Kadir van Lohuizen

The Human Consequences of Rising Sea Levels

Lannoo

For my dad

THE SEA IS COMING

Kadir van Lohuizen

It was 2011, and for a visual survey on land migration, I had travelled from the southernmost tip of South America to Alaska. One of my stops was the beautiful San Blas Archipelago on the Caribbean side of Panama. Actually, it's just called 'Guna Yala', because it's the land of the Guna people, one of the original populations of Panama. I interviewed people there for my story, and to my surprise, they all talked about an upcoming evacuation. I didn't really understand what they meant. When I asked, they said, 'The sea is coming'. Still, the penny didn't drop. 'You know, the sea is coming, it destroys our houses and we can't grow crops anymore and our drinking water gets salty.' They spoke of rising sea levels.

Fortunately for the Guna people, their region also includes a higher coastal strip. This is where everyone is moving to. 'Next year', they assure me. When I return four years later, no one has moved, and only a school and a clinic have been built. Funding was committed by the Panamanian government in 2017, but there have since been delays due to additional budgetary requirements as well as COVID-19. It does seem that the relocation is finally happening.

Even though I'm Dutch and live below sea level myself, until then, I'd never suspected that rising sea levels were already threatening certain parts of the world. Several months earlier, I'd been in the Ganges-Brahmaputra-Meghna Delta of Bangladesh and had witnessed the vulnerability of the coastline. I had seen people doing everything they could to protect their country from the sea.

I decided to research areas of the world where rising sea levels are an urgent problem now, not just one for the next generation. But how do you visualize a problem that often isn't yet visible? My list gradually grew: Bangladesh, Guna Yala, Kiribati…

It takes a different way of working than the one I'm used to as a photographer. Normally I use the light as a guide, but now my guide is the tides. The first time I visited a family on the coast in Bangladesh, the man said, 'You should have been here two weeks ago, the water was in the house'. I looked at the sea, which was very far away … and realised that if I wanted to show the impact of the sea, now and later, I had to time my visits to coincide with high tide or, rather, spring tide, which occurs once or twice every month. When you see the amount of rising water, it's easy to picture sea levels rising by another 2 m or 3 m.

My first time in Bangladesh, I travelled with the NGO Displacement Solutions. They help people in Bangladesh and other areas of the world who, as a result of the climate crisis, need to build a new life elsewhere. I photograph in black and white, as I do almost always. I realise that aerial photography often captures the issue of rising sea levels better, but the drones may be too big and too expensive. Someone in the Netherlands helps me set up a system to mount a small camera on a kite. I can't see what I'm doing, but I can programme the camera and use the wind direction to find the right angle. The disadvantage is that if there's too much or too little wind, it doesn't work. But in most cases, the results are satisfactory.

I am in Bangladesh two years after Cyclone Aila, and the delta still bears the traces of devastation. Sparing no effort - in other words, thousands of people - the dykes are being reinforced and rebuilt. But unlike previous cyclones when the water receded after the storm, in many places the water remains, turning the land saline and the drinking water brackish.

Research is always an important basis for my projects, especially for a topic like this. I'm aware that I'll need to do a thorough job, in part to avoid giving climate sceptics ammunition. Fortunately for me, the *New York Times* gets onboard at a fairly early stage. Based on my research, together we decide which regions I will visit. It is important that I look for a geographical spread and also look closer to home, with my own country, the Netherlands, as the last destination.

Until recently, the warming of the oceans was the main cause of rising sea levels. When water warms up, it expands, which causes a rise in the water level known as 'thermal expansion'. Meanwhile, the main reason for rising water levels is the melting of the Greenland and Antarctic ice sheets. If the entire Greenland ice cap melts, sea levels will rise by 7 m, and if Antarctica melts, by 86 m.

I go to Greenland, perhaps one of the few countries that is relatively happy with the climate crisis. The melting ice sheet makes raw materials accessible, creating better conditions for farmers in the south of Greenland. They even grow strawberries there these days. This local production makes it cheaper for its inhabitants and makes Greenland less dependent on 'Motherland' Denmark.

I get to go with a group of scientists to the East Greenland Ice-core Project (EGRIP), a research station in the middle of the ice sheet. A Lockheed Martin C–130 Hercules from the U.S. Air Force is the only connection to the outside world. The aircraft has skis and lands on the world's longest runway. I'm at an altitude of 2,700 m, of which 2,400 m is ice. A few years ago, ice flows - in fact, gigantic frozen rivers - were discovered in Greenland. EGRIP involves drilling right through the ice layer to the land beneath. Drilling is performed at the source of the ice flow, and here the entire ice sheet moves 15 cm per day (!) towards the ocean. The researchers believe this is another reason that sea levels are rising even faster than expected.

A long corridor takes me to the actual research station a few metres below the surface, where the temperature is always −18 °C. Every hour, a bar of ice rises. When I arrive, they're at a depth of 1,500 m, and the ice is about 20,000 years old. It is in fact an incredible history book of the world's climate. Scientists can tell when there was an ice age, when temperatures were warmer and so on. A colour difference indicates a large volcanic eruption, when the ice sheet was covered with a layer of volcanic ash.

Kiribati is an island state in the Pacific with more than 100,000 inhabitants. It is located between Hawaii and Fiji and is about the size of India, but it consists, for the most part, of ocean. The land is only 1.5 m above sea level. Here you can feel how vulnerable people are.

From a distance, with its white beaches and turquoise lagoons, it looks like paradise. But up closer, in the capital, there is a different picture. The population is very dense; many people have already moved here from the other islands in the hope of finding work and a 'safe' place.

The situation is quite dire - there is actually no adequate sewage system, which means that the lagoon is heavily polluted, and the beaches have already been mostly washed away. The coral reefs are dying. As a result, they no longer function as a natural buffer, and the coast is rapidly eroding.

It's spring tide when I'm there, and the government hands out sandbags that people can use to protect their homes, yet there are large areas that are already submerged. This makes good drinking water a problem here too, and people can barely grow crops.

Kiribati previously had President Anote Tong, who seized every opportunity to call attention to the fate of his people and the country as a result of the climate crisis. Kiribati is in danger of disappearing, a fate that is likely to await other countries as well, such as Tuvalu, the Marshall Islands, the Maldives and other island states. The islands cannot be protected. There is no funding and no time, and as one resident put it, 'We have the smallest carbon footprint, but we pay the highest price'.

President Tong decided to take action himself: The government of Kiribati bought a piece of land from the Anglican Church in Fiji, and a press statement was issued saying that Kiribati would eventually move its inhabitants to Fiji. When the Fijian government heard about this, they were not amused. Since then, the official story is that the land is used for agriculture and the crops are exported to Kiribati.

It certainly highlights the problem: There is no focus on countries disappearing and what is going to happen to their people. The term 'climate refugee' is used, but you aren't given asylum if you are fleeing because of the climate crisis.

In 2019, I was supposed to go back to Kiribati, but now there's a new president, who's called 'the Trump of the Pacific', because he claims the climate crisis is fake news.

As I mentioned earlier, I also wanted to show that the problem is closer to home, too. On the United States East Coast, for example, there is almost no coastal protection. In 2005, I was in New Orleans right after Hurricane Katrina hit. What little coastal protection there was had been breached, and a tidal wave had flooded the city and the large surrounding area. Hurricanes are common, but now they're intensifying and becoming more frequent. If sea levels rise and there is hardly any coastal protection, you can guess what the consequences will be.

Miami occupies a unique position. The city is built on limestone, which is porous. By now, every Dutch expert has visited. The city was hoping for a master plan that would protect the city. But everyone has come to the same conclusion: Building a seawall makes no sense, as the water will just run underneath it. This means that, as things currently stand, Miami Beach will have to be evacuated around 2050, soon followed by the rest of the city. Only the poor part, 'Little Haiti', slightly higher above sea level, will hold out longer.

At the monthly spring tide, part of the streets of Miami Beach are covered. Seawater rises through the sewer system. To prevent this, the city council has installed 600 pumps in recent years, but it's a temporary solution. Moreover, most of these pumps are electric, and many malfunction during the first heavy storm. Meanwhile, Miami is still building one residential tower after another. It feels like dancing on a volcano.

New York, on the other hand, had a wake-up call in 2012: Superstorm Sandy. Much of Manhattan was flooded and left without power. A master plan is supposed to protect the city. A large dyke should surround Manhattan, but the height of the dyke has been calculated to prevent a second Sandy and hasn't factored in the rising sea level. In addition, neighbourhoods such as Brooklyn, Queens, the Bronx and Rockaway aren't covered by the plan. Jersey City, on the opposite bank, isn't included either.

The Netherlands had its last wake-up call in 1953, when a severe storm combined with a spring tide collapsed the dykes in the southwest and inundated a large area. Nearly 1,900 people lost their lives. The Delta Plan was developed in the years that followed - estuaries were closed and the dykes were raised to delta height, so that a disaster like the one of 1953 should statistically occur only once every 10,000 years.

I was born 10 years after the flood, but the fear of the sea is very real, and we visit the Delta Works on school trips. In primary school, you have to be able to draw how a dyke is built.

It took 40 years to complete the Delta Works, and, in this small country that is mostly below sea level, we imagine we're safe. We no longer fear the sea. But is that justified?

The Netherlands is a delta: Large rivers flow into the North Sea. We are sometimes referred to as the drain of Western Europe. I went to high school in Gouda, a town in the west of the country, known for its cheese. It is a city built on peat, like most of the west of the Netherlands, and, thanks to land subsidence, roads and gardens must regularly be raised. With the current prolonged droughts, this

is worsening by the year, and the quays are often inundated. The city recently decided on the drastic step of lowering the water table so the city won't actually sink in the future. It highlights the problem of a low-lying delta.

But what will happen if there's a marked rise in the sea level? Everyone agrees that the Netherlands can handle a 1 m sea level rise, but what is the worst-case scenario? In 2018, the Ministry of Infrastructure and Water Management and the Delta Programme Commissioner asked the research institute Deltares to investigate.

The findings are alarming. By the end of the century, the sea level could rise by 1 m to 3 m; it just depends on whether we stick to the agreements of the climate summit in Paris: There should be a rise of no more than 2 °C, preferably no more than 1.5 °C. We are now heading for a 3 °C to 4 °C increase.

The Dutch are world-famous for their coastal protection and water management expertise, and knowledge transfer has become a major export product. Many believe that technology and our expertise will always protect the Netherlands. But is that really so?

The question is not *if* the sea level will rise by 3, 4 or 5 m, but when. With 2 or 3 m, the rivers can no longer drain their water naturally and the dykes won't be high enough, certainly not during a heavy storm. There are now plans, sometimes wild plans, to continue to protect the Netherlands, but the question is, how much will this cost and how long do we have left?

Every year, the Netherlands spends vast amounts of money on coastal protection and spares no effort to assure the country's safety. And yet one question remains unanswered: Why doesn't the Dutch government want to look at the worst-case scenario? After all, this is precisely what we do when it comes to safety in aviation, on the road or, more recently, when dealing with COVID-19. We take the measures, and the worst-case scenario isn't the best; it is leading. The wordst-case scenario should be the leading scenario when it comes to the planning for the inevitable.

When it comes to the climate crisis, we always seem to think, 'Things will work out all right in the end'. But everything indicates that it's not going to be easy and we're leaving the real problems for the next generations. They'll be angry. And rightly so. •

A schematic illustration of the climate- and non-climate-driven processes

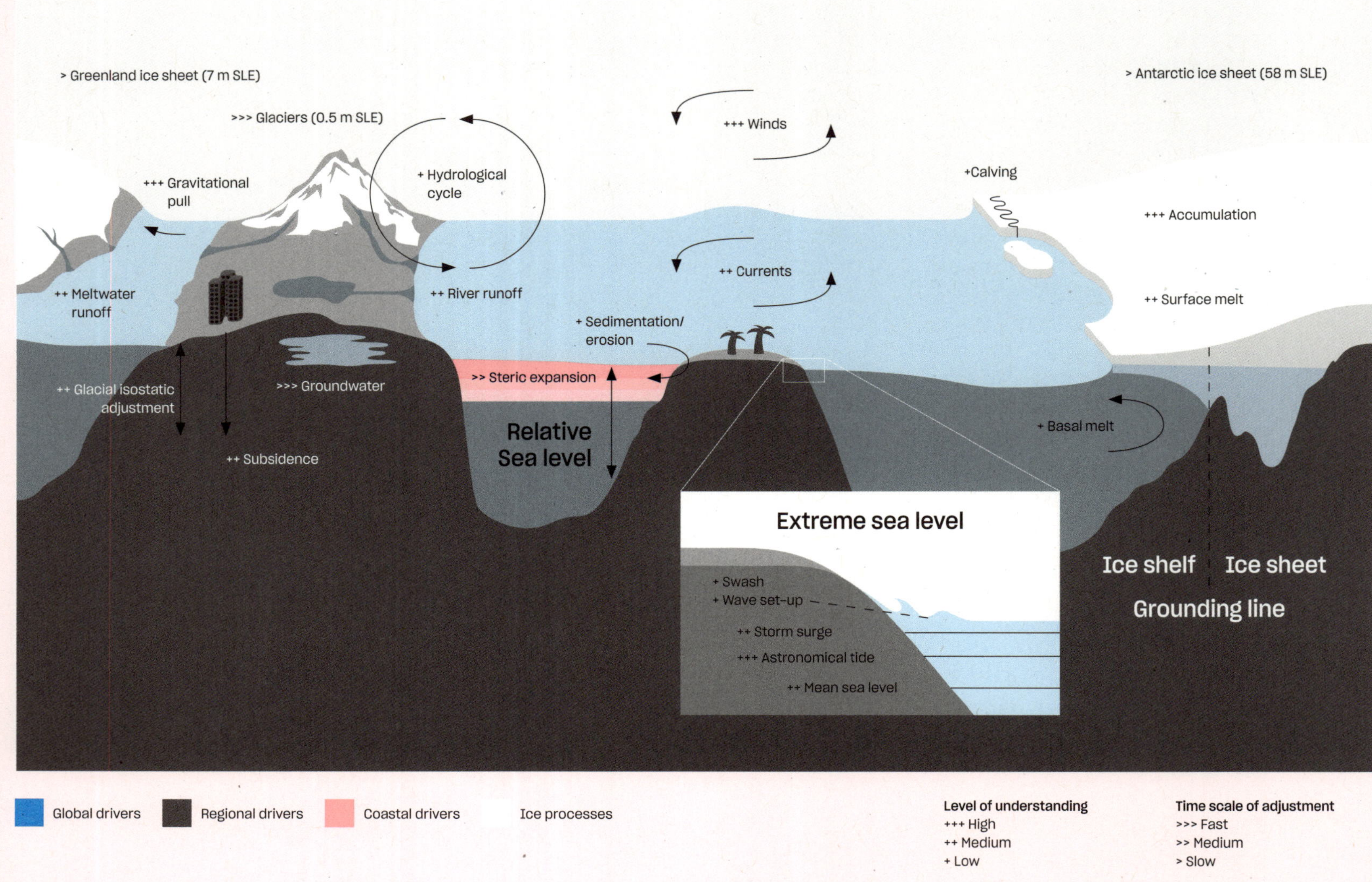

A schematic illustration of the climate and non-climate driven processes that can influence global, regional (grey), relative and extreme sea level (ESL) events (pink and red) along coasts. Major ice processes are shown in purple and general terms in black. SLE stands for sea level equivalent and reflects the increase in global mean sea level if the mentioned ice mass is melted completely and added to the ocean.

Source: Oppenheimer et al., in press

WATER
Henk Ovink

Water and the climate crisis are directly linked. We know this by default and from disaster. The climate crisis is a water crisis. Nine out of 10 natural disasters are water related. Between 2001 and 2018, droughts, floods, landslides and storms caused over US$1.7 trillion in damage worldwide, according to the UN, impacting over 3.4 billion people, the majority of them in Asia (UNU–INWEH & EM–DAT, as cited in UNESCO World Water Assessment Programme, 2020). Without water, there is no energy and no food. But too much water and ever-increasing 'extremes' also go hand in hand with far too *little* water - periods of drought align with the flow of refugees and increased conflicts. We are depleting our natural water supplies at a ruinous rate, and sea level rise is jeopardizing our cities and deltas.

PUTTING WATER CHALLENGES ON THE WORLD MAP

Starting in 2015 as special envoy for international water affairs, I wanted to put all water challenges literally on the map of the world, to spark a better understanding of water's complex challenges, their interdependencies and the forthcoming opportunities. *The Geography of Future Water Challenges* - developed together with the Netherlands Environmental Assessment Agency and spearheaded by Willem Ligtvoet with a consortium of global scientists - states that: 'Water security is related to three water-related challenges: water scarcity (too little water), water pollution (dirty water) and flood risk (too much water). In the coming decades, these challenges and their impact on people's daily lives are expected to increase due to population growth, economic development, increased agricultural production and the climate crisis, in turn affecting water availability, sea level rise and weather patterns. In order to secure water resources, now and in the future, an understanding of the complexity of water-related challenges and the existence of possible gaps is essential, as a basis for the development of sustainable strategies that can adequately reduce risks for the population, economic development, ecosystems, and water associated migration and conflicts.'

The 2020 *Global Risks Report* (World Economic Forum, 2020) agrees, as it lists water crises - time and again - as one of the top global risks. Water is linked to the economy, geo-politics, the environment, the climate crisis and more. The report reiterates a painful song, played over and over again: water scarcity, which already affects a quarter of the world's population, will only increase. Crop yields will likely drop in many regions, undermining the ability to double food production by 2050 to meet rising demand. The way we grow food, produce energy, dispose of waste and consume resources is destroying nature's delicate balance of clean air, water and life that all species - including humans - depend on for survival. The climate crisis not only dries out our lands and waters and floods our coasts, destroying our economies, but it is also 'the greatest threat to global health in the 21st century' (World Health Organization, n.d.). With extreme weather conditions putting populations around the world at risk of food and water insecurity, today's children face a future of increasingly serious climate-related hazards: less-nutritious crops, air pollution exacerbated by burning fossil fuels, rising average temperatures and other weather-related disruptions to livelihoods.

WATER AS A CATALYST FOR CHANGE

As I have learnt over the past decade, working around the globe, in every region, on every continent, with different people and partners everywhere, it is water that can drive us apart, cripple our lives, destroy our environments and our economies and strengthen the impacts and origins of the climate crisis. But it is also water that enables us to come together and do better; it catalyzes the changes we need and is the true inspiration for sustainable development, lasting partnerships and transformative climate action.

REBUILDING AFTER SUPERSTORM SANDY

Working in the United States after Superstorm Sandy, as senior adviser to President Obama's Hurricane Sandy Rebuilding Task Force, water was the main driver of our work: its devastating impact, and its transformative capacity not only to rebuild better, but also to leapfrog towards a more resilient future. Water informed the solutions for moving ahead, looking not back but towards the future. It capacitated and empowered communities at risk, the people and their institutions. Water helped inspire all to come together, despite distrust in government and within society, despite different needs and interests, despite different opinions.

One of the task force's initiatives was a design competition called Rebuild by Design, aimed at quickly generating implementable design solutions. The competition brought the region together in developing transformative action for a resilient recovery. At the presentation of the results to the project partners, Damaris Reyes, from the community partnership Good Old Lower East Side, said, 'Rebuild by Design made us sit on the same side of the table, but we can easily sit back on the other side again'. In other words, trust - much needed for transformative action that lasts - comes by turtle and leaves by horse. Investing in people is not easy, but it is a necessity ánd amazing. It comes with a cost: never let go, never turn your back. We need a process, an approach, of continuity, consistency and commitment - we need it through an approach driven by 'science and solidarity'.

A COMMITMENT TO INTERNATIONAL WATER AFFAIRS

In 2015, I was appointed as the Netherlands' first special envoy for international water affairs, a first for the Netherlands, but also a first for the world. In this capacity, I have worked to support communities at risk, to inform better decision-making, to help raise societal and political awareness, to enable finance and investment opportunities, to build coalitions across stakeholders and to inspire transformative action and innovative approaches, while remaining grounded in my personal convictions and principles of leaving no one behind. Starting with the people, not the projects or the politics. Integrating holistically all needs, opportunities and challenges. Looking at the future while learning from the past, but not replicating our past mistakes. Working towards sustainable and catalytic action, embraced by all. From a common yet ambitious shared ground, to a sustainable, equitable and resilient future. A future that can continue to inspire and help drive the change so needed for our world, our planet and our communities.

I often travel from disaster to disaster, to inform better practices and to provoke a rebuilding approach not in response to the disaster, but in overcoming that perspective and investing in the future. With the world at risk and disasters more complex, interconnected and interdependent, impacts are seen not only from the damage caused, but also from our future vulnerability. Replicating the practices from the past only makes us more vulnerable tomorrow.

In my travels, I have met and worked with experts, community leaders, children and politicians and alike. All with different backgrounds, needs and interests. Through water, I have managed to ignite a conversation, a partnership even, a process leading towards increased awareness and understanding, enabling actions that matter. Water empowers people and institutions; it helps to better capacitate them for challenging tasks. Water inspires this collaborative process to spur novel ideas, to identify opportunities and projects to work on. With water, we work collectively from the ground up, to invest together in a better, more sustainable, more resilient and more inclusive future.

WATER AS LEVERAGE FOR RESILIENT CITIES

Water as Leverage for Resilient Cities Asia is the programme I initiated to spur this collaborative action (World Water Atlas, n.d.), to make it concrete, to identify needs and opportunities while building partnerships across all layers of society, across all institutions and their silos, across everything and everyone. Water as Leverage is living proof of both the need for action and the opportunities we can implement, if only we drive our actions inclusively, holistically and sustainably.

Water as Leverage brought me to Khulna, Bangladesh, in the fall of 2019, just weeks after Cyclone Bulbul had made landfall. My partner for action, Mr. Abir-ul-Jabbar, the city's chief planning officer, told me, 'The mangroves saved Khulna City!'. Khulna, Bangladesh's third-largest city, sits at the convergence of the Bhairab, Rupsa and Mayur Rivers, serving as a port and an important gateway on the northern edge of the Sundarbans. This magnificent mangrove forest, twice named to the UNESCO World Heritage List, is treasured as the home of amazing flora and fauna. The mangroves are beloved by local communities for providing livelihoods and sustaining the region. While the Sundarbans suffered damage during the cyclone, the mangroves in fact slowed wind velocities, sparing inland cities from devastation. Khulna City is a testament to the long-term value of investing in water and nature, not only to sustain daily life, but also to dull the catastrophic force of climate disasters.

HOTSPOTS OF THE CLIMATE CRISIS

Asia is the hotspot of the climate crisis, where climate disasters, economic and urban growth and people's vulnerability converge. This is where the complexity and interdependency of our vulnerability is exposed. This is also where these hotspots, these converging places of needs, become places of opportunity, if only we are able to use our capacities, fulfill our political and societal responsibilities and use the insights gained from science to better inform our decisions and investments, to spur action for a better future, to progress towards sustainability and resilience.

Asia is not alone in being a continent at risk. Vulnerable places and vulnerable communities in the context of the climate crisis and sustainability challenges are places of opportunity, if only we deliver on our promise, if only the world can act with 'science and solidarity', if only we will show that we care.

VULNERABLE NATIONS

Small islands and development states, the Middle East and Africa have all been battered by the climate crisis, natural disasters, famine, social inequalities, political oppression, geo-political tensions, wars, conflicts and terrorism. The most vulnerable are hit hardest and have the hardest time getting back on their feet. Inequality and insecurity cannot be easily overcome by a pilot project, a one-off, by doing good for a day. Cultural change for sustainable development means geo-political and multilateral cultural change. Global action means exactly what the words tell us: action by all, collectively across the planet.

From Peru, to Chile and Mexico, to Canada and the United States, my journey of water took me from coast to cities, from rivers to wells; from governments and businesses, to communities, schools and NGOs. Water connected my travel, work and actions. It helped inform new post-disaster resilience practices, better inclusive decision-making and innovative and preventive action.

Working in Chile in the Elqui River Basin, on a regional water resilience strategy, using data, with all stakeholders, aimed at all levels and layers of society. Informing mayors, miners and farmers how to do better, working with schools and kids, politicians and scientists. In Peru after the massive El Niño floods in 2017. In the United States after every flood and storm, but more often before the disaster, working towards real resilient communities and institutions to prevent the climate impacts and increase societal capacities. Across Africa, in the Middle East, in places where conflicts drive people away, finding spaces, time and hope through water conversations and actions.

LOCAL ACTION AND CAPACITY

No matter where in the world, in Afghanistan, China, Vietnam or Bangladesh; in South Africa, Mozambique, Egypt or the Middle East; in Europe or in the Americas. Water is life - it helps build a better future and inform sustainable actions, and it helps bring us together. Local action, local capacity and local needs must be leveraged with global commitments, with indigenous knowledge and cultural capacity contributing to reducing social vulnerability. The understanding, skills and philosophies developed by societies with long histories of interaction with their natural surroundings inform decision-making about fundamental aspects of life, from day-to-day activities to longer-term actions. This knowledge is integral to cultural complexes, which also encompass language, classification systems, resource use practices, social interactions, values, rituals and spirituality. 'These unique ways of knowing are important facets of the world's cultural diversity and provide a foundation for locally-appropriate sustainable development' (UNESCO, n.d., para. 3).

GLOBAL COMMITMENTS AND AGREEMENTS

In 2015, the world agreed on the *2030 Agenda for Sustainable Development* and the 17 Sustainable Development Goals (SDGs), not to cherry-pick from but as a holistic, comprehensive agenda for sustainable development. Social, economic, cultural and environmental challenges and opportunities are all interlinked. These interdependencies determine the way we live and thrive, and the way we must invest. Investing in water, sanitation and hygiene (WASH) is the first line of defence and the first step towards a sustainable recovery. Never has the sixth SDG, 'Ensure access to water and sanitation for all', been more vital for saving and protecting lives. Even better, investing in water has a trickle-down effect across all of the SDGs. But to deliver on our promise of meeting the SDGs, we need collective commitment, programme continuity and consistency of ambition.

The High Level Panel on Water (HLPW) was founded in 2016 with a core focus on the sixth SDG. Comprising 11 heads of state and government, under the leadership of Ban Ki-moon, Secretary-General of the United Nations, and Jim Kim, president of World Bank Group, the HLPW has travelled the world forging partnerships, developing understanding and securing commitments for water action. The HLPW agreed on the three principles for water action across the *2030 Agenda for Sustainable Development*: understand, value and manage water (better). Three pillars that are foundational for any sustainable and transformative water action. Only then can water be the enabler we need it to be, the leverage for catalytic, sustainable and inclusive action.

INVESTING IN OUR FUTURE

Together, we must leapfrog ahead and invest more and better in water capacity, land management and infrastructure - blue, green and grey. It is time to scale up our investments in integrated, inclusive and sustainable water programmes and projects. Doing so pays off, according to the World Health Organization and UN-Water (2014): Every US$1 invested in safe drinking water in urban areas yields more than US$3 in saved medical costs and added productivity. For every US$1 invested in basic sanitation, society earns back US$2.50. In rural areas, US$7 is gained or saved for every US$1 invested in clean drinking water. So far, we have failed to seize this opportunity. We continue to invest in infrastructure projects from the past, taken off the shelves, to fill economic stimulus packages. Focused on jobs alone for fast economic recovery, these projects offer no added value for integration, inclusion or sustainability. The *2030 Agenda for Sustainable Development* and the 17 SDGs should lead the way for recovery, really preparing us for the challenging future ahead. Investing in water across the 2030 agenda is the added-value enabler we so urgently need.

While we all know preparedness pays off, in terms of climate resilience, preparedness offers a return on investment of five or 10 times or more. And this is counting only the losses prevented and risks reduced. If we take into account the investment opportunities and added value - from better health, increased security, improved ecology, a decreasing gender gap and strengthened youth capacity - the benefits are numerous. Why shy away from sustainable investments, increasing resiliency and opening up our portfolios for more and a much wider range of opportunities?

While we have great and inspiring examples, we lack a steady flow of sustainable investments. Our promises compete with outdated infrastructure investments. If we continue replicating the past, we'll end up more vulnerable, less equal and more fragile than before. Our commitment is challenged by vested interests in past mechanisms. We need to overcome these vested interests, grounded in the past, single focused and aimed for despair and a disastrous future. We need to accelerate and expand our promises and our commitments, by science and through solidarity. Investing across the *2030 Agenda*, in a pipeline of blue and green opportunities, means investing in people, across the world. We must practise what we preach.

STARTING AT THE SOURCE

The choice between prevention and repair is false. Both are essential. We need to start at the source: reduce greenhouse gasses and make efficient and careful use of our planet and all of its resources. Yet at the same time, we need to prepare boldly, comprehensively and inclusively for tomorrow's extremes. Our human-made systems are simply not fit for that future. Our cities are built on hard structures. No capacity to hold the rains overflowing our stormwater facilities, no parks or green roofs. No sewage systems that can stand up to these extreme events. All over the world, our cities and communities face these shocks and stresses. And everywhere, the impact shows our vulnerability. We have destroyed our natural systems, too. Our rivers are channelled up, urbanized or even covered by infrastructure and buildings. While these natural systems used to meander, shrink and grow depending on their flows, they are now stuck in human-made barriers, designed and engineered according to outdated standards. This kind of infrastructure paid off, but only in the very short term and only from a financial perspective. It is devastating for climate mitigation and resiliency, with disastrous impacts on marginalized communities and our biodiversity system, wrecking our food security and economy in the longer run. When will we learn to do better? To mitigate and adapt, to prepare before responding and to invest everything we have in a sustainable future, leaving no one behind? We have no choice, but lessons learned are costly. Yet we have every opportunity to change course now, with water as the leverage for sustainable development and climate action; tackling social, economic, cultural and ecological challenges.

The availability of clean drinking water safeguards health, education and development, equal opportunities and inclusive sustainable growth. Preserving our ecosystems and natural resources ensures the resilience of our planet and society. By taking a preventive approach on our coasts and deltas and in our cities, we can avert the most serious problems and prepare ourselves and our world for a sustainable future that is strong and resilient. Water and water narratives can unite people around the world - politicians and scientists, city dwellers and country dwellers. We have to come up with new solutions to tackle our future challenges, since the solutions of the past will make the world a worse place tomorrow. By being proactive, we can understand our future and build resiliently. Our policies are based on our understanding of yesterday and not our understanding of tomorrow. Innovation also involves the task of helping us change our policies and practices.

A NEW APPROACH

For this we need a new approach, one that is rigorously inclusive, innovative and comprehensive, with everything and everyone working together from beginning to end. A mechanism through which future understanding becomes an inspiration and drives innovation forwards, and which includes everyone in the process - bankers and investors are as much a part of this as policymakers and politicians, community leaders, NGOs, academics and the businesses that develop these solutions. Because with a better collective understanding of the future, we can gain a better idea of how to fund innovations arising from that understanding. These are the millions we need to invest to secure the billions for the projects that will really make a difference and prepare our society and planet for our challenging future.

This approach demands a free zone, a safe space, or soft space, as Allmendinger and Haughton (2009) describe: 'areas where deliberate attempts are made to introduce new and innovative ways of thinking, especially in places where there is considerable resistance to cross-sectoral and inter-territorial governance.' And in my own book, *Too Big: Rebuild by Design: A Transformative Approach to Climate Change* (2018), co-authored by Jelte Boeijenga, in reflecting upon my work for then-president Obama's task force, I call on the need for 'a different culture of working together. The enormous complexity calls for room to experiment, reflect and innovate. Time and space where we can step outside existing interests and frameworks, legitimized by the huge challenges ahead. Challenges which can never be met by our current approach and conditions, our agreed-upon procedures and institutions, rooted as these are in the past. We need room to experiment, in which we tie government responsibilities to the strength of initiatives in our communities, in academia, by activist groups and by our businesses and investors. Room to take risks, make mistakes and learn, to set change in motion. This experiment requires a safe place that frees us from the "not allowed" and "won't work" attitudes, based on yesterday's experiences. A place where we can collaborate, where it's all about people, not positions, on a level playing field, space where we can get away from solidified relations and accepted roles. A place where everyone can be vulnerable, where government and communities or citizens and corporations are not on opposite sides, where everyone can change roles and positions. A place where disagreement and even conflict can grow into understanding and better solutions. Here we can organize a truly inclusive and transparent process, rooted in trust and collaboration, a process where individual vulnerability is respected and can grow into a collective force for change'.

DOCUMENTING THE CLIMATE CRISIS

Kadir van Lohuizen finds himself in this context: challenged by humankind's failures and vested interests, he documents the decline and degradation of the planet, and of the places we live.

Yet his images document not despair but the fine line between the power of nature and human hope. Yes we can, but we can go either way, and right now, he tells us, everything is going to waste. We must change course. We *can* change course. There is no time to waste if we want to achieve our climate and sustainable development goals and thus safeguard our planet and our future. For this, we need big and small successes.

I met up with and partnered with Kadir while I was travelling the world in my quest for water security for all. Knowing Kadir from his past project, *Where Will We Go?*, I was inspired to dive deeper into the challenges we face, and to ask better questions. What are the mechanisms behind our actions? How can journalism and documentary photography help increase our understanding? How can we build awareness and understanding and strengthen the capacity to stand up, to act and provoke the future instead of continuing to linger in the past? Kadir's provocation matches with my own experience and ambition and my push to rapidly and massively increase our understanding of the complexity of our challenges through science and data, and to develop new and transformative action based on facts and through inclusive partnerships. We must act now, together - we have no time to waste! •

Global mean sea level rise

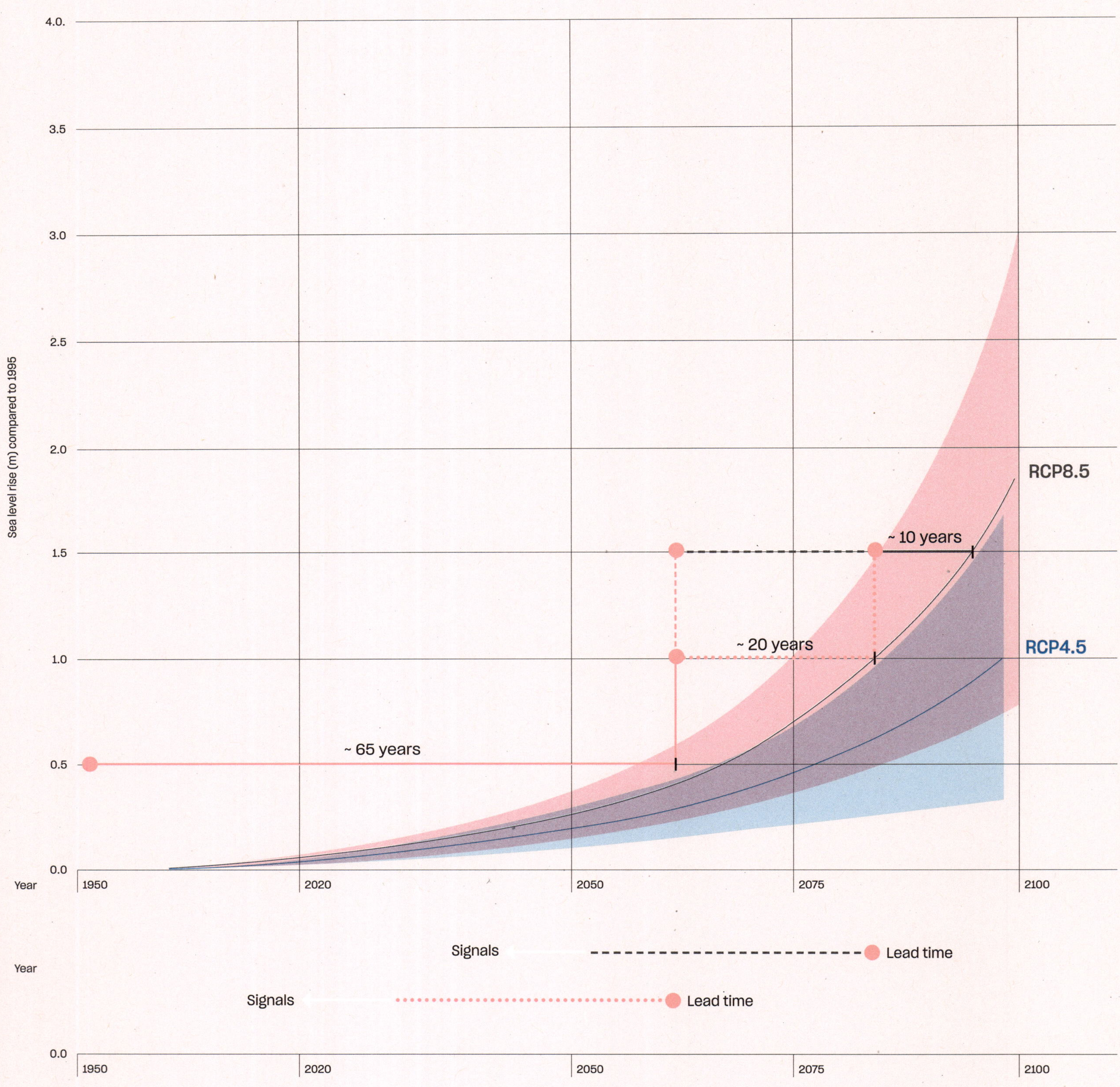

Global mean sea level rise including accelerated mass loss from Antarctica for RCP 4.5 (blue) and RCP 8.5 (grey) and the potential consequences for coastal adaptation. The bandwidth presents the 5th-95th percentile. The horizontal bars present an adaptation pathway existing of a sequence of measures for 0.5 m sea level rise and the functional lifetime of these adaptation measures in the event of an accelerating sea level rise according to the median AA value for RCP 8.5 or the 95th percentile for RCP 4.5. The lower panel shows a lead time of 30 years for planning and implementation, and required timing of signals and decision. An alternative pathway could start with the 'orange' measure and switch directly to grey (dashed lines).

Source: Haasnoot et al., 2020
Le Bars et al., 2017aa6512

Amplication factor of historical 1-in-100-year extreme sea level event

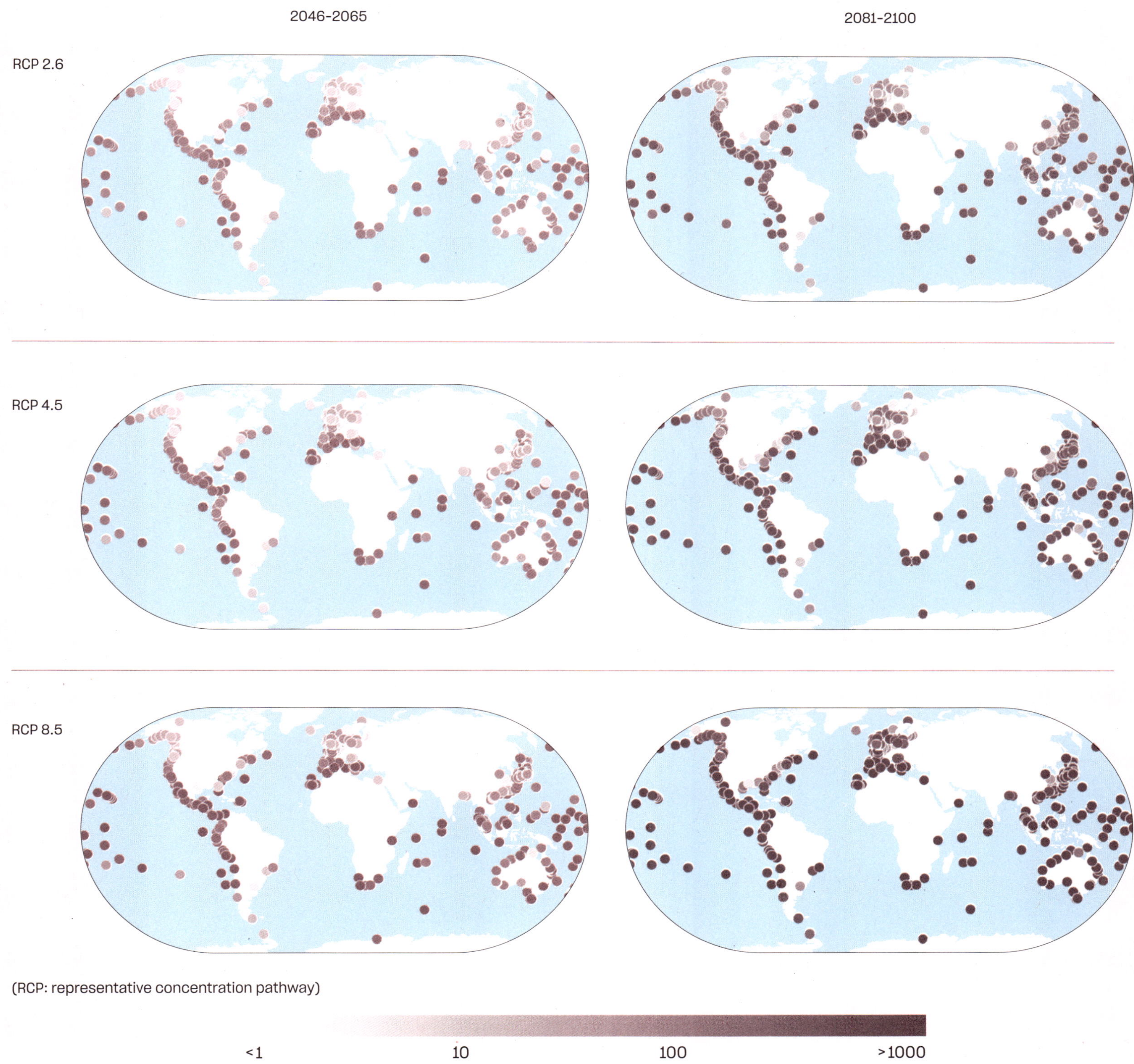

The colours of the dots express the factor by which the frequency of extreme sea level (ESL) events are expected to increase in the future for events which historically have a return period of 100 years. Hence a value of 50 means that what is currently a 1-in-100-year event will happen every two years due to a rise in mean sea level. Results are shown for three RCP scenarions and two future time ranges as median vaues. Results are shown for tide gauges in the GESLA-2 database

Source: Oppenheimer et al., in press

GREENLAND

UNITED STATES

PANAMA

INDONESIA

PACIFIC

BANGLADESH

NETHERLANDS

AFTER US THE DELUGE

GREENLAND
Dorthe Dahl-Jensen

One of the major causes of the world's rising sea levels is the melting of Greenland's ice sheet, which is occurring at an unprecedented rate. For some of the people in the south of Greenland, this is good news - due to the absence of snow and ice, they can now grow crops for the first time. In Greenland's north, it is a completely different story.

The Arctic is warming at a much faster rate than the rest of the world. This warming is causing the large ice sheets in Greenland to melt and retreat, permafrost to thaw, sea ice to reduce and smaller glaciers to thin. Arctic communities are heavily concentrated in coastal areas, where accelerating sea level rise threatens both the infrastructure and the way of life. The impact of rising sea levels on coastal regions will depend on our ability to mitigate and adapt to its consequences. This in turn will depend on our ability to accurately project the rate of sea level rise.

My research is mainly in understanding how the climate has changed in the past and how the Greenland ice sheet has changed in volume. We know that there have been large, abrupt changes in the past, even without human influence. This knowledge is an important tool in understanding how the climate and the ice in the Arctic are changing now, and it shows us that the climatic system can react strongly to changes. We need to know what we can expect in the future.

THE GREENLAND ICE SHEET AND ICE STREAMS

The Greenland ice sheet is the second-largest body of ice on Earth, close to 3 million km³ in volume. If all of this ice melted, it would equate to a 7 m rise in global sea level. While the Antarctic ice sheet is 10 times larger, the Greenland ice sheet's location in the Arctic means it is presently experiencing a much greater increase in temperature. The Arctic's ice sheet, ice caps and glaciers are by far the most dominant component of melting ice worldwide. The present global mean sea level rise is 3.5 mm each year. Of this, Arctic ice contributes 1.5 mm per year, while the remainder is from heating of the ocean, loss of land water and melting of terrestrial ice in the rest of the world.

As the temperatures around Greenland rise, we see more and more melting along the margin of the ice sheet, both from increasing melt near the coast and as a result of the area experiencing melting migrating to higher elevations on the ice sheet. The ice melt from the Greenland ice sheet is mainly from this marginal melt and discharge of ice from the ice streams surrounding Greenland. The contributions from these two sources are similar in size. The second part of this mass loss, from the ice streams surrounding Greenland, is what I will focus on here.

Ice streams are areas with high surface velocities; they can be seen as ice rivers cutting their way through the slower-moving ice in the ice sheet. There are ice streams all around Greenland, but two are exceptional. The Jakobshavn Isbræ on the west coast of Greenland, near Ilulissat, has a surface velocity of 14 km/year and drains 70 GT to 80 GT (80 km³ to 90 km³) of ice each year. This represents 0.2 mm of sea level rise just by itself. The Northeast Greenland Ice Stream (NEGIS) covers the largest area and discharges 30 GT (34 km³) of ice each year.

Most of the ice streams surrounding Greenland have accelerated since 2000 and are thus discharging more ice into the sea. As an example, the velocity of the Jakobshavn Isbræ has doubled from 7 km/year prior to 2002 to 14 km/year after 2004. The changing velocities are believed to be connected to the warming of the ocean, an increase in surface meltwater penetrating the ice streams to the bed beneath and an increase in snowfall on the top and centre of the ice sheet. We are not yet able to fully model the ice streams and their changing velocities due to lack of understanding about the process.

EAST GREENLAND ICE-CORE PROJECT

Since I was a student, I have been involved in the drilling of deep ice cores through the Greenland ice sheet, from the top to the bedrock, 2,500 m to 3,100 m below the surface. Ice cores contain layers upon layers of annual snowfall, becoming older and older with depth. They can thus be used to learn about the past climate. Near the bedrock, the ice layers are more than 150,000 years old. The projects are international and involve researchers from Japan, China, the United States, Australia, New Zealand and Europe. We establish camps that can house around 30 scientists during the summer months on very cold and remote sites on the Greenland ice sheet.

At present, we are drilling the East Greenland Ice-core Project (EGRIP) in the centre of the NEGIS, not only to understand the past climate, but also to learn how ice streams flow, as they are such important, yet poorly understood, contributors to sea level rise. EGRIP started in 2015, when we moved the previous camp, North Greenland Eemian Ice Drilling (NEEM), 440 km across the surface of the Greenland ice sheet. The traverse included moving the 55 t main building, 'the Dome', on skis, at a rate of 10 km/hour. The ice core drilling and the first measurements on the ice cores are done in subsurface snow trenches, where temperatures are cold and not influenced by the surface weather. In these trenches, international teams measure ice and climate properties and cut samples for further analysis in more than 100 laboratories around the world.

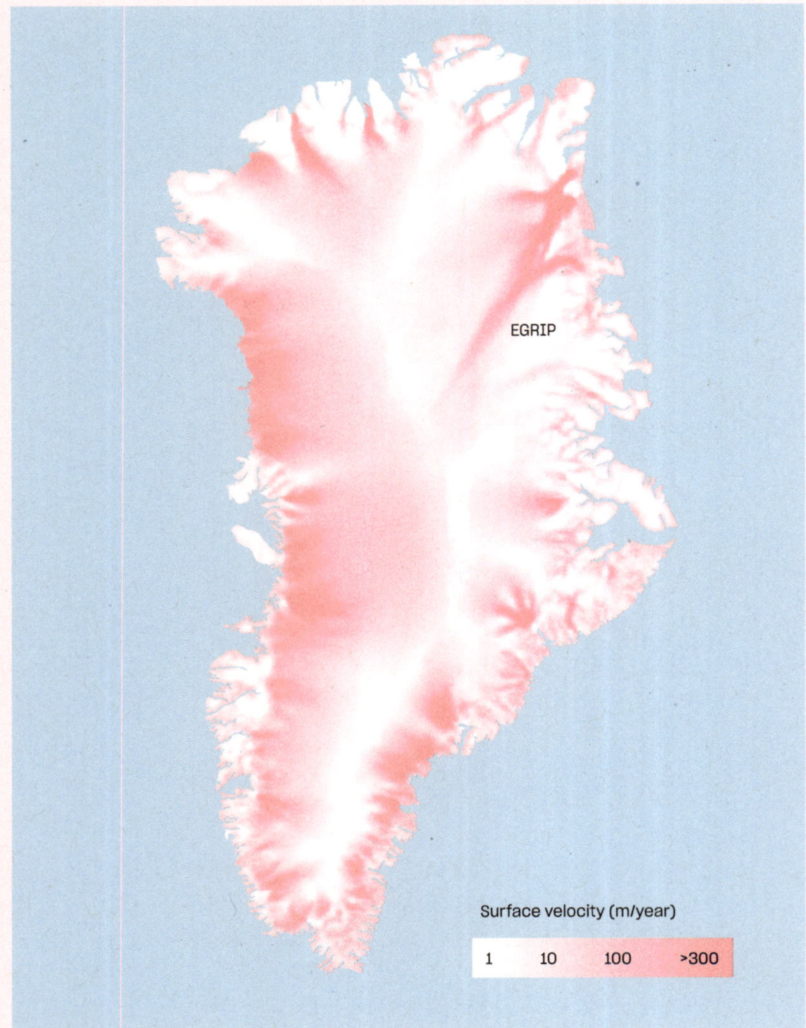

Source: Joughin et al., 2016

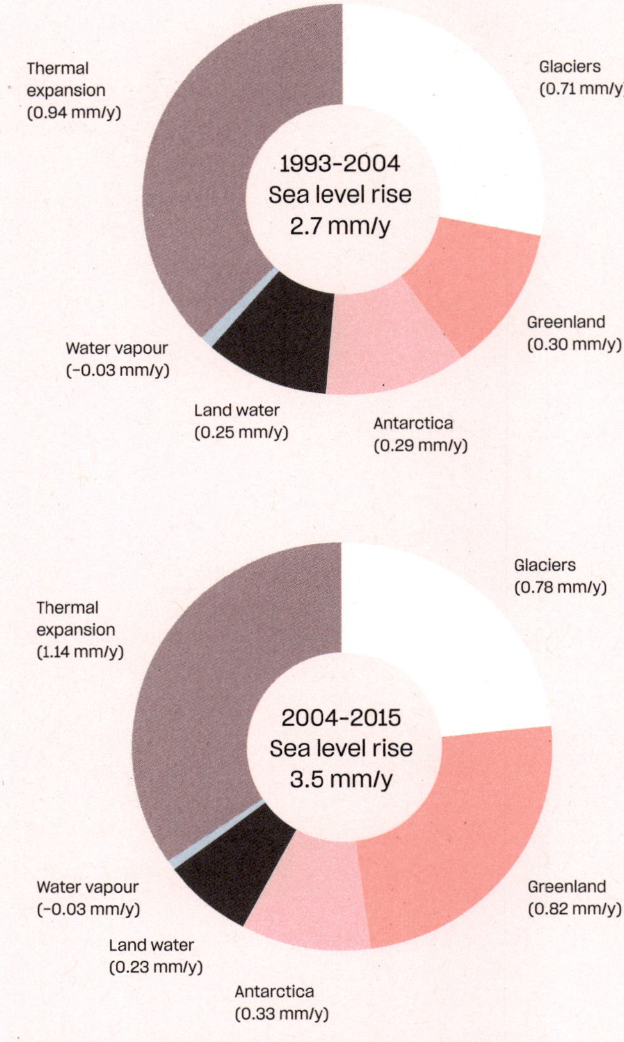

Source: World Meteorological Organization, 2018

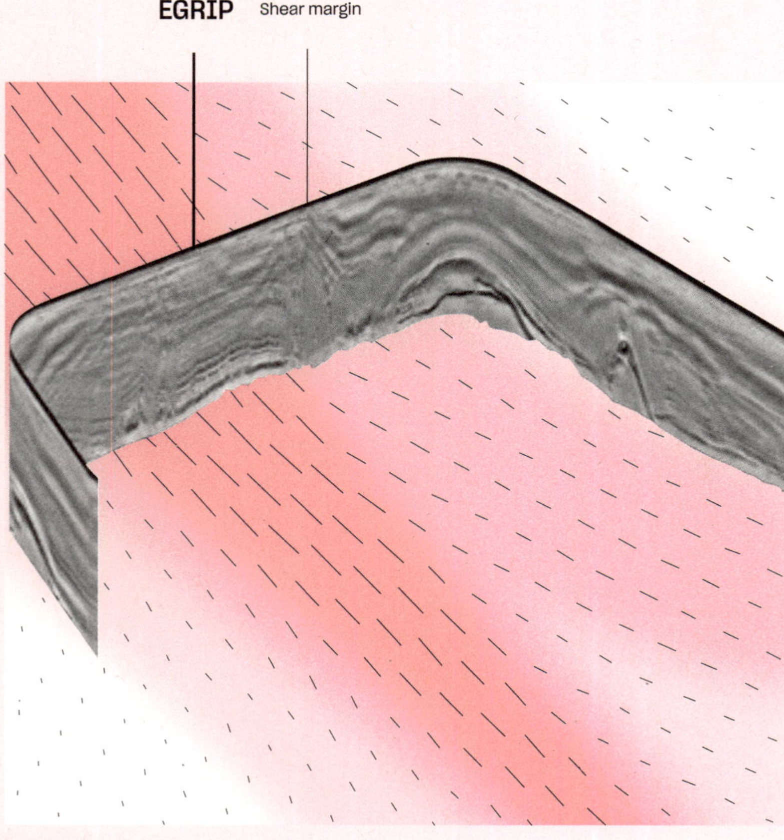

Image of the NEGIS near the EGRIP site. The fast-flowing ice is shown in red and is flowing towards the north. A radio-echo diagram from NASA's Operation IceBridge mission is shown crossing the ice stream at the EGRIP site.

Source: NASA, n.d.

FIRST FINDINGS FROM EGRIP

We established the camp and started making subsurface trenches at EGRIP in 2016. During the field seasons since the camp was established, we have been drilling the ice core in the centre of the NEGIS. The ice is 2,650 m deep here, and on the surface, the ice and the camp are moving at a rate of 57 m per year. Our last field season was in 2019, where we reached a depth of 2,120 m. We still have 530 m to drill, and this is the ice that will really tell us about flow in ice streams. How does the ice deform near the bed? How much water is flowing at the bed? Is the water in the bed itself or in veins in the ice? At the present depth of 2,120 m, where the ice is 45,000 years old, we see very different ice than we have seen in other ice cores. The ice crystals tell a story of ice that has expanded due to the accelerating flow in the ice stream. The result is ice with very different flow properties than ice we have seen elsewhere, and it is important for understanding the velocity of the ice flow. From our investigations of the main ice core, we have made very detailed radio-echo sounding measurements from the surface of the ice in the area around the EGRIP camp. The depth-penetrating radar maps the bedrock and the internal layers of the ice.

The radio-echo sounding image shows great disturbance in the internal layers while crossing the margin of the ice stream left and right of the EGRIP camp. The shear margin of the ice stream is where the ice changes from slow-flowing ice (dark pink to red) to fast-flowing ice-stream ice (white to light pink) and clearly shows how the ice is sheared apart. Analysis of the shear margins show that the margin area is depressed by 15 m to 20 m, forming a valley following the margin. A short ice core drilled in the shear margin also clearly shows very deformed ice right to the surface of the ice sheet where the snow still has not been compressed into ice. We are really looking forward to being able to have another field season, hopefully in 2021, where we can drill to the bedrock and continue our surface investigations. Although we have started observing a very special flow at EGRIP, the real results from the deep ice are still waiting to be unveiled.

OUR FUTURE AND SEA LEVEL RISE

The knowledge gained from using ice cores to understand the flow of ice has taught us lessons about the special flow of ice-stream ice. From our ice core studies, we can see that it is natural for the climatic system to change, to wax and wane between glacial and interglacial periods, and to have rapid unstable events lasting up to 2,000 years during the glacial periods. The present warming in the Artic has clearly been caused by

As humanity unavoidably has to adapt to sea level rise, an important question will be, how fast will it happen? This uncertainty in sea level predictions is worrying. How do we prepare for the inescapable future of sea level rise?

human activities, leading to increased concentrations of greenhouse gasses in the atmosphere. All evidence from the past shows that increasing temperatures in the Arctic causes ice to melt, resulting in rising sea levels. In less than a hundred years, atmospheric temperatures in the Arctic will very likely be 5 °C warmer than those we experienced between 1950 and 1980, before the largest contributions to human-caused warming started. This brings us to the same temperatures we had during the last interglacial period, 130,000 years ago, where we know that Greenland lost 20% of its volume and global sea level rose by 5 m to 9 m. As humanity unavoidably has to adapt to sea level rise, an important question will be, how fast will it happen? This uncertainty in sea level predictions is worrying. How do we prepare for the inescapable future of sea level rise?

The sea level rise in the Arctic will in general be less than the global average, because large land areas are still rising after the termination of the last glacial period 11,700 years ago. However, the Arctic ice that is melting right now will lead to large sea level rises in the southern hemisphere. The studies at EGRIP to understand the ice stream flow behaviour are urgently needed to improve our predictions of ice loss, and thus sea level rise. Inland studies must be connected with studies on the interactions of ice streams and ocean. How does the ice flow into the ocean, and how do icebergs break and calve from ice streams? How does the warming ocean influence the acceleration of ice streams, and what will happen when the ice steams retreat and lose their 'wet feet'?

To me, it is clear that the best strategy to avoid accelerated loss of mass from the Greenland ice sheet, and thus serious increases in sea level, is to reduce global emissions of greenhouse gasses. This would slow the increase in atmospheric temperature, and thus the increase of melting ice. It is a global responsibility to reduce emissions, and while the effects of the climate crisis are strongest in the Arctic, which is thus exposed to humanity's greatest changes in living conditions, sea level rise is something that impacts everyone on Earth.

MY WISHES FOR A BETTER FUTURE

My research has focussed on understanding the fast-flowing ice streams on the Greenland ice sheet. The ice streams, with their discharge of freshwater into the ocean, are very important for fisheries. Freshwater, with nutrients from melting ice, is an important source for the primary production. Knowledge of the freshwater discharge in the future will thus be crucial for the communities in the North. While we can observe from satellite and direct measurements that most ice streams are accelerating as the ocean warms, we also see that the Jakobshavn Isbræ has slowed down in very recent years. We really need to focus on understanding the ice streams, as they are so important for both communities and sea level rise globally. I hope that the research at the EGRIP site on the NEGIS will help us improve our understanding of ice streams. In this way, I hope to contribute to our future.

Also, I hope to be able to contribute to Inuit-managed research programmes in northern Baffin Bay, between Canada and Greenland, with the intention of forming the first Inuit-managed area. Pikialasorsuaq is an obvious place to enhance the observations of ice streams and the impact on living resources around the ice sheet margins, where the ice streams deliver freshwater to the ocean. Such programmes will benefit local communities and contribute valuable data for global sea level predictions, which are important for all communities close to the sea. •

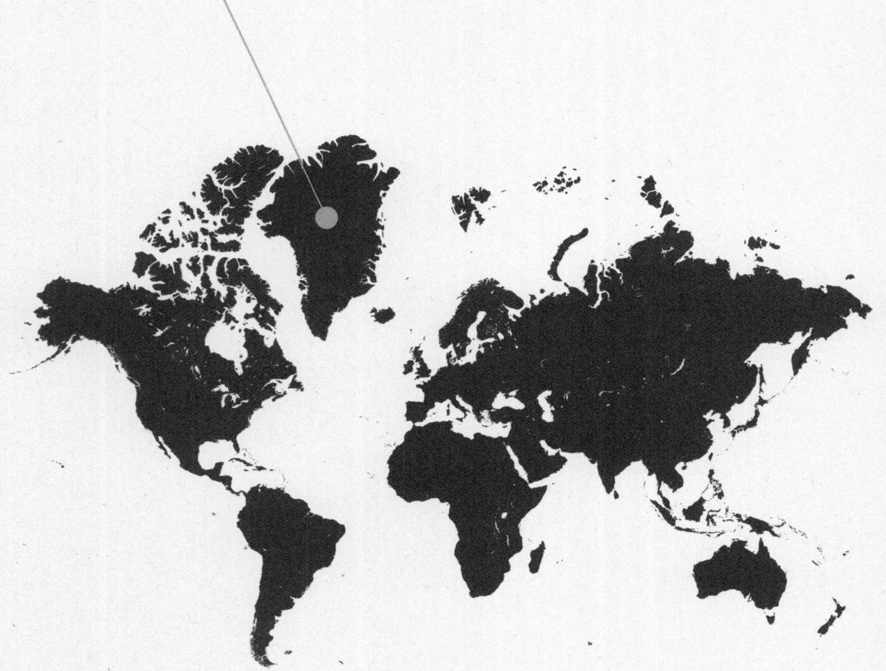

The main reason for the world's rising sea levels is the melting of Greenland's ice cap and glaciers, which is occurring at an unprecedented rate.

For some of the people in the south of Greenland, this is perceived as good news. Due to the absence of snow and ice, they can now grow crops for the first time: potatoes, onions, lettuce and even strawberries. Residents hope this will make them less dependent on Danish imports and that food prices will decrease.

In the north of Greenland, it is a different story. Here, sea ice is disappearing, and glaciers are retreating at an alarming rate. In 2012, there was an extreme melting event, where almost all of the ice cap melted on the surface. On average, the summer melt lasts 70% longer now than it did in the early 1970s. If the entire Greenland ice cap melts, oceans will rise by 7 m on average. For traditional hunters, life has become difficult, and sea animals are facing difficult times trying to survive in this changing environment.

As goes Greenland, so goes the world.

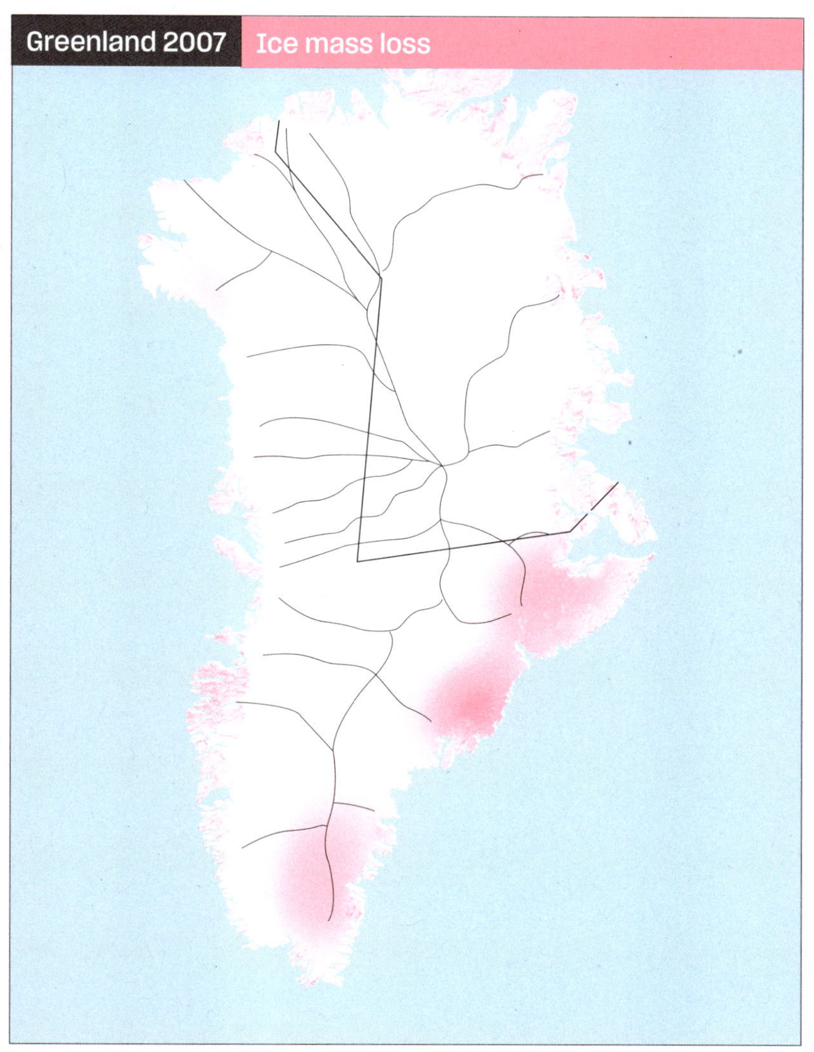

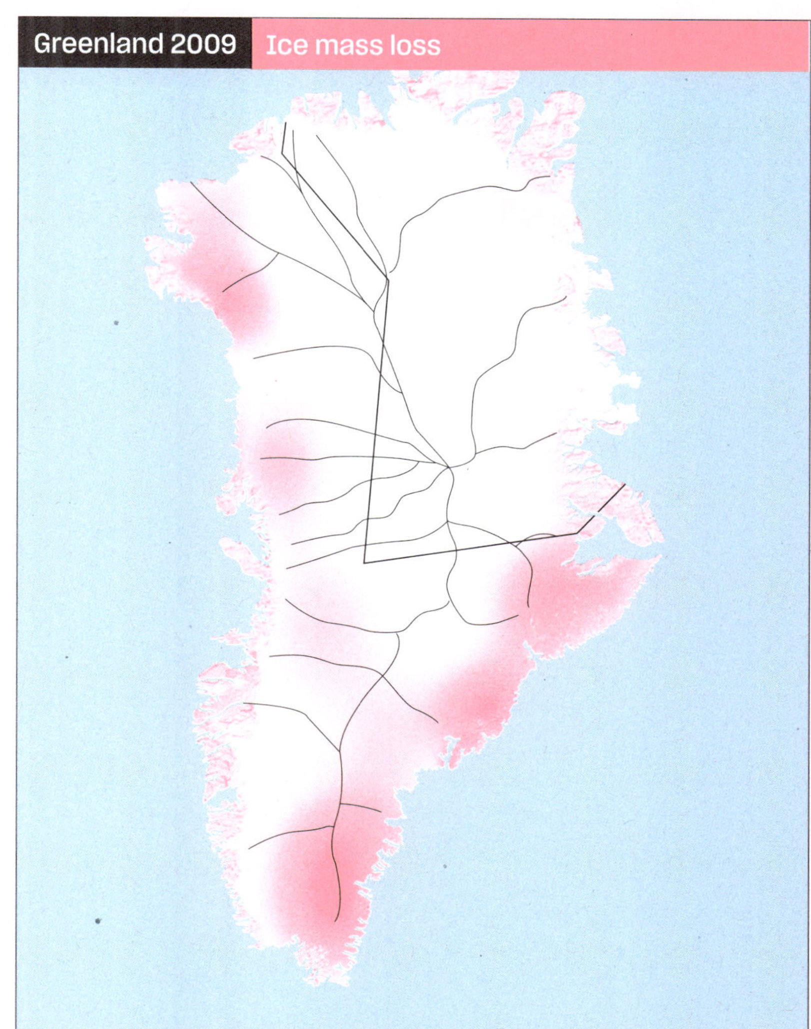

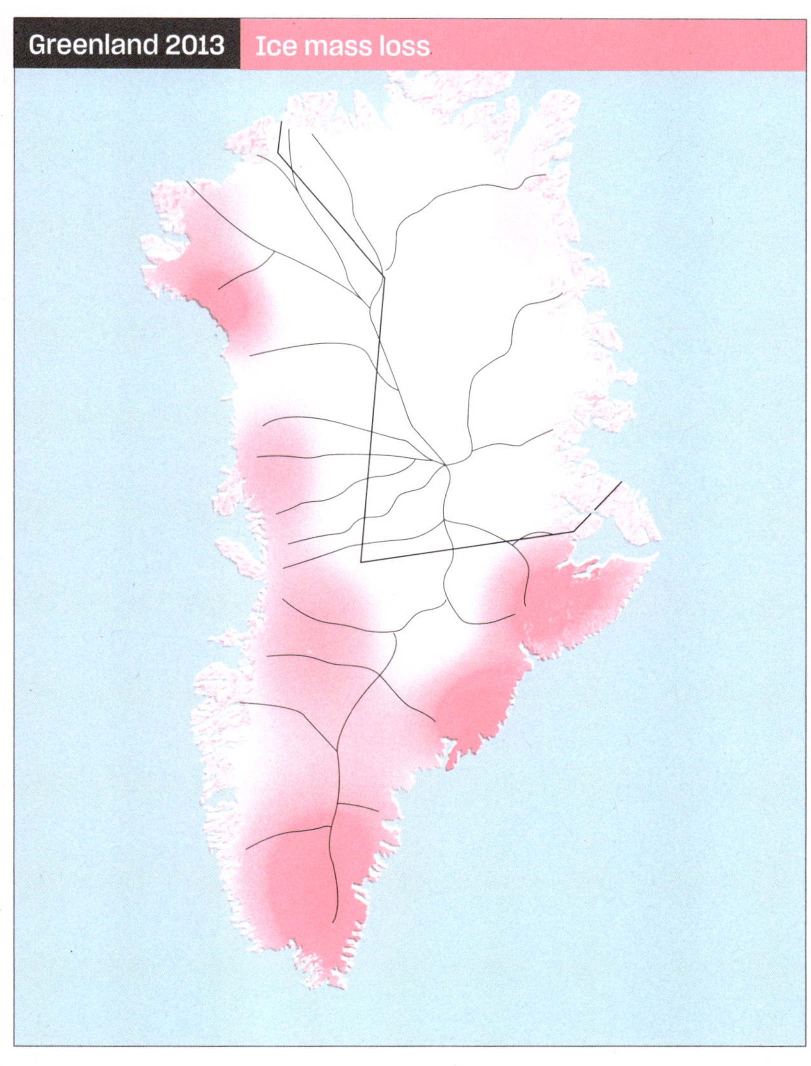

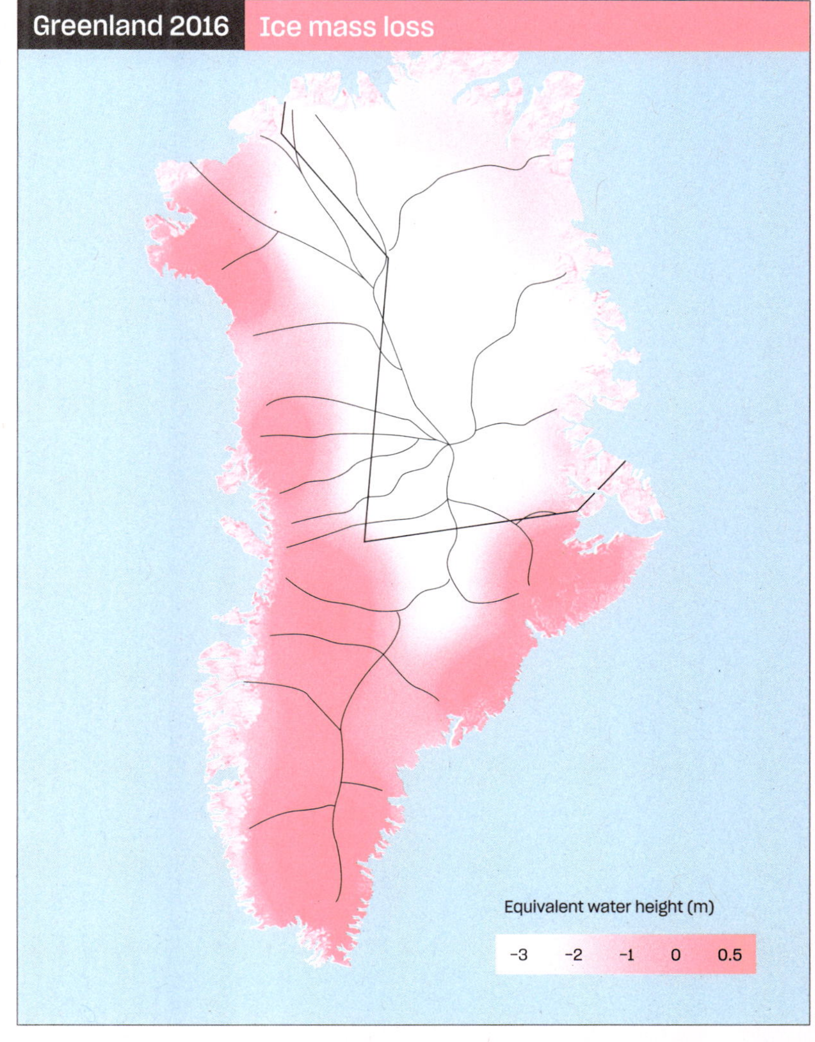

Source: NASA Jet Propulsion Laboratory, n.d.

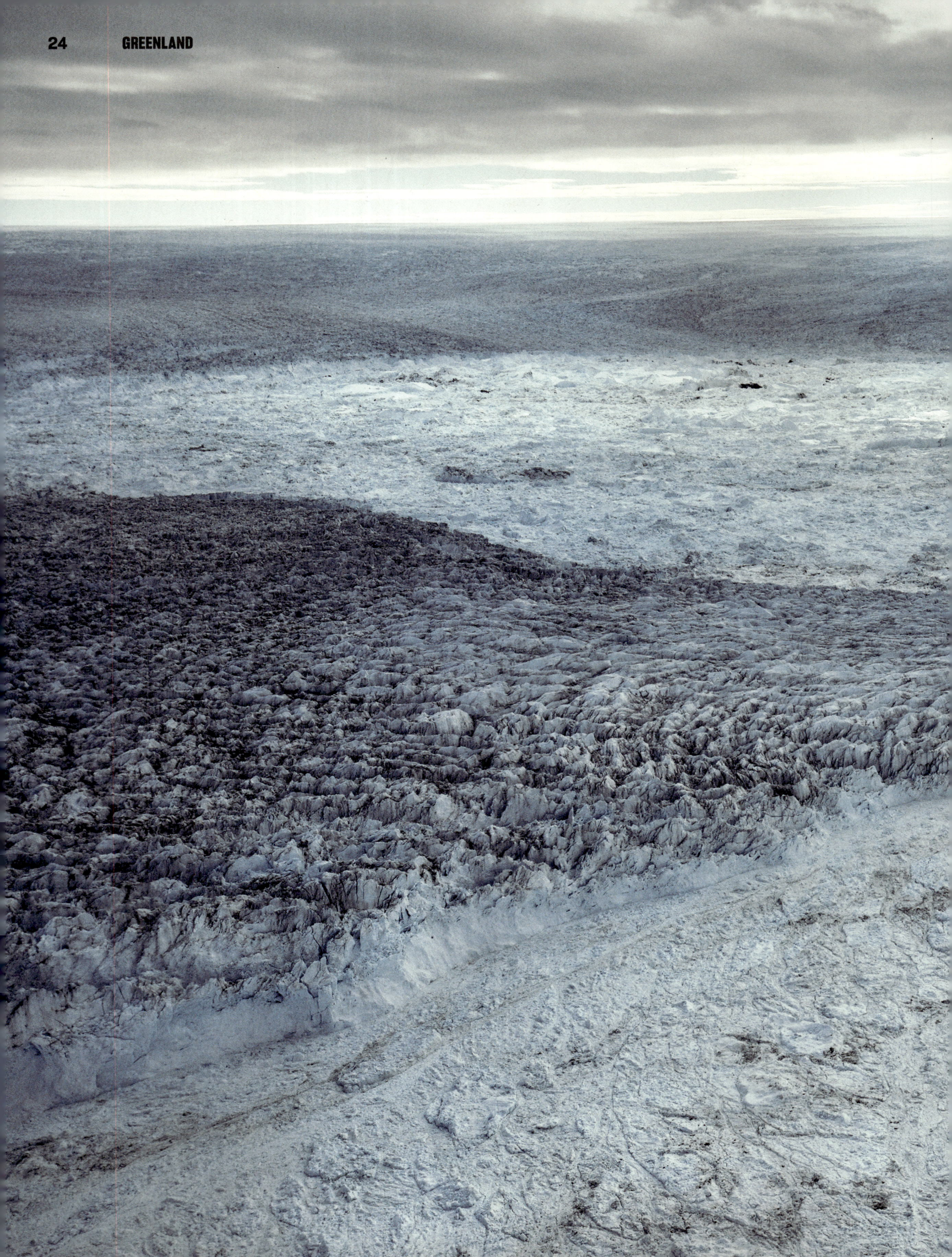

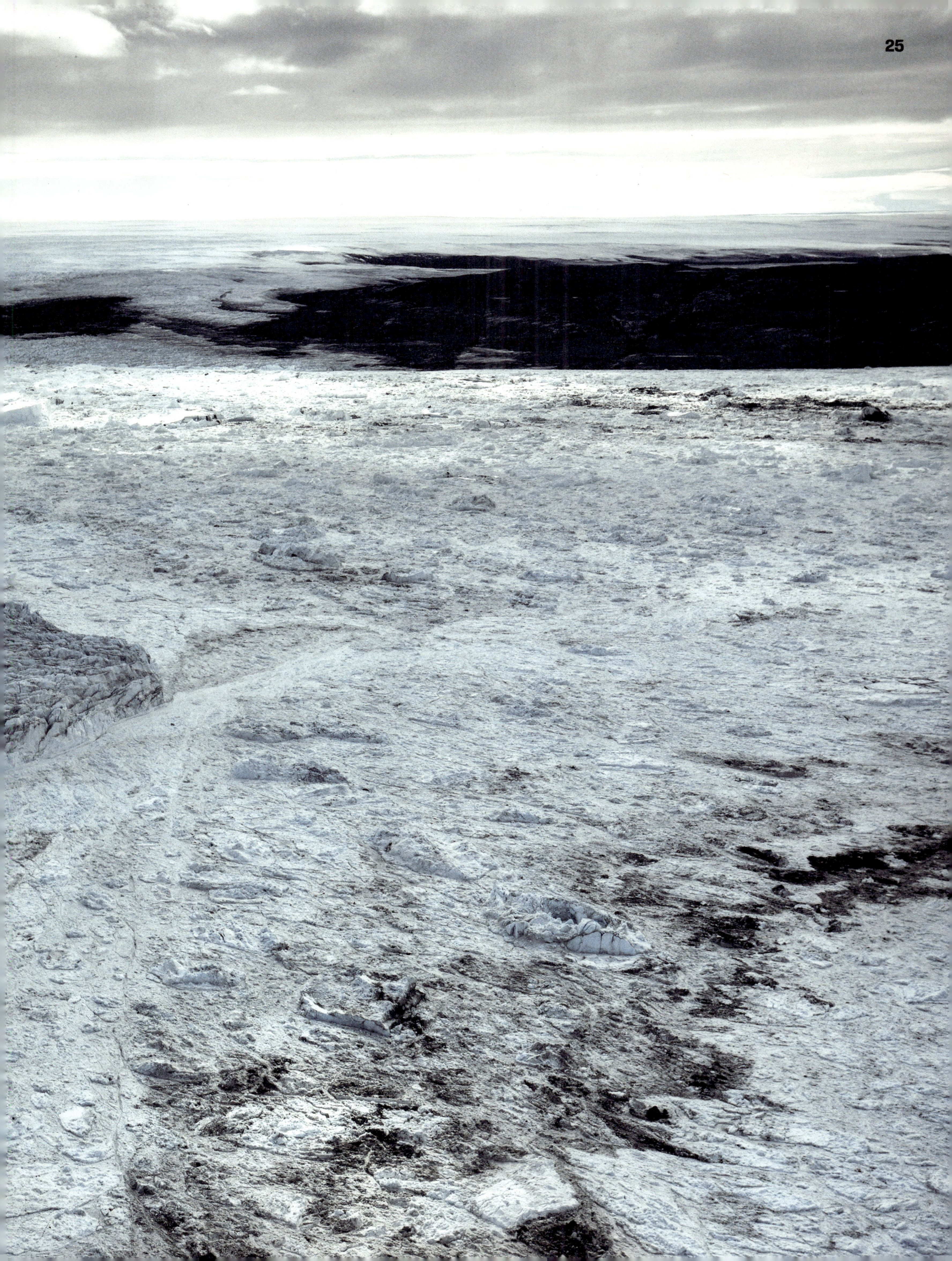

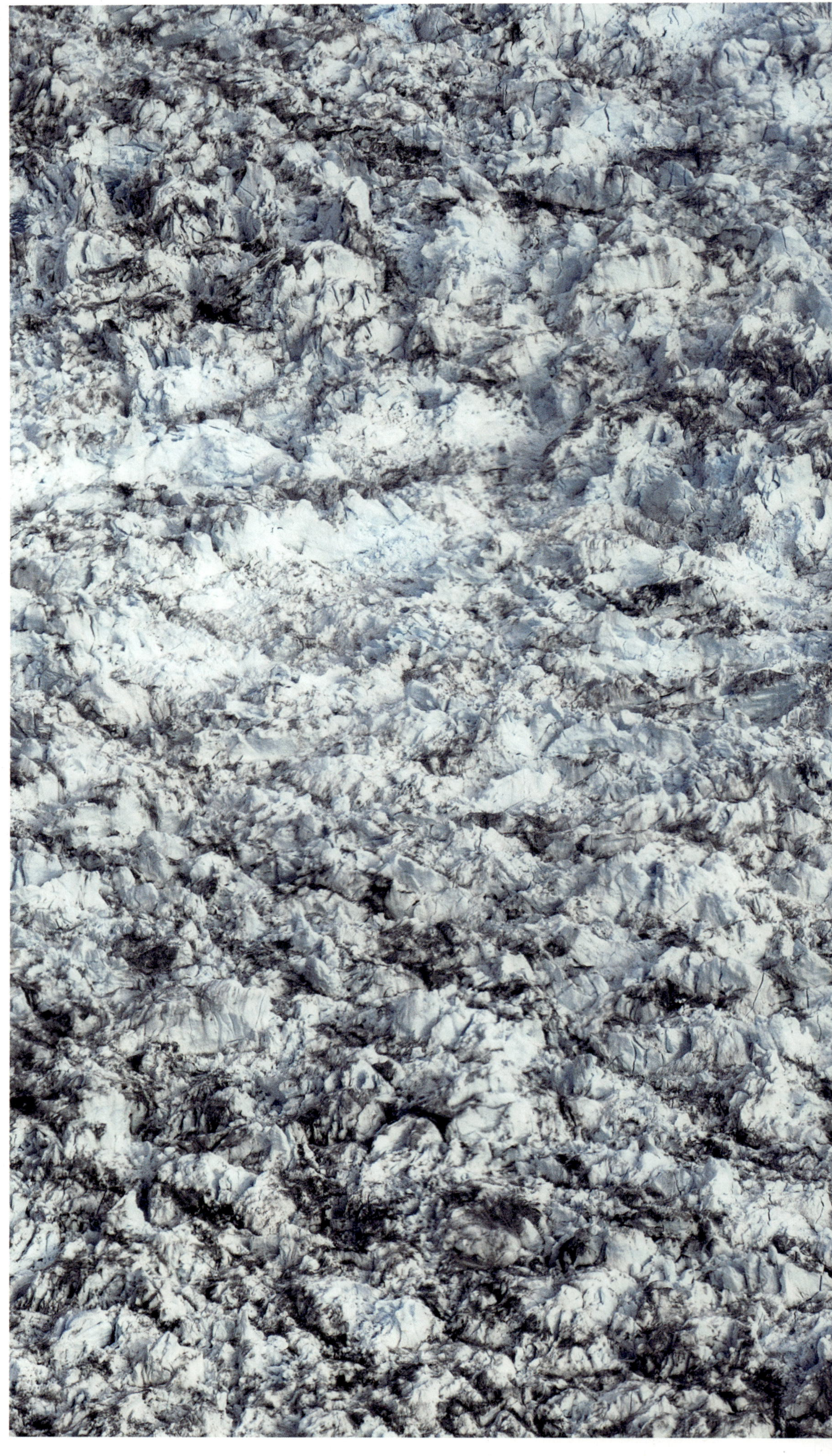

< Pages 24-25: The Sermeq Kujalleq glacier (Greenlandic), also called the Jakobshavn Isbræ (in Danish). The glacier is close to the town of Ilulissat, on the west coast of Greenland. It's one of the fastest-moving (and fastest-retreating) glaciers in the world, moving about 20 m a day. It is therefore a large contributor to rising sea levels.

A very large piece of ice breaking off of the Sermeq Kujalleq glacier.

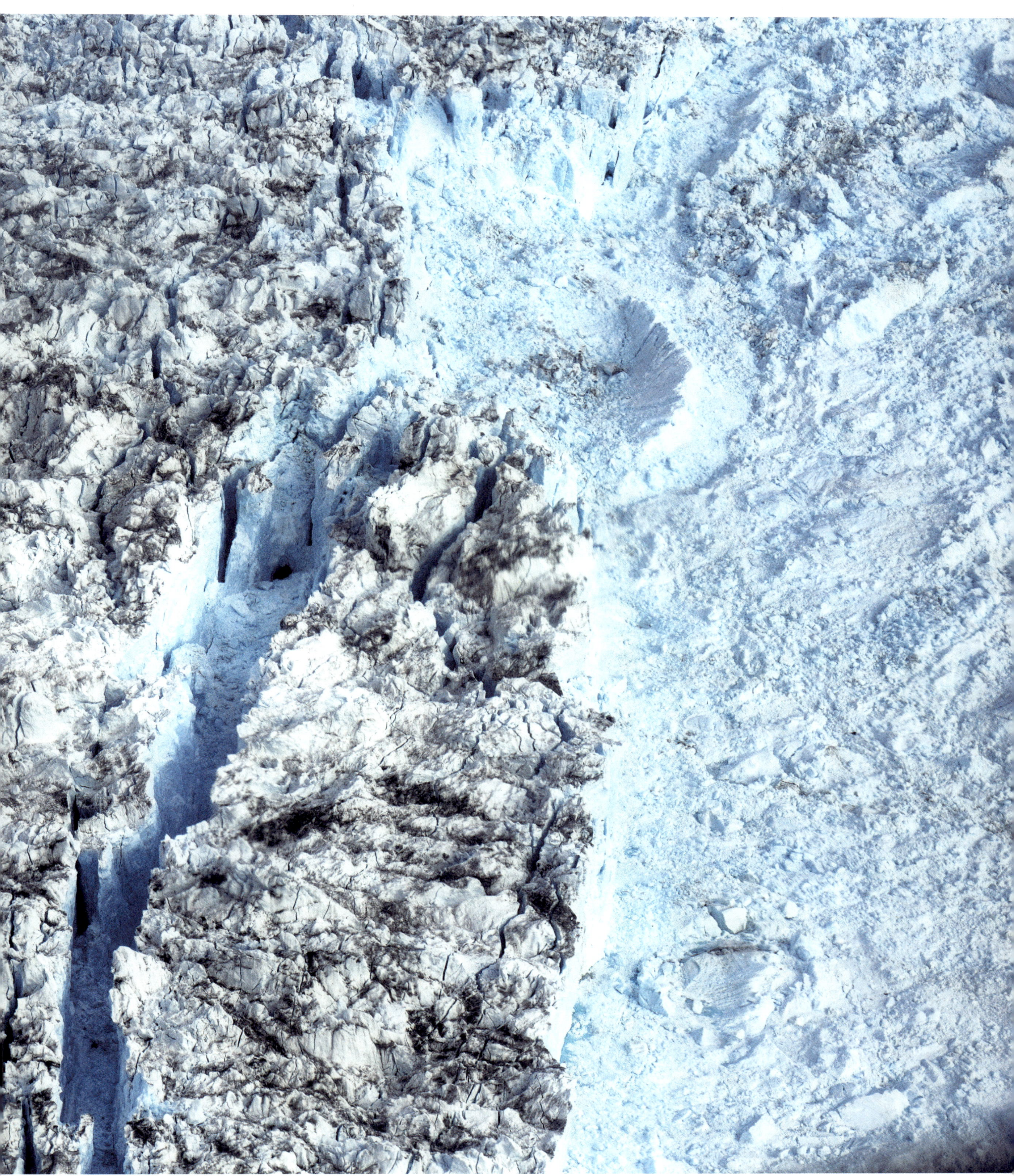

Melt rivers on the ice sheet, close to the edge near Kangerlussuaq. Due to the climate crisis, glaciers retreat at a rapid rate and the ice sheet itself also melts, forming melting streams, reservoirs and underground rivers.

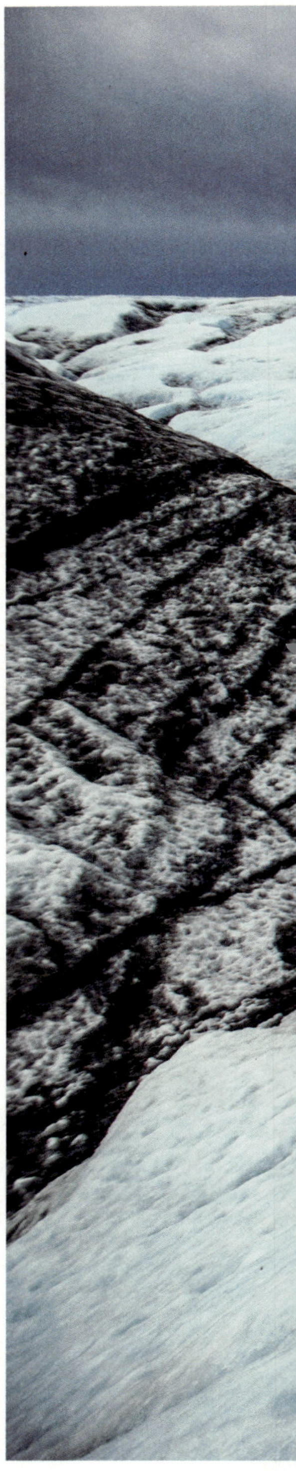

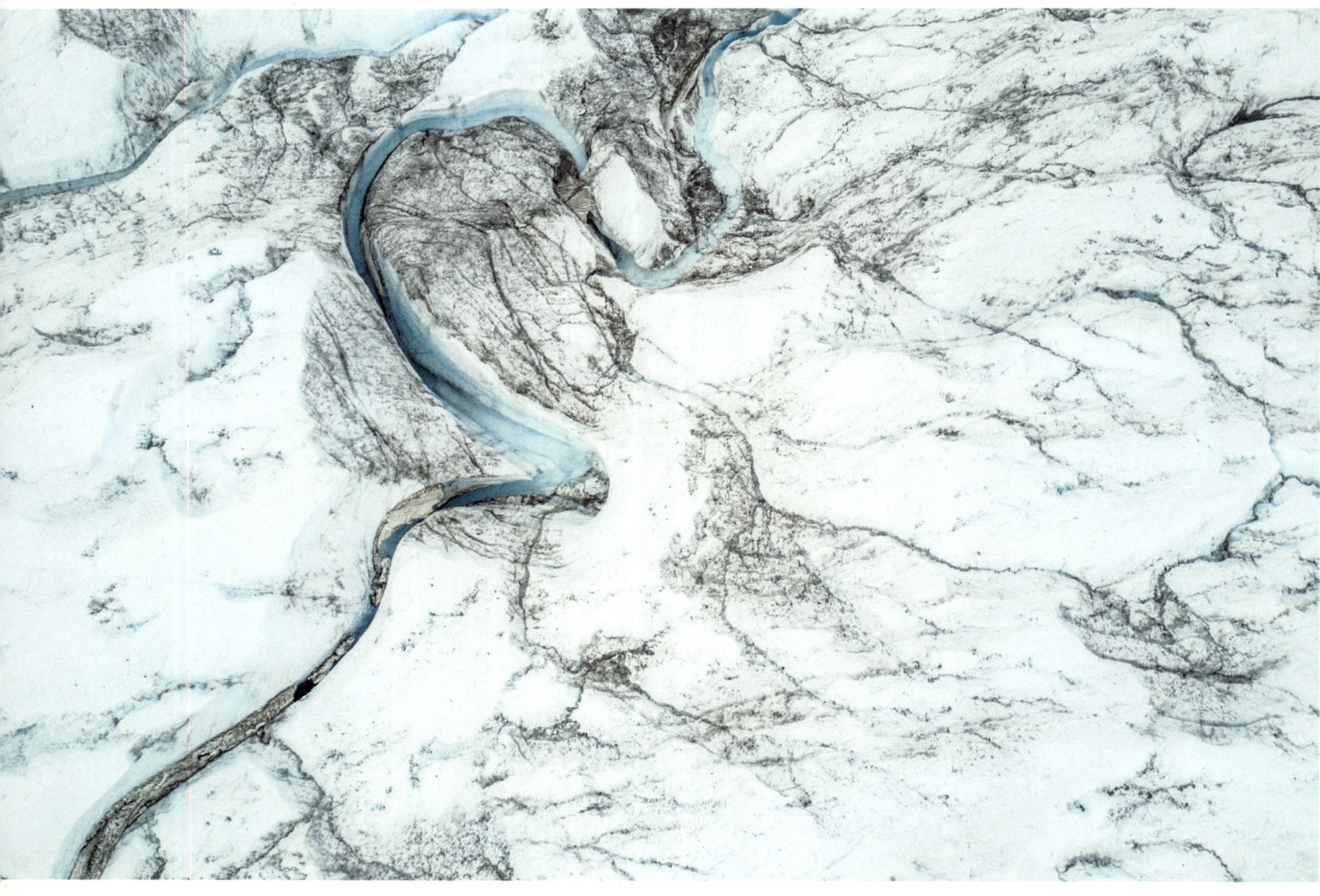

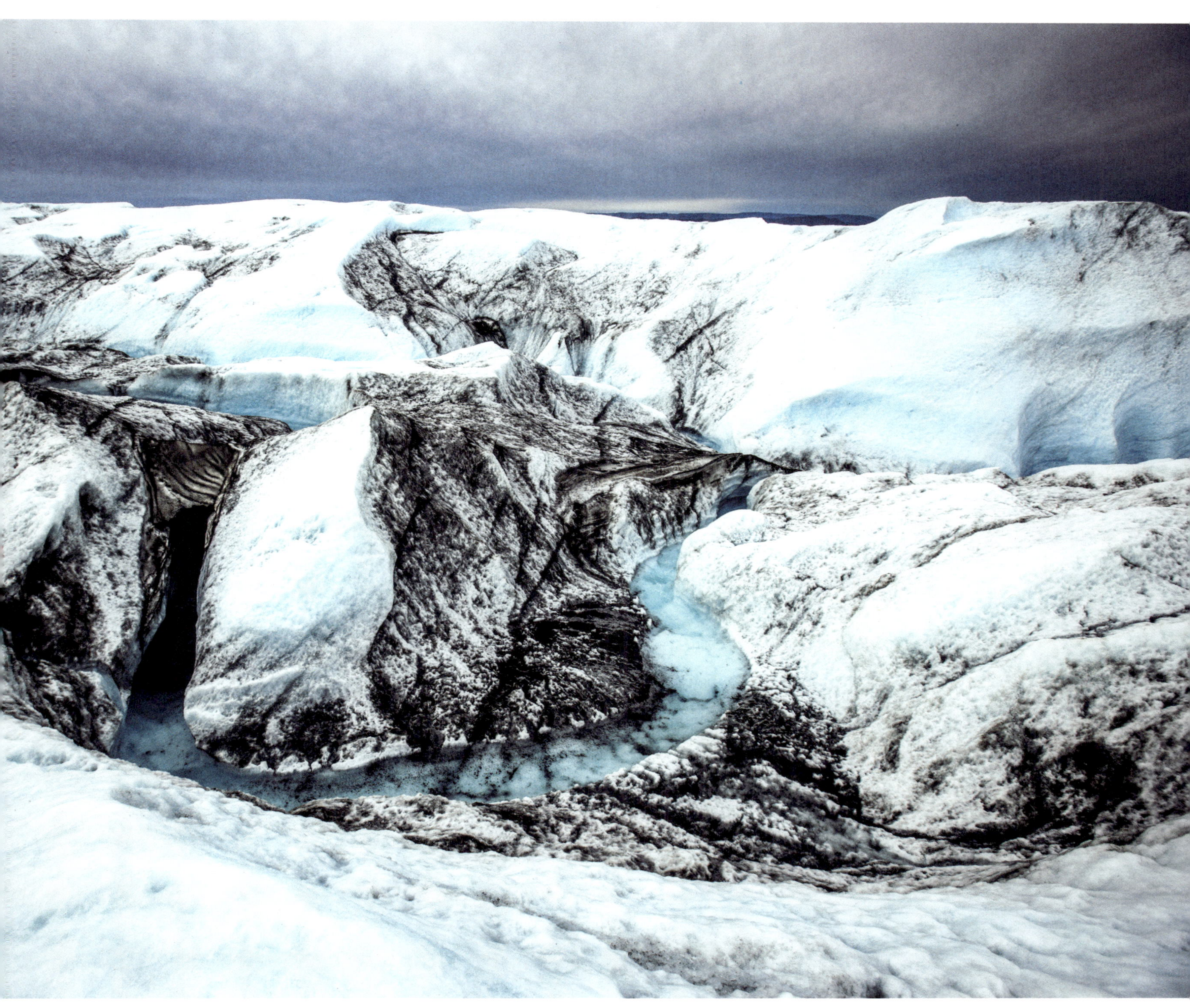

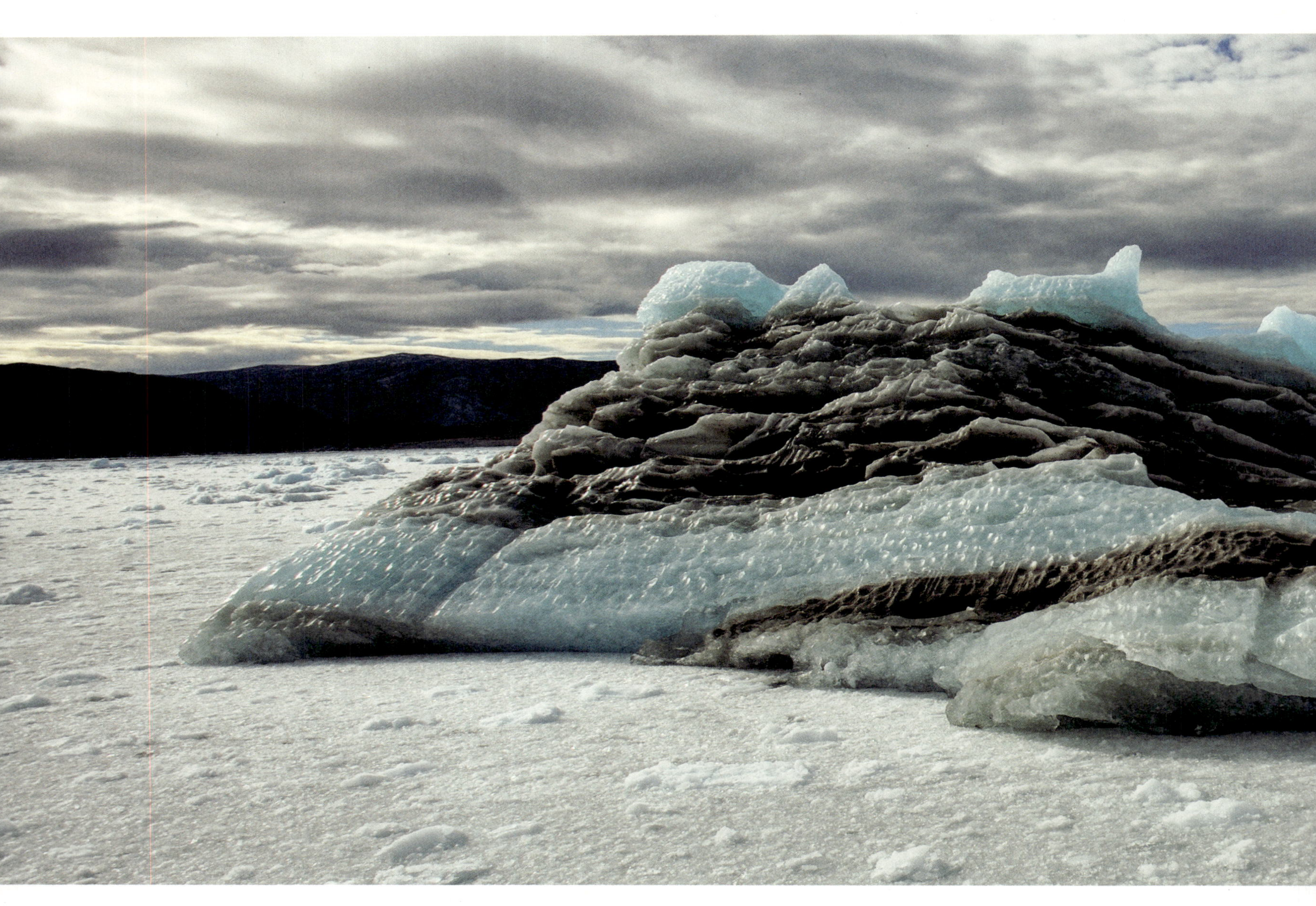

An iceberg floats in front of the
Eqip Sermia glacier. Together with
the Sermeq Kujalleq glacier, it is the
most active glacier in the Northern
Hemisphere.

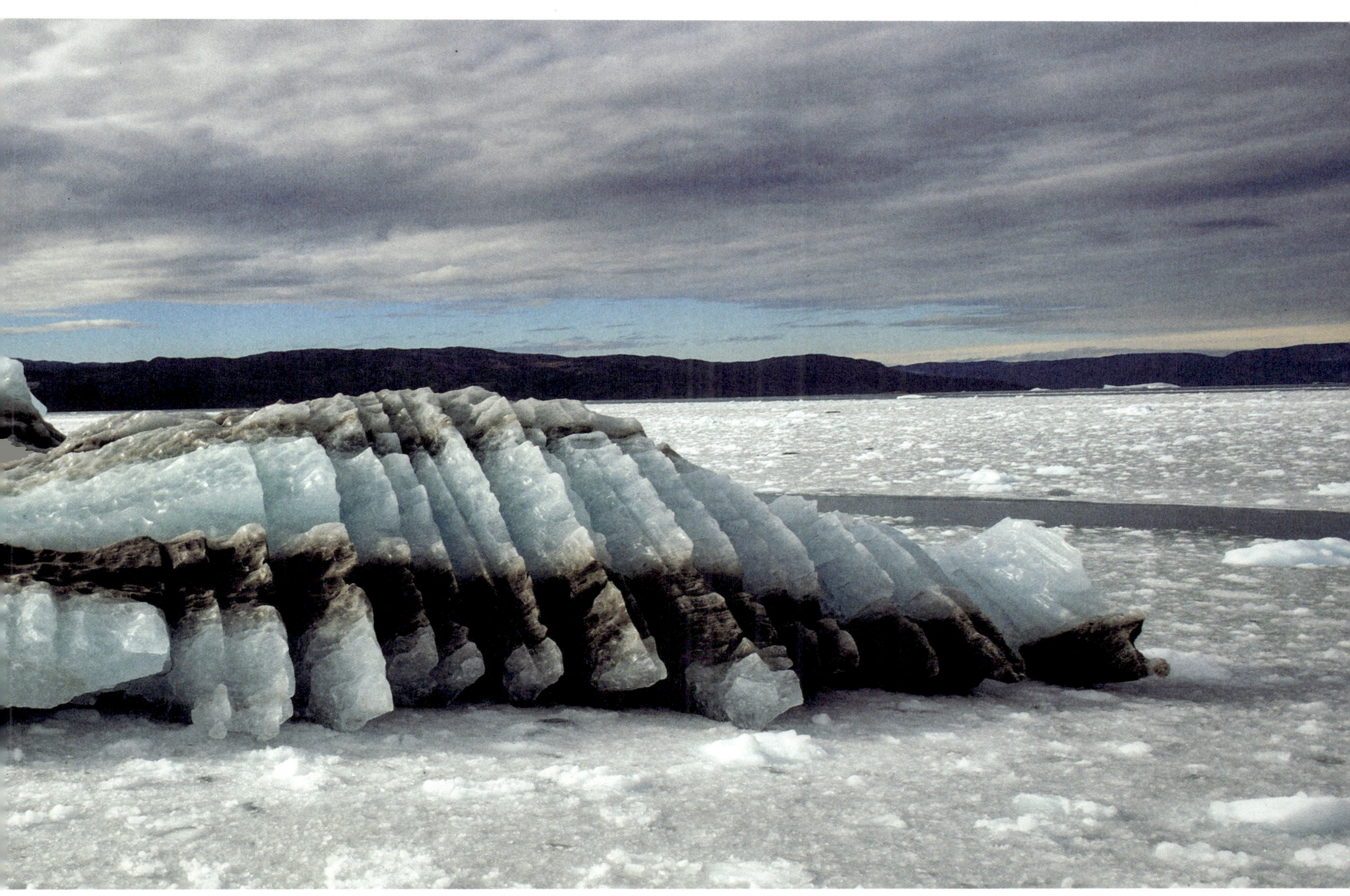

Meltwater from the ice sheet
near Kangerlussuaq.

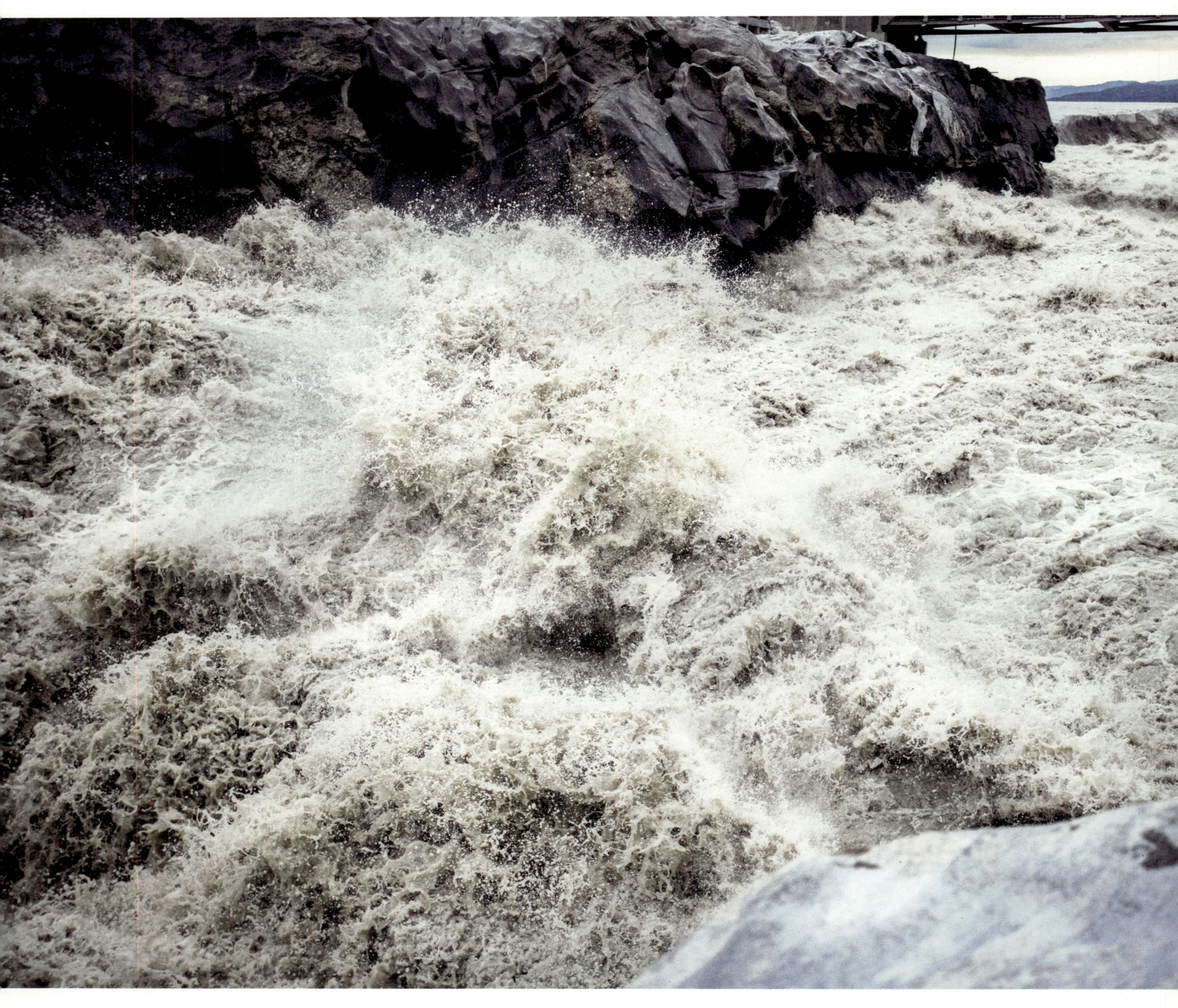

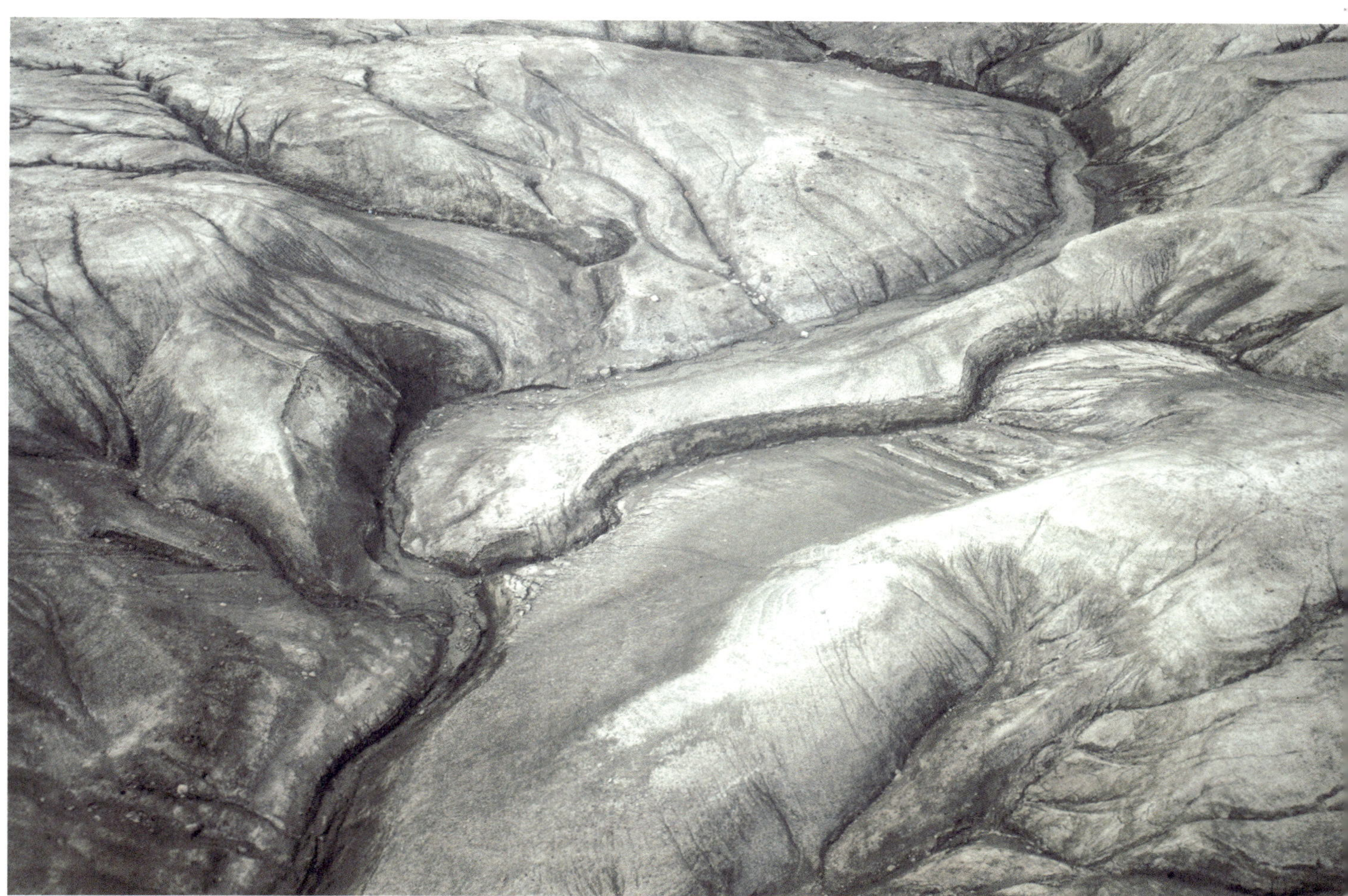

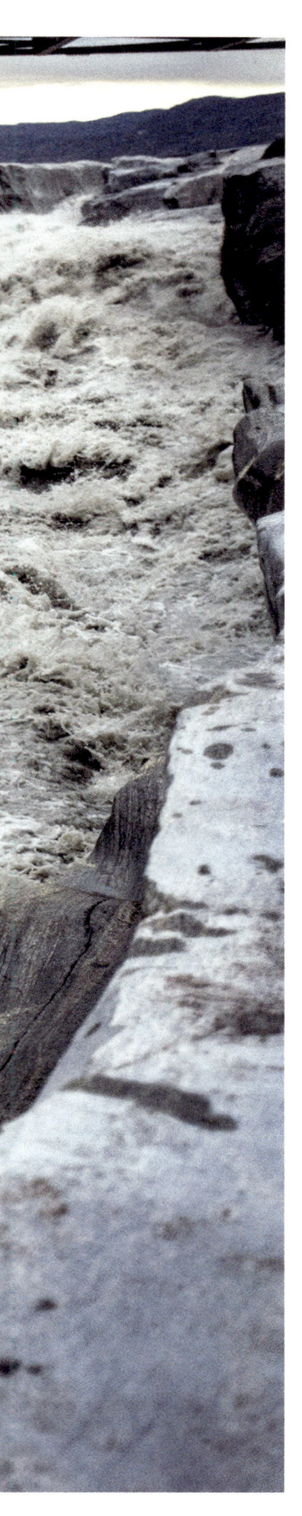

Landscape near Kangerlussuaq.

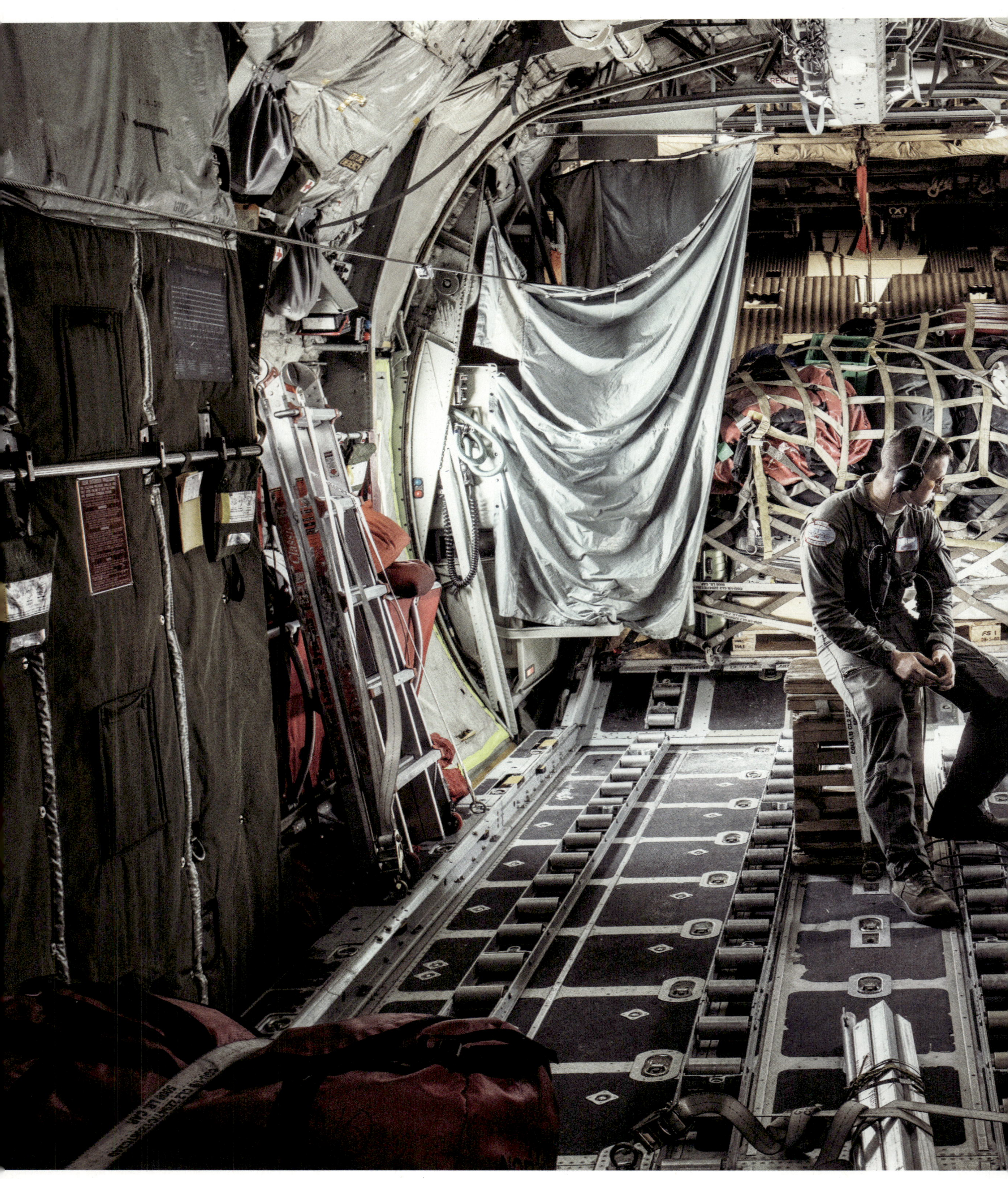

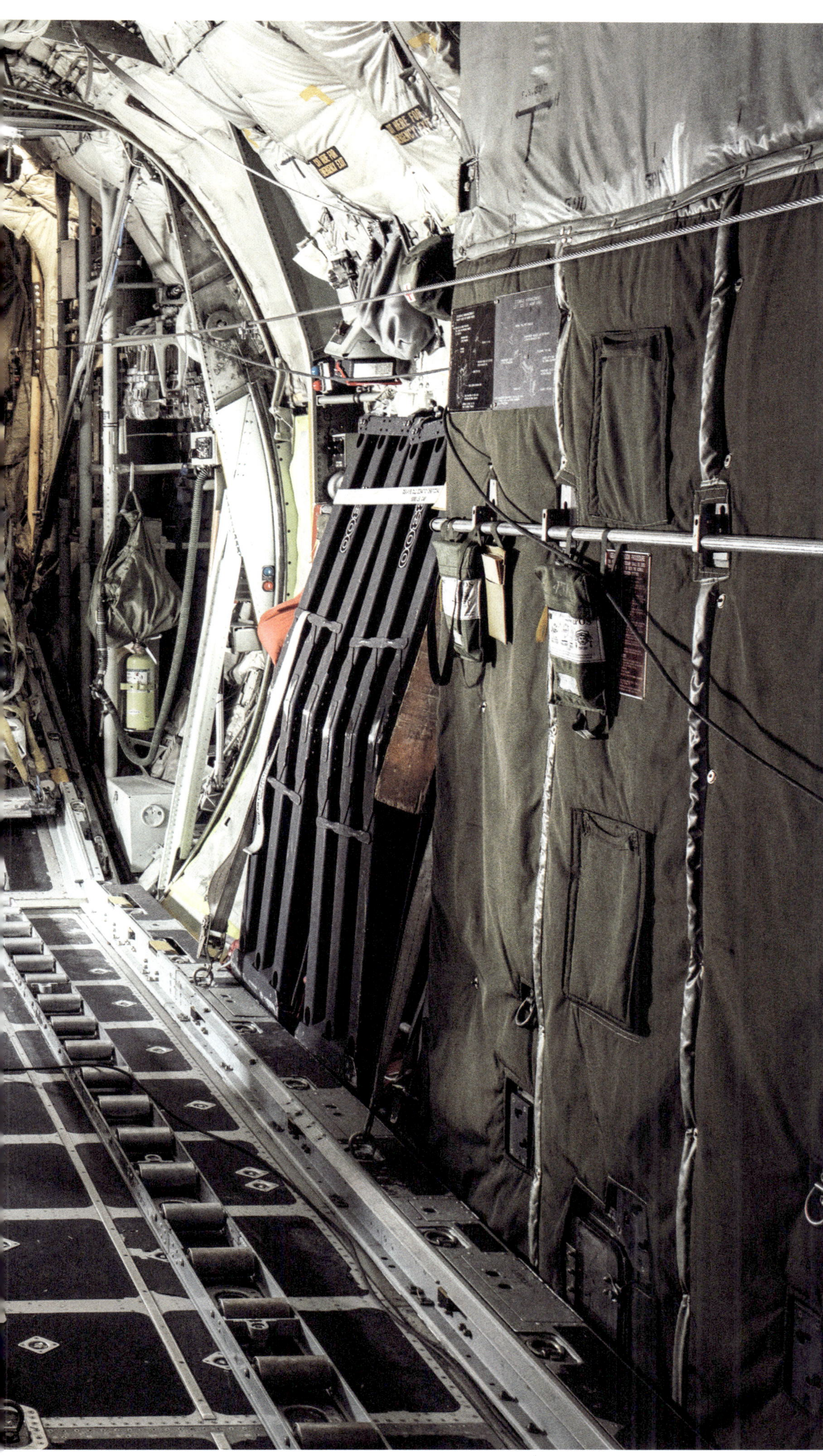

A Lockheed Martin C-130 transport plane takes scientists to the East Greenland Ice-core Project (EGRIP). The flight from Kangerlussuaq to the camp takes about 3.5 hours. The plane lands on skis.

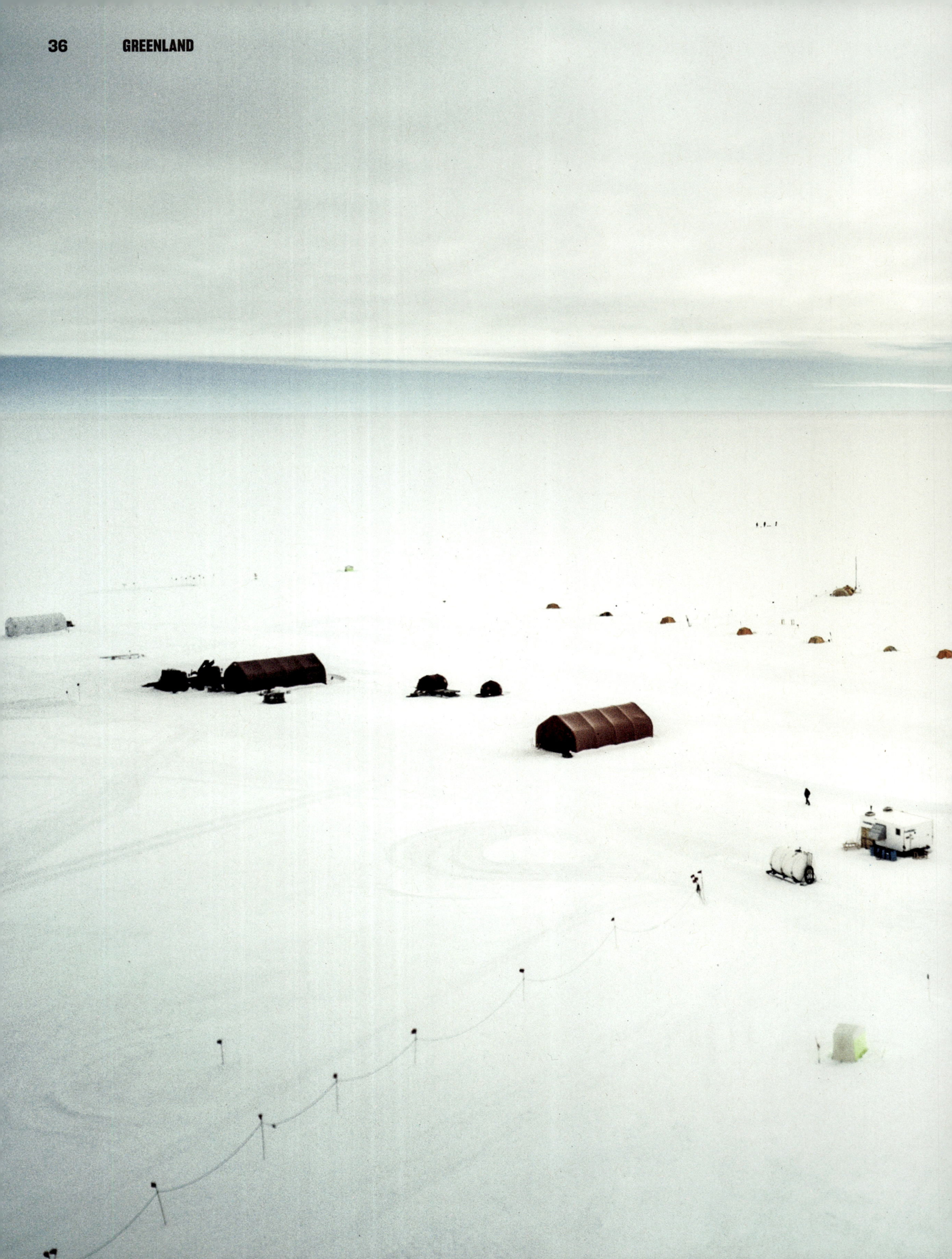

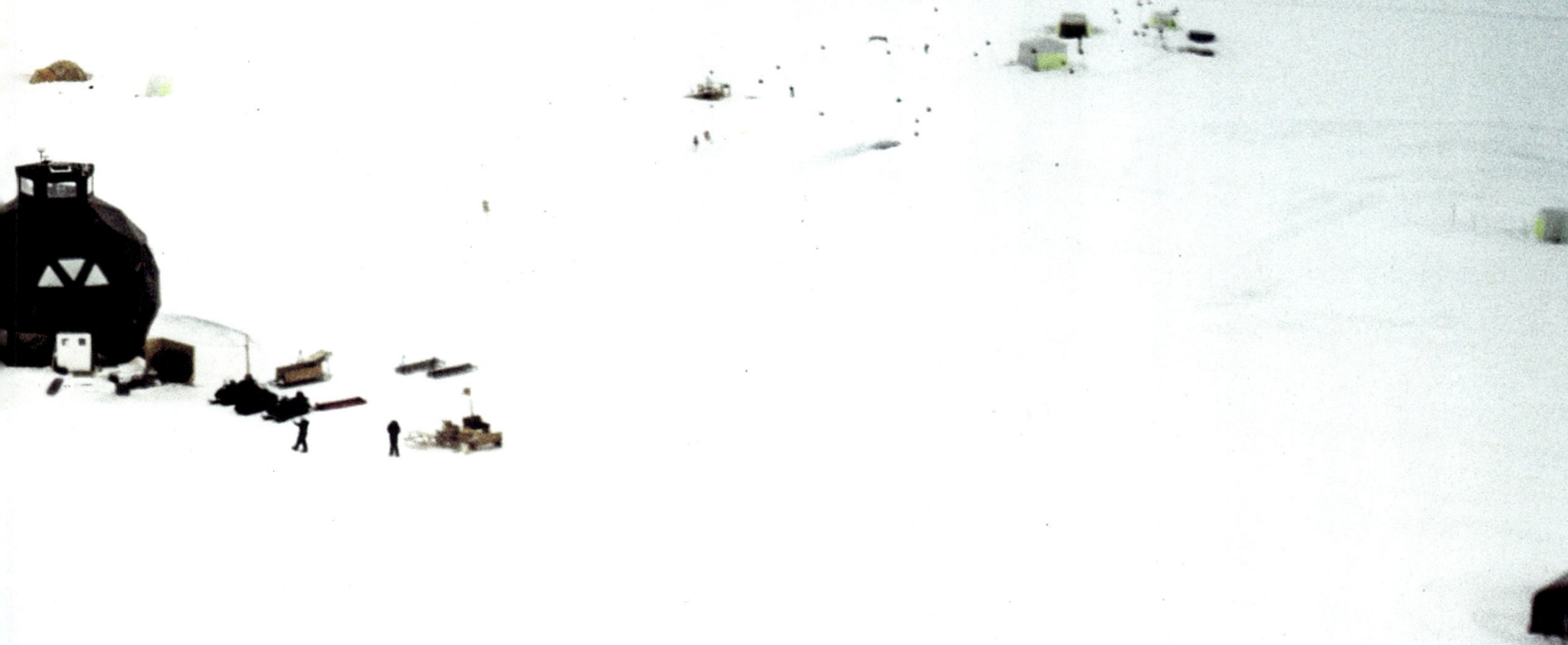

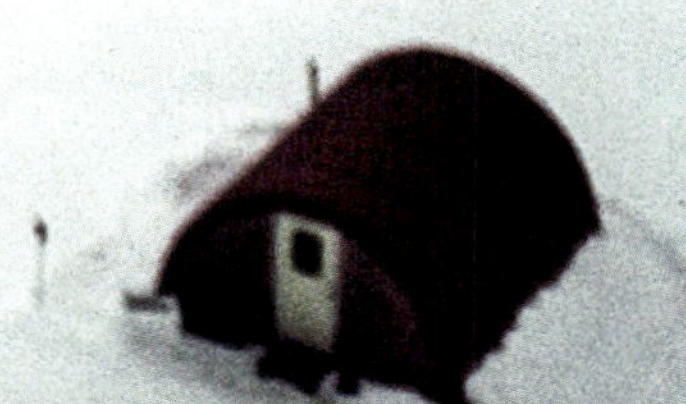

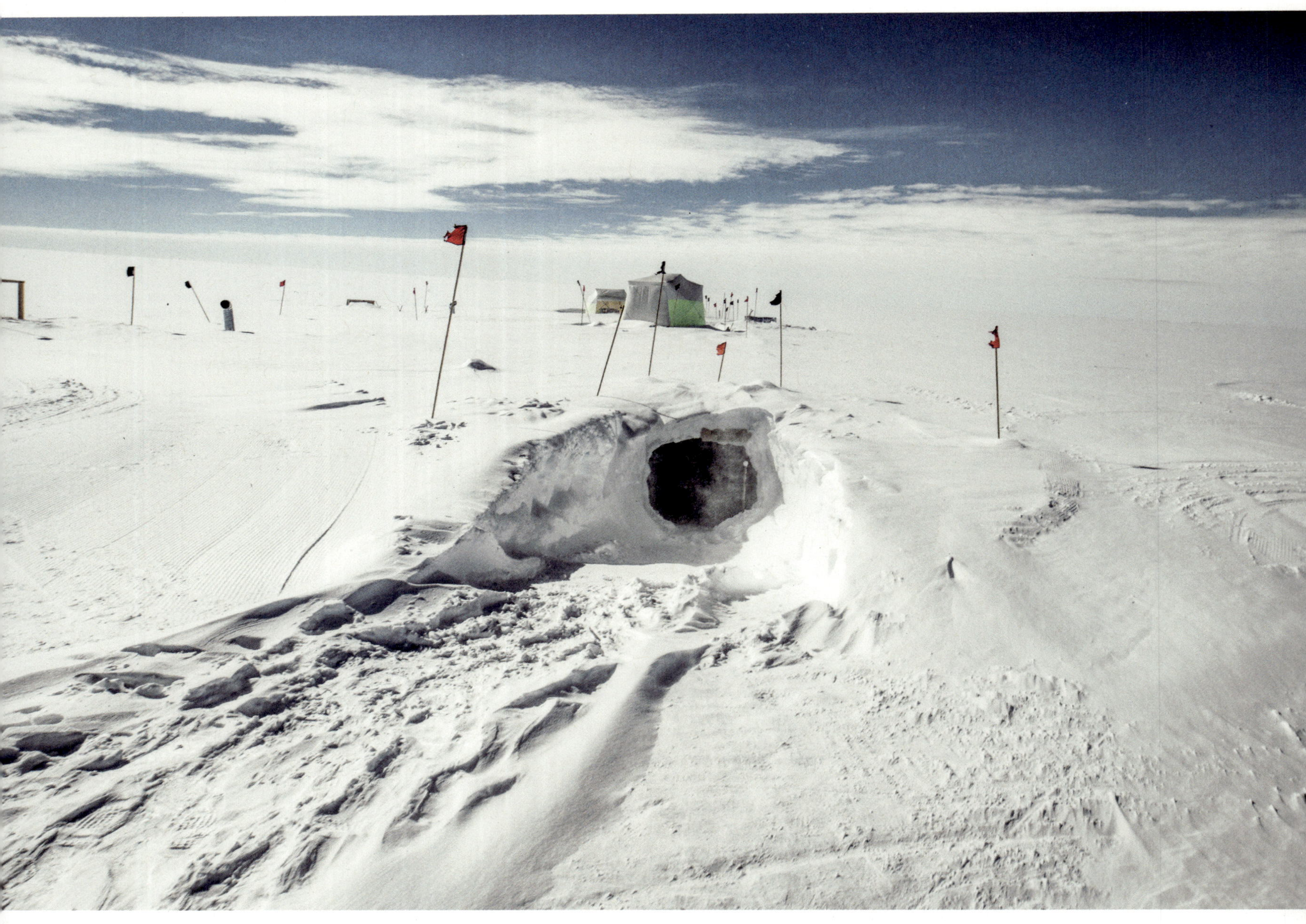

< Pages 36-37: The East Greenland
Ice-core Project (EGRIP) camp on
the Greenland ice sheet.

The East Greenland Ice-core Project
(EGRIP) camp is located at the
northeast of the Greenland ice sheet.
A couple of years ago, ice flows were
discovered on the Greenlandic ice
sheet, essentially frozen rivers which
move to the ocean.

Therefore, scientists drill to the bedrock to determine the extent these
flows contribute to rising sea levels.
The elevation of the camp is 3,000 m,
and the ice sheet is 2,700 m thick. The
ice sheet is moving towards the ocean
at a rate of 15 cm a day, which could
prove to be a major contribution to
rising sea levels.

Ice cores being stored at the East
Greenland Ice-core Project (EGRIP).
The ice is from a 1,500 m depth and is
approximately 22,000 years old.

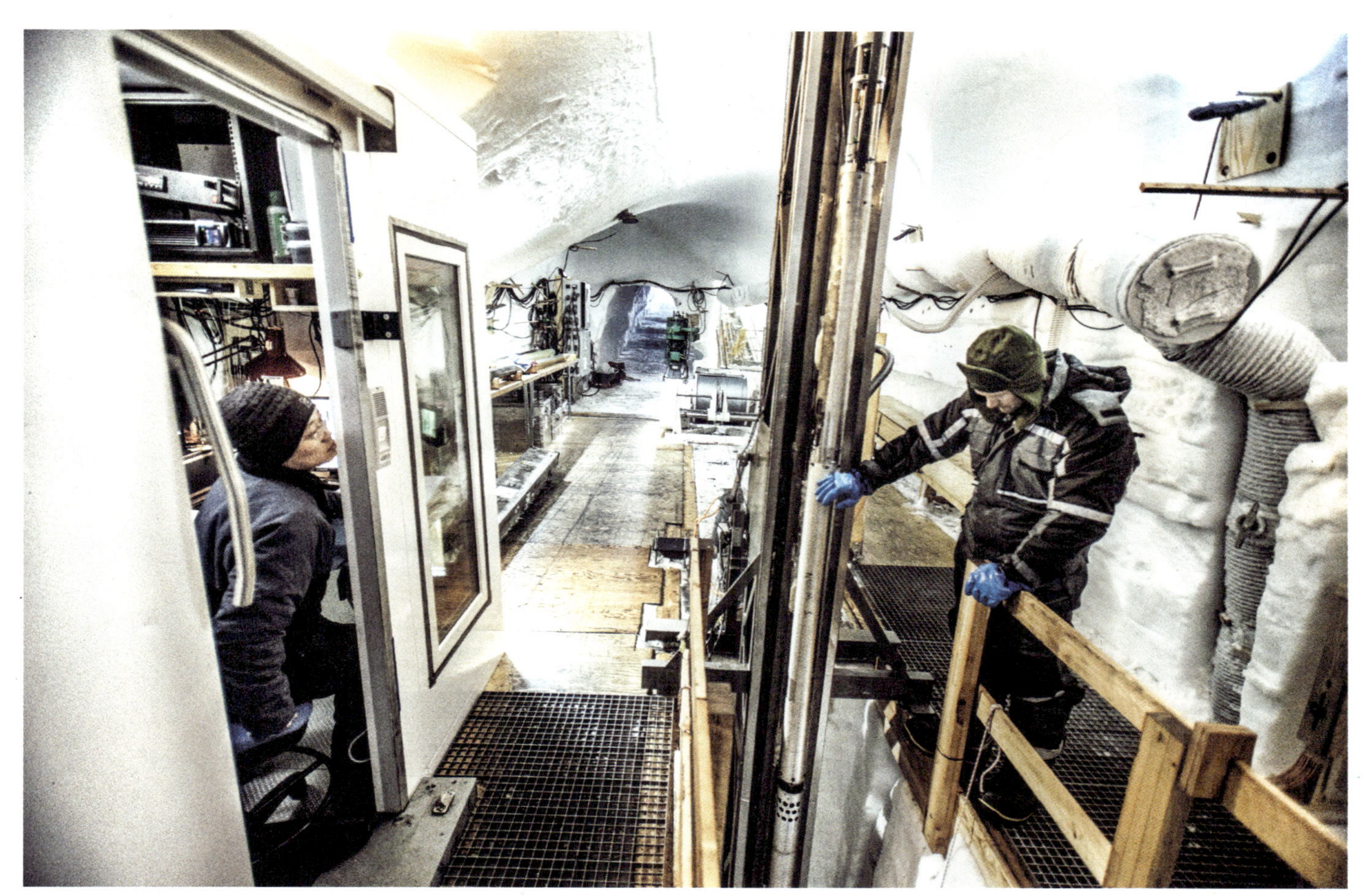

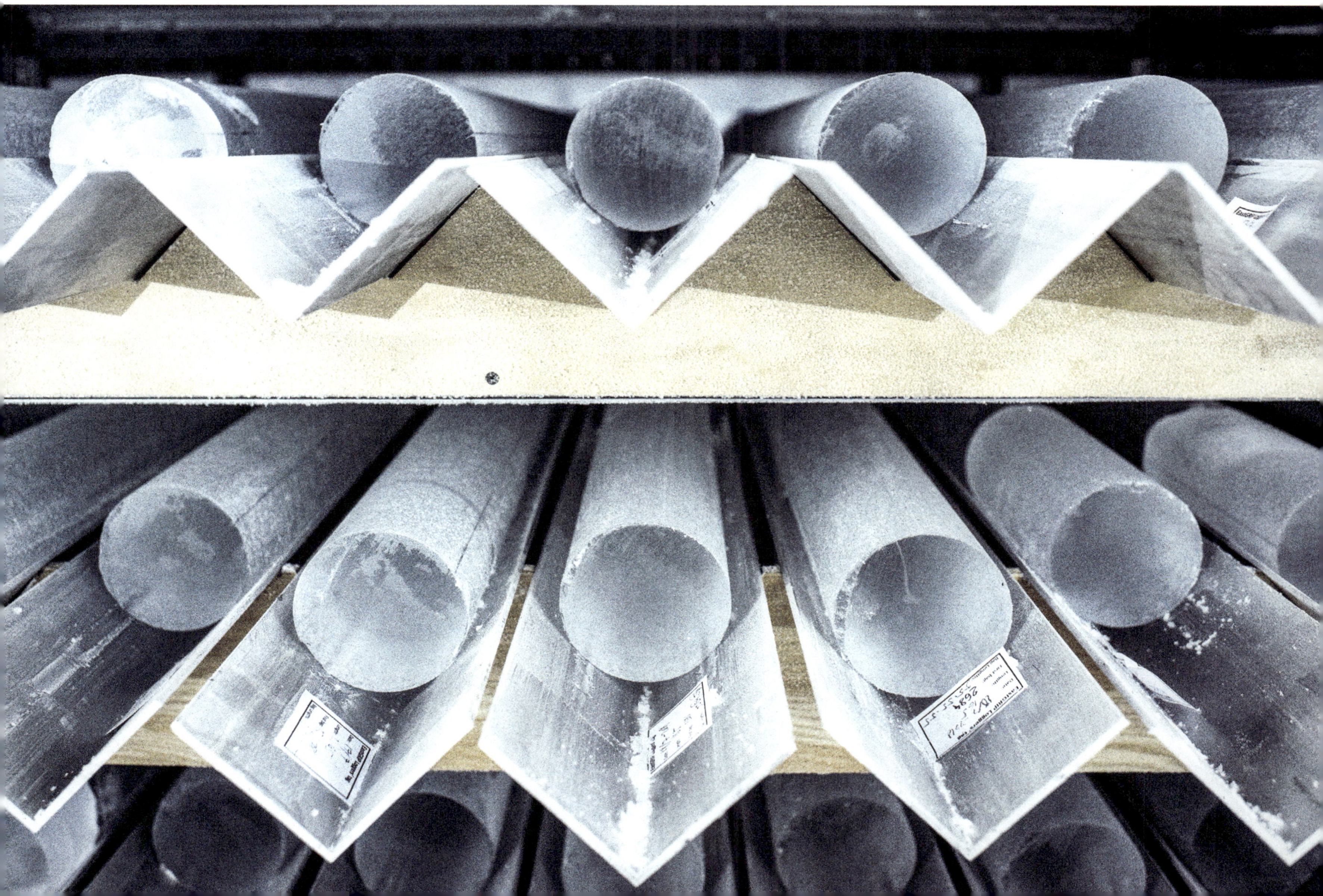

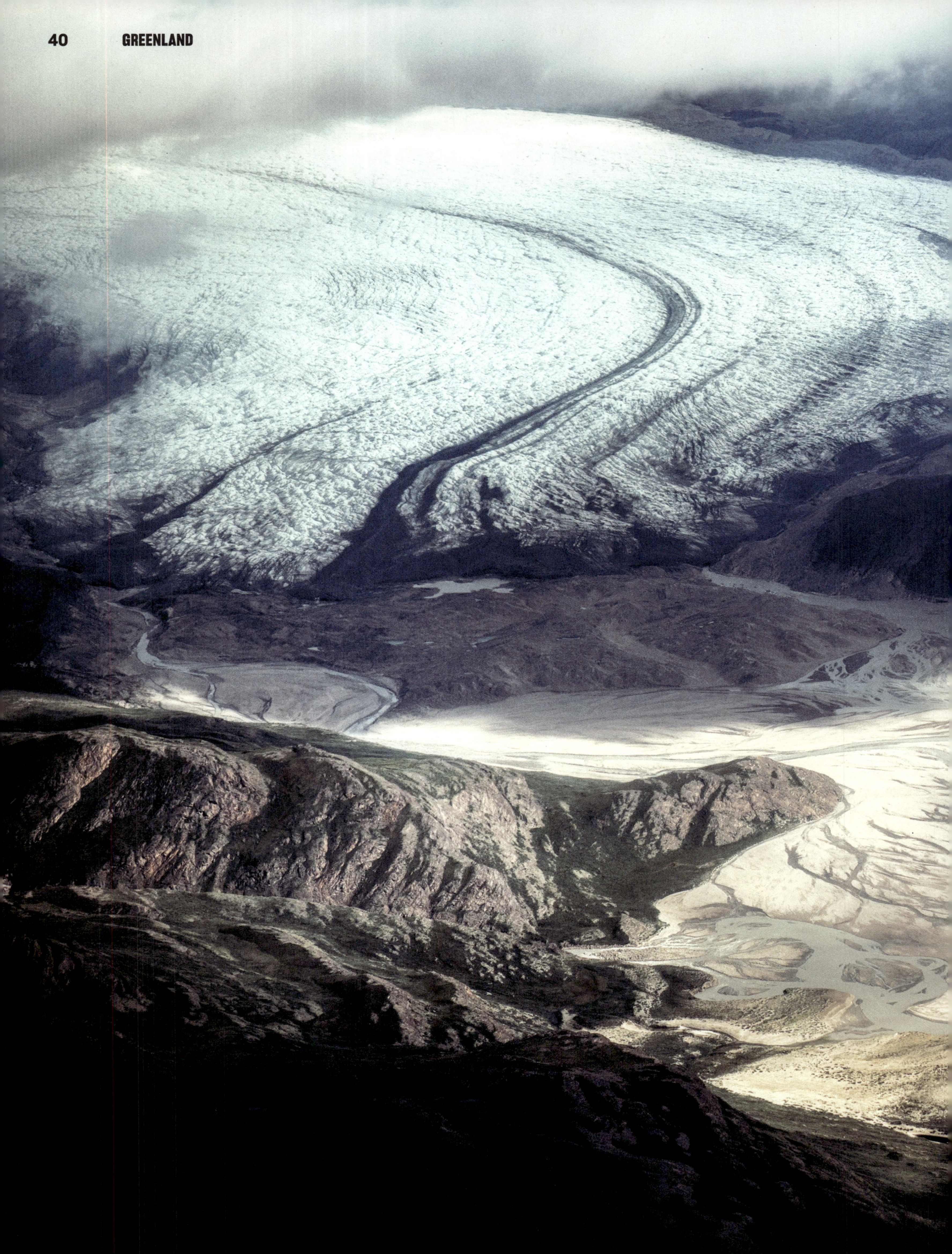

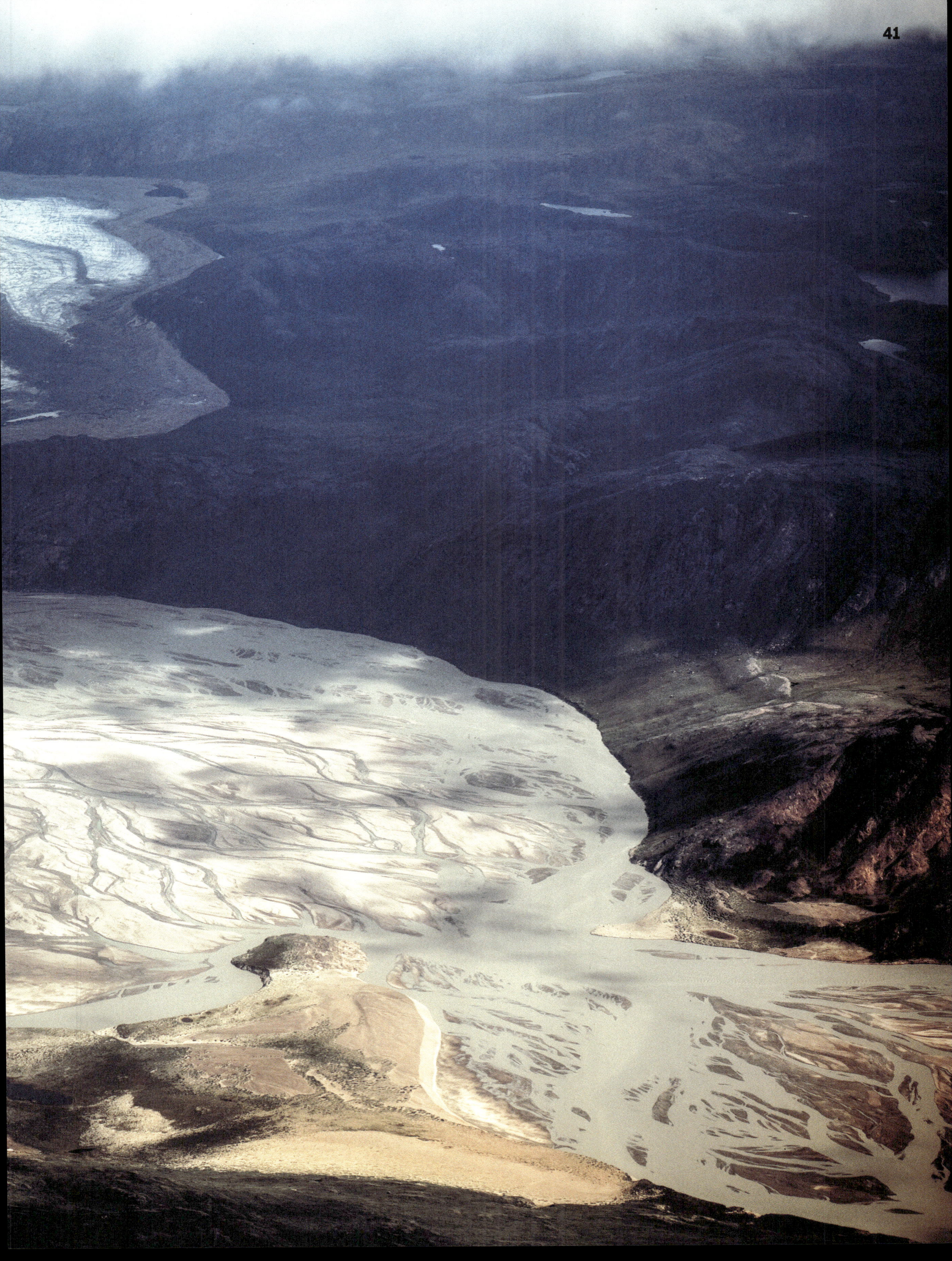

< Pages 40-41: The edge of the
ice sheet and melt rivers near
Kangerlussuaq.

The Eqip Sermia glacier and the
Sermeq Kujalleq glacier, the most
active glacier in the Northern
Hemisphere. The difference in colour
shows where the glacier used to
be. Greenland's immense ice sheet
is melting as a result of the climate
crisis.

Icebergs in northeast Greenland
seen from a height of 12 km.

The edge of the ice sheet near
Kangerlussuaq, with a melt river in
front.

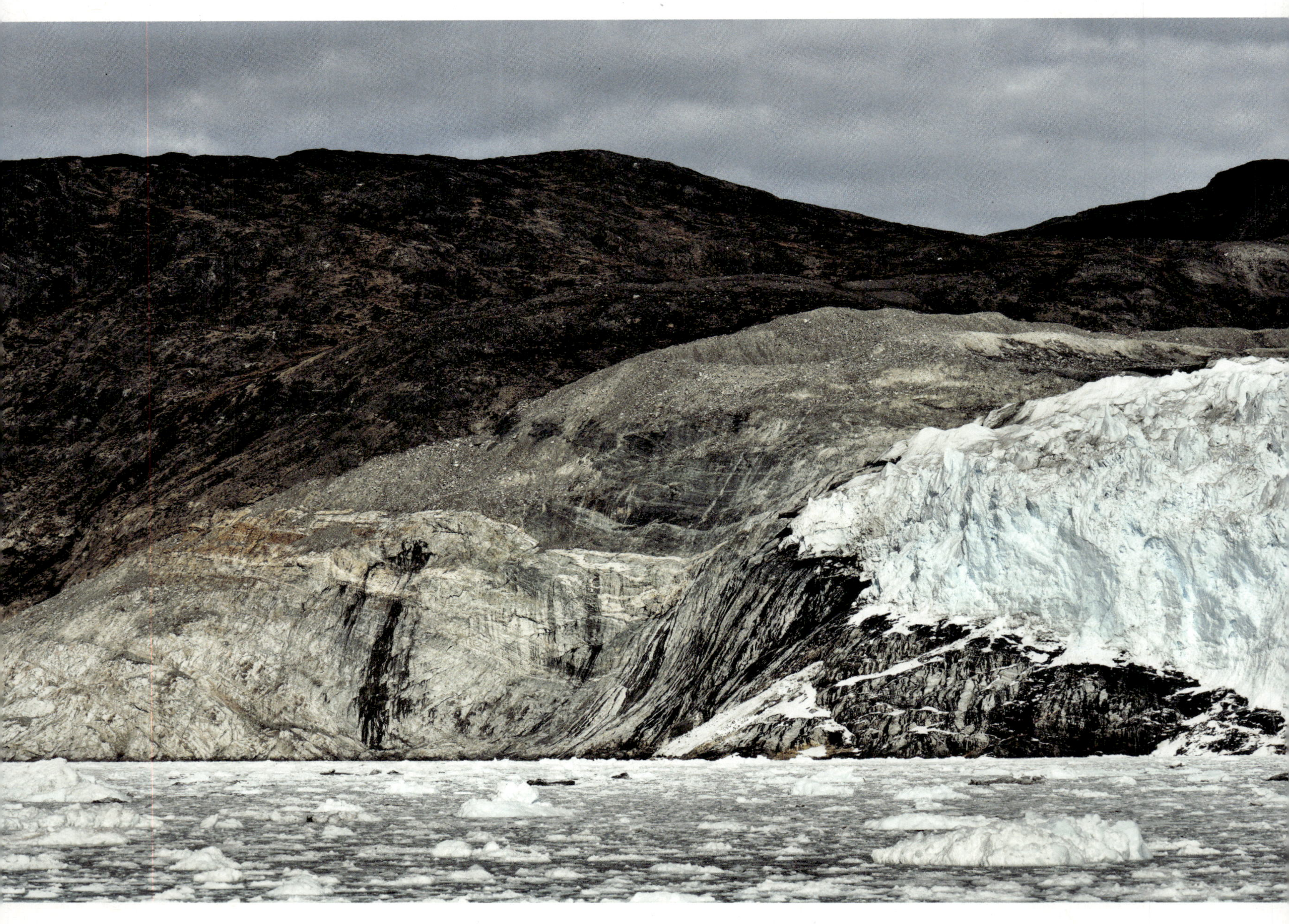

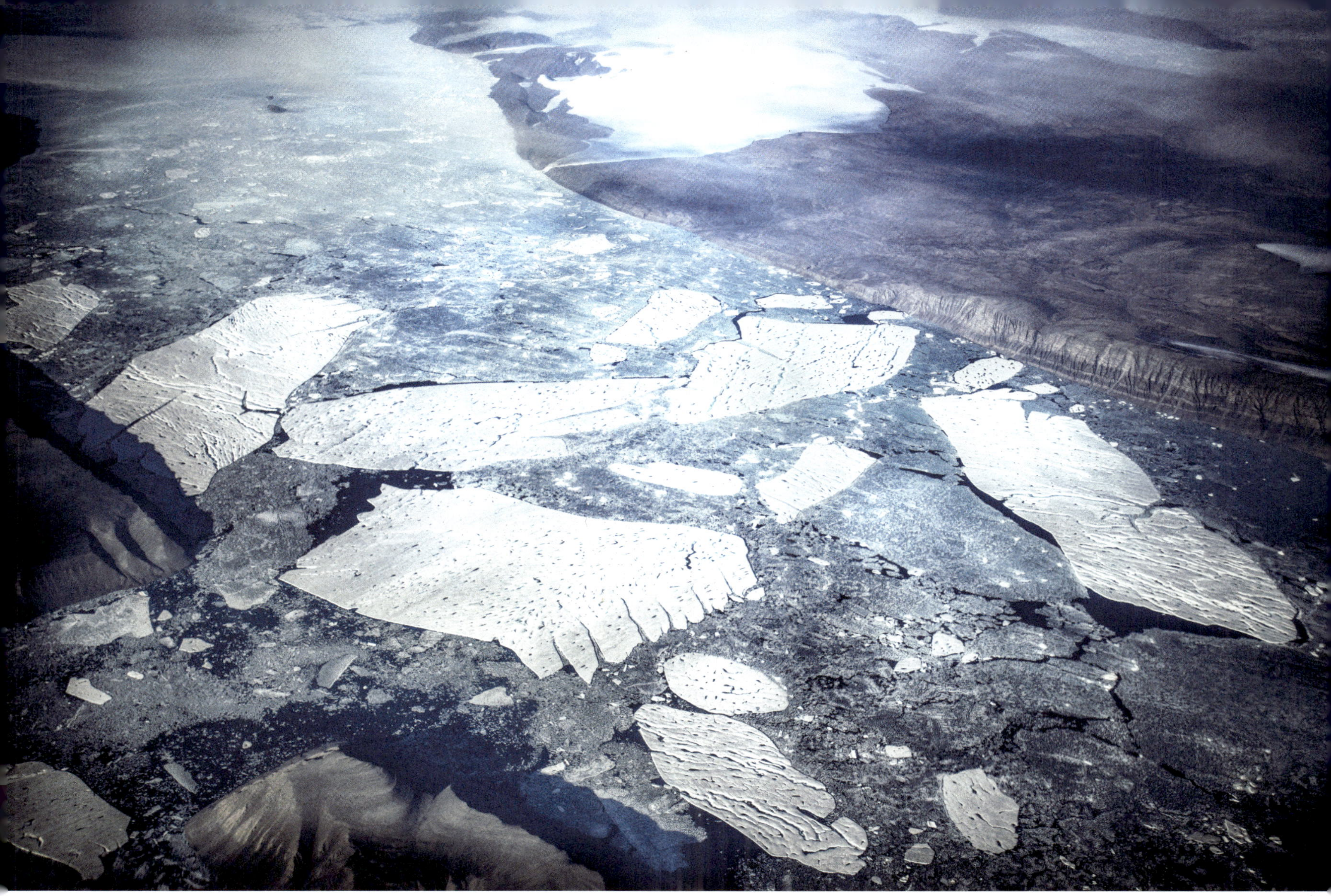

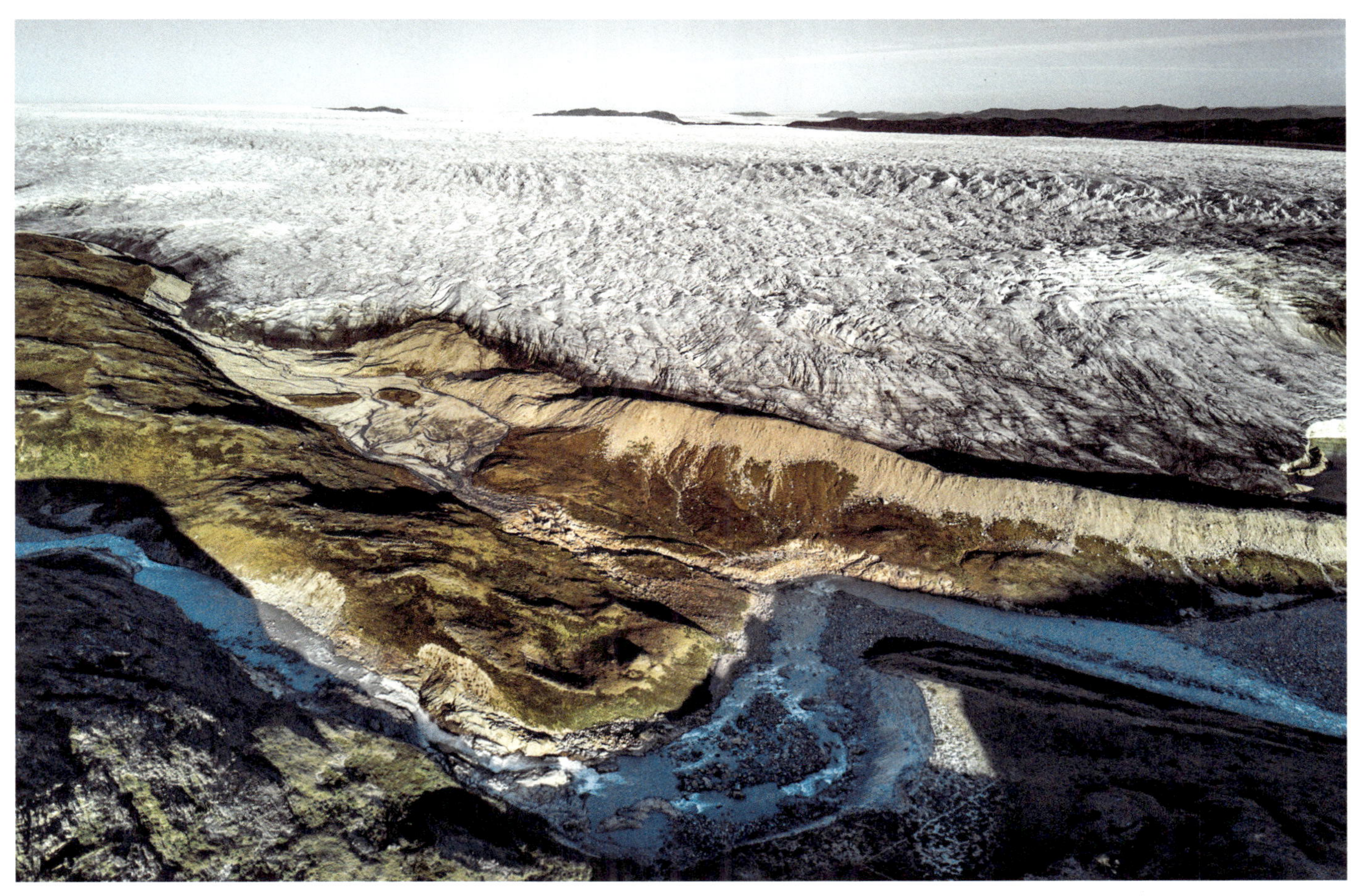

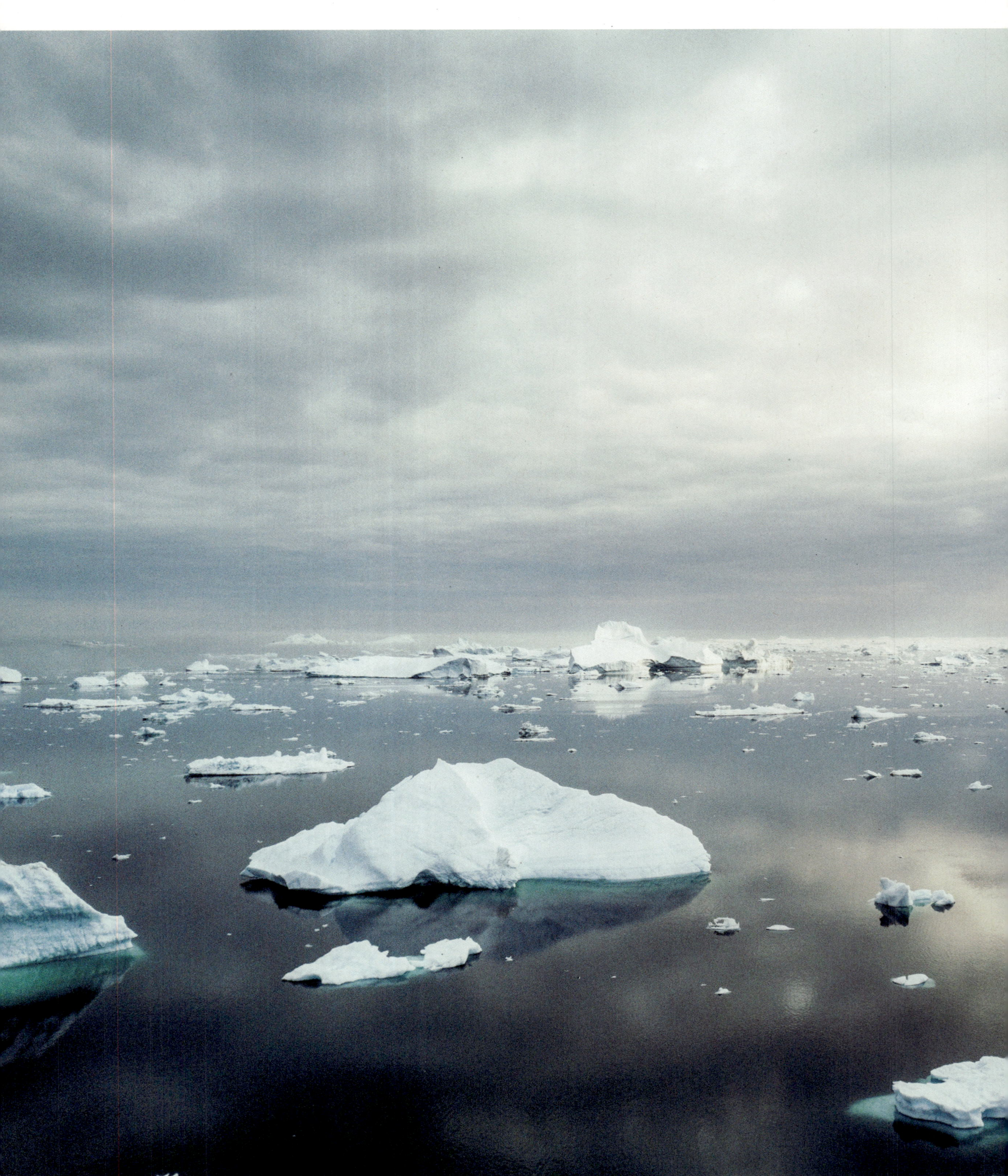

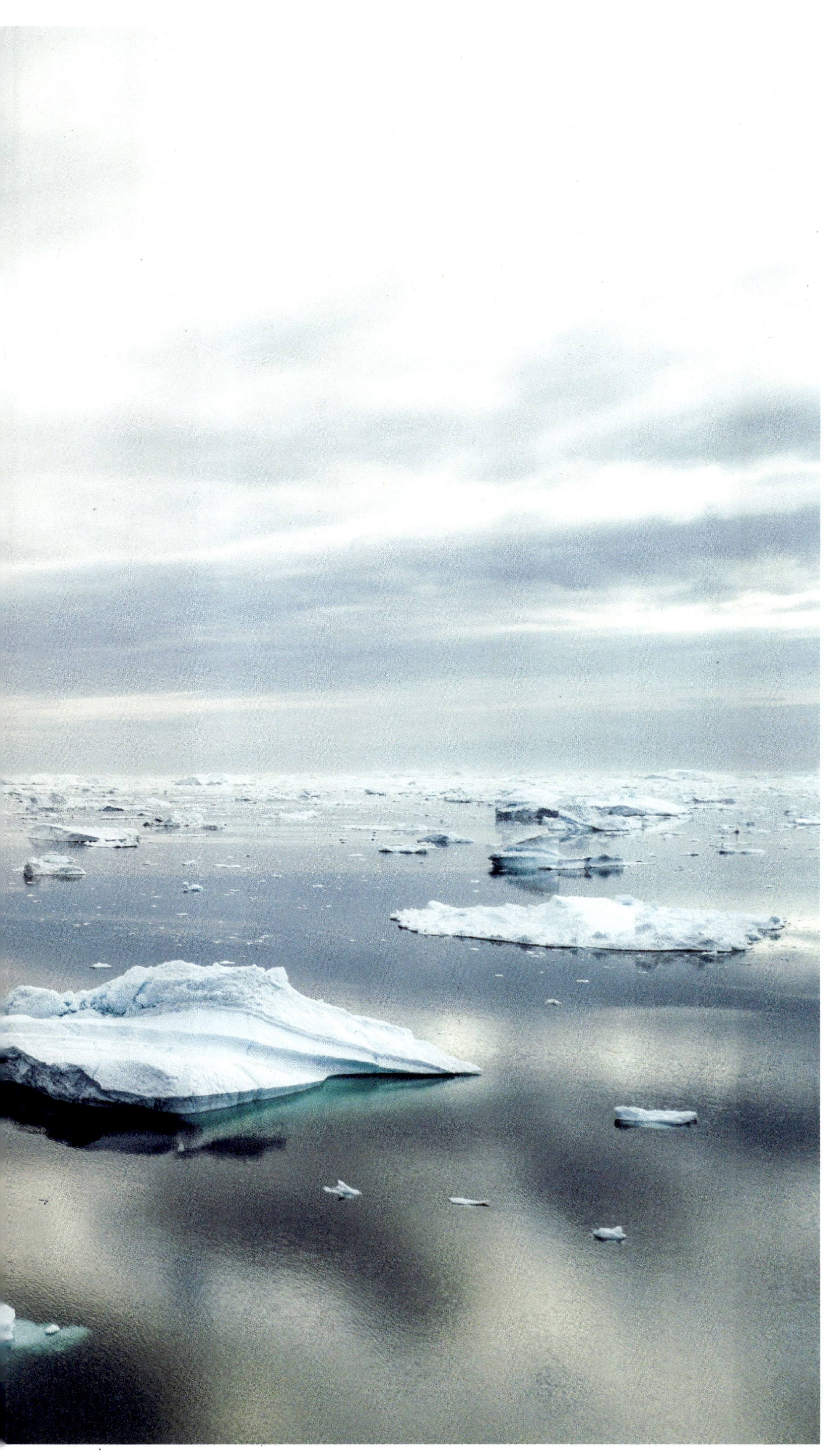

Icebergs in front of Oqaatsut. Only 50 people live in the community, most of whom used to be hunters. Nowadays, due to the instability of the sea ice, it's much more difficult to hunt for sea mammals like whales and seals.

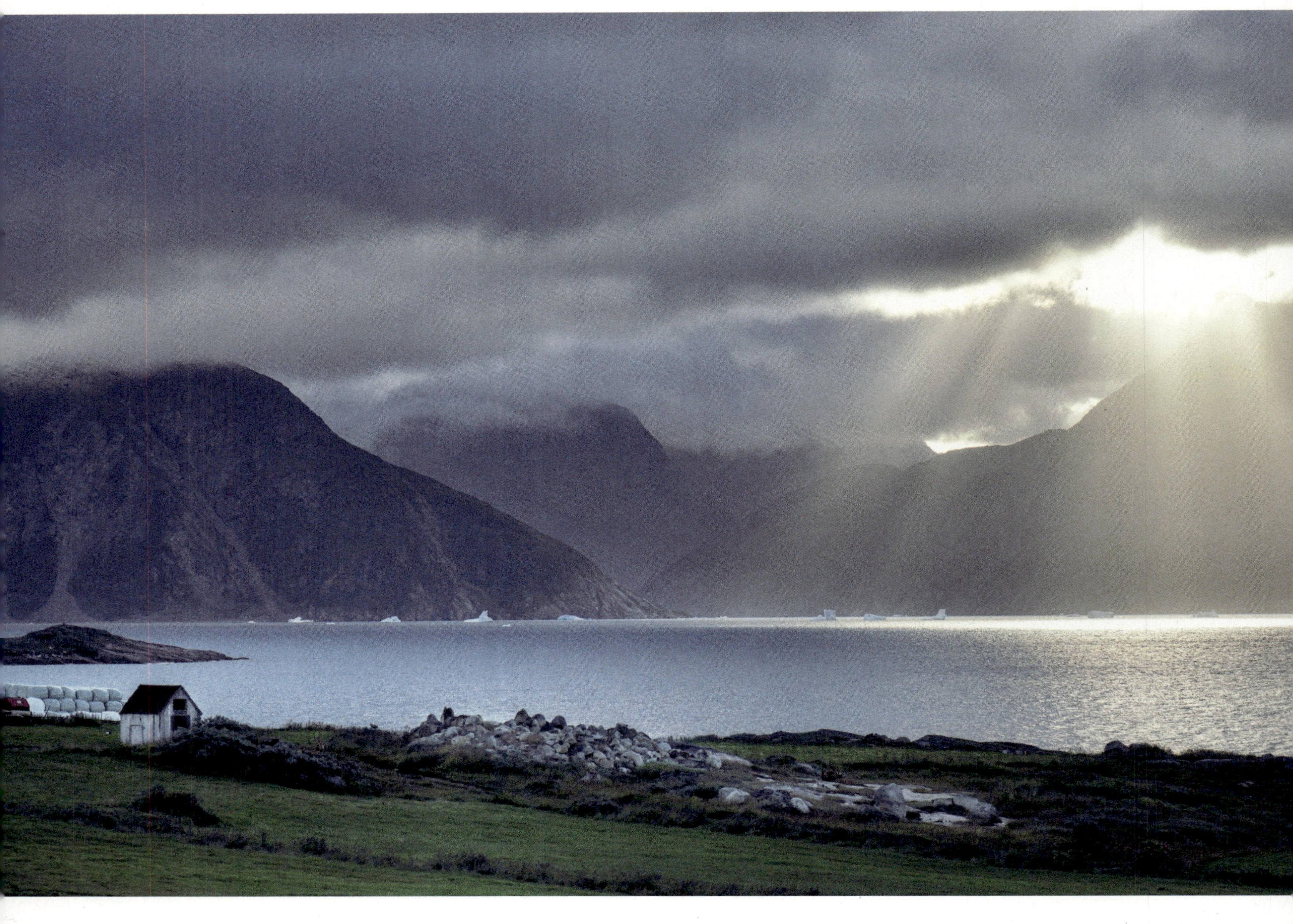

The channel with icebergs floating
by at Sillisit.

Fishers in Ilulissat.

A hunter has caught a seal close to
Oqaatsut (formerly Rodebay, a Dutch
whaling station in the 18th century).
The seal meat is used for consumption
and to feed the dogs. The skin is used
for clothing.

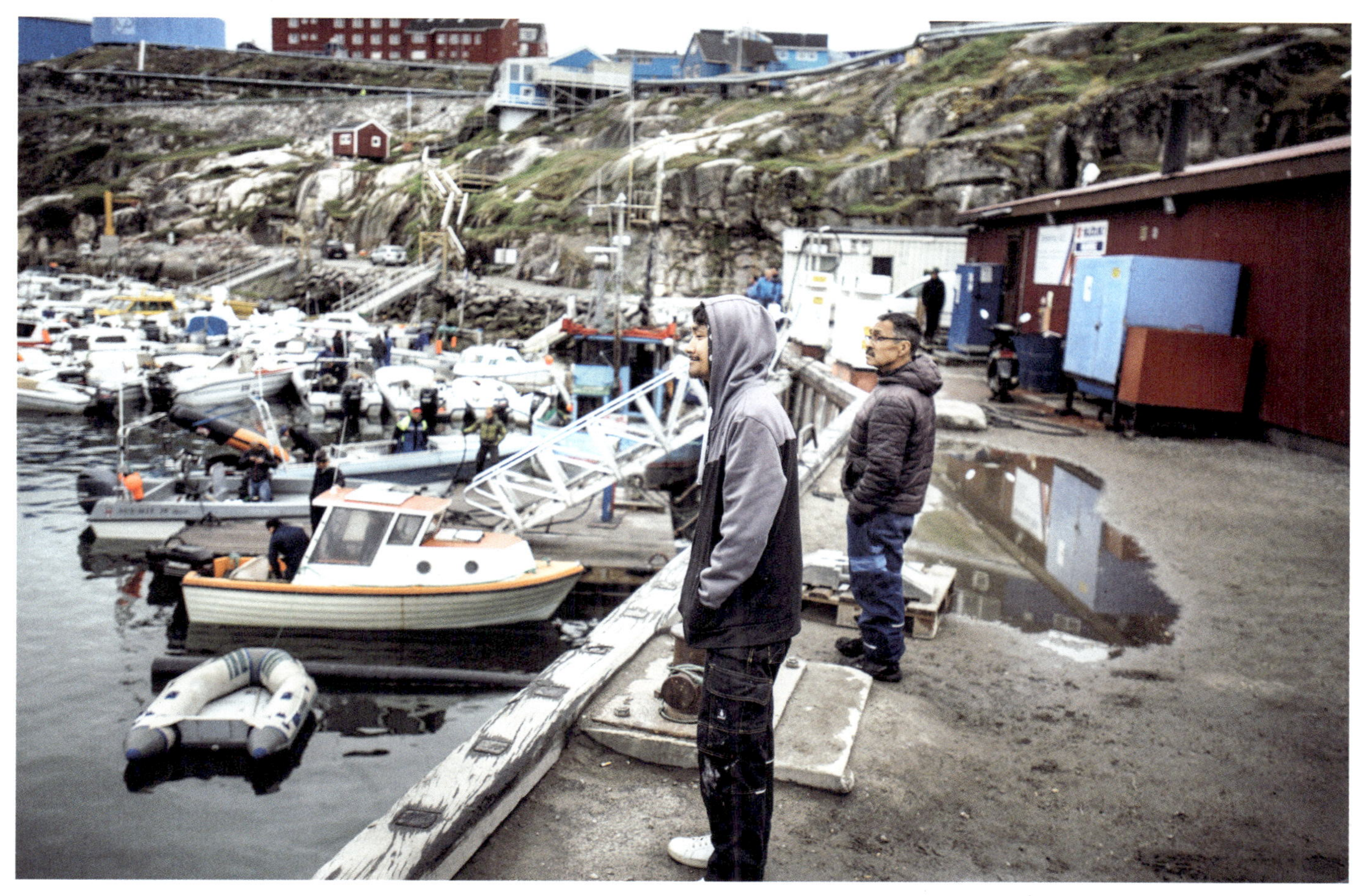

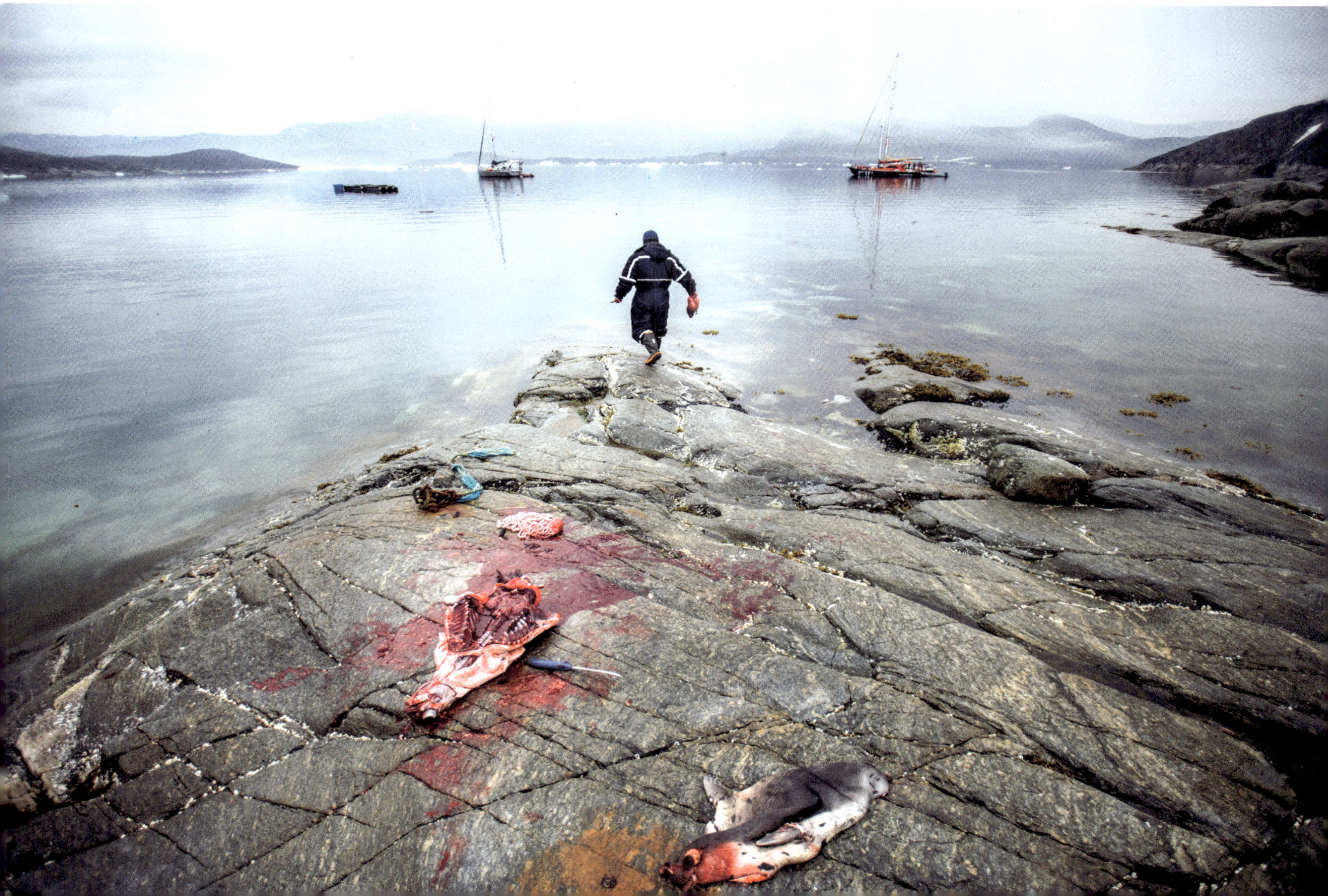

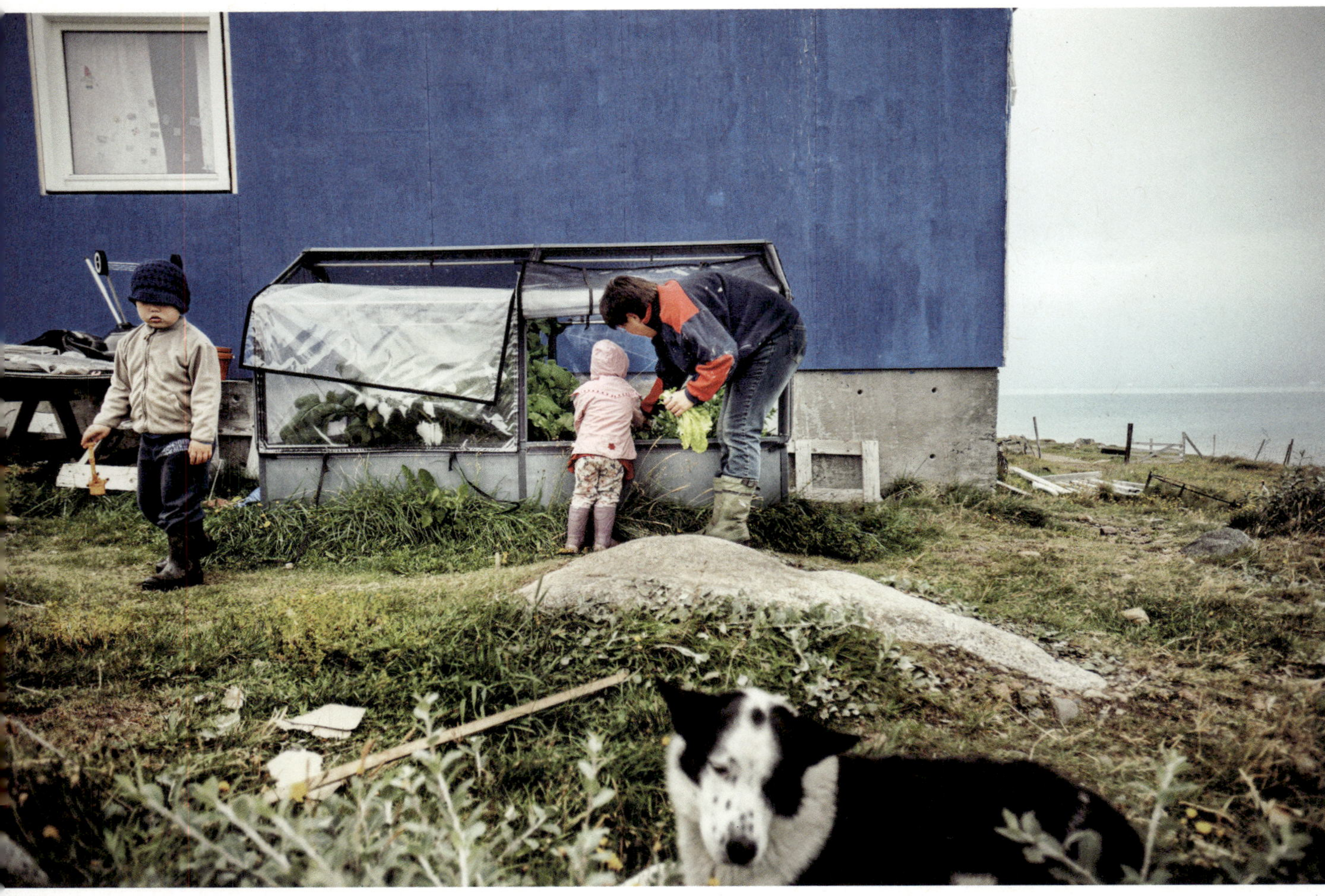

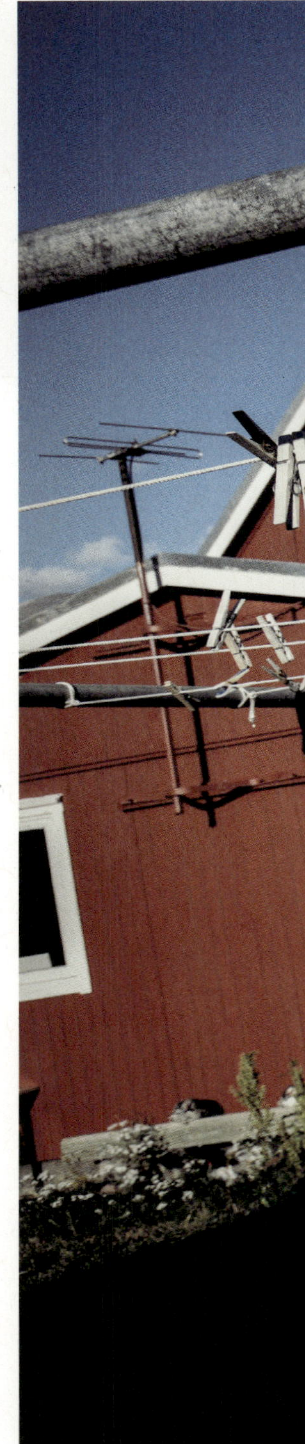

Elna Jensen with her children at their small greenhouse in Sillisit. Nowadays, the people of southern Greenland can grow all kind of crops due to higher temperatures and the lack of snow.

> Malene Egede at home in Igaliku.

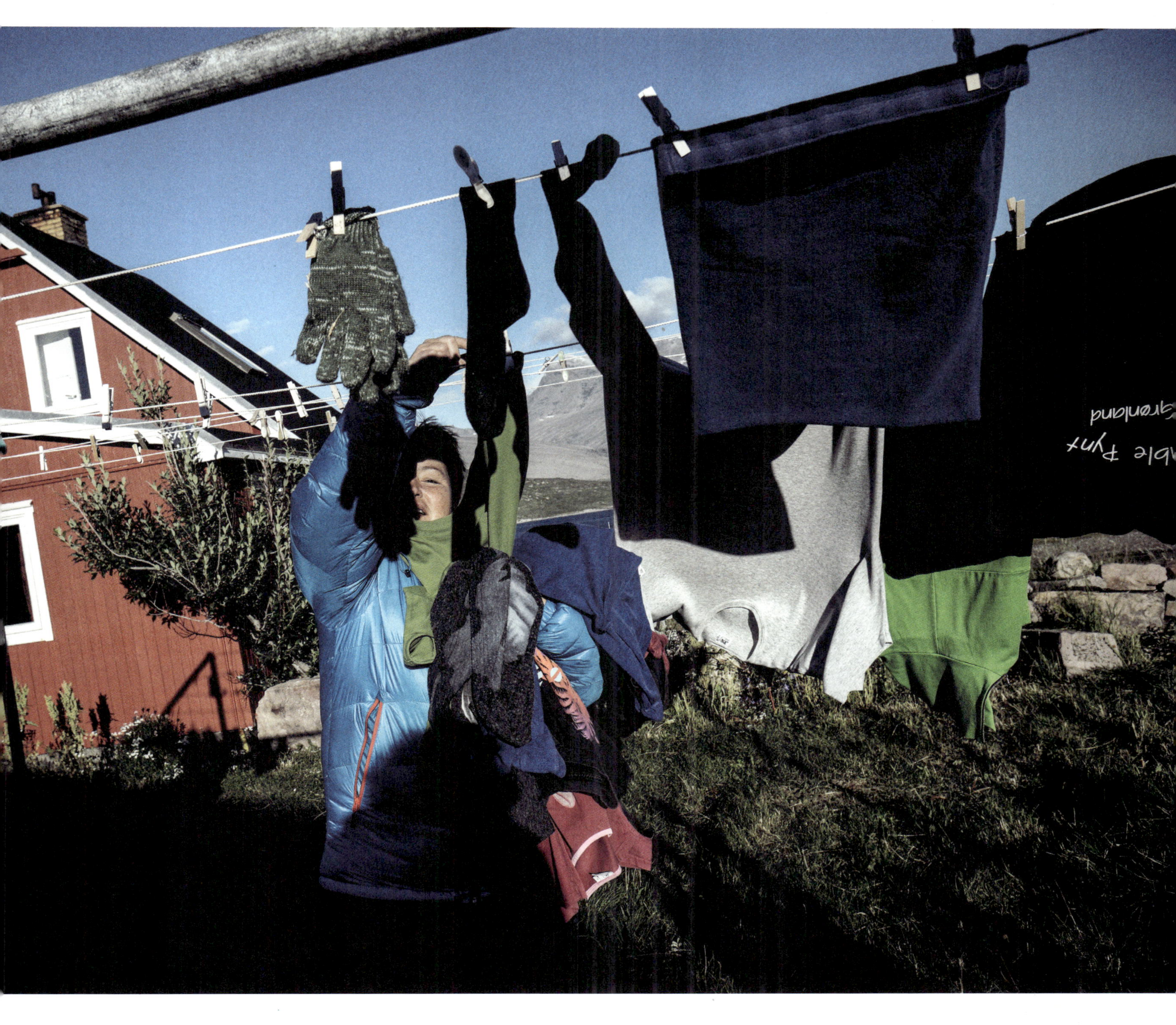

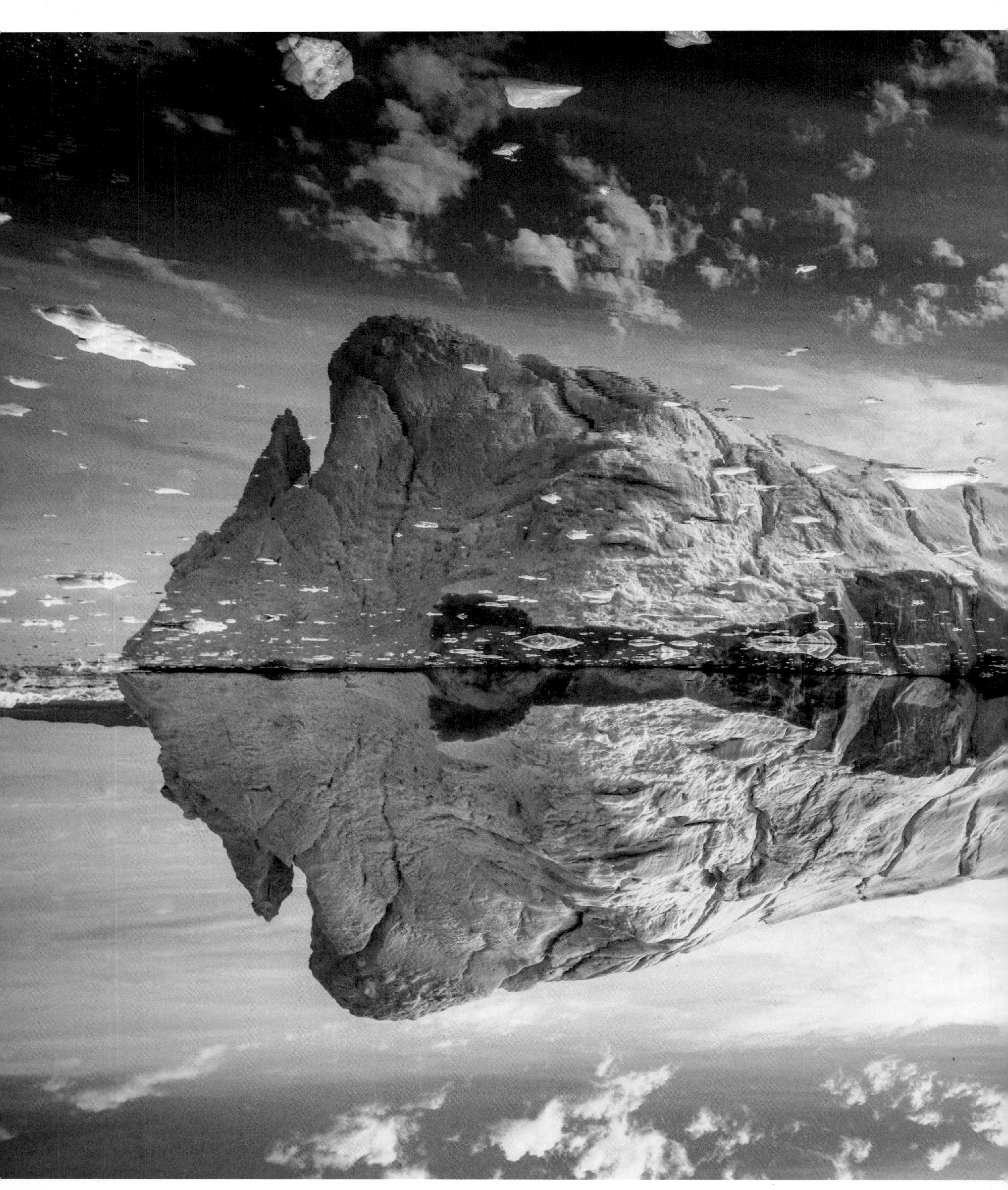

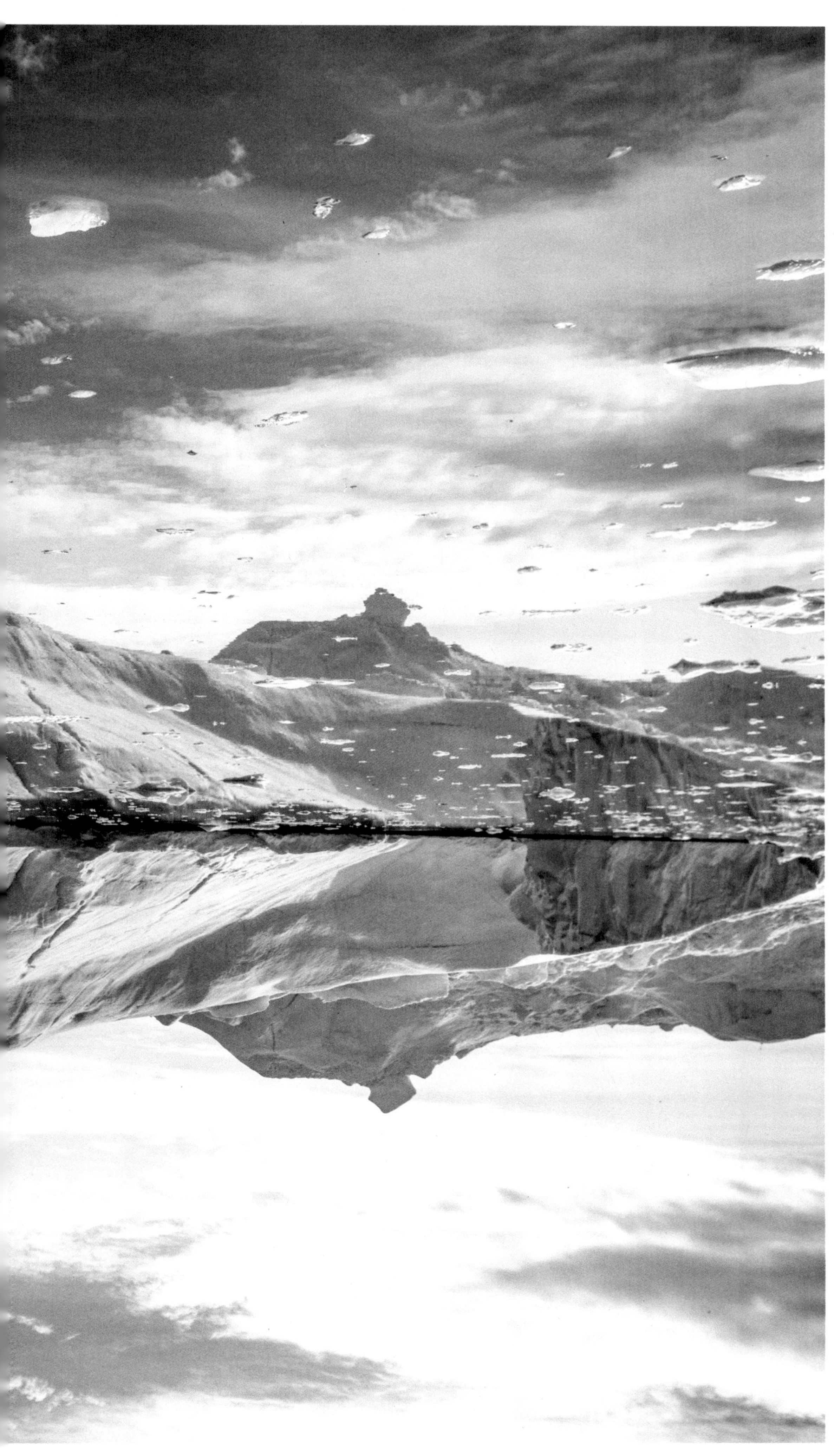

An iceberg in Disko Bay. Disko Bay is often advertised as the birthplace of icebergs. The ice originates from the Sermeq Kujalleq glacier (also called the Jakobshavn Isbræ), one of the fastest-retreating glaciers in the world. Smaller pieces of ice resulting from a quicker melt often block waterways and ports.

> Pages 52-53: Tourists in Disko Bay near Illulisat.

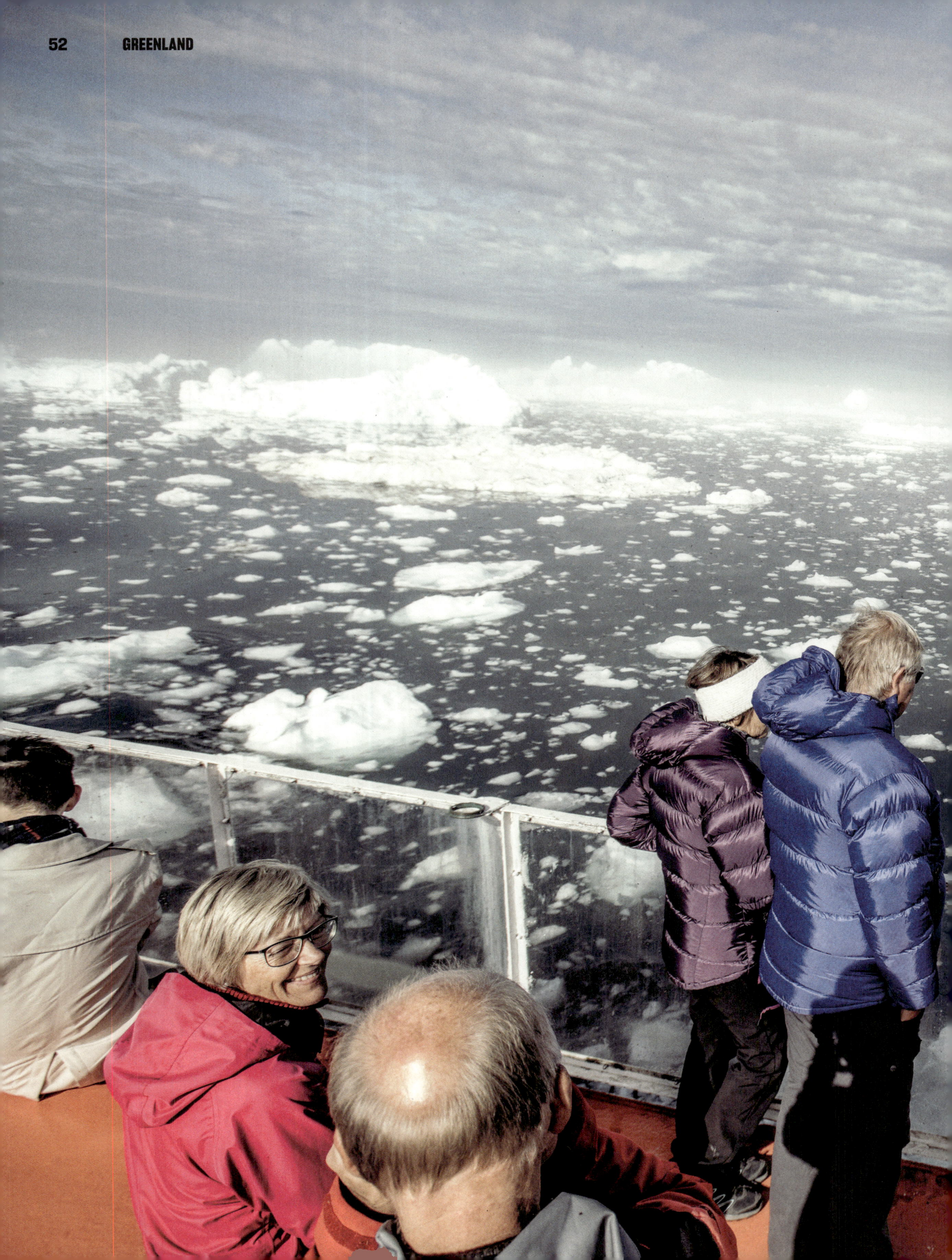

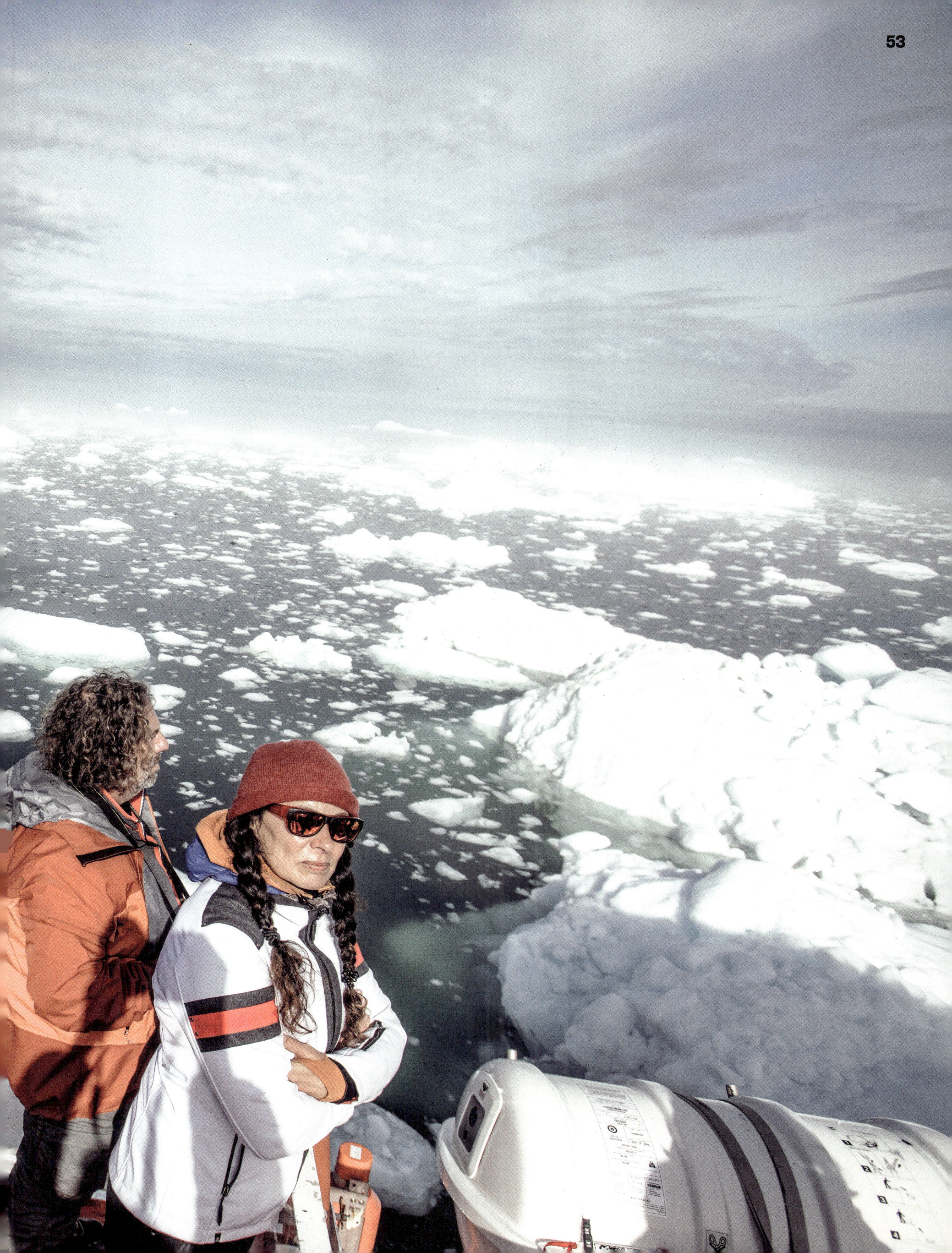

THE UNITED STATES: MIAMI AND NEW YORK

Jeff Goodell

The East Coast of the United States is experiencing rising sea levels at a rate three times the global average. Nearly 40% of the US population lives in coastal zones, mostly unprotected, with these areas generating half of the US GDP. If nothing is done, the US economy will be affected on a scale never seen before. While New York may survive, Miami is built on porous limestone and likely can't be saved.

MIAMI

Miami is an American fantasy, a city that glints and sparkles in the tropical sunshine, where the rich go to party and drug cartels go to launder their cash, a city of white-sand beaches and Art Deco hotels, a front porch of democracy for Cuban and Haitian immigrants, a buzzing financial hub that connects the United States with Latin America. Above all, Miami is a triumph of modern engineering, a city that has risen to heights no city should, a place where asphalt and air conditioning have made nature irrelevant. It is a great city built on a swamp, a technological miracle no less remarkable than the first human footprint on the moon or the iPhone in your pocket.

But Miami's glittery future is imperiled by rising seas. There's a long list of cities that are vulnerable to sea level rise, including London, Boston, New York, Mumbai, Jakarta and Shanghai, as well as low-lying nations like Bangladesh and the Philippines and islands like the Maldives and Bermuda. But South Florida is uniquely screwed, in part because about 70% of the 9 million people in South Florida live within 2 miles (3.2 km) of the coast. And unlike most cities, where the wealth congregates in the hills, southern Florida's most valuable real estate is right on the water. The Organisation for Economic Co-operation and Development lists Miami as the most vulnerable city worldwide in terms of potential property damage, with more than US$416 billion in assets at risk from storm-related flooding and sea level rise.

TOPOGRAPHY AND GEOLOGY

When it comes to protecting itself from rising seas, South Florida has two big problems. The first is its remarkably flat topography. Half of the area that surrounds Miami is less than 5 feet (1.5 m) above sea level. Its highest natural elevation, a limestone ridge that runs from Palm Beach to just south of Miami, is a scant 3 m above sea level. With just 1 m of sea level rise, more than a third of southern Florida will vanish, displacing more than 800,000 people; at 2 m, more than half of the region will be submerged. If the seas rise by 3 m, the only things in southern Florida that will poke above the water are the tops of buildings and a 60 m high pile of garbage on the outskirts of Miami known as Mount Trashmore. And the waters won't come in just from the east - because the region is so flat, rising seas will come in nearly as fast from the west, too, through the Everglades.

Even worse, South Florida sits above a vast and porous limestone plateau. 'Imagine Swiss cheese, and you'll have a pretty good idea what the rock under southern Florida looks like', Glenn Landers, a senior engineer with the U.S. Army Corps of Engineers, told me. This means water moves around easily - it seeps into yards at high tide, bubbles up on golf courses, flows through underground caverns and corrodes building foundations from below. Conventional sea walls and barriers are ineffective in this kind of environment - the water just seeps around them. Another problem: the city of Miami Beach already has about 100 km of sea walls on the island. The vast majority of them are on private property. How do you force people to raise them higher? Do you pass a law requiring everyone whose property includes a sea wall to spend US$100,000 or so to upgrade it? Does the city pay for it? In America, we can't even get everyone to wear a mask during a pandemic to protect their health, and the health of others. Good luck passing a law that requires them to build an expensive sea wall.

Like every city, the impacts of rising seas won't be felt equally in Miami. When waterfront property values decline due to flooding and other problems, the jet-setters who own high-end real estate in Miami Beach can easily move elsewhere. But the region's middle-class residents - who have most of their savings tied up in their homes - face the risk of generations of wealth being wiped out. And low-income communities, which were pushed away from prime waterfront property during Miami's initial expansion and often segregated onto the region's high ground, now face a fight to hold on to that elevated land as developers seek property situated away from rising seas.

A CITY UNDER THREAT

Already, Miami streets flood with every high tide. And it will just get worse. Just 20 cm of sea level rise will threaten the viability of the regional drainage system, which keeps Miami from returning to the swamp it once was. Adapting this system is expected to cost some US$7 billion. Hundreds of thousands of residential septic tanks will become inoperable, because the tanks don't work when groundwater tables rise along with sea level. Miami-Dade County has 108,000 properties on septic systems, most of them owned by middle-class residents. The County estimates that it would cost about US$3 billion to build out a sewer system that reaches everyone.

In the last few years, the fight to save Miami has begun in earnest. The city of Miami Beach has earmarked about US$1 billion to raise roads and install more than 50 stormwater pumps to keep the city dry. In addition, the U.S. Army Corps of Engineers has proposed a controversial US\$4.6 billion plan to build a barrier in Biscayne Bay to help protect downtown Miami. But there is still no coordinated regional plan, and no real thinking about serious infrastructure problems, such as how to protect access to the airport, which is in a particularly low-lying section of the city.

'Saving Miami is a huge undertaking', one city engineer told me. 'The costs - and complexities - are endless.' Just raising a road is an impossibly complicated task. 'When you raise the road by even a few inches, what happens to the water?' the engineer asked rhetorically. 'It runs off the road into the buildings and homes beside it. So you have to raise those too. Or else you have to just figure out a way to live with the water.'

PLANNING FOR THE FUTURE

John Stuart, the head of the architecture department at Florida International University, has spent years thinking about what South Florida's future might look like. 'It's pretty clear that we are not going to be able to stop the water from coming in, so how will we live?' Stuart is inspired by Stiltsville, a collection of structures built on pilings in Biscayne Bay during the 1940s and '50s by Miami residents looking for a place to party beyond the easy view of the law. (Although they are abandoned now, a few of the Stiltsville structures still survive in the bay.) Stuart and his colleagues are trying to imagine what a city in the water would look like. How do you get electricity? Who provides emergency medical services? 'It is really unlike anything humans have tried to do before', Stuart says. 'How do you build a floating city?'

Stuart is clearly energized by the creative challenge of thinking about this. And if sea level rise happens slowly enough and Miami doesn't get slammed with a hurricane and the drinking water supply doesn't go bad and the real estate market doesn't crash and the beaches aren't washed away, the city may well have time to transform itself into a modern-day Venice.

But in the long run, the water will win. Stuart compares Miami to Baiae, an ancient Roman resort town in the Bay of Naples that was once a playground for the super-rich, including Emperors Nero and Caligula. Today, because of ground subsidence connected with nearby volcanoes, the ruins of Baiae are under water. 'This is what humans do', says Stuart. 'We inhabit cities, and then when something happens, we move on. The same thing will happen with Miami. The only question is, how long can we stick it out?'

For Stuart, who lives in Miami Beach, the city's risky future doesn't diminish his love for the place. 'That's the thing about Miami', he says. 'You'll want to be here until the very end.'

We inhabit cities, and then when something happens, we move on. The same thing will happen with Miami. The only question is, how long can we stick it out?

Rising sea levels on the East Coast of the United States are due in part to the fairly recent discovery that ice has its own gravitational pull that propels water towards it. Consequently, when ice melts, it contributes to the already rising sea level. Because western Greenland's glaciers and pack ice are melting more quickly than ever before, the eastern seaboard of the US is grappling with these sharp increases in sea levels.

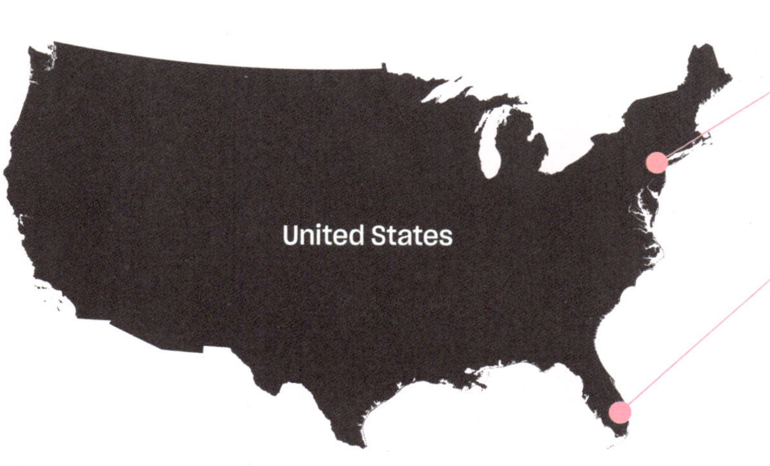

New York

Population US: 331,000,000
Population New York: 8,623,000
In 2012, New York was hit by Superstorm Sandy. It caused major flooding, a blackout in large areas of Manhattan, and a huge bill for damages. It was a wake-up call for the city, as it showed what could happen if the sea level were to rise permanently.

Miami

Population US: 331,000,000
Population Miami-Dade County: 2,795,000
Miami is one of the cities currently most at risk from sea level rise, as it is built on porous limestone. It is believed that Miami and Miami Beach will need to be evacuated by 2060. Nevertheless, there are currently plans to build 32 residential towers in Miami Beach.

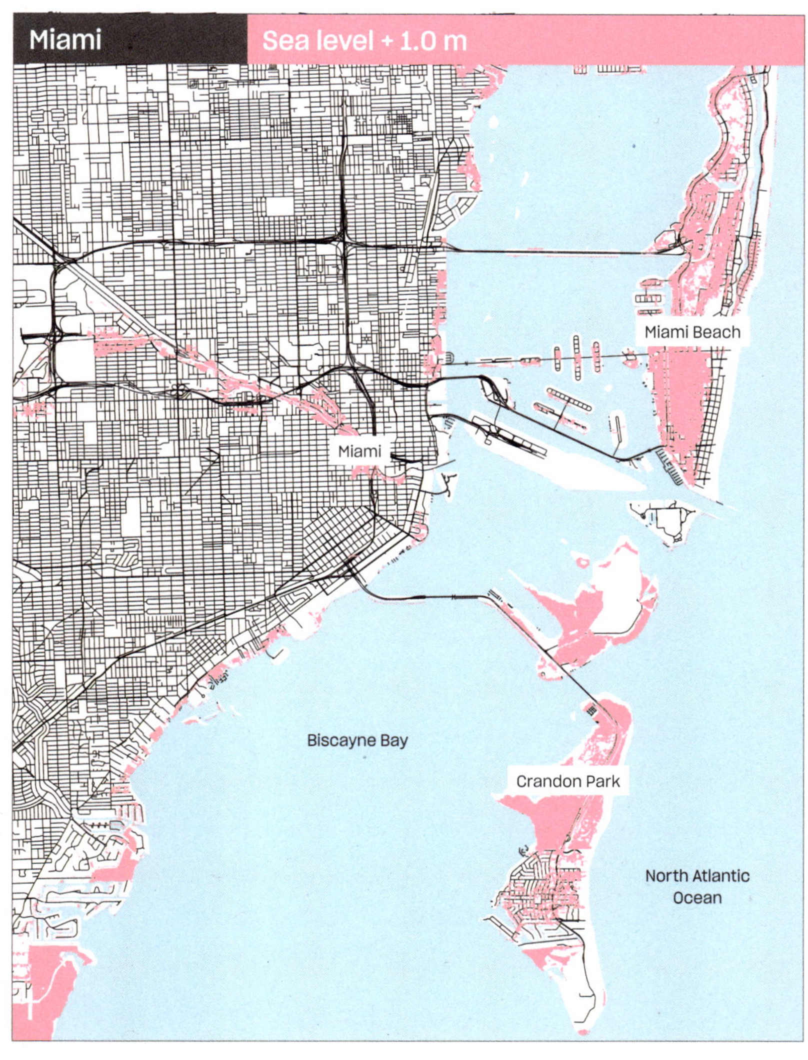

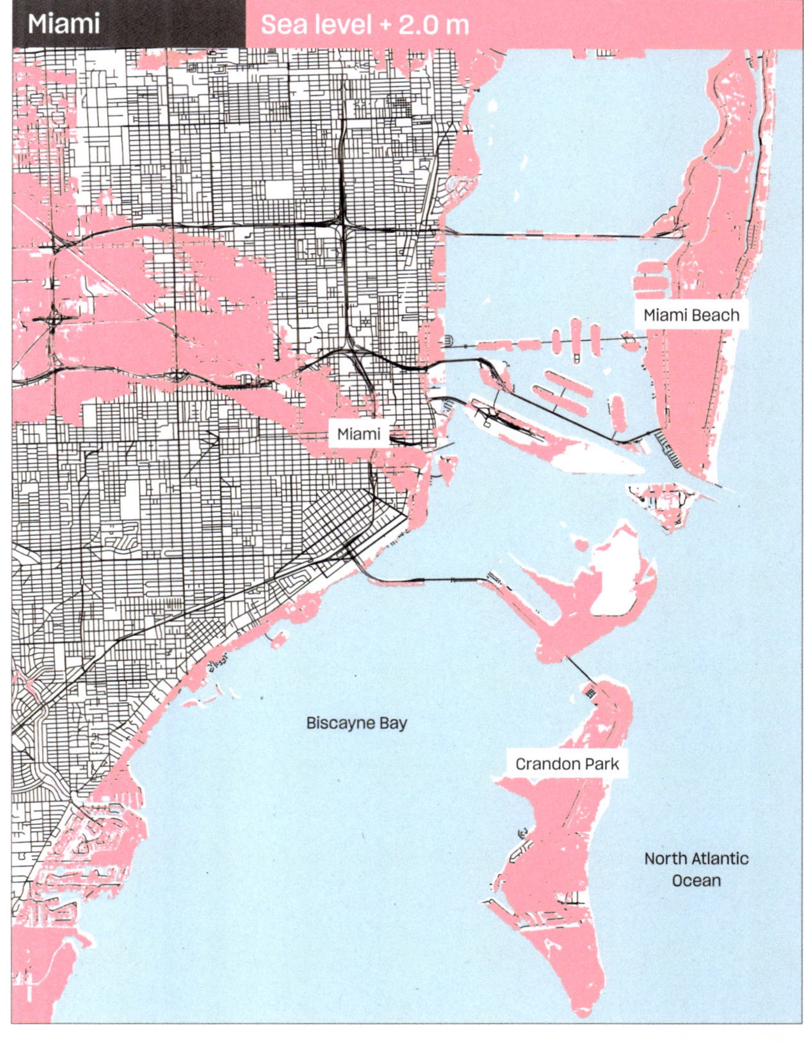

Source: Climate Central (www.climatecentral.org)

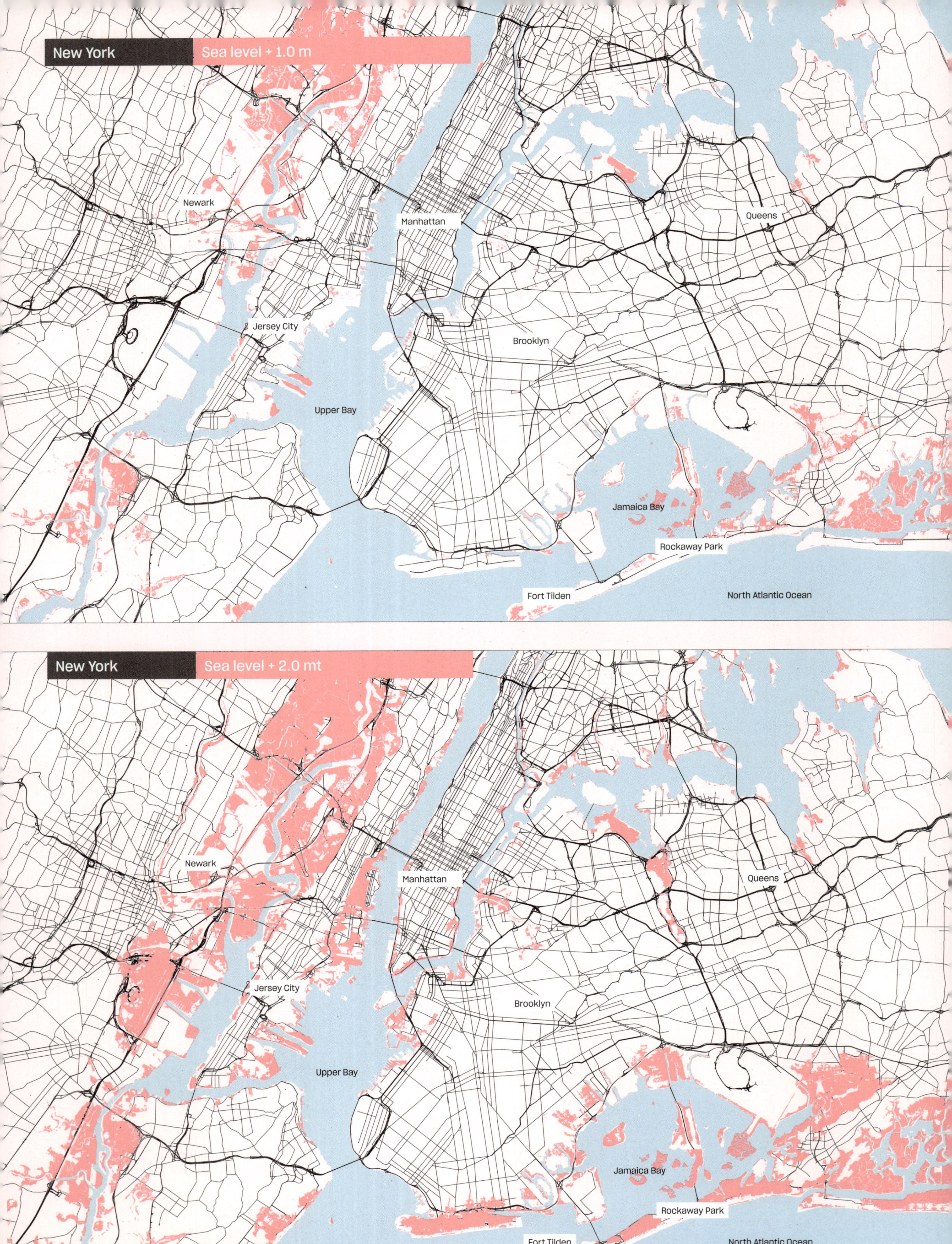
New York
Sea level + 1.0 m
Newark
Manhattan
Queens
Jersey City
Brooklyn
Upper Bay
Jamaica Bay
Rockaway Park
Fort Tilden
North Atlantic Ocean
New York
Sea level + 2.0 mt
Newark
Manhattan
Queens
Jersey City
Brooklyn
Upper Bay
Jamaica Bay
Rockaway Park
Fort Tilden
North Atlantic Ocean

NEW YORK

In a world of rapidly rising seas, New York is better prepared than many coastal cities. As anyone who has seen the rock outcroppings in Central Park knows, much of Manhattan is built on 500-million-year-old schist, which is impervious to saltwater. There is plenty of high ground, not just in Upper Manhattan, in Washington Heights, but also along a ridge that runs diagonally through Queens and Brooklyn, including places like Park Slope and Jackson Heights. Finally, the city has brains, money and attitude - New York is not going to go down without a fight.

But in other ways, New York is surprisingly vulnerable. First, it's on an estuary. The Hudson River, which runs along the west side of the city, needs an exit. So, unlike a harbour city like Copenhagen, you can't just wall off the city from the rising ocean. Second, there are a lot of low areas in Brooklyn, Queens, and, most importantly, Lower Manhattan, which has been enlarged by landfill over the years. (If you compare a map of the damage from Hurricane Sandy in 2012 with a map of Manhattan from 1650, you'll see that they match pretty well - almost all of the flooding occurred in landfill areas.) The amount of real estate at risk in New York is mind-boggling: 71,500 buildings worth more than US$100 billion stand in high-risk flood zones today, with thousands more buildings at risk with each foot (0.3 m) of sea level rise. In addition, New York has a lot of industrial waterfront, where toxic materials and poor communities live in close proximity, as well as a huge amount of underground infrastructure: subways, tunnels, electrical systems. And because of changes in ocean dynamics, as well as the fact that the ground beneath the city is sinking as the continent recovers from the last ice age, seas are now rising about 50% faster in the New York area than the global average.

LESSONS FROM HURRICANE SANDY

In the aftermath of Hurricane Sandy, a barrier system known informally as the 'Big U' was proposed that would loop around the entire bottom of Manhattan, from 42nd Street on the east to 57th Street on the west. The project, like many projects of this size and scale, ran into financial and legal problems. Instead, a smaller barrier system on the east side of Manhattan, known as the East Side Coastal Resiliency Project, is moving forward. The US$1.5 billion project is an undulating 10-foot-high (3 m) steel-and-concrete-reinforced berm that will run about 2 miles (3.2 km) along the riverfront. There are plans in the works to build other walls and barriers in the Rockaways and on Staten Is-land, as well as in Hoboken, New Jersey, across the Hudson River. And then there is the idea of walling New York off from the sea entirely whenever a storm approaches. Recently, the U.S. Army Corps of Engineers proposed a massive 6-mile-long (9.6 km) retractable barrier in New York's outer harbour to protect the entire city from Hurricane Sandy-like surges. The project, which would cost US$119 billion and take 25 years to build, has already run into fierce opposition from New Yorkers who point out that the barrier would be ecologically devastating to New York's harbour, trapping sewage and toxins within. And given how fast sea levels are rising, and the uncertainty of what's to come, there is a good chance the barrier would be obsolete by the time it was completed.

BEYOND BARRIERS

In any case, fortifying New York will require more than just walls - it will require a radical rethinking of the relationship between the city and the people who live in it. If the central role of government is to keep people safe, what happens when people realize they are not safe? What is the government's role in keeping people out of harm's way? How does the government compensate people whose properties are under water? Adriaan Geuze, a Dutch landscape architect who has done as much thinking about how to live with water as anyone, compares sea level rise to other transformative catastrophes such as the Dust Bowl, a partly human-caused natural disaster that profoundly changed the geography of America and also expanded the role that government plays in ensuring the long-term welfare of even the most vulnerable people. 'Dealing with sea level rise is going to require a rethinking of the social contract in America', Geuze told me.

In a world of rapidly rising seas, New York is fortunate. It has enough money and enough high ground to ride out whatever comes in this century. The question is, what kind of city will it be? Will it be a safe, livable place, alive with art and commerce, inspiring to the world? New York has always defined our idea of what a city is and can be. Now, as the water rises, New York may well define our idea of urban survival. 'I have the frame of 100 years', Geuze told me. 'Maybe 8, 9 feet (2.4 m to 2.7 m) of sea level rise. We can deal with that. But there will come a moment when no matter what you do, even a rich city like New York won't be able to do anything to protect itself. When is that moment? I don't know. But it is coming. What Mother Nature is telling us right now is, we are not in control.' •

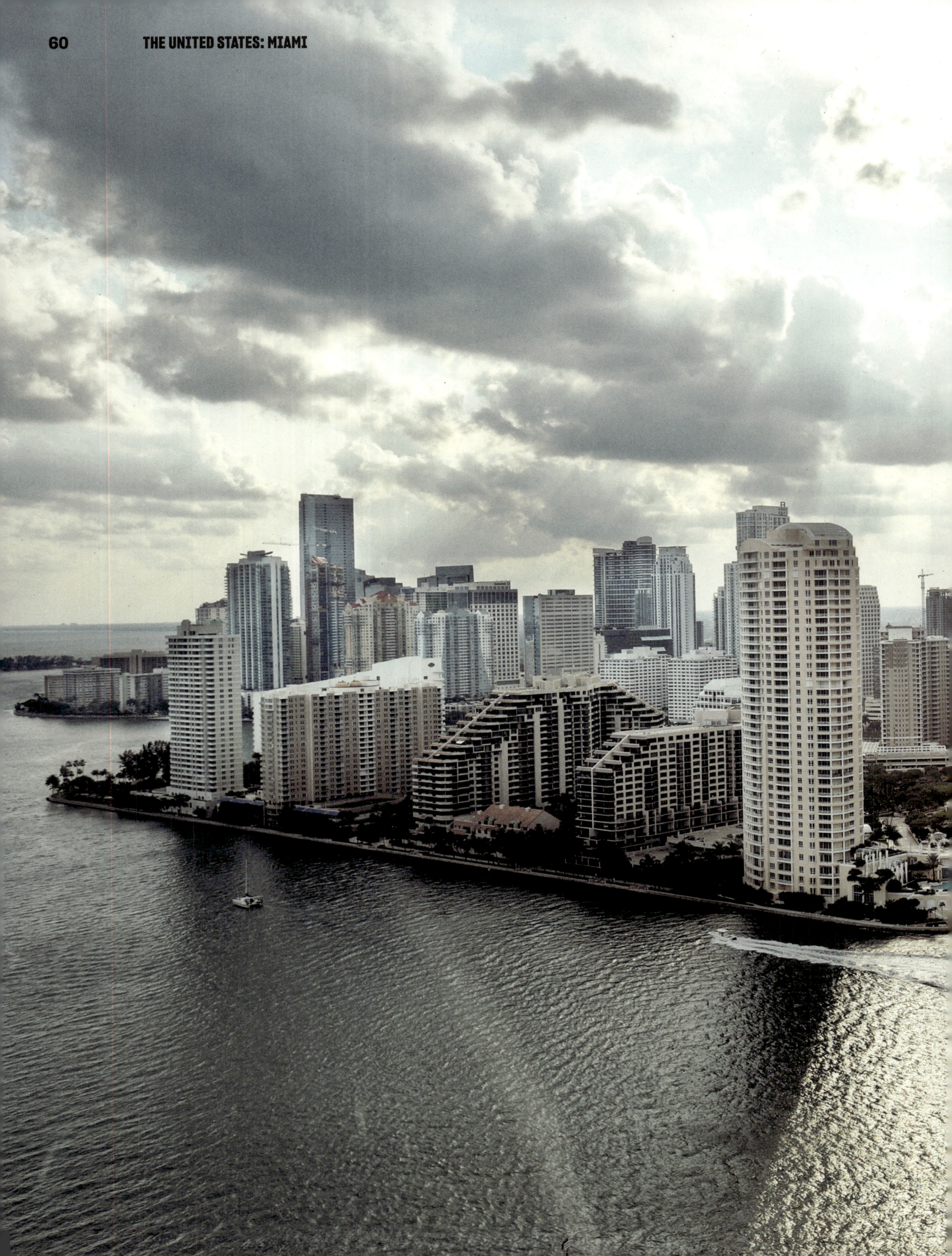

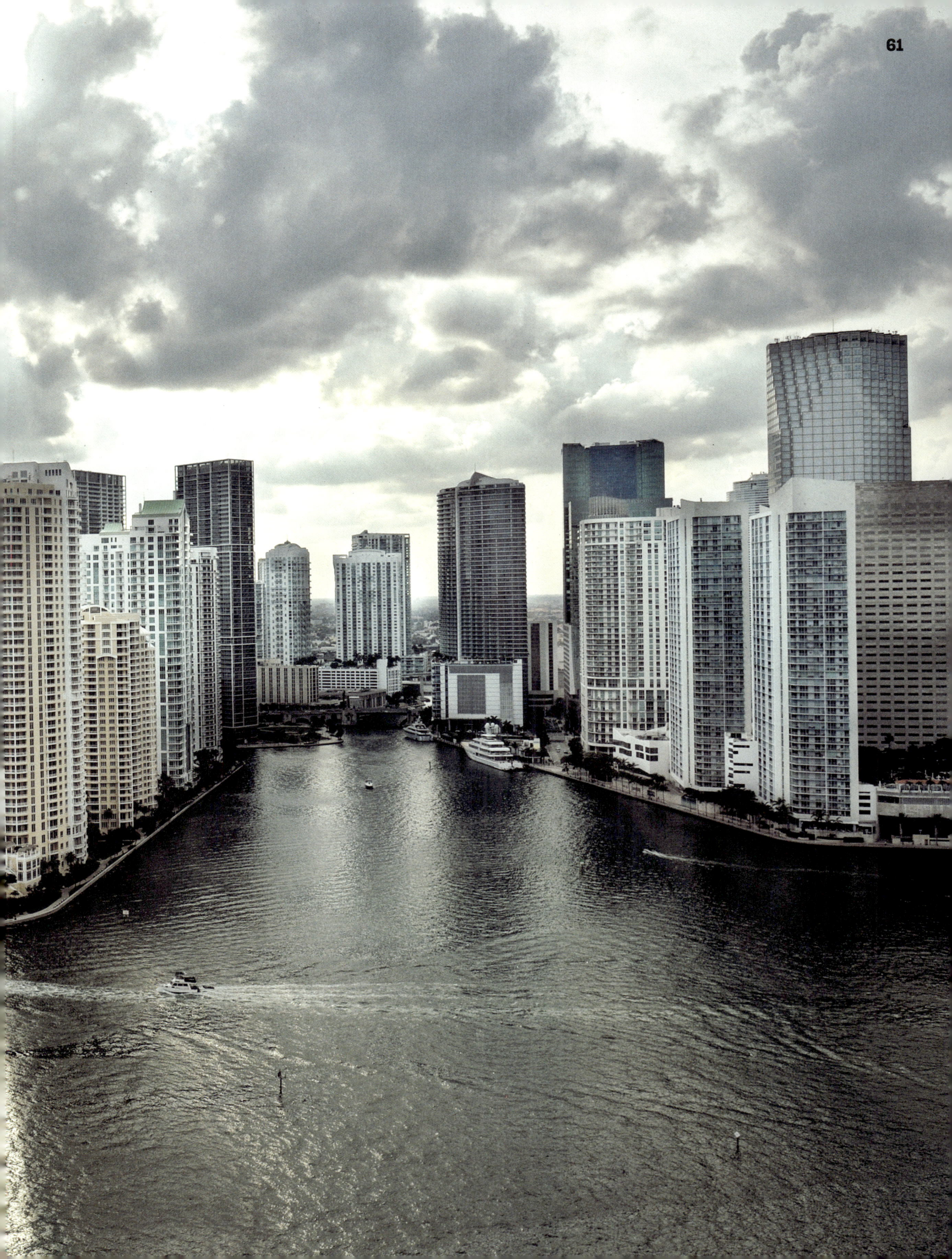

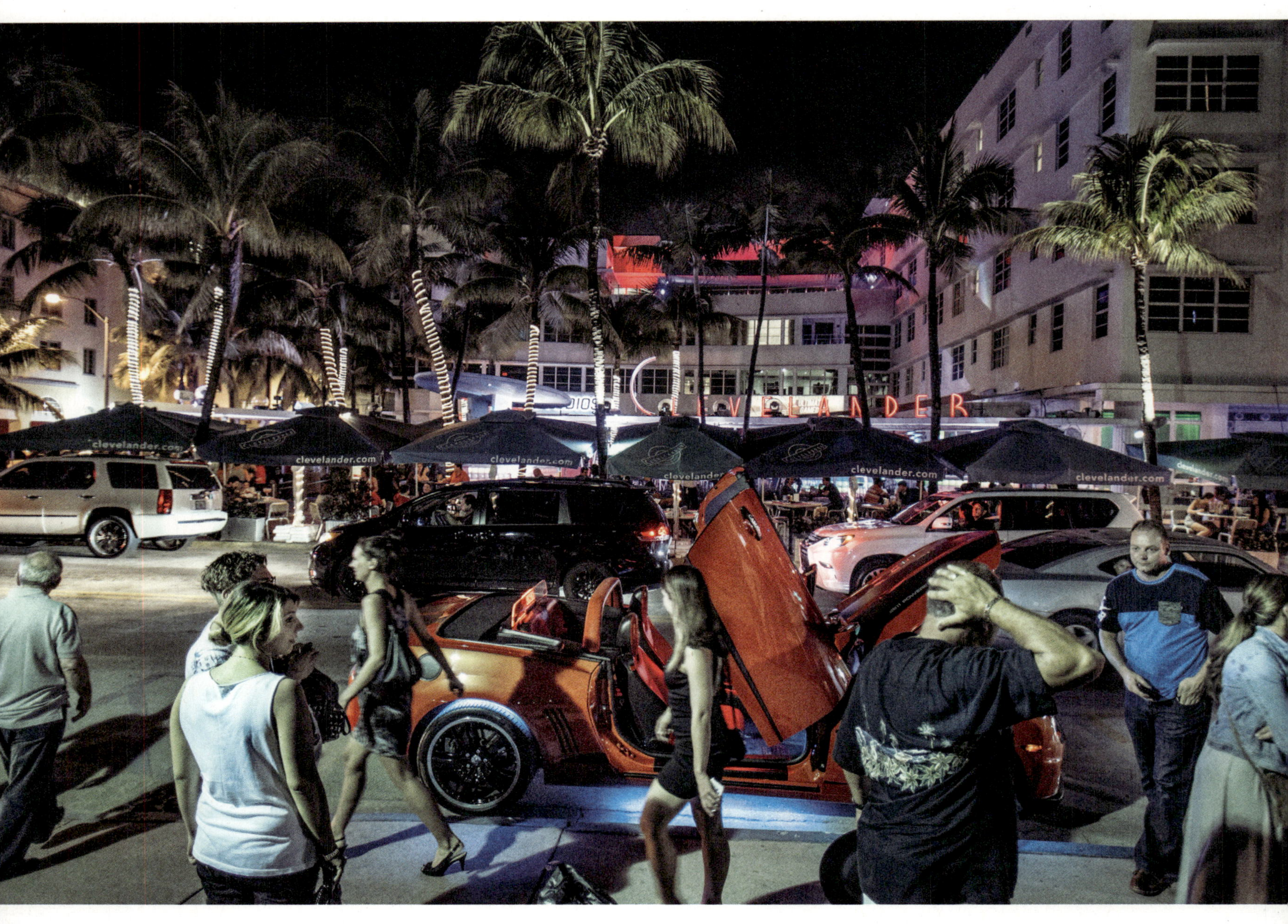

Friday night on Ocean Drive, Miami Beach.

< Pages 60-61: Downtown Miami. It's been predicted that by 2060, Miami Beach and the Bay Area will need to be evacuated, and by 2100, 60% of the city will need to be evacuated, including the downtown area. So far, experts think that Miami cannot be protected due to the fact that it's built on limestone.

The Porsche Design Tower is being built in Sunny Isles Beach. The tower is home to very expensive apartments where you can park your Porsche (included in the price) in your living room.

The boulevard at Miami Beach is protected by a rather low sea wall. According to experts, Miami cannot be protected due to the fact that it's built on limestone. Therefore, it's believed that Miami Beach and the Bay Area will need to be evacuated by 2060.

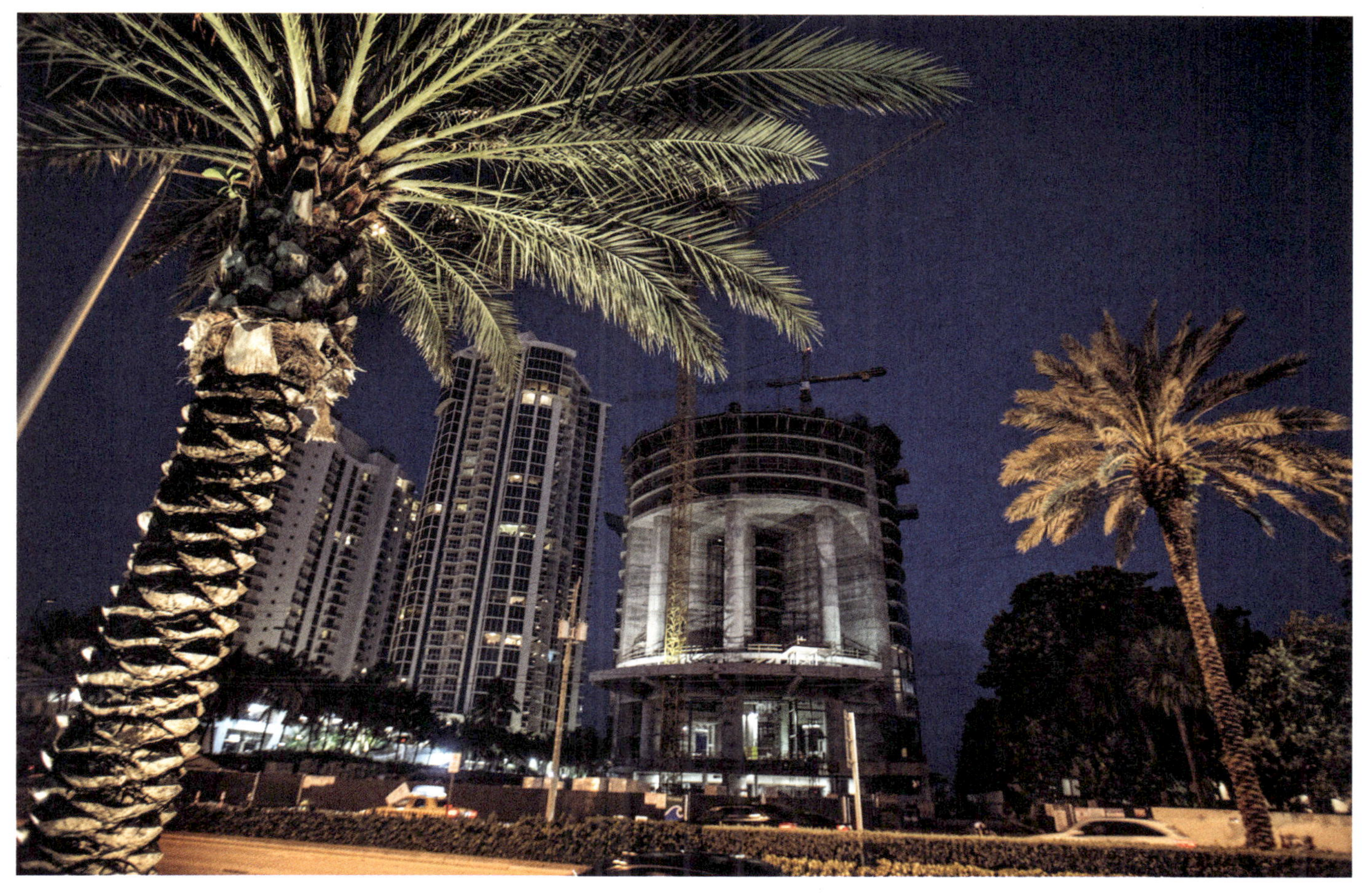

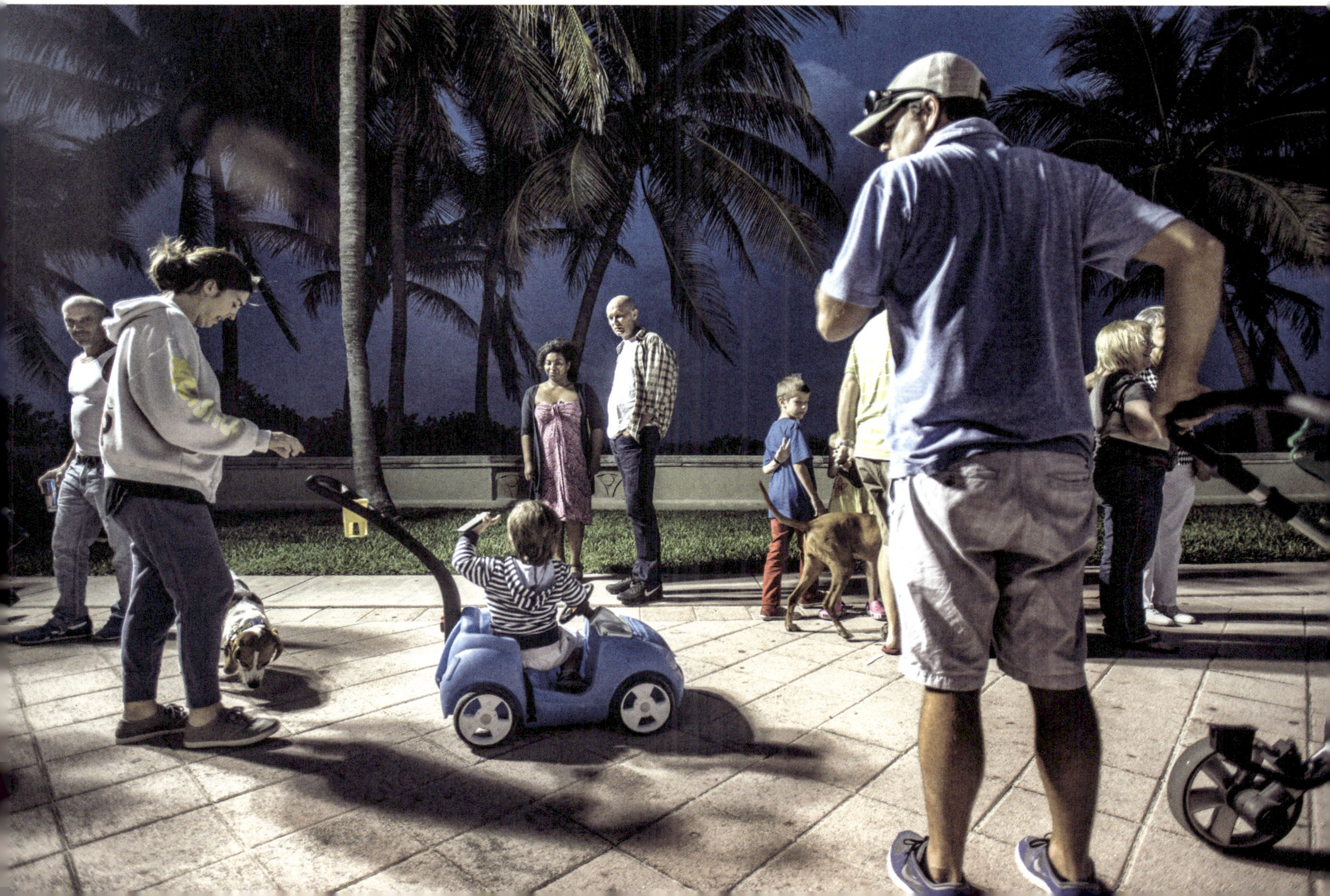

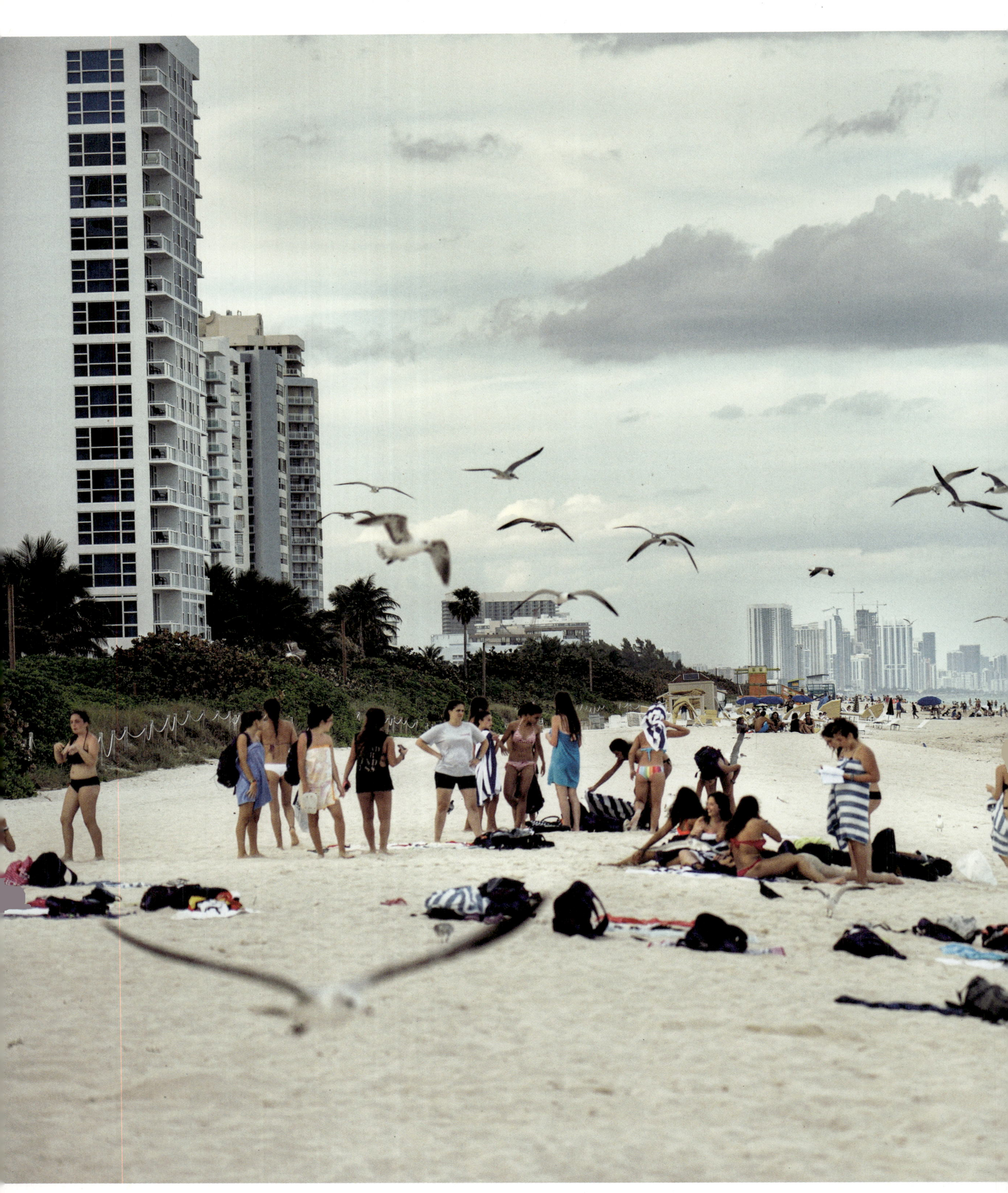

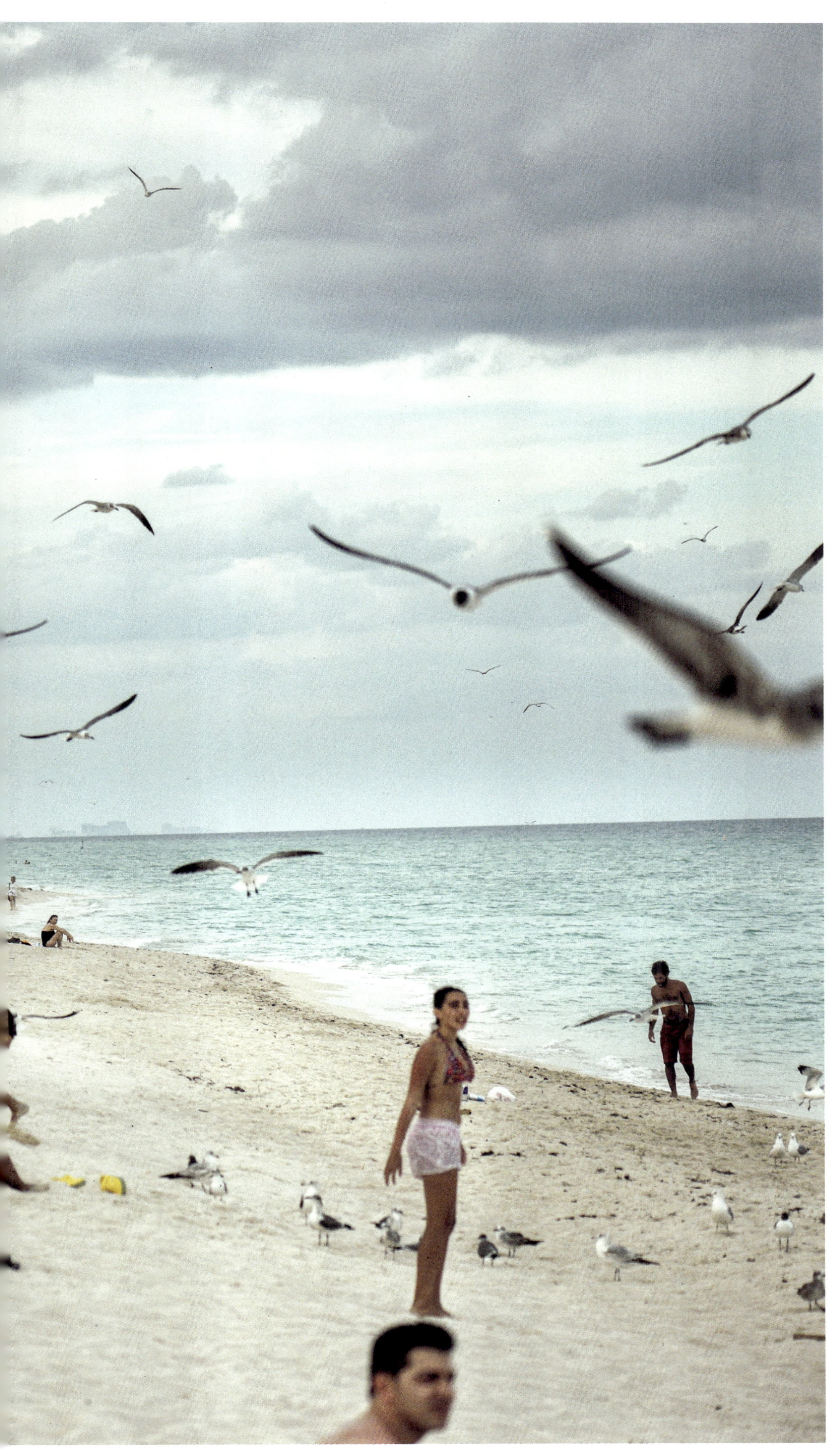

New condo towers on Miami Beach. According to experts, Miami cannot be protected due to the fact that it's built on limestone. Therefore, it's believed that Miami Beach and the Bay Area will need to be evacuated by 2060.

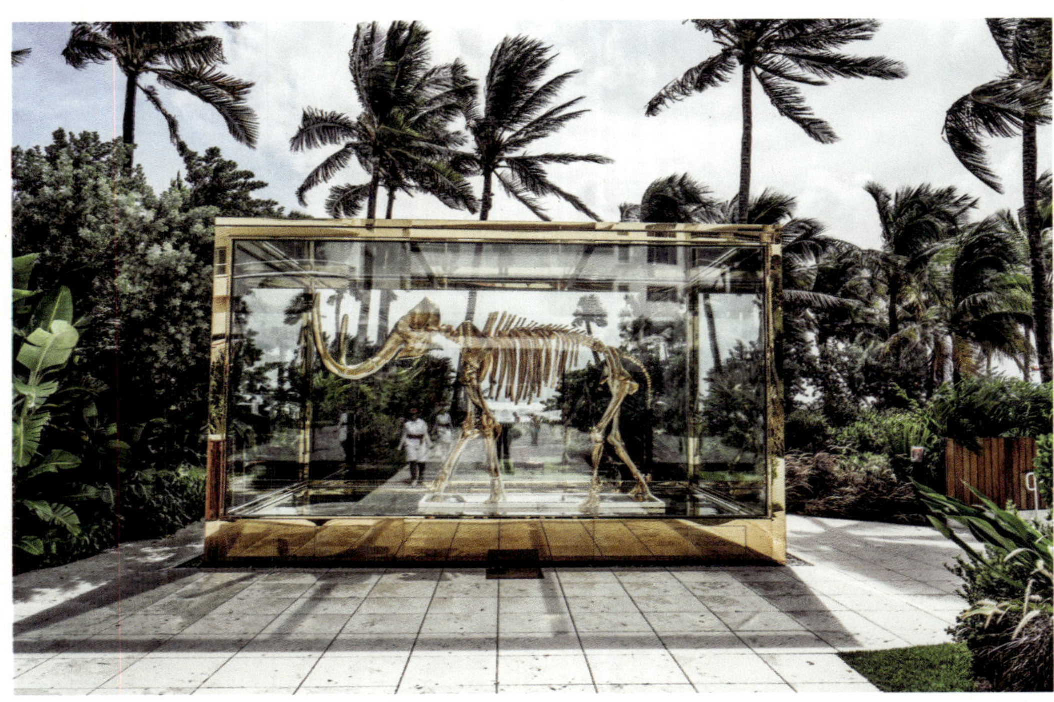

< Jeff Koons' gold-plated prehistoric relic at the newly opened Faena Hotel in Miami Beach.

> Sunday afternoon in the Miami Bay Area.

Miami's South Beach and the entrance to the port of Miami. Several cruise ships leave for the Caribbean every day.

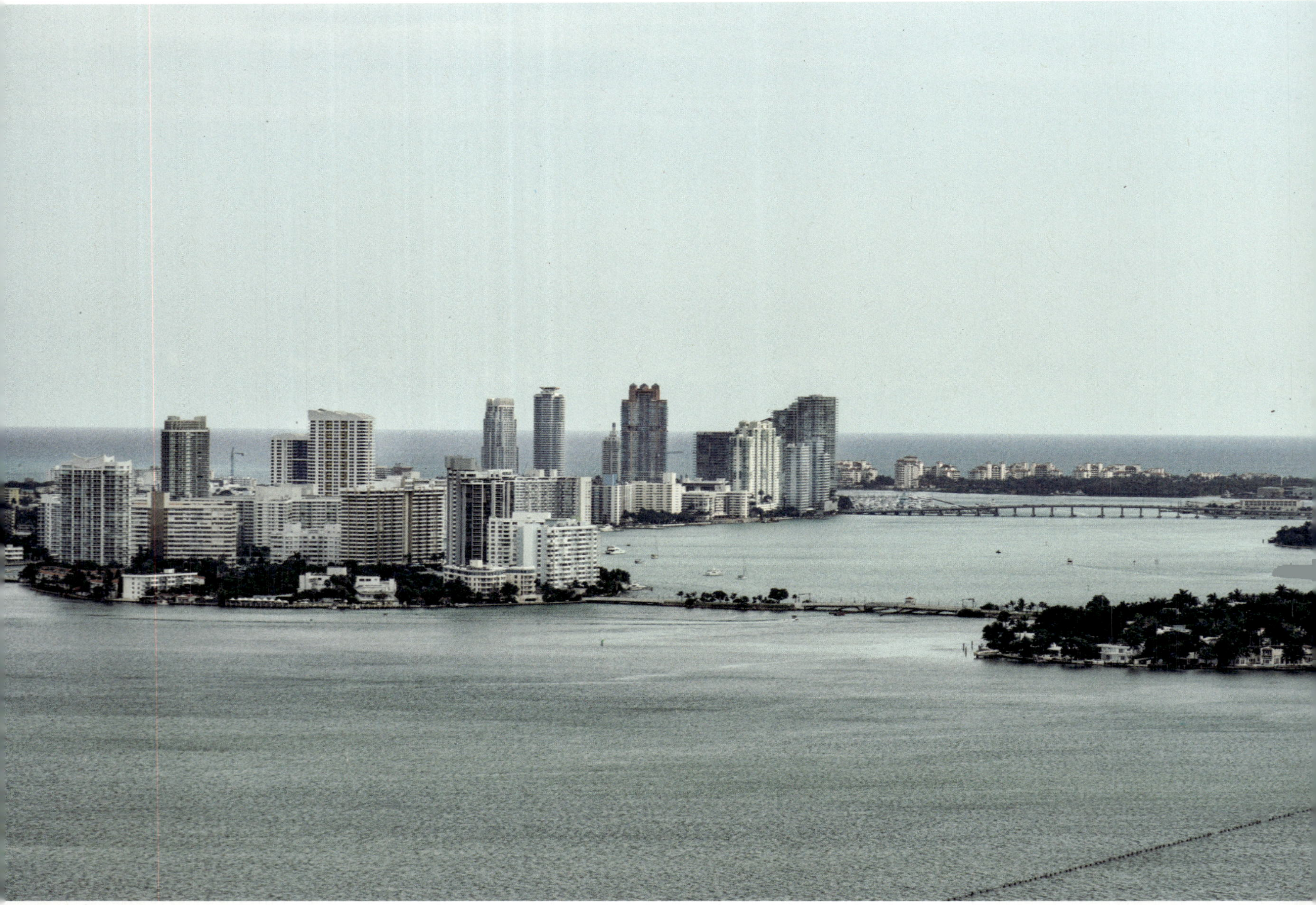

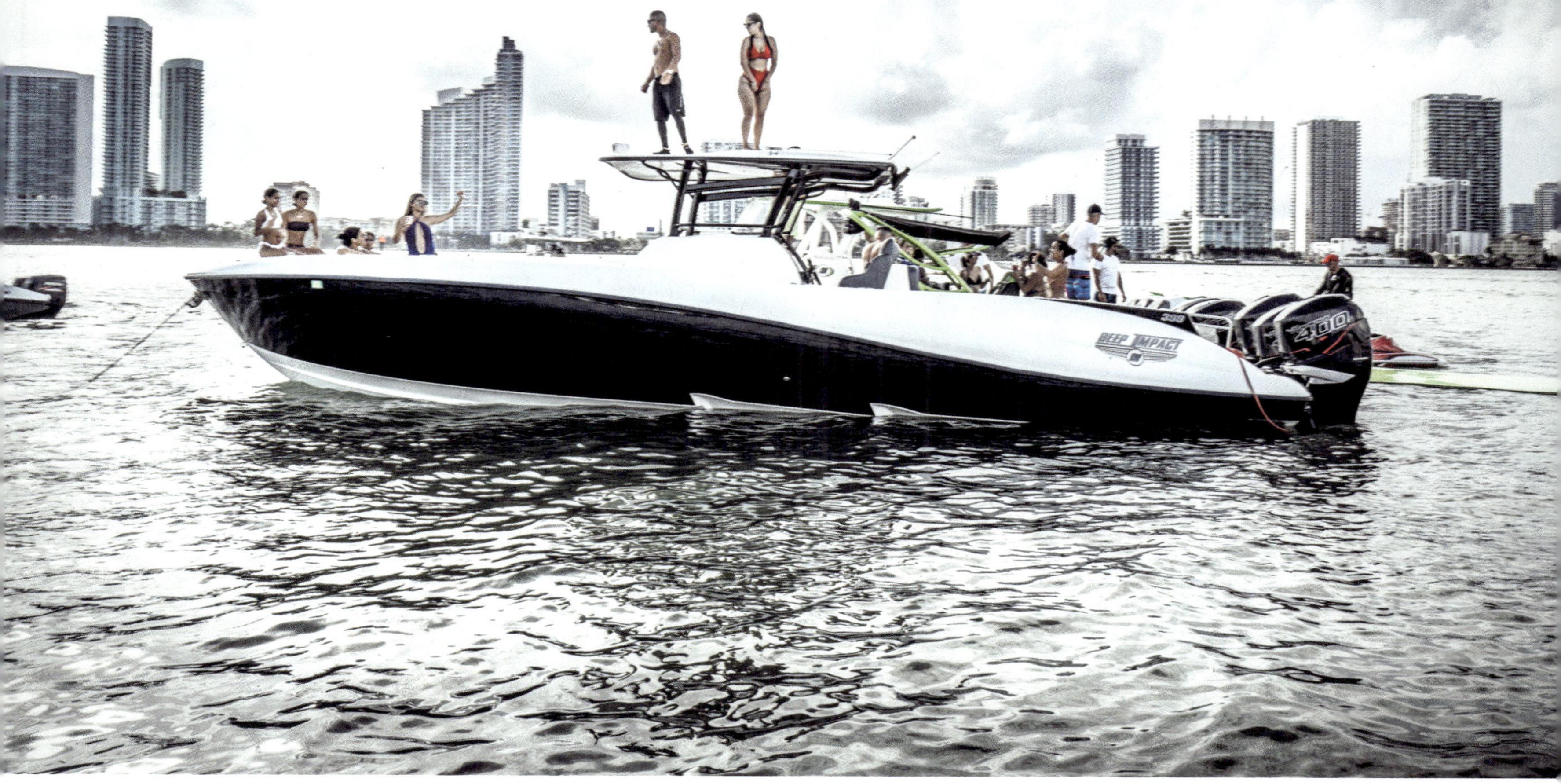

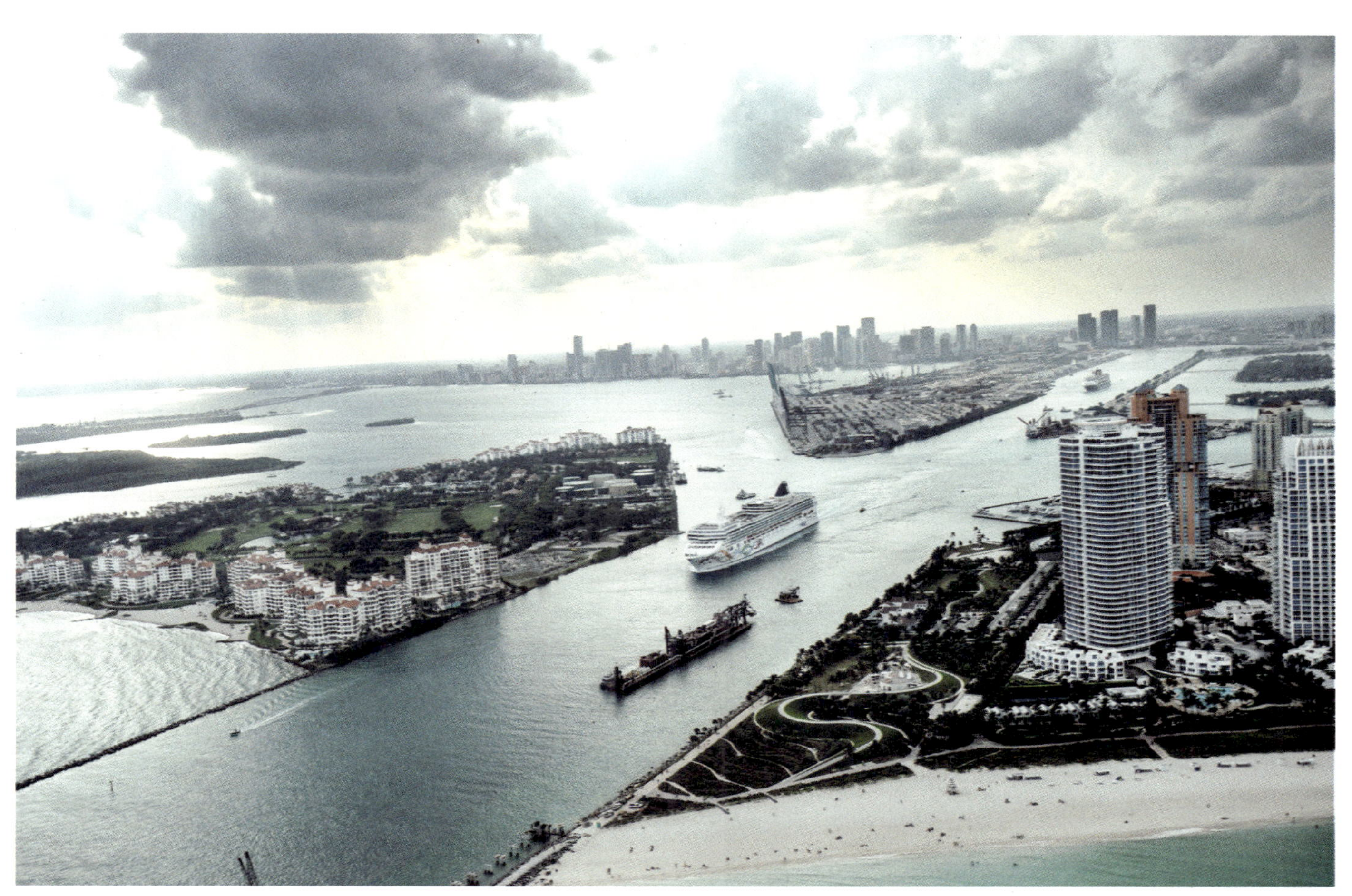

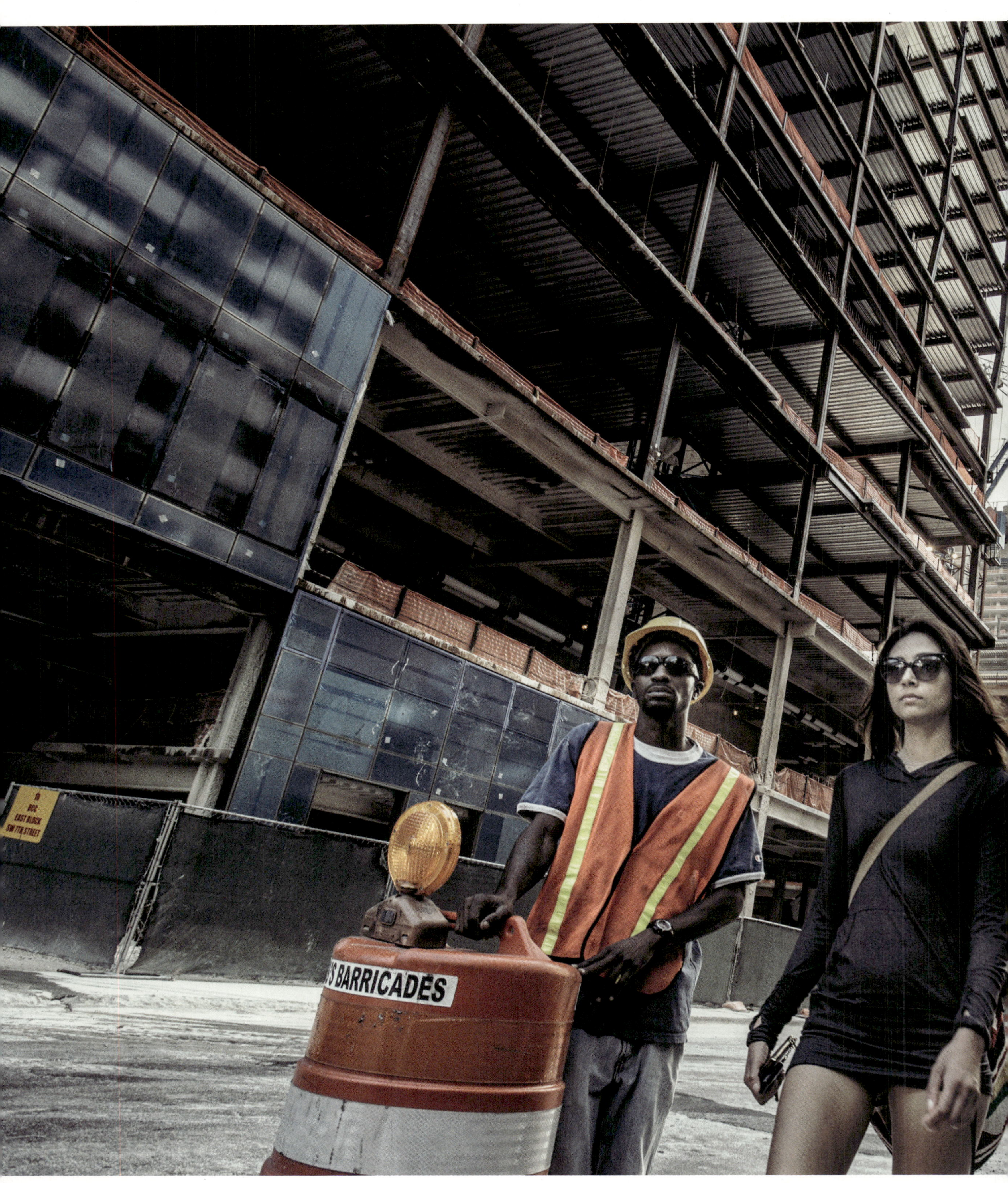
S BARRICADES

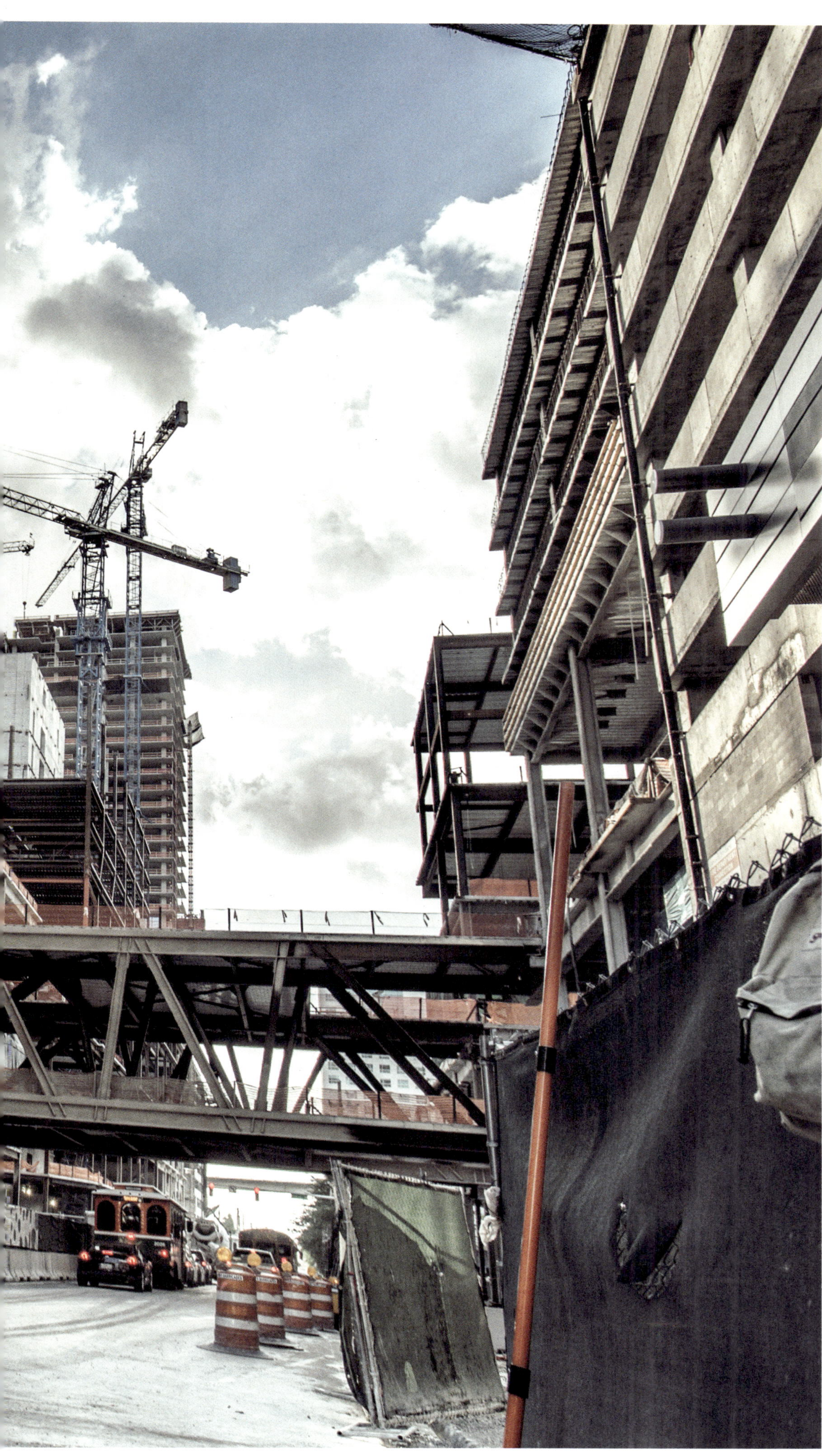

The Brickell City Centre in downtown
Miami under construction. The project
consists of three condo towers,
a hotel, shopping malls, theatres,
cinemas and more.

The authorities of Miami Beach are installing 600 pumps to prevent the streets from flooding during high tide.

King tide at Miami Beach. The seawater in the street comes up over the poorly maintained sea wall at Indian Creek and through the drainage system.

Star Island is one of the most expensive zip codes in the United States. The islands are very vulnerable to the rising seas.

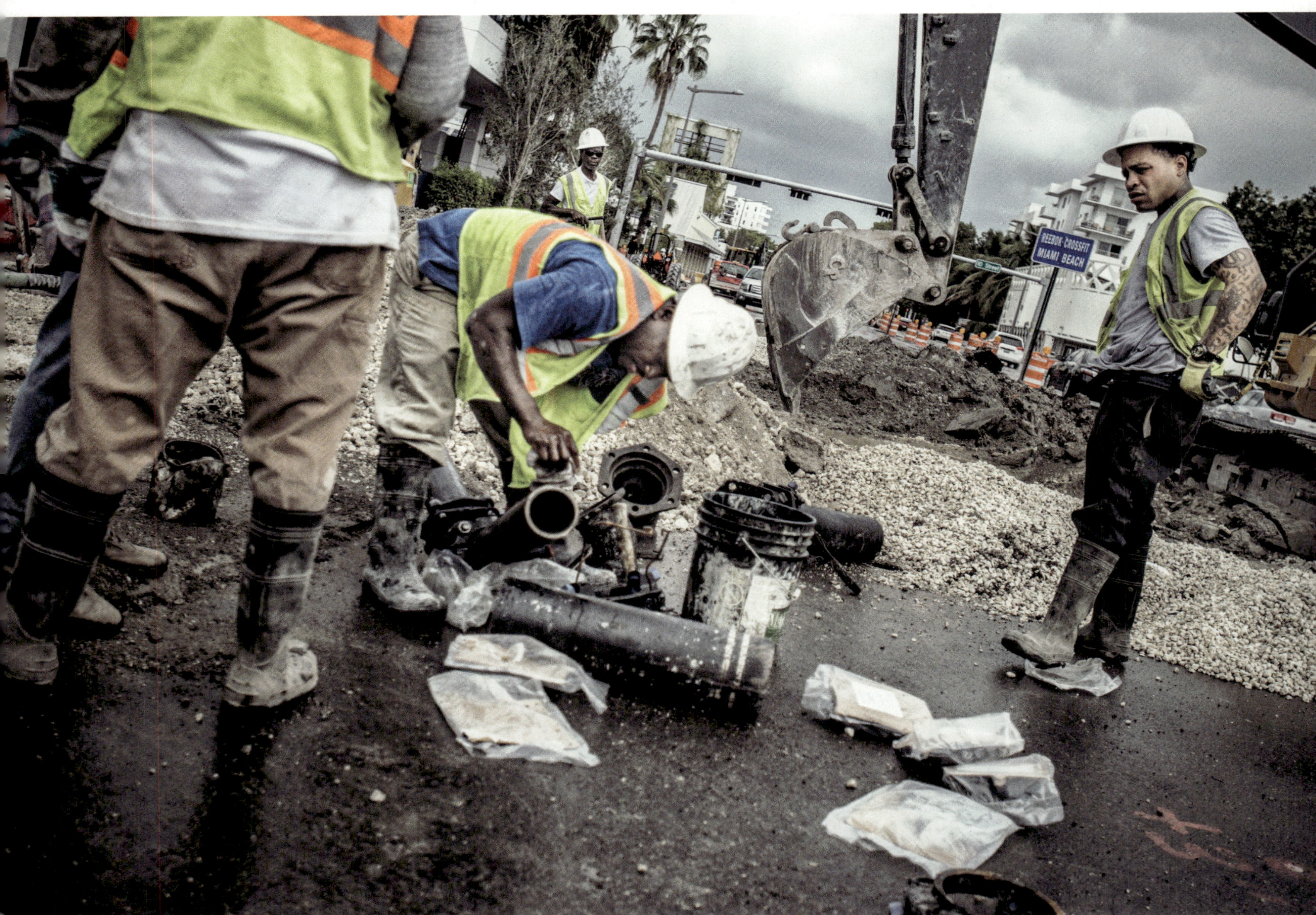

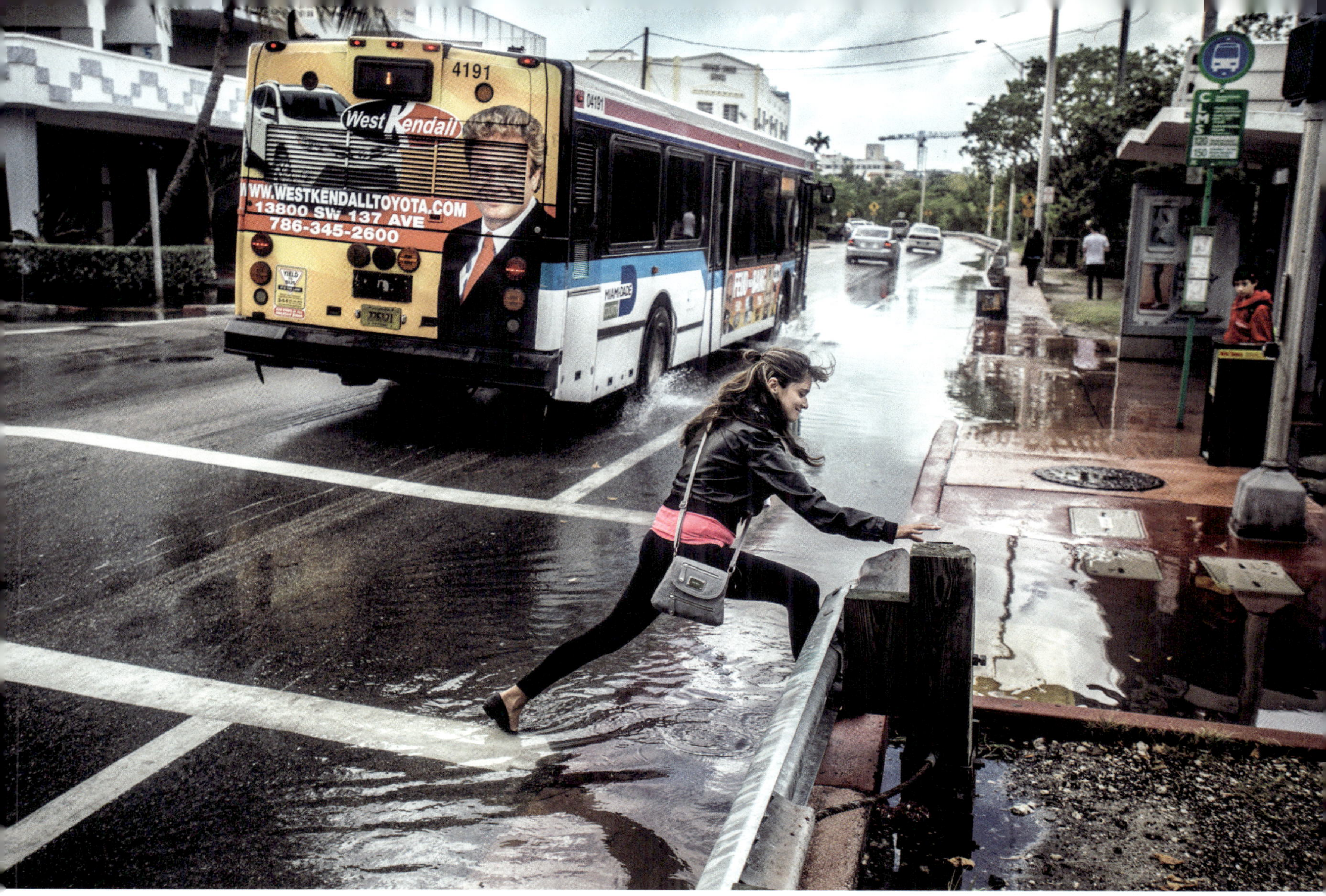

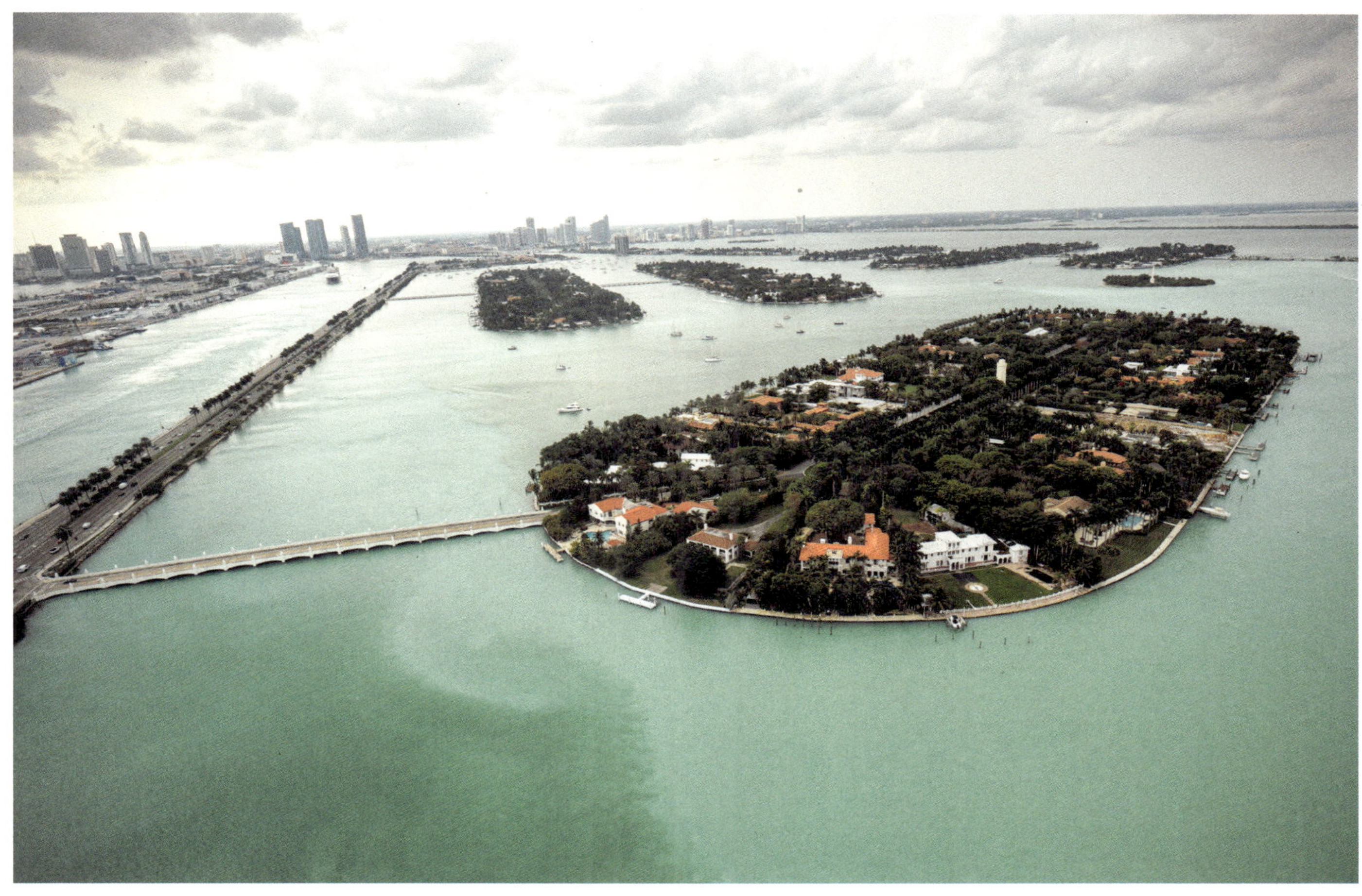

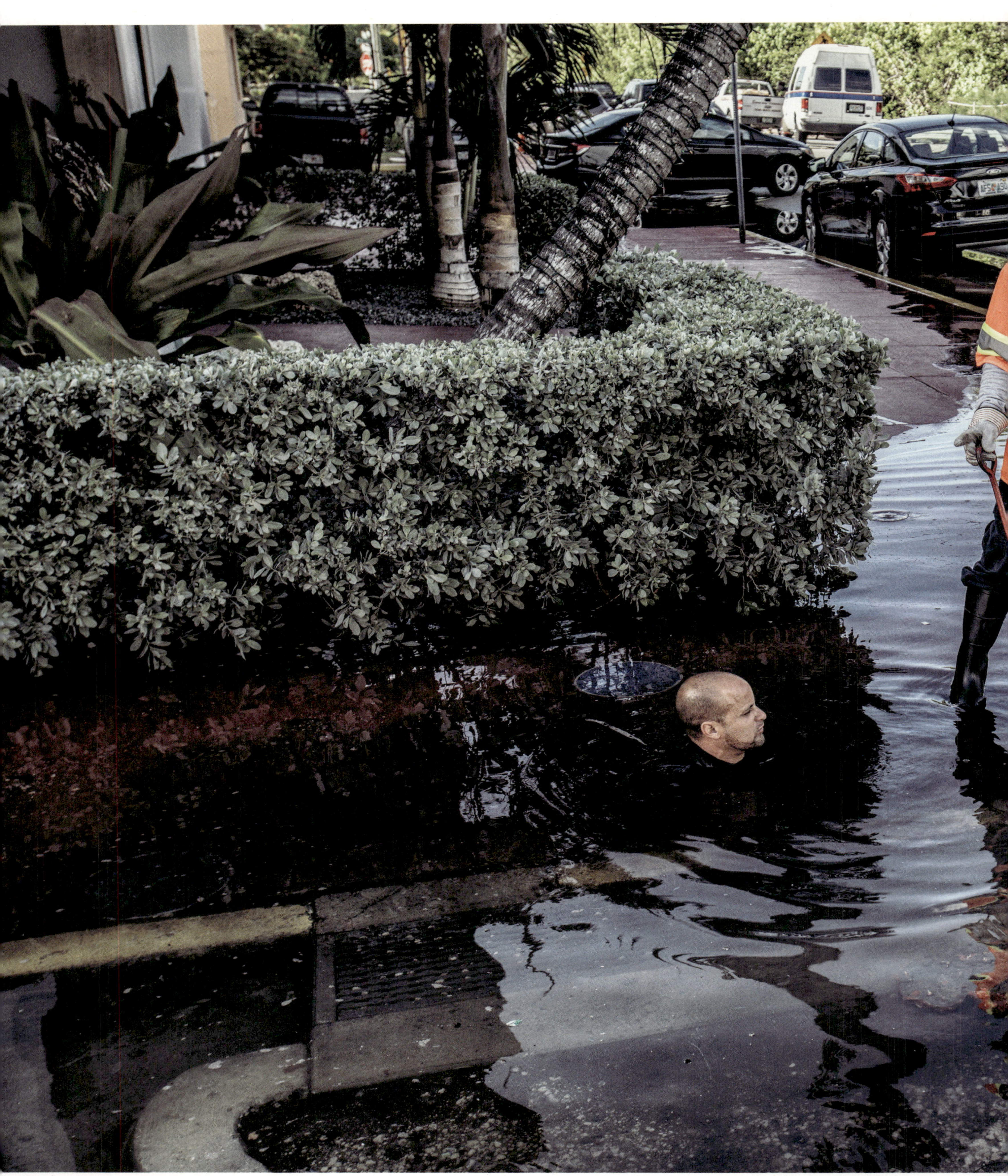

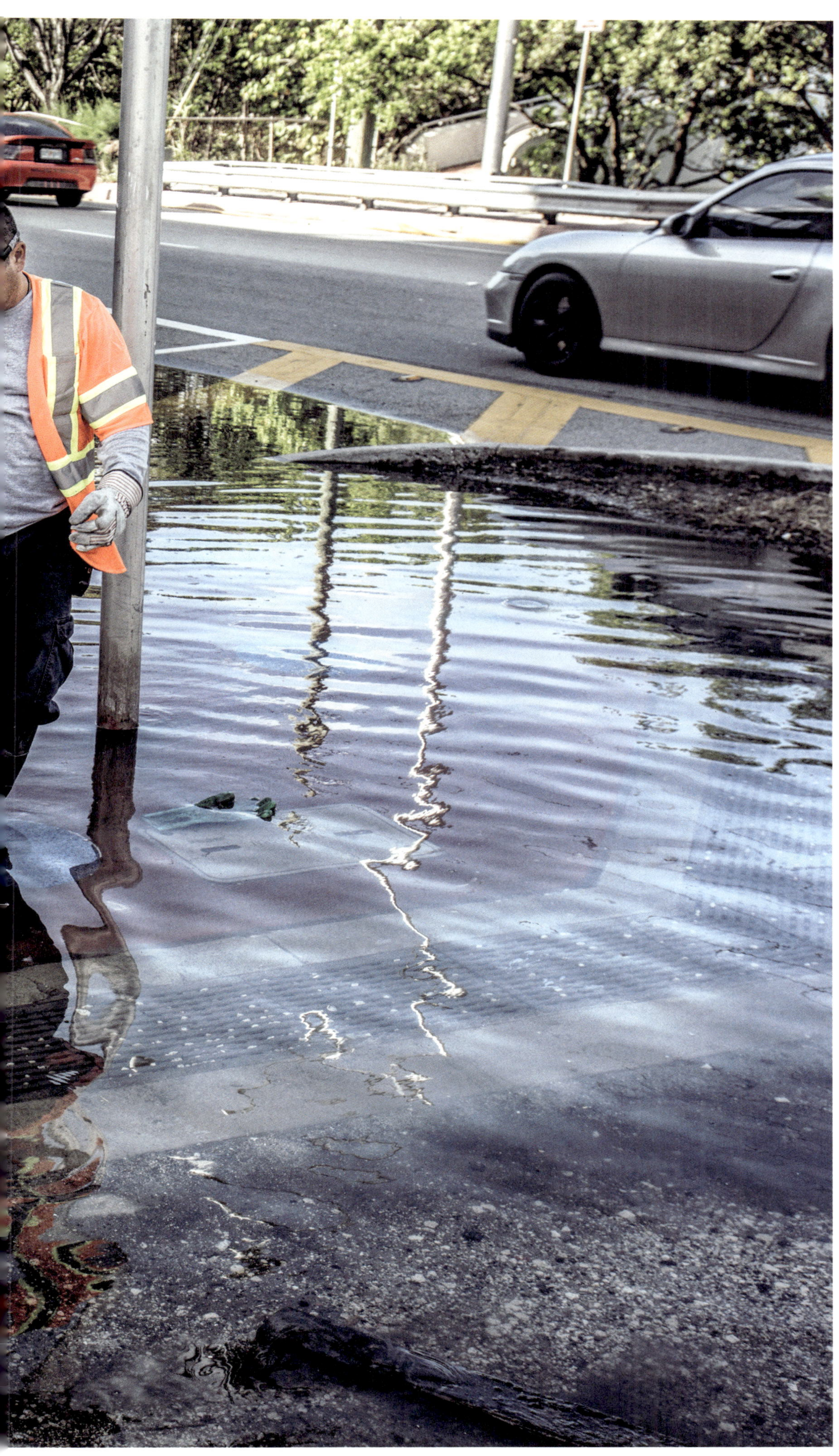

King tide at Miami beach. Workers are checking if the drainage system is blocked, but they concluded it was seawater.

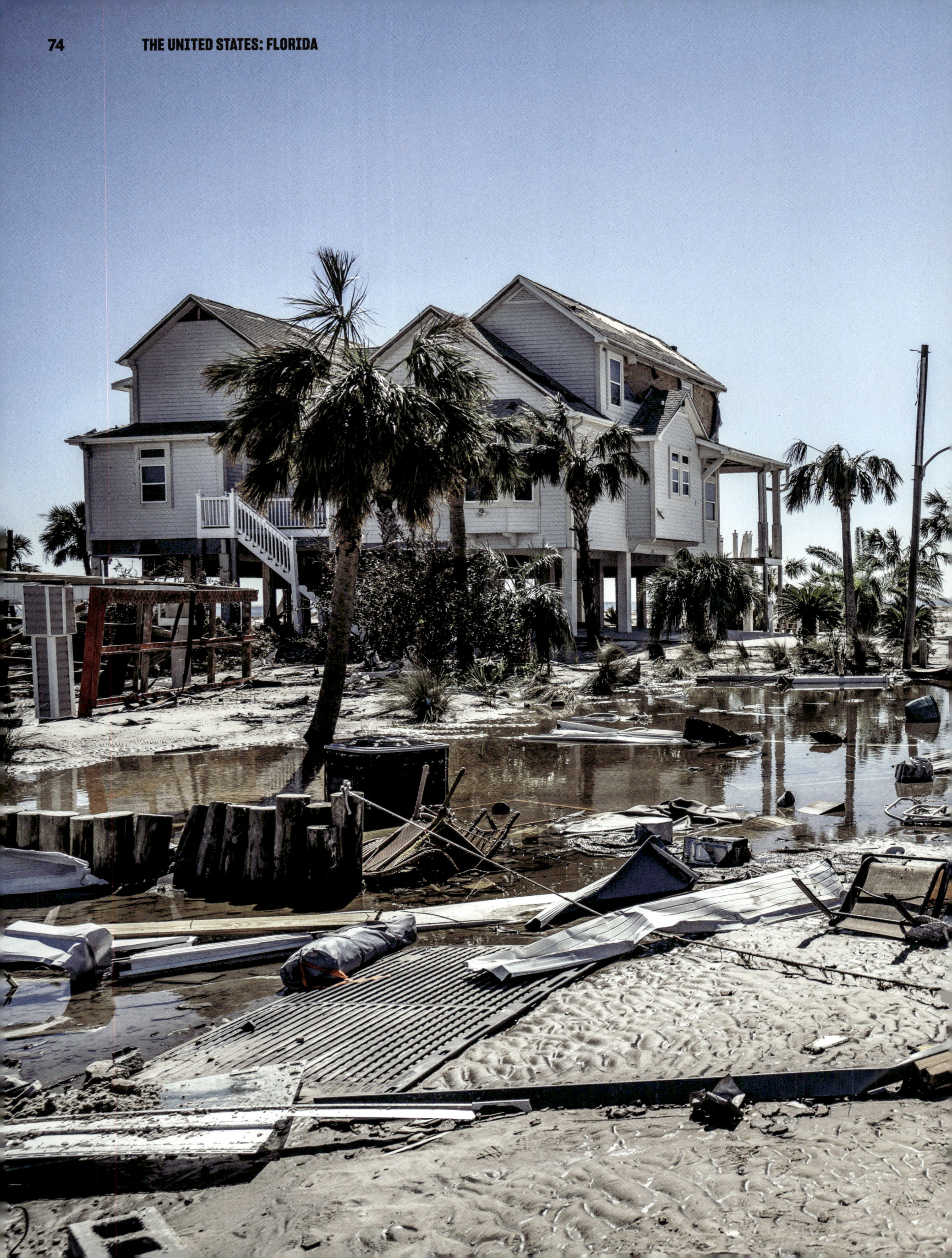

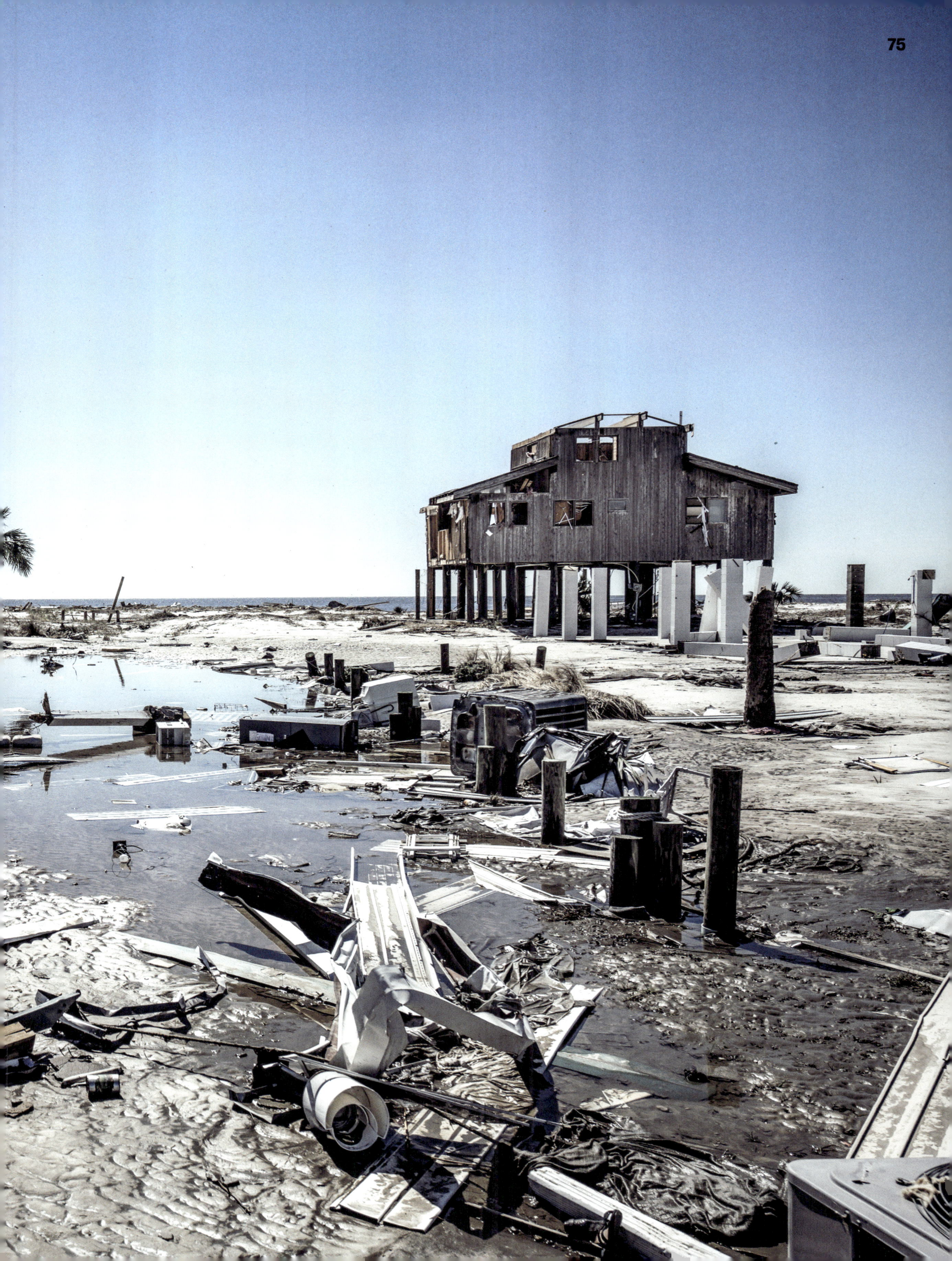

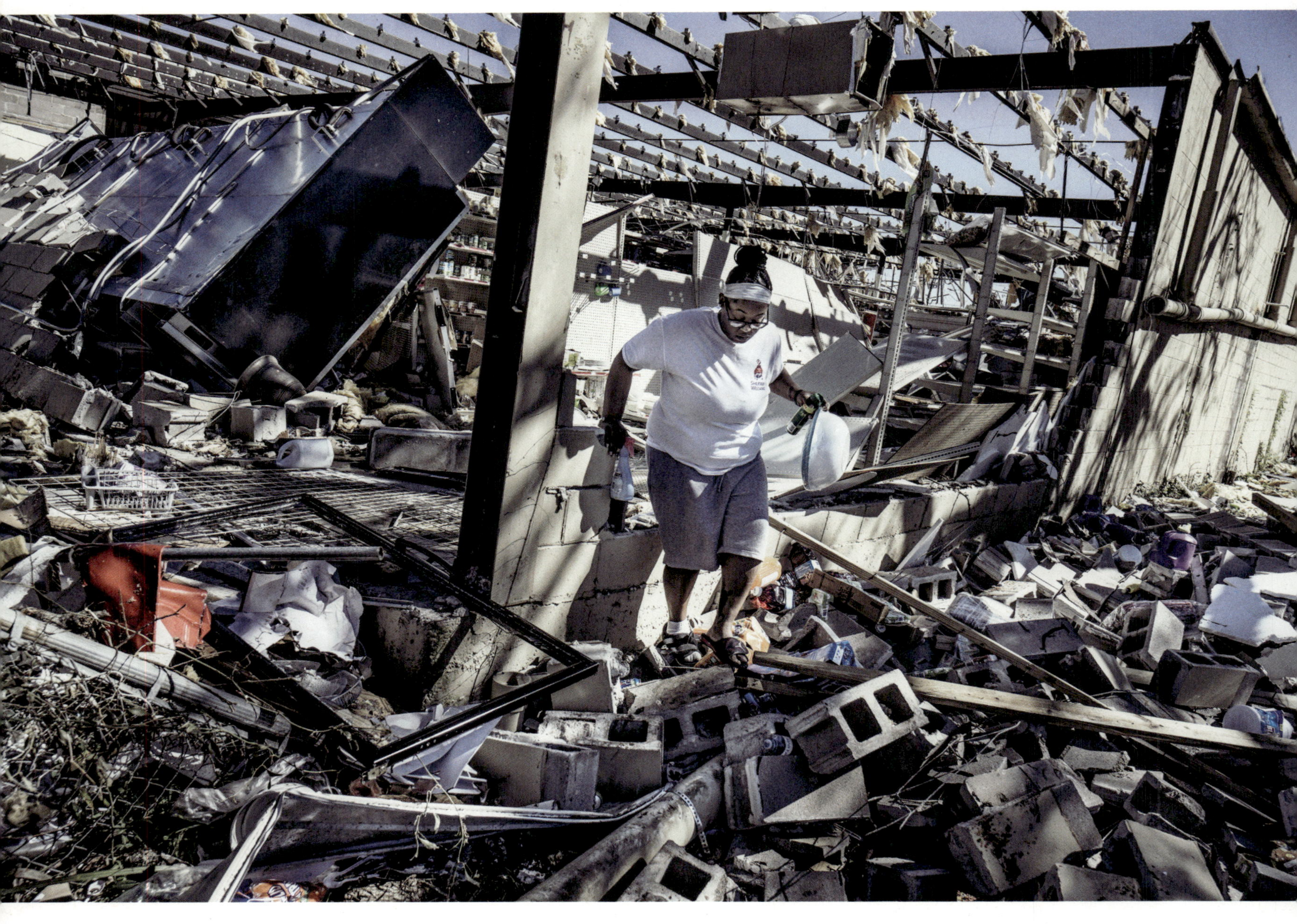

The aftermath of Hurricane Michael. Residents of Panama City Beach are finding water and food three days after the hurricane.

< Pages 74-75: Hurricane Michael hit Florida in October 2018. Mexico Beach was largely flattened by the storm surge. The town had 1,000 inhabitants; around 280 stayed for the hurricane.

> A house has been swept off of its foundation by the storm surge, landing a few hundred metres away, largely undamaged.

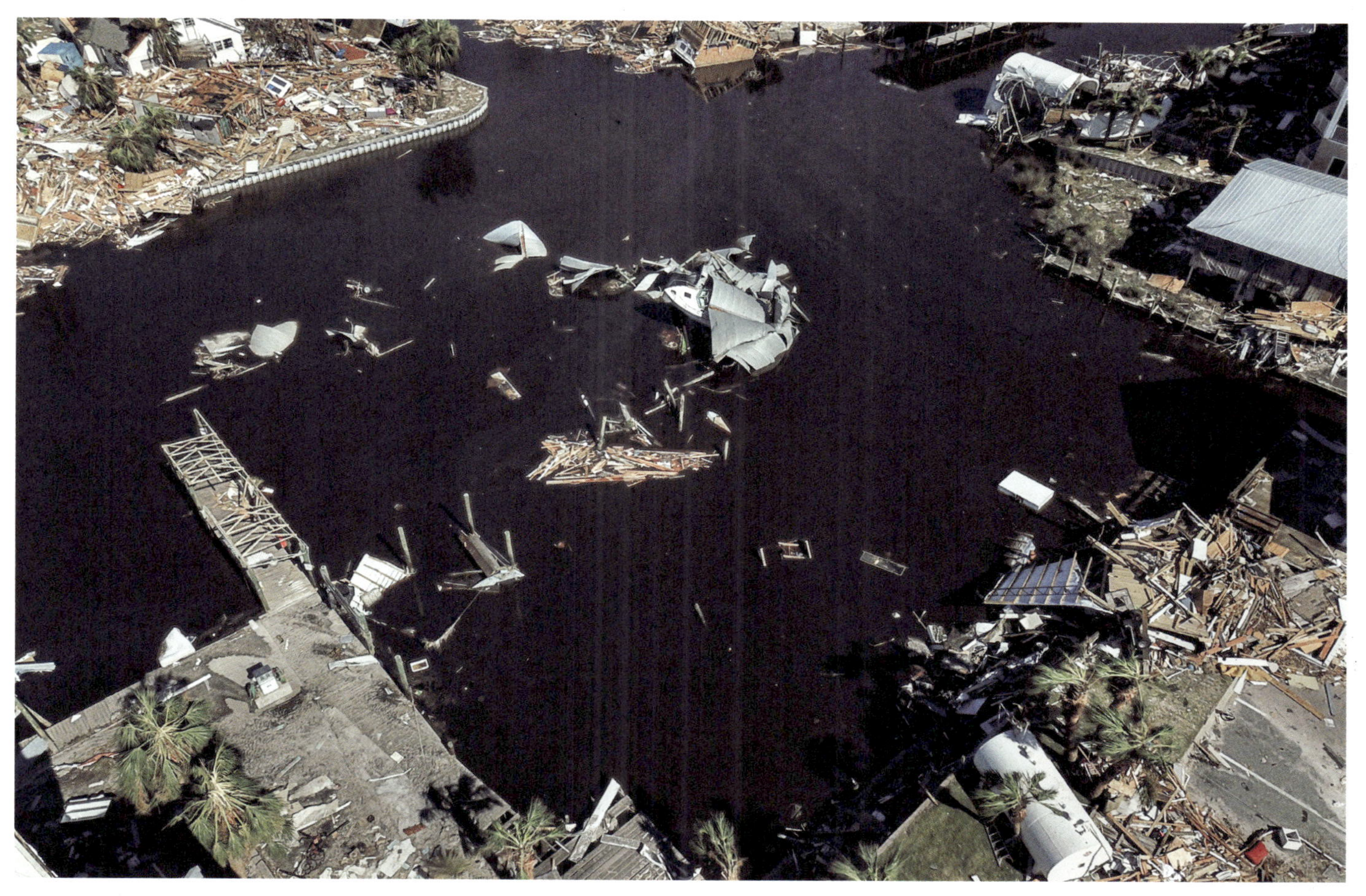

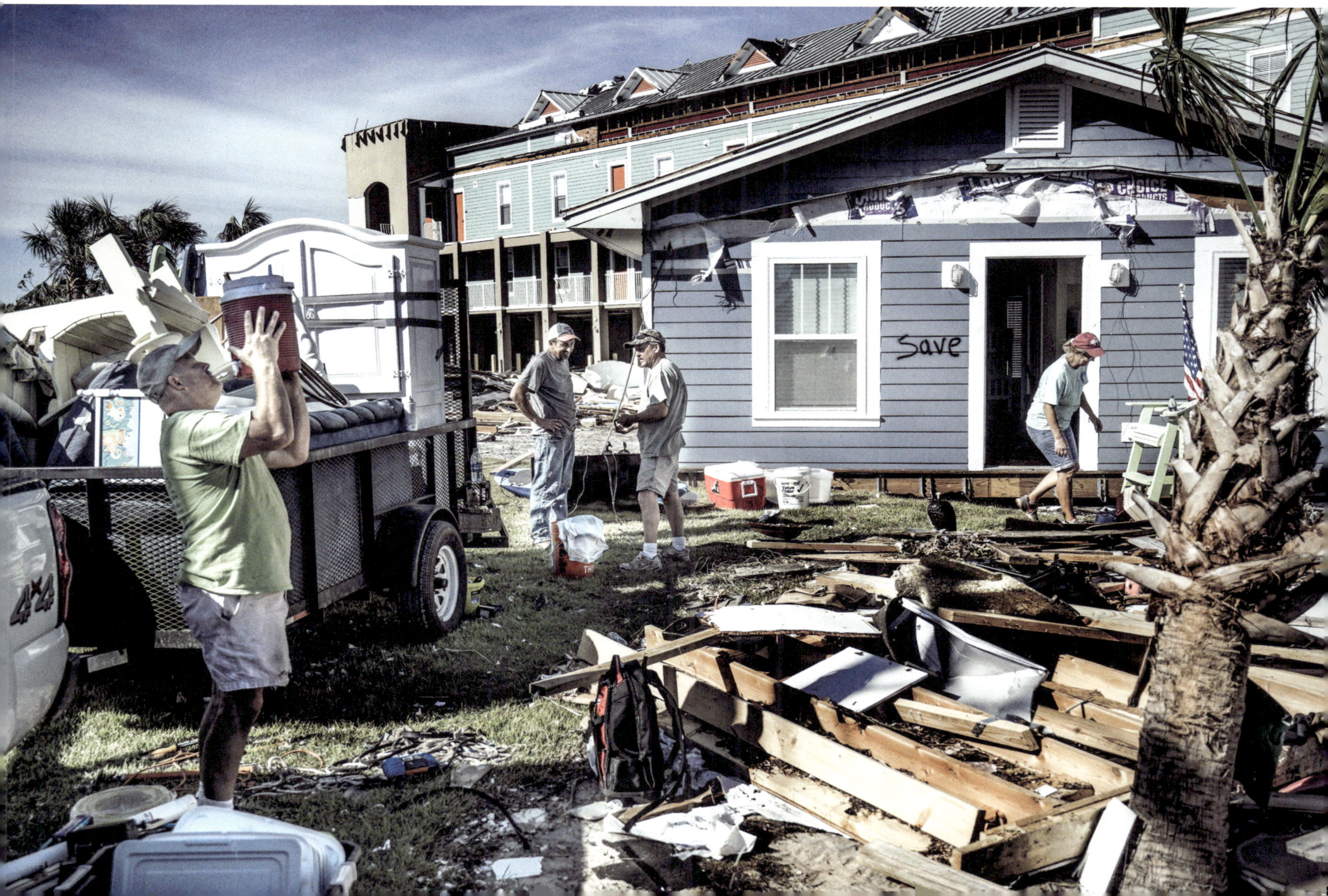

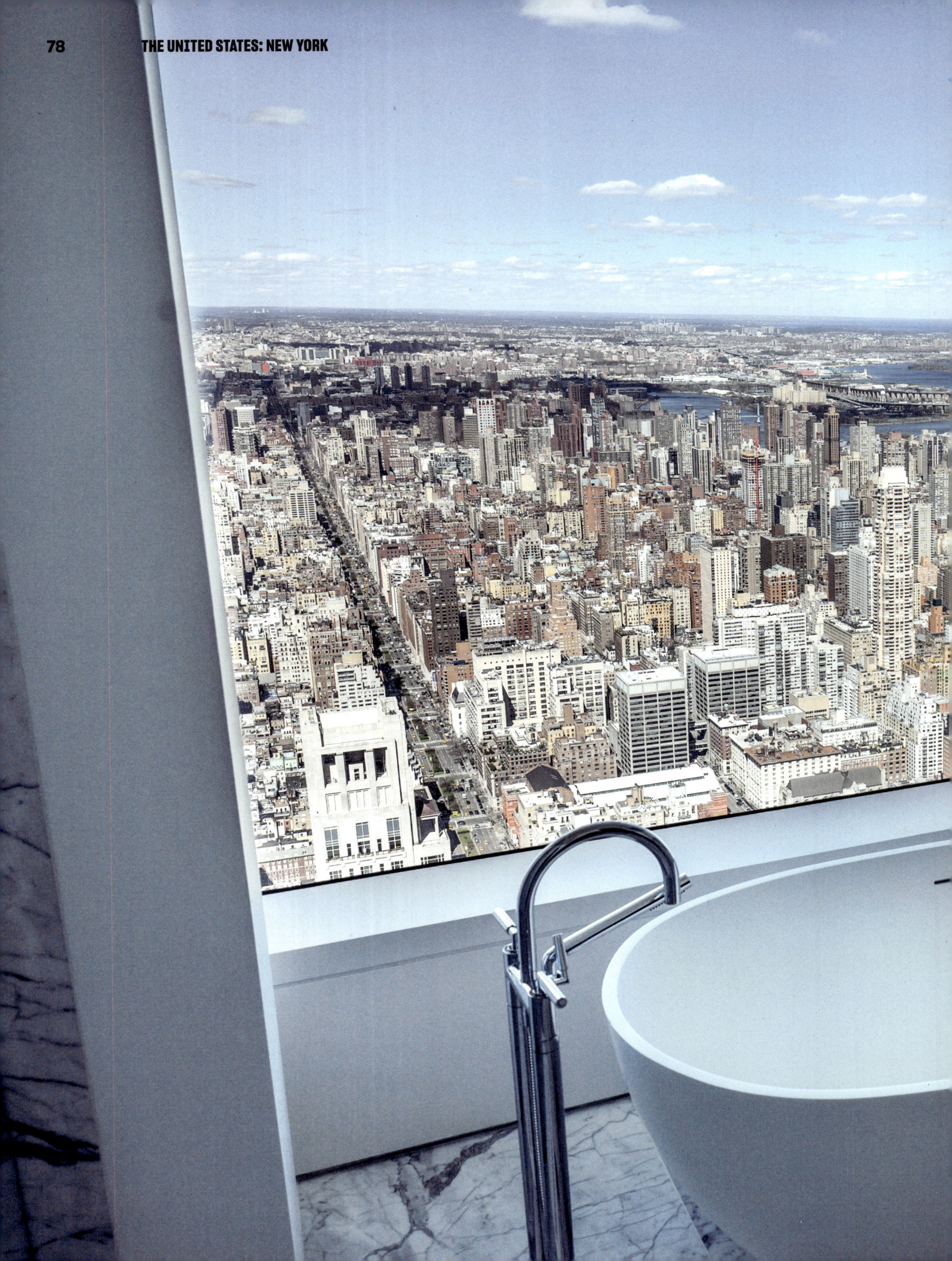

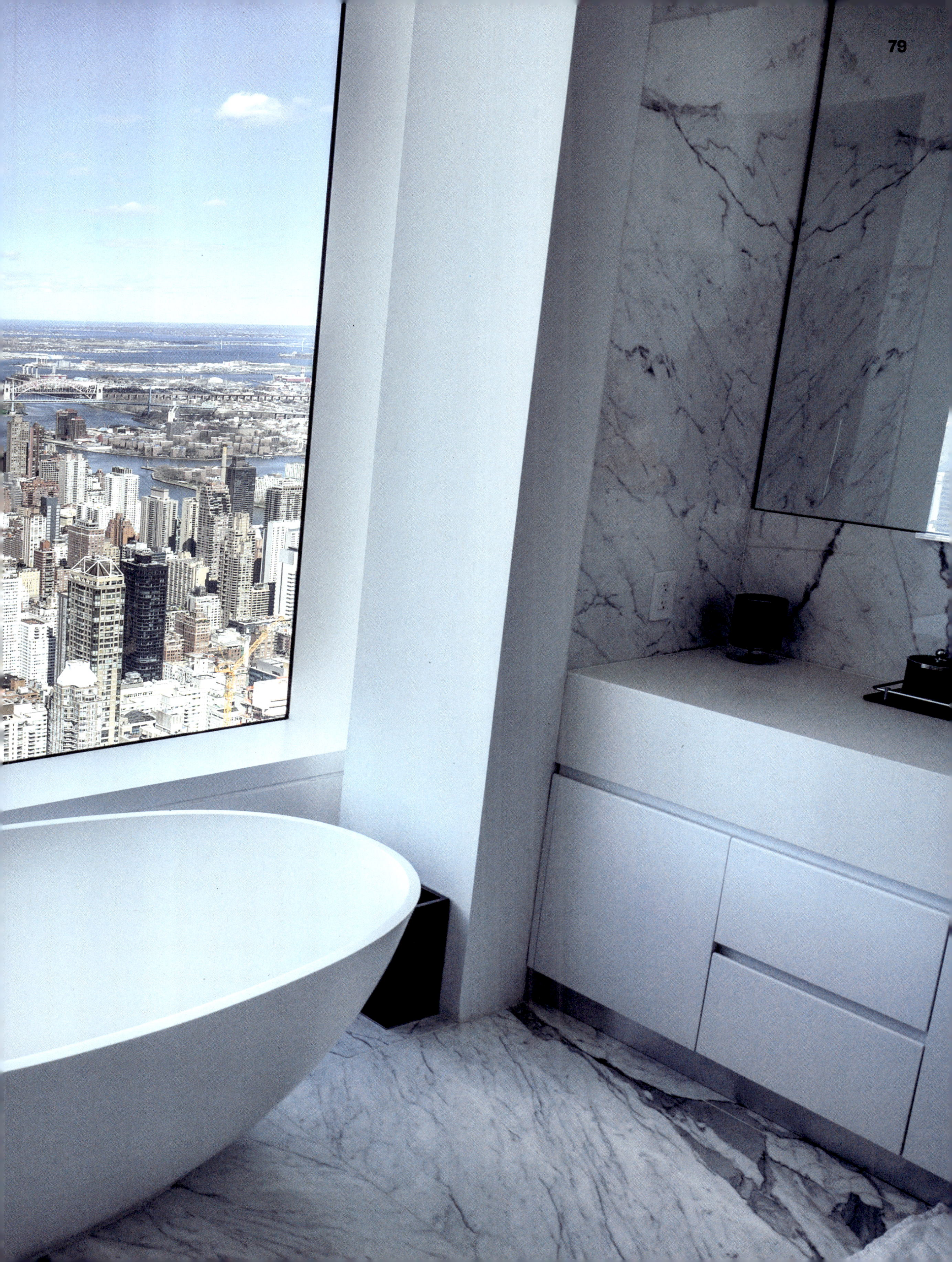

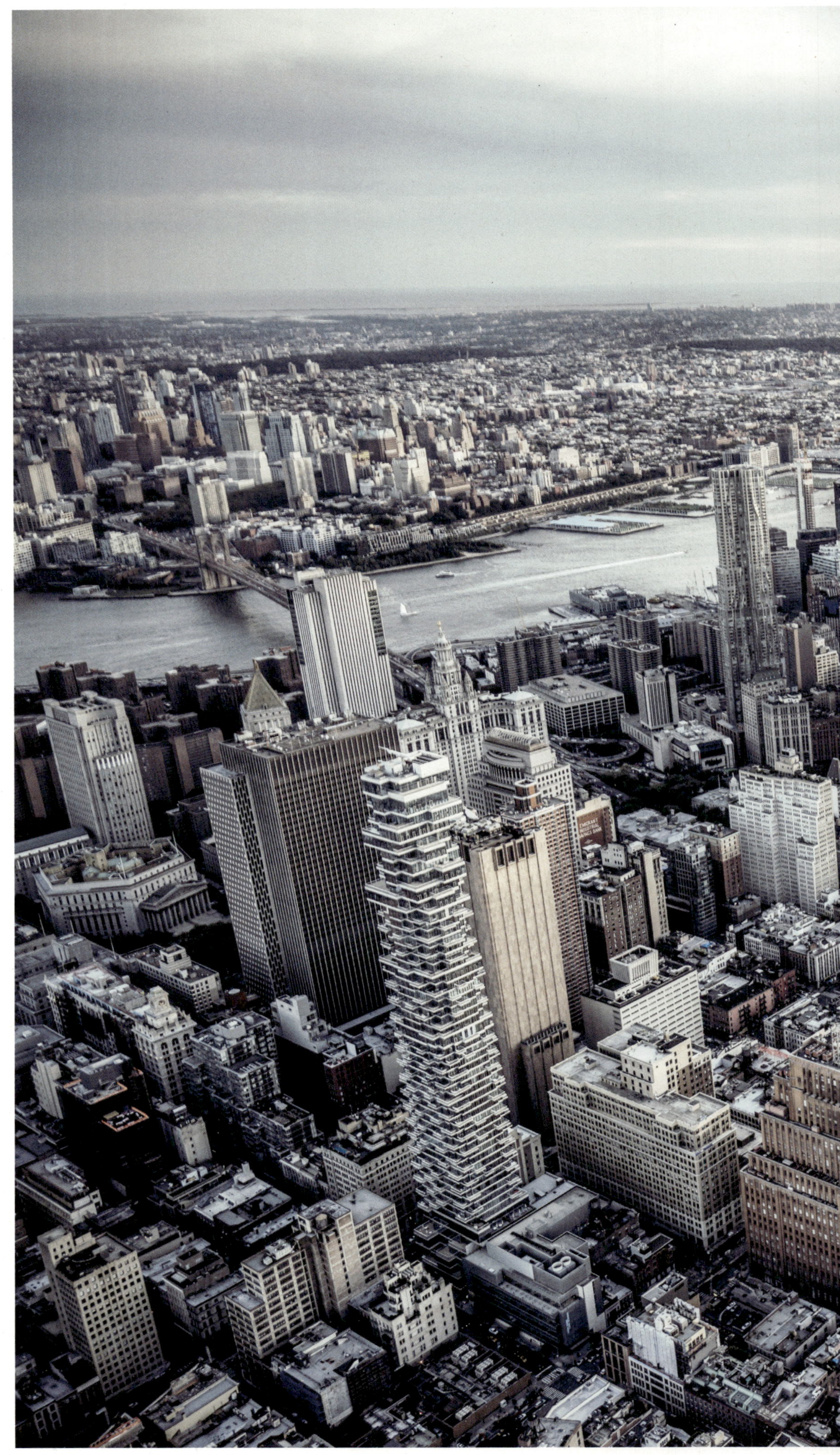

< Pages 78-79: A new apartment
building on Park Avenue.

Downtown Manhattan.

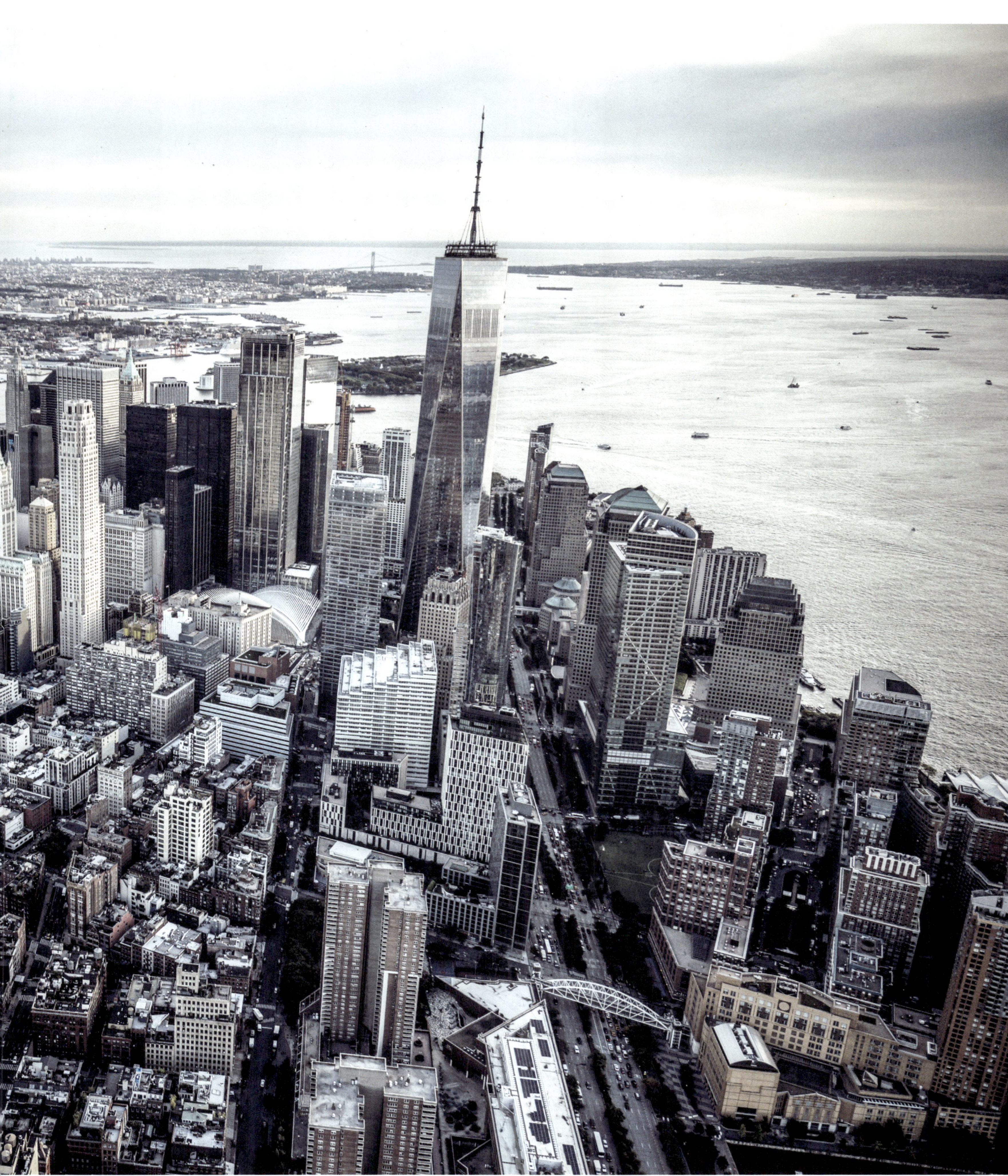

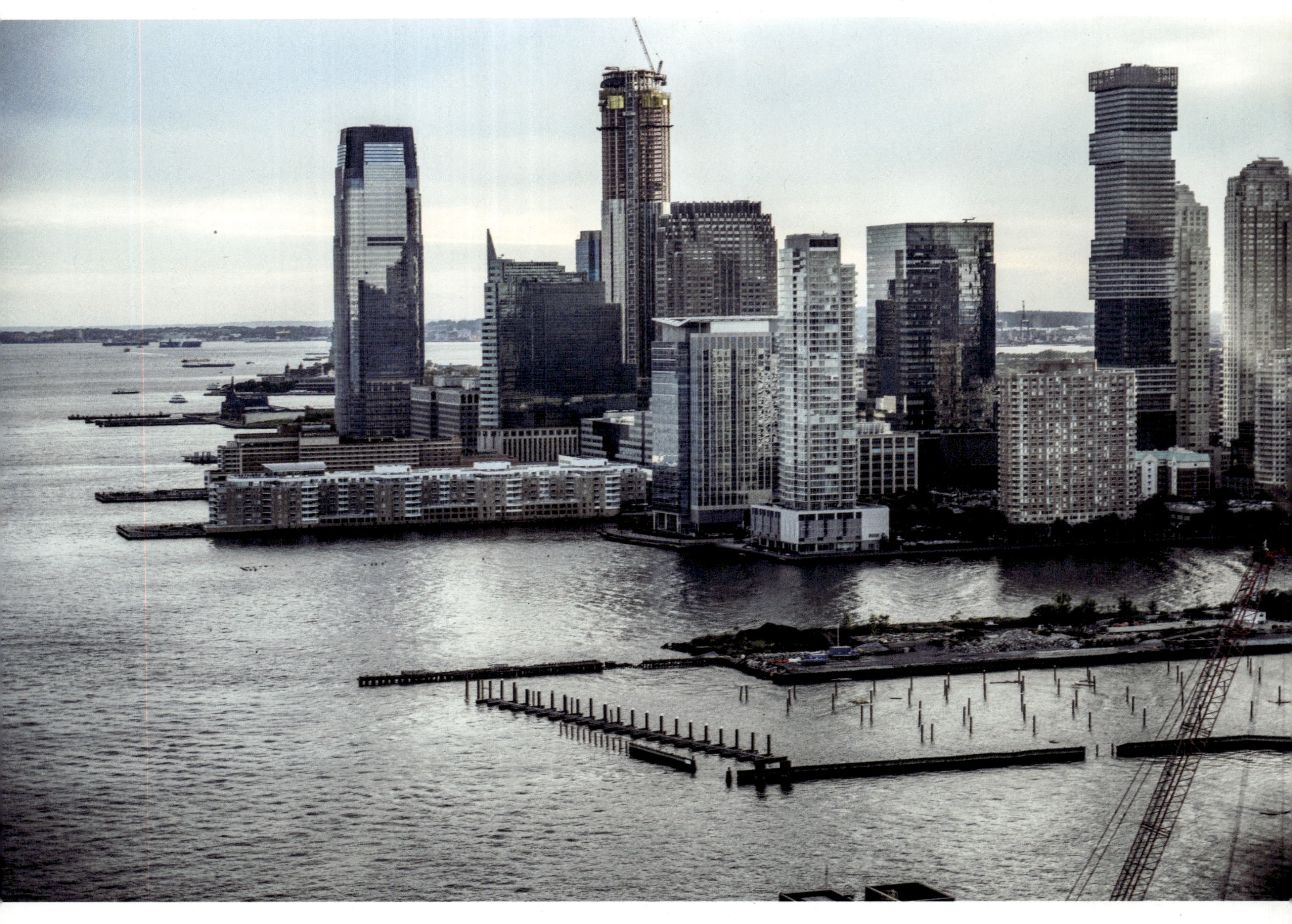

New buildings in Jersey City are
unprotected from rising sea levels.

A commemorative mural at a res-
taurant on Staten Island, which was
severely hit by Superstorm Sandy.

The boardwalk on Coney Island.

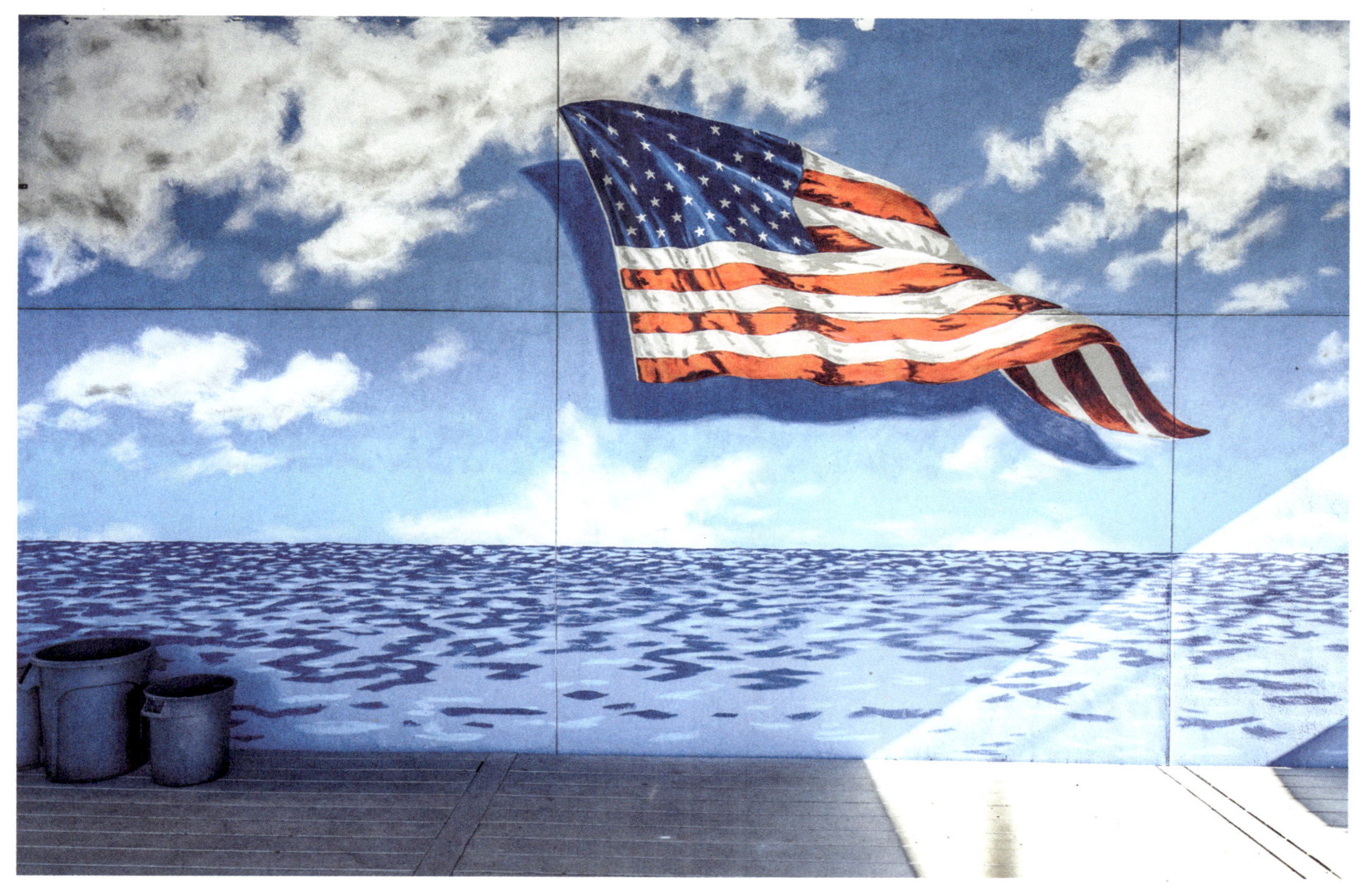

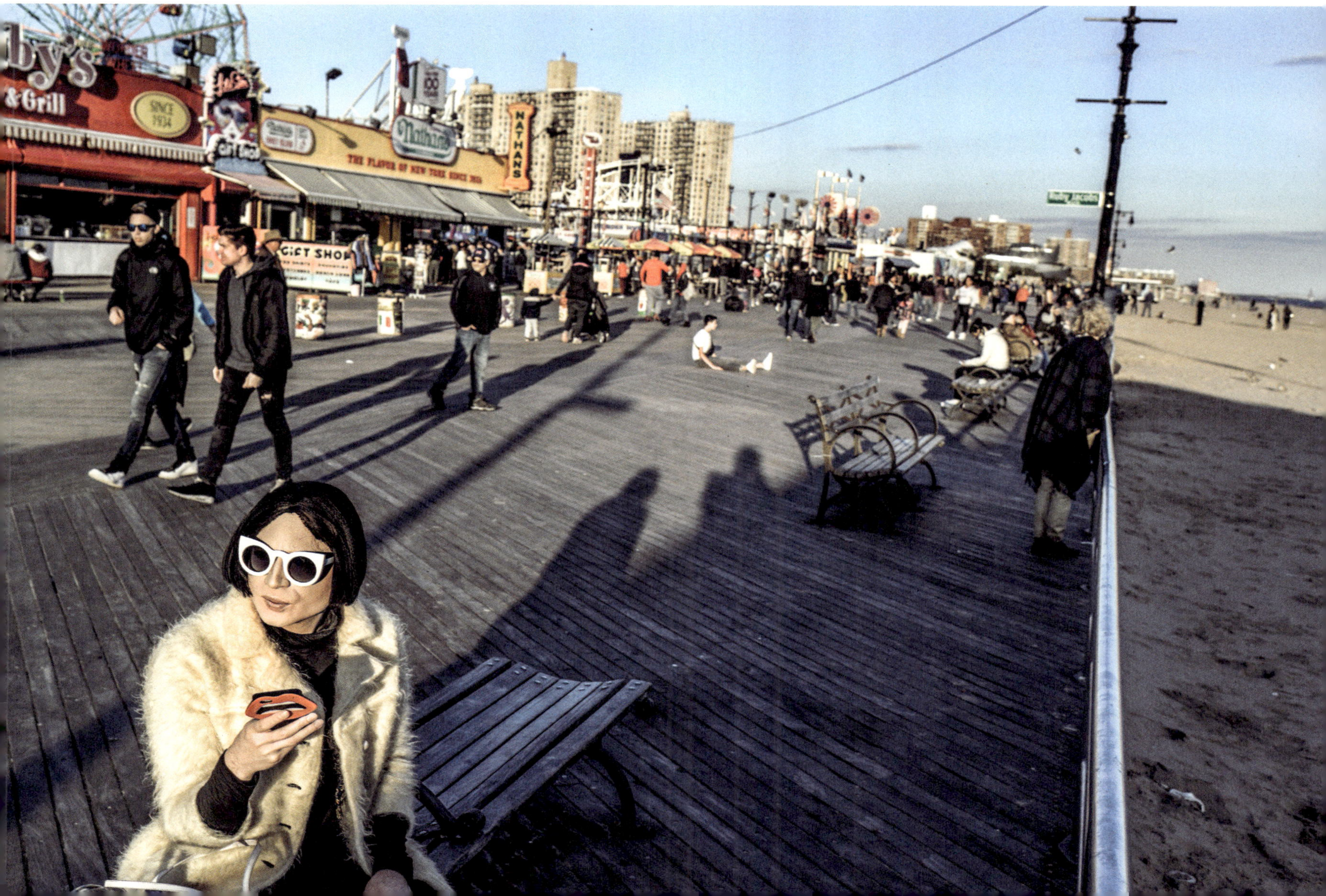

New York, as seen from the marshes around the Hackensack River in New Jersey.

Schoolchildren plant reeds to protect the shores against erosion in Jamaica Bay.

Pelican Island, between the barrier islands of Fire Island and Smith Point.

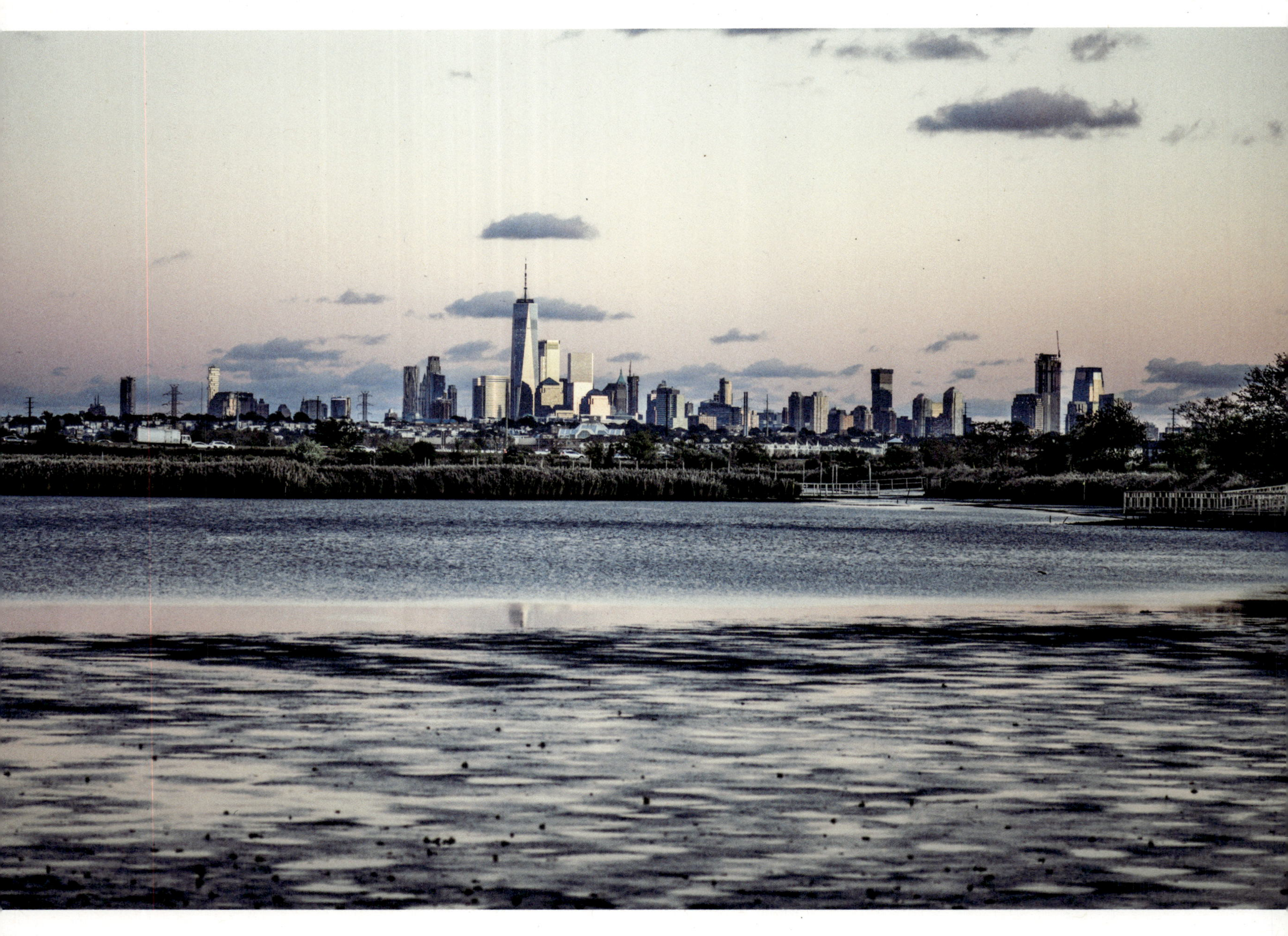

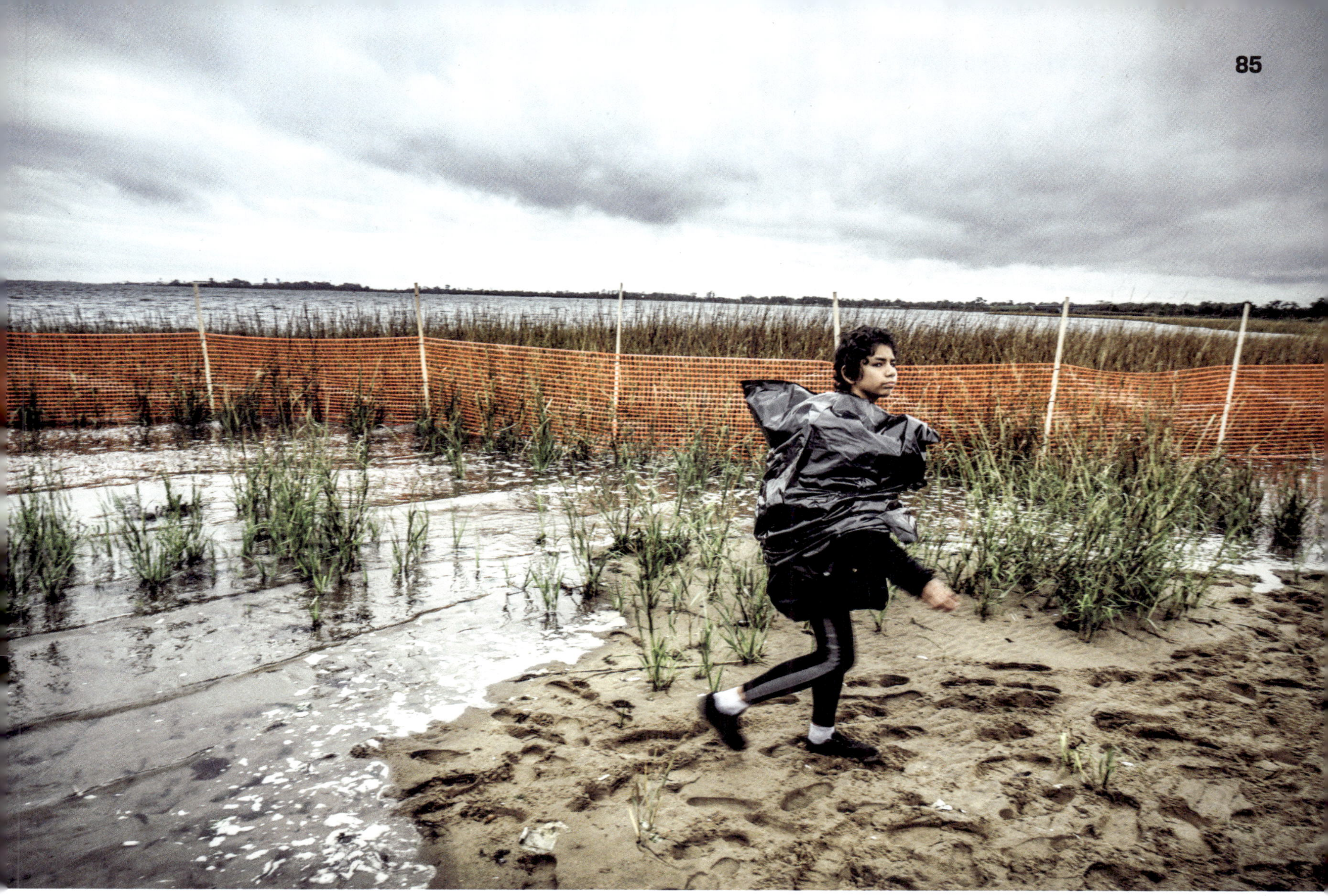

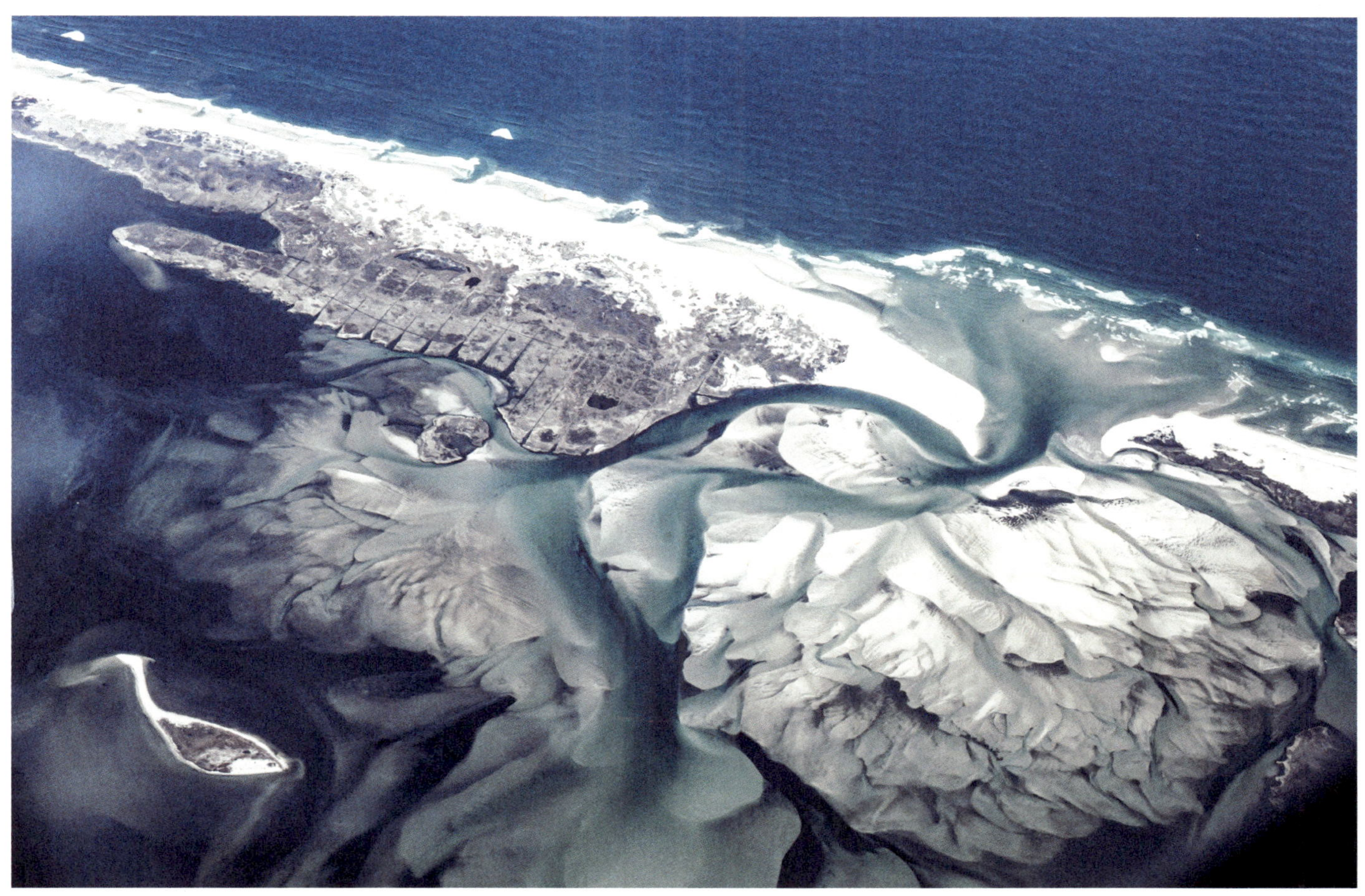

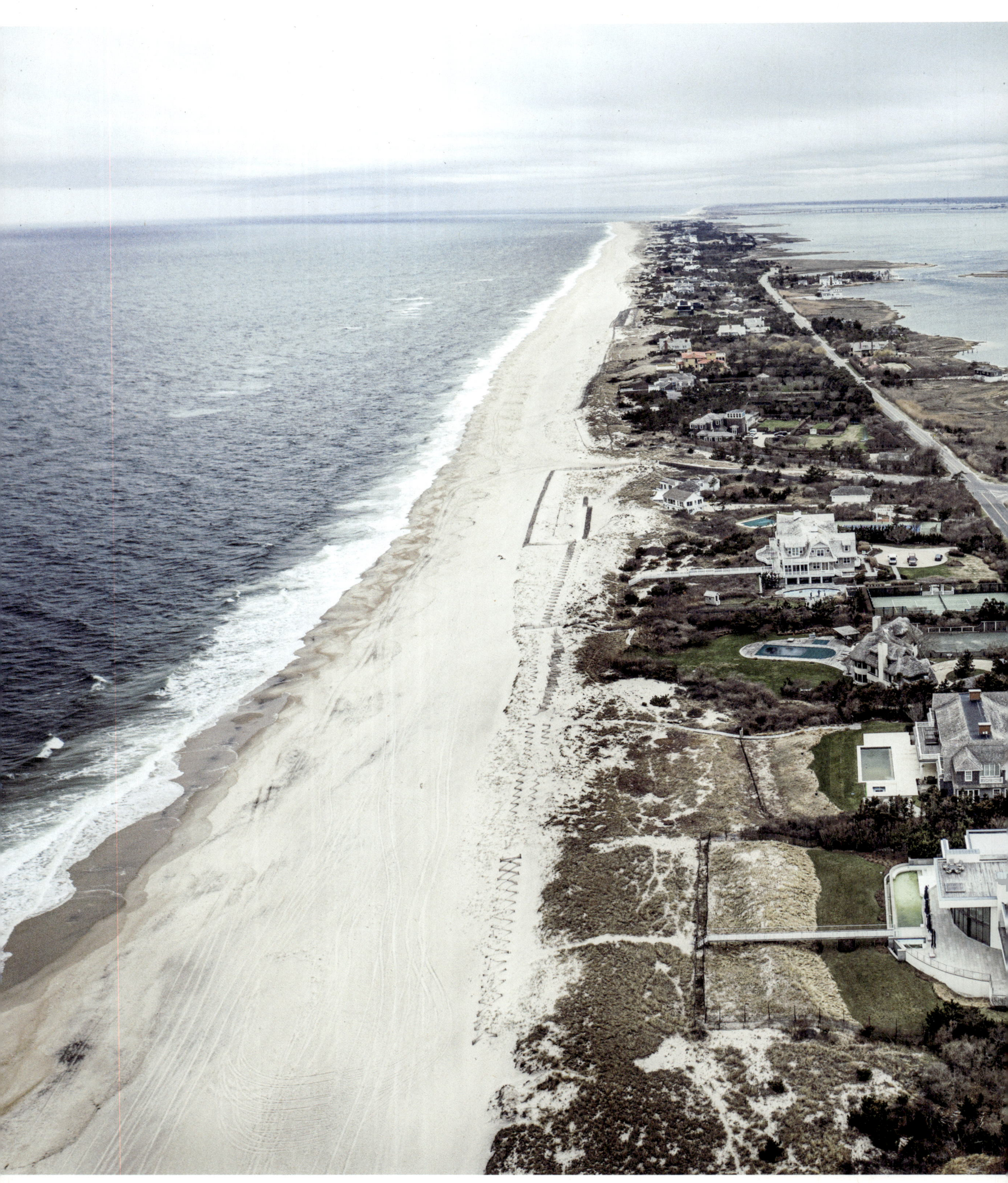

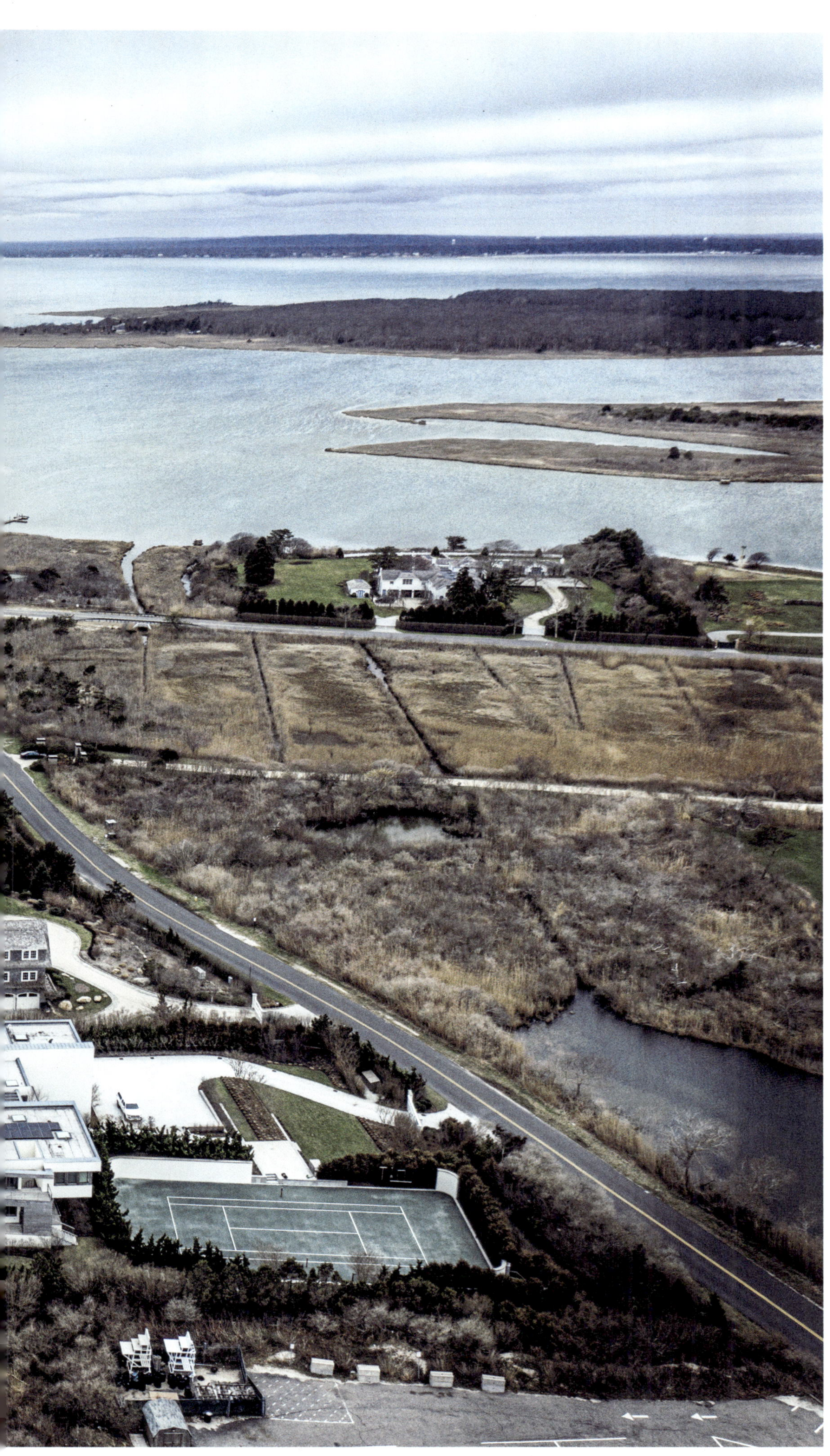

Extremely expensive real estate at the Hamptons, on a barrier island in front of Long Island. The villas are vulnerable to storm surges and sea level rise.

> Pages 88-89: New York, as seen from Jersey City.

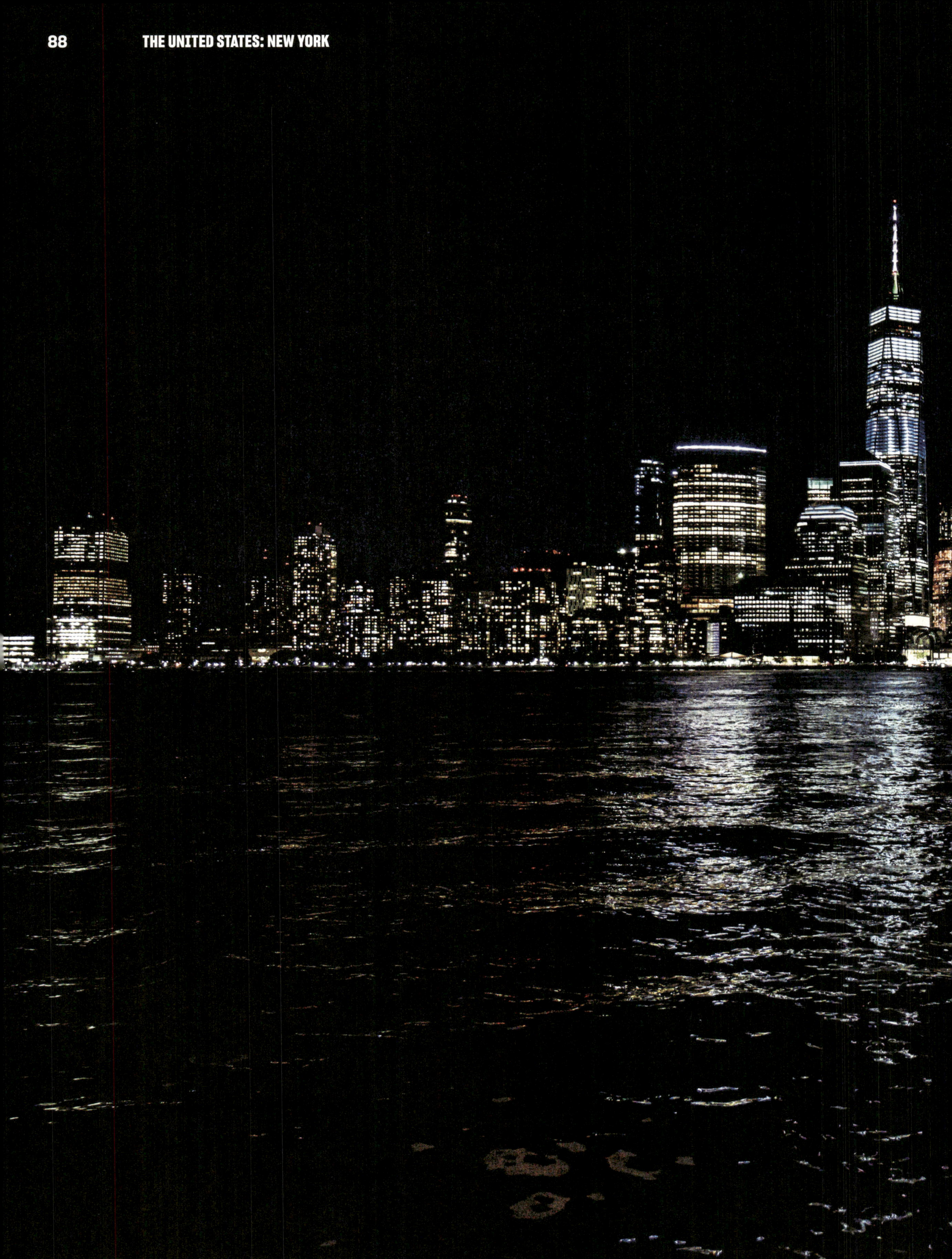

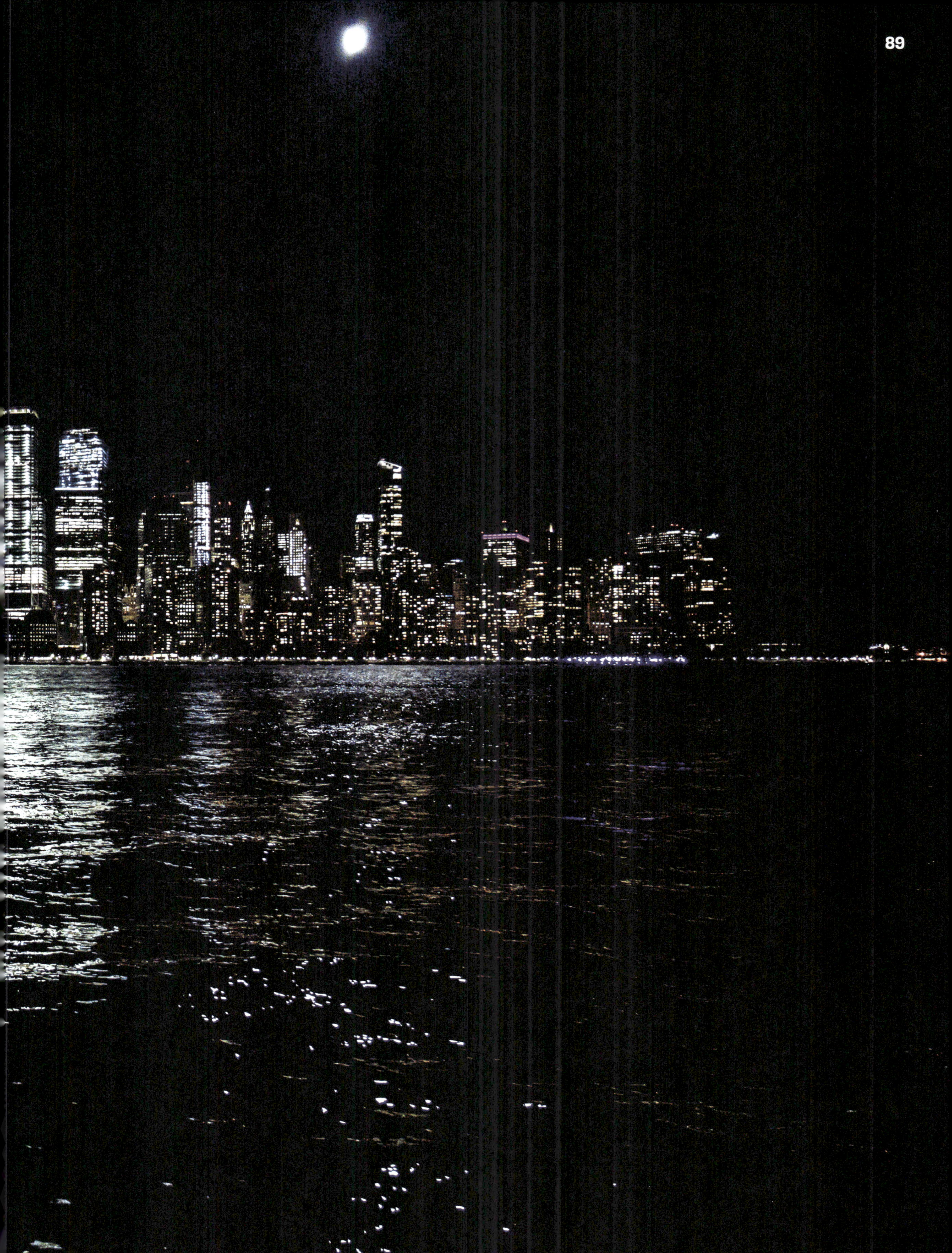

PANAMA: GUNA YALA
Elliot Brown

In 2017, the government of Panama approved funding to evacuate four of Guna Yala's 300 islands, but the project was delayed due to budget constraints and then COVID-19, so the people still remain where they are, in ever-worsening conditions. Originally, the people were supposed to have been evacuated in 2012.

The Guna Yala region, in the northeast of Panama, is made up of four sub-regions distributed in 52 communities, of which 39 are low-lying islands, spread over an area of 2,358.2 km². Its head is the administrative island of Gaigirgordub. Located at the western end of the Guna Yala region, Gardí Sugdub is one of the most densely populated communities, with a population of 1,013, according to the 2018 local health center census. The population shares a common language, Guna.

Gardí Sugdub also played a key role in the Dule Revolution of 1925, under the leadership of the Sagla Olonibiginya, to confront the system that wanted to oppress the existence of our culture.

PLANNED RELOCATION

In 2008, while on summer vacation from university in Panama City, I visited relatives in my community of Gardí Sugdub, in the Guna Yala region. I started hearing the elderly talk about moving to the mainland. Reasons for this included population growth, vulnerability to extreme weather conditions, flooding in the Cartí River, strong winds and other events related to the climate crisis.

Having to move from one place to another is not new for the inhabitants of Gardí Sugdub, whose origin is in Dupgandí (Bayano), east of Panama City. Several centuries ago, our ancestors were forced to move to the north coast, and 163 years ago, they moved from the banks of the Cartí River to live on the islands, specifically to an island called Aspandup, to escape mosquito-borne diseases and other evils that were decimating the population. Unfortunately, on September 7, 1882, a tsunami struck the entire Guna Yala region. Aspandup was swept away by the onslaught of the waves, leaving only a coral bass. The surviving settlers moved to a nearby island, Yandub, and eventually began to populate the other islands, including one they named Sugdub ('crab island').

RISKS AND VULNERABILITIES

Currently, Gardí Sugdub is vulnerable to natural disasters such as storms, hurricanes and tsunamis. This was evident in early 2020, when strong hurricane-force winds coincided with high tide, causing damage to both traditional and concrete houses in several Guna Yala communities, including Gardí Sugdub, where seawater flooded some houses. In addition, the homes in these communities lack a sanitation network, drinking water, sewerage and a waste collection system.

At the beginning of 2015, after finishing my university studies in environmental biology at the University

Not everyone in Gardí Sugdub wants to move to the mainland. Many older adults, who have lived on the island all their lives, do not want to move; they prefer to stay on the island no matter what.

of Panama and returning to Gardí Sugdub, I became interested in the issue of resettlement, in adaptation processes and in how this altered the socio-ecological system. For example, the relocation of our ancestors to the islands inadvertently triggered a maladaptive process through the mining of coral to build barriers, expand the island's surface and protect homes from ocean waves and tides, which has unfortunately been shown to be damaging in the long term. In fact, by removing coral, our ancestors unknowingly increased their vulnerability to storms, because coral reefs play a role as natural protective barriers against ocean currents. The current process of adapting once again to living on the continent must be undertaken with caution to avoid unwanted maladaptive processes, practices and behaviours and to design appropriate mitigation strategies. Relocating any community involves many risks, including its division (Lazrus & Arenas, 2020).

COMMUNITY RESISTANCE AND CONCERNS

Not everyone in Gardí Sugdub wants to move to the mainland. Many older adults, who have lived on the island all their lives, do not want to move; they prefer to stay on the island no matter what. Some even wonder if the climate is currently changing, as, according to their observation, the climate has been changing for many years and they do not see a sense of urgency. As a result, there is a risk that the community will split in two. On the other hand, some people, especially young people, have a vision for community development, as it makes life more comfortable - for example, being able to have 24-hour power on the coast of Cartí, a school and a hospital under construction, the idea of a regional headquarters for the University of Panama and the

Guna Yala is an autonomous indigenous region of Panama with a population of 31,500. Guna Yala consists of an archipelago of more than 300 islands (the majority of which are uninhabited) and a strip of mainland. Because of rising sea levels, the incidence of storm surges and flooding has increased dramatically in recent years, leading to life becoming increasingly risky in some of the overpopulated islands.

In response, the government planned to evacuate the first four islands in 2013, but the project was delayed due to budget constraints and then COVID-19, and the people still remain where they are, in ever-worsening conditions. When the indigenous Guna Yala do eventually relocate, they will have to learn to become sedentary farmers, whereas now they work primarily as fishers.

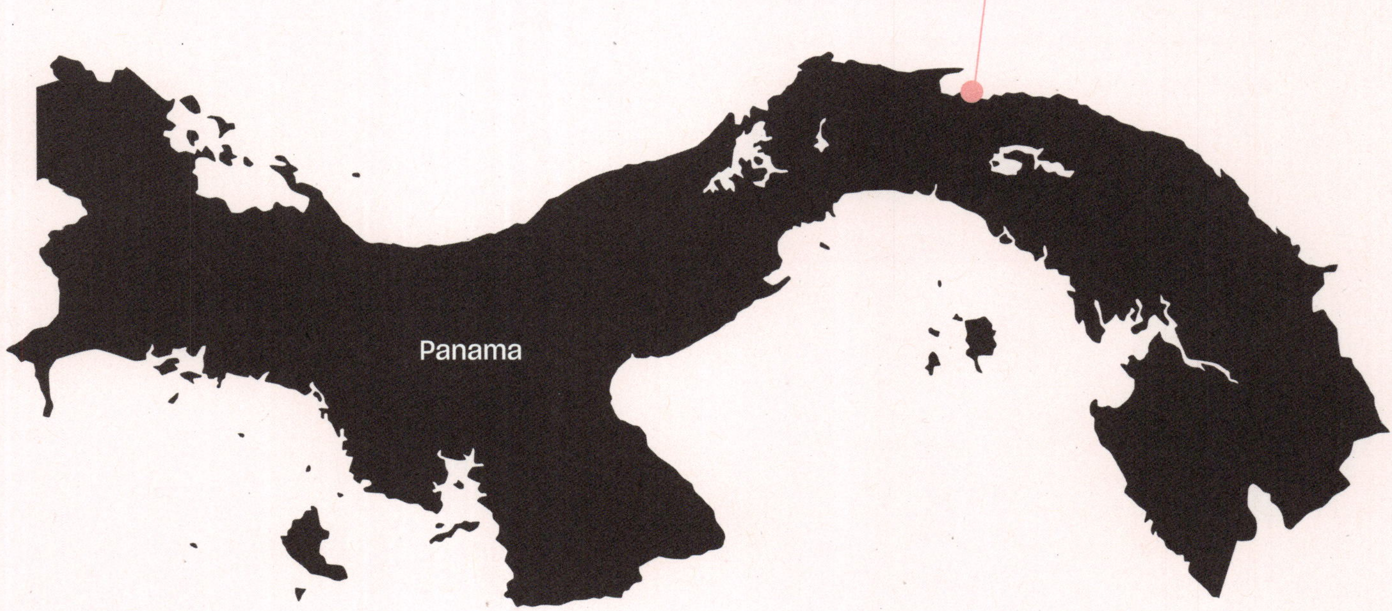

construction of houses with concrete infrastructure and paved streets, including parking lots. However, this vision of accelerated development does not consider other realities, such as a higher cost of living, adverse impacts on the environment and the loss of ancestral knowledge. Our culture is rich and vibrant, and we are proud of our history of fighting to uphold our customs of protecting Mother Nature. The Guna people are ruled by the sages, who are the guardians of our ancient knowledge and traditions. Each community has a leader, called 'Sagla', chosen by the community. Guna Yala has three regional leaders, called 'caciques', who are in charge of managing relations with the Panamanian government. It is common for caciques to be accompanied by a small group of younger advisers, usually with a university degree. However, the re-

ality is that it is not always easy for a young person to be heard and for their voice to be taken seriously in the community, so we always face the challenge of deciding where to focus our energy, inside or outside of the community. There are other young professionals, including me, who are interested in contributing meaningfully to the community.

GETTING INVOLVED

In 2017, I began to think more about the issue of community relocation, which motivated me to convey the correct information to my community, knowing that many developers and construction companies were not concerned about the environment or the culture. Environmental problems were not new to me. Prior to this, I had studied jaguars ('assu barbad' in Guna) in the

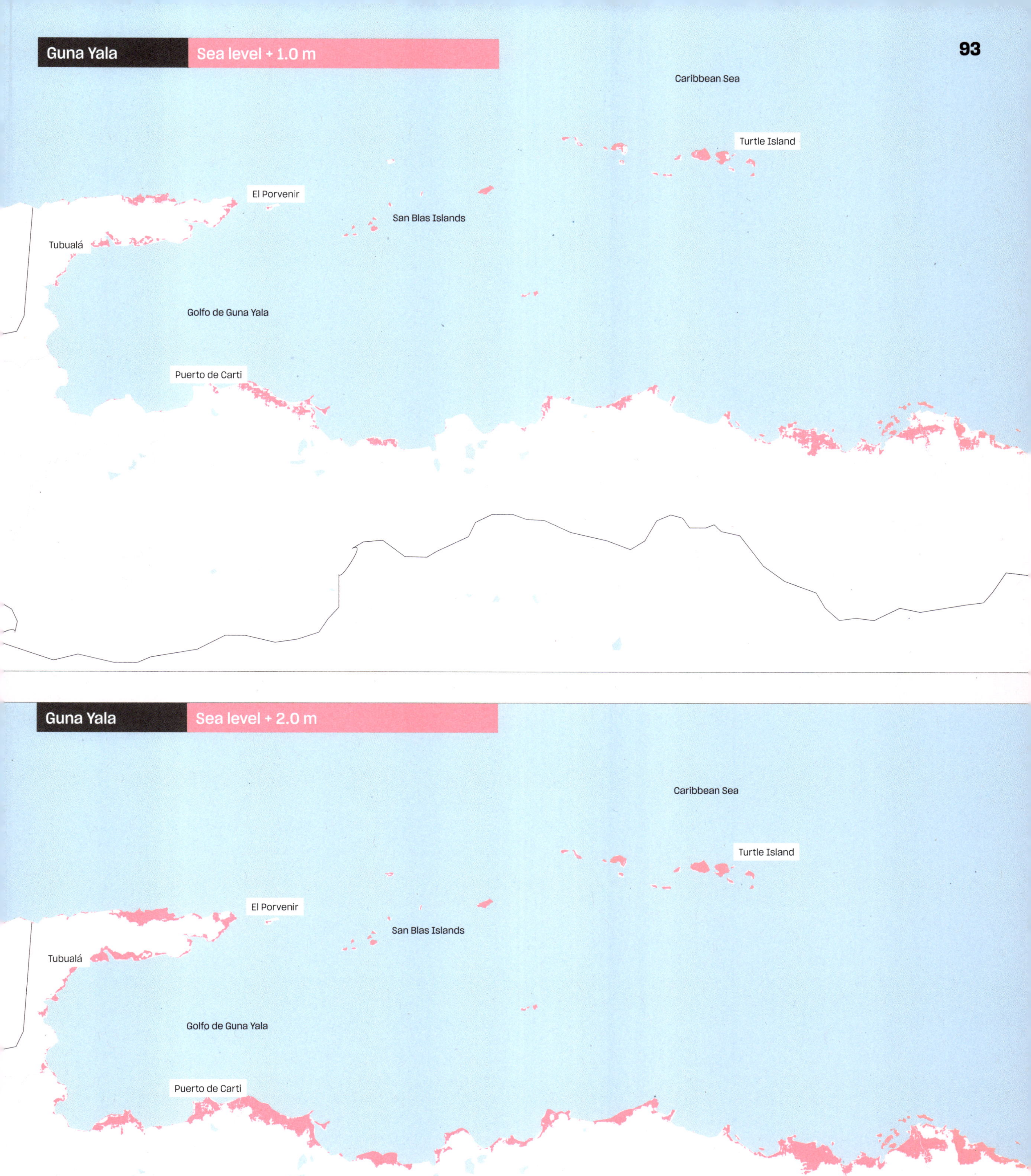

Source: Climate Central (www.climatecentral.org)

It is fair to say that relocating a community is not a sprint race but a relay race. But despite all of the delays, obstacles and challenges, the people of Gardí Sugdub are proud to pioneer relocation to Guna Yala.

protected Narganá Wilderness Area in Guna Yala, using camera traps, in addition to various educational activities. Currently, I am an associate researcher at Yaguará Panamá, a conservation foundation, and previously, I was a volunteer for the Ministry of Environment of Panama and international NGOs to protect jaguars from extinction.

In fact, that year I began a master's degree in management and evaluation of environmental impact studies at a university in Panama City, in order to be more prepared and thus help the community. It was then that I felt it was time to be at the forefront in Gardí Sugdub, and I began to get involved in the commission that the community had formed, to ensure that the interests of the community were protected. The formation of a neighbourhood commission was a very good initiative, since community professionals were empowered with the issue, contributing multidisciplinary experience in fields such as education, health, economics, the environment, sociology, law, carpentry and the media.

PROJECT APPROVAL AND START-UP

In 2017, the Panamanian government of President Juan Carlos Varela finally supported our request to build 300 houses to relocate the community. In addition, the Inter-American Development Bank (IDB) supported the project with financing for technical assistance for the relocation process. This allowed the hiring of several national and international consultants to advise and document the process and offer the community access to some of the funds related to the activities to move

the project forward. The IDB's participation was a great impetus for the relocation process and revitalized the community. Three Guna professionals from the community, one of whom was me, were selected to work in collaboration with the external consultant charged with guiding the drafting of the relocation plan and monitoring the entire process.

In 2018, while we were at a committee meeting, news emerged about the order to continue the project. At that time, an assembly was organised in the congress house to communicate and inform the community of the good news: construction of the 300 houses was going to begin. Once the construction of the houses was approved by the Panamanian government, the main obstacle to the project was the environmental impact assessment (EIA) necessary to move the project forward. At that time, it was the responsibility of the group of Guna professionals from the community to accompany the external consultant in the visit to the different government agencies, to pressure them and provide more information and clarity.

The project's EIA was delivered to the Ministry of Environment at the end of 2018; however, it was rejected for non-compliance with the requirements, one of them related to the Guna Yala waste collection system. The community had to pay a visit to the town of Chepo on Christmas Eve to meet with the mayor and support our request to collect waste, especially from the project for the new Gardí sweep. The meeting was not successful in the end, as we did not receive help from the mayor, but more than US$300 was spent on the visit. In the end, the EIA was delivered again with the pertinent adjustments, supported by the community commission and accepted in January 2019.

COMMUNITY REVIEW AND PARTICIPATION

But another process followed, that of citizen participation, which had to be completed within the term established by Executive Decree 123, which regulates environmental impact studies in Panama. A valuable contribution from the community was the observations and recommendations included in the environmental impact study. Next, the community was invited to participate in a meeting organised by the environmental impact studies department, in which the developer, the construction company and the environmental consultant and the director of environmental impact studies of the Ministry of Environment also participated. At the meeting, we formally delivered the changes and improvements that the consultant needed to make to the project's EIA. All of the points raised by the community were included.

Once the study was approved, in June 2019, the construction company was authorized to start work on the project. However, the bureaucratic processes continued and now there was no funding for the compensation for ecological damage that the construction company had to pay to the Ministry of Environment. That process delayed the project by a few more months. Gardí Sugdub tirelessly supported the process to keep the relocation project moving forward, despite so much uncertainty about the project. These processes did not stop the community meetings to address the issue of who had the right to housing in the new neighbourhood, during which the requirements that each family had to meet were detailed. The list of beneficiaries of the houses was approved by the assembly in December 2019.

DELAYS AND CHALLENGES

Early 2020 finally saw the arrival of the first trucks, tractors and heavy equipment for the project. The area where the community will be relocated was levelled, and the foundation construction for the houses began in late February 2020. Unfortunately, the COVID-19 pandemic halted the project for several months. The only relocation-related project that has been completed so far is a 55 km power line bringing electricity to Guna Yala for the first time.

Meanwhile, Gardí Sugdub's leadership has had to focus on facing the challenges of COVID-19, educating the community to overcome the health emergency. During that process, I had to be on the front line of the fight against the virus, where I tested positive for COVID-19 but remained asymptomatic. Unfortunately, some of the other community members who were very active in the relocation effort passed away, as well as four other community members. Now, Gardí Sugdub is ready to get back to work on relocation and apply all of the lessons we have learned in facing the pandemic, including new ways to combine and complement the traditional wealth and modern knowledge that are also present in the community. It is still necessary to adopt the norms of coexistence behaviour and adapt them to the new reality of having biosafety protocols in health matters.

Now I understand that, by nature, any resettlement or relocation process is very complex, and as a result, it is a long-term process. It is fair to say that relocating a community is not a sprint race but a relay race, as Gardí Sugdub's experience illustrates. Despite all of the delays, obstacles and challenges, the people of Gardí Sugdub are proud to pioneer relocation to Guna Yala, which will eventually require the relocation of around 28,000 people throughout the region (Displacement Solutions, 2014). We hope that by sharing our experience, we can contribute to initiatives that bring similar people from all over the world together to learn from each other. In all cases, local communities are, or should be, the main driver of any relocation process that occurs in our current times, characterized by a changing climate. •

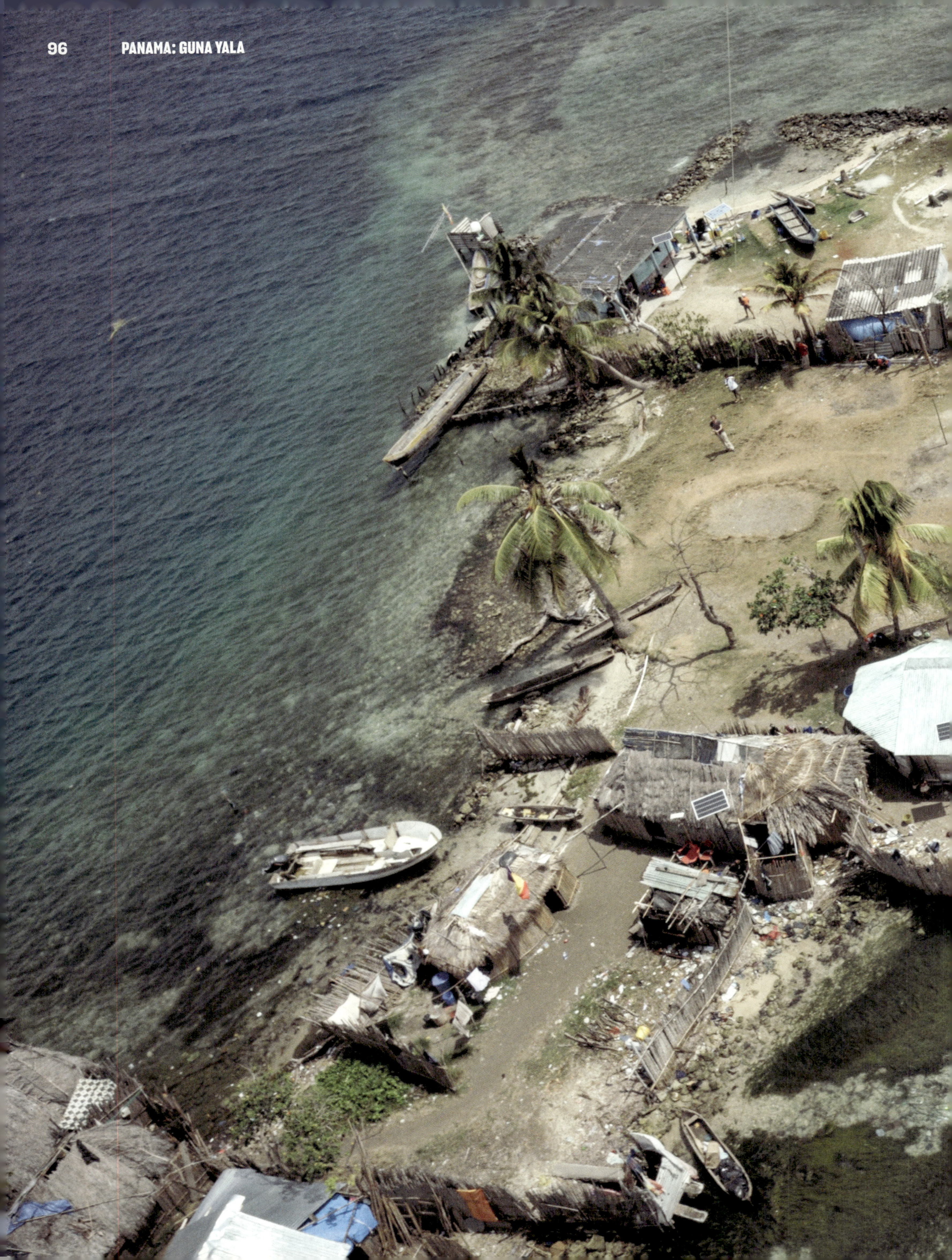

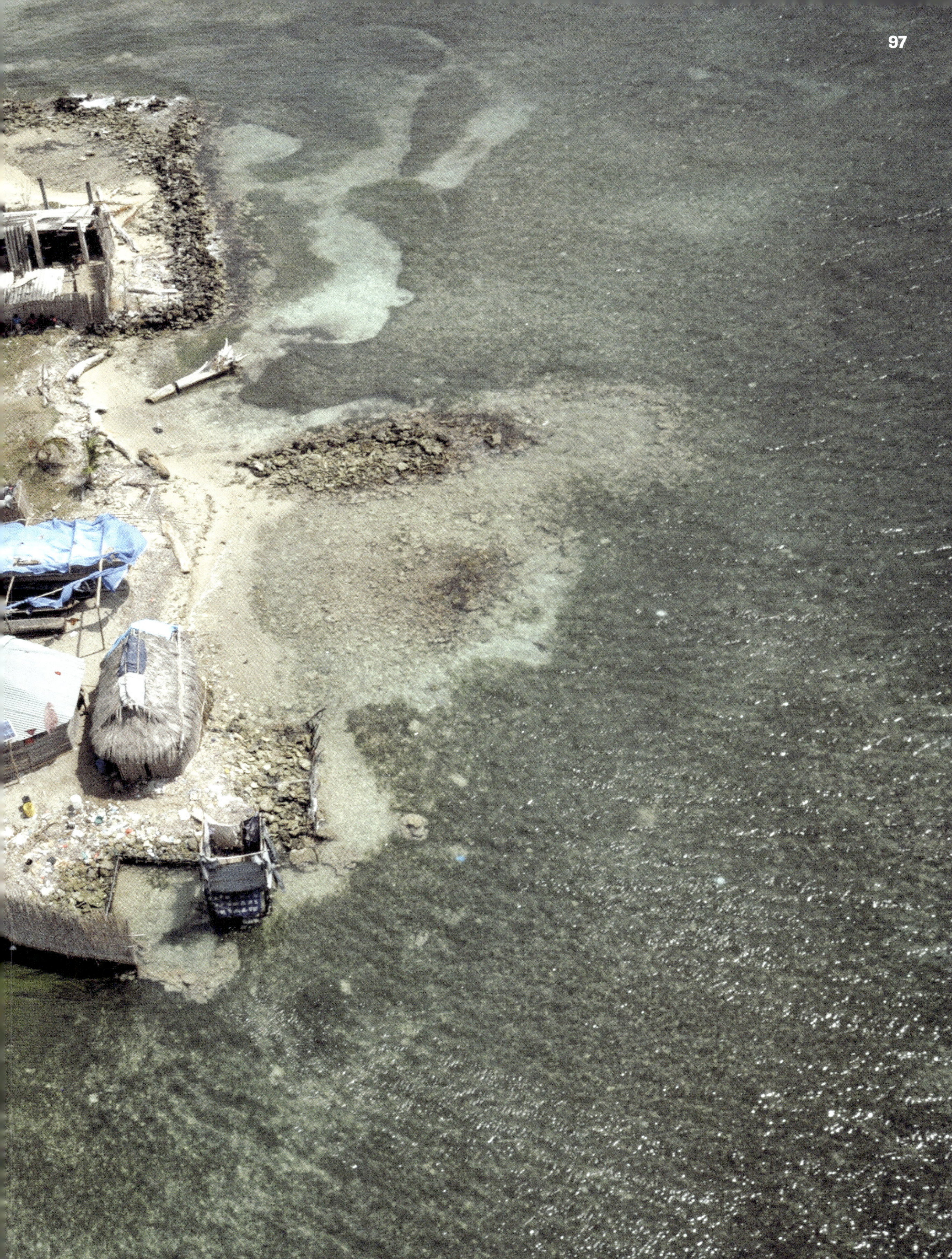

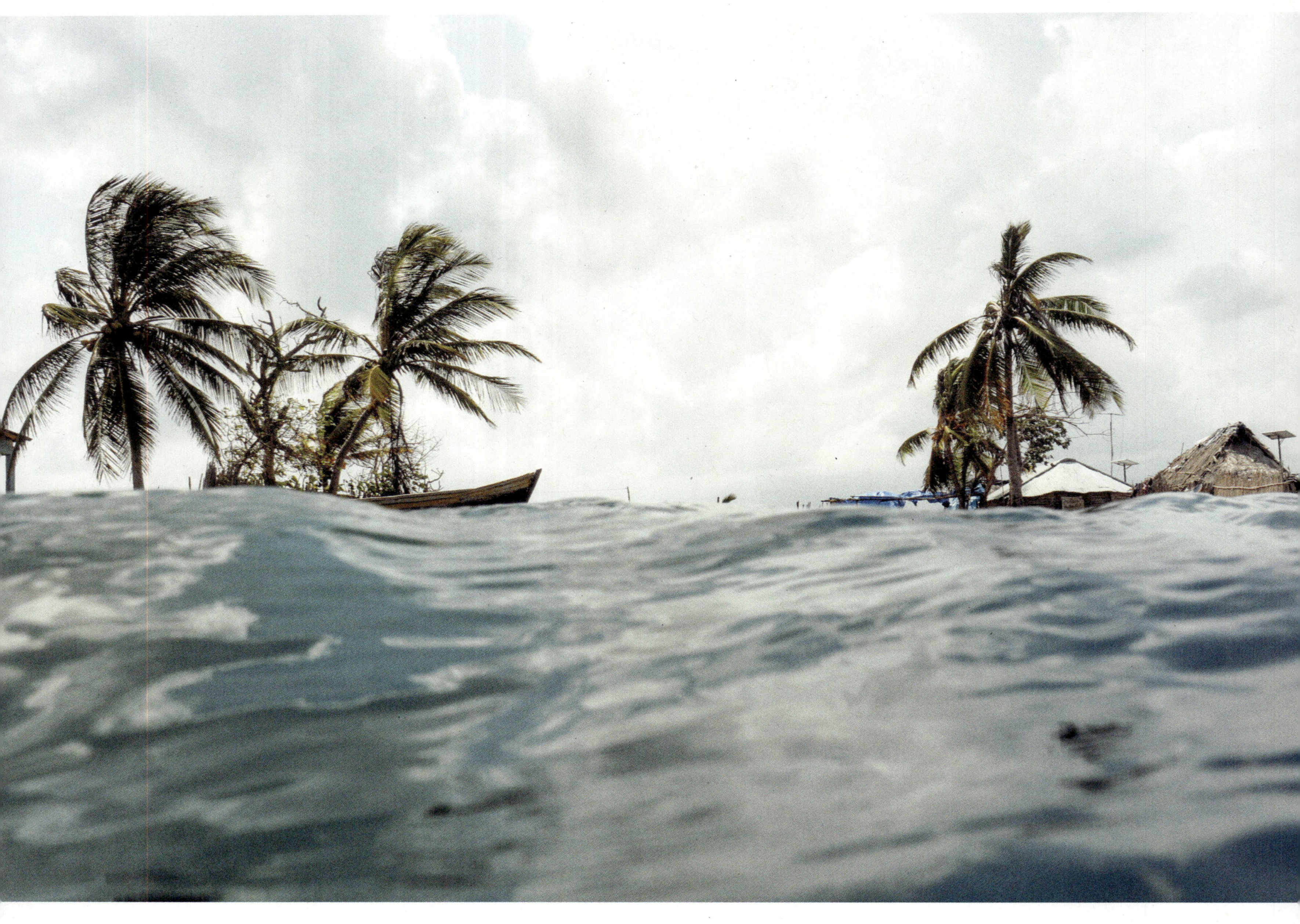

< Pages 96-97: Cartí Cohabita is one
of the most vulnerable islands; the
population tries to protect the island
from the sea by building seawalls from
coral.

The shoreline of Corazón de Jesús
is eroding quickly. Inhabitants try to
protect it using dead coral.

Mulatupu, like all islands in the Pacific,
is eroding quickly. Sea walls built with
coral bring ample protection.

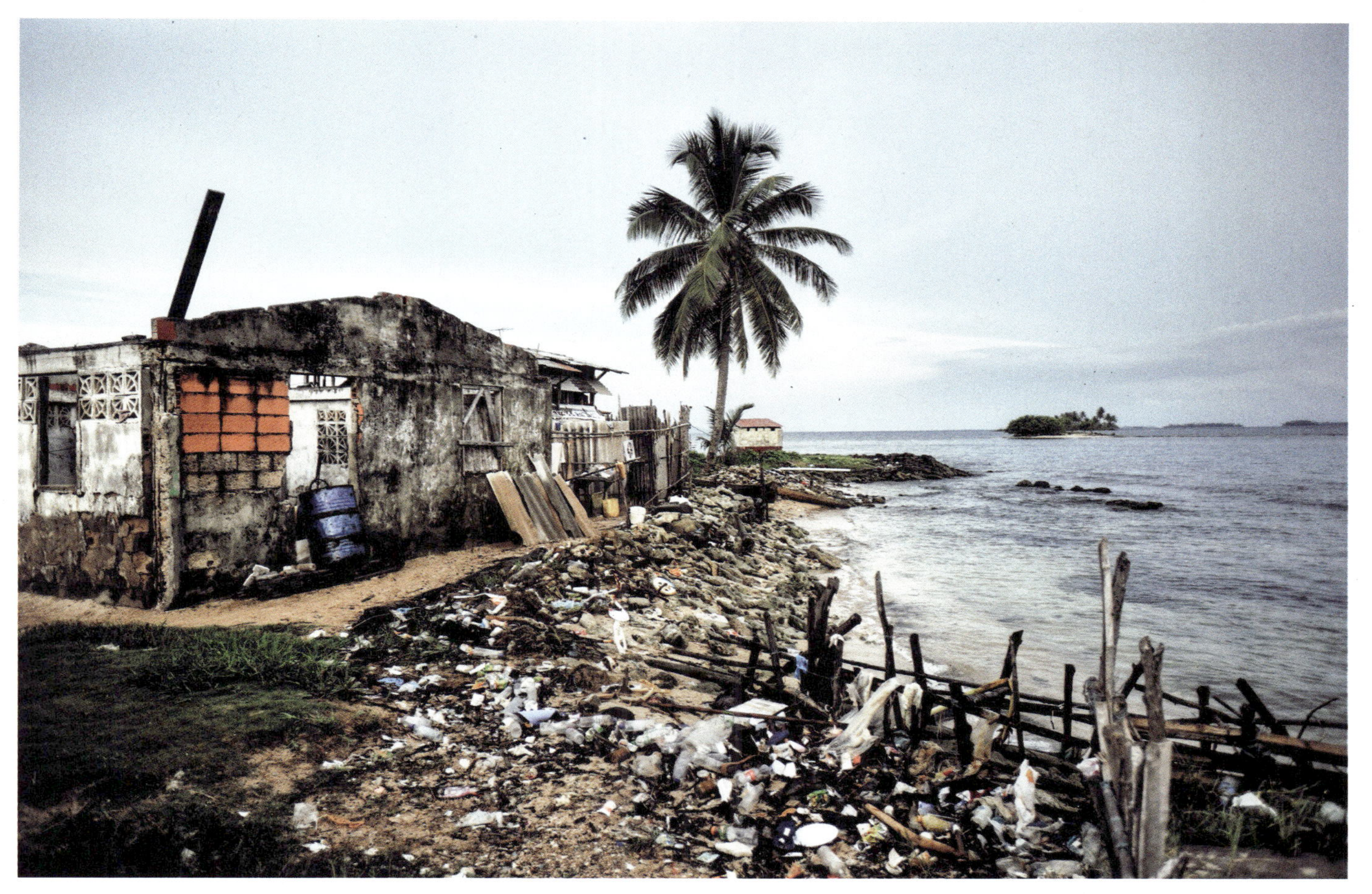

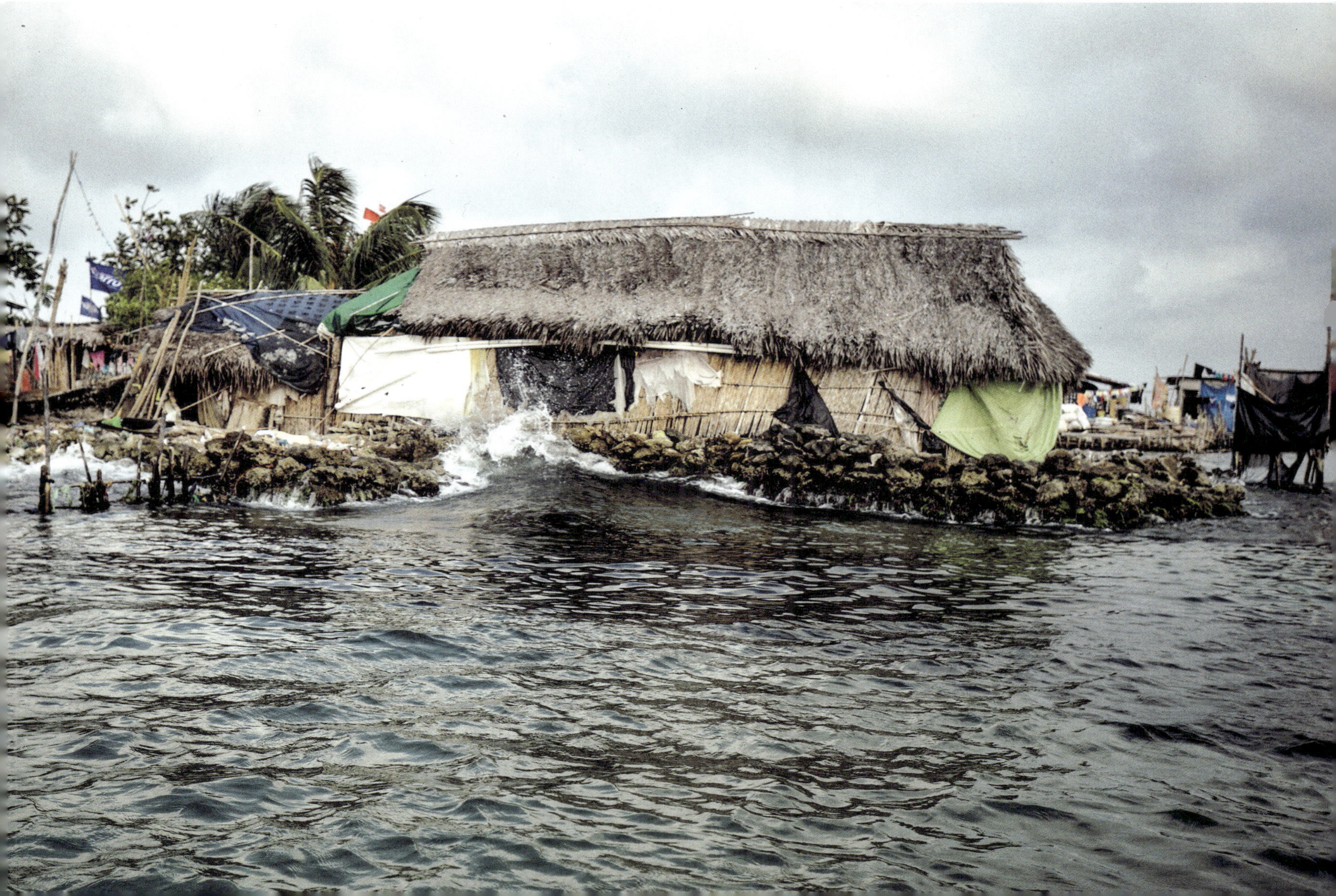

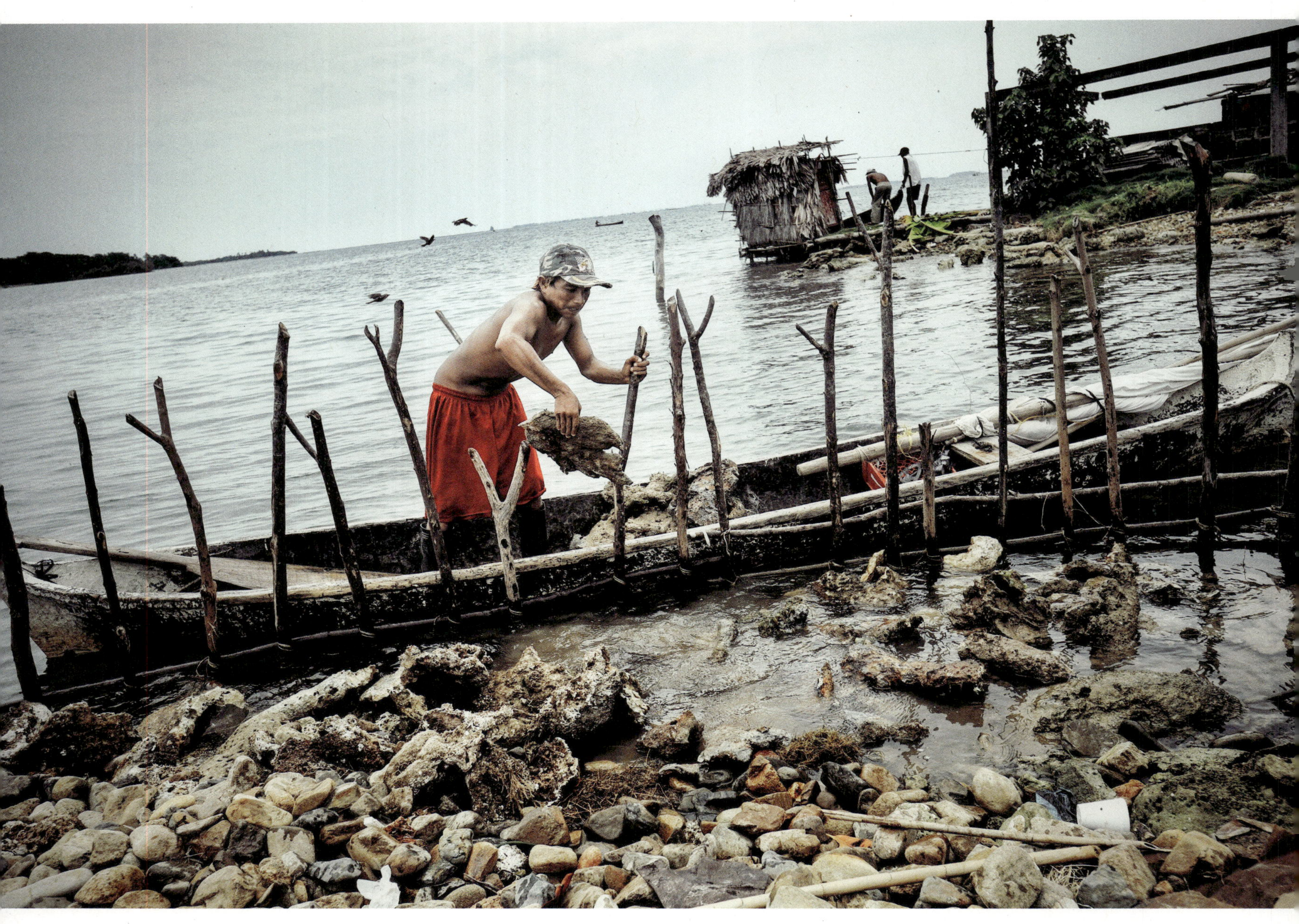

A worker on the island of Río Azúcar dumping dead coral to protect the island.

Students wait at the pier of Ustupo to be brought back home by boat. Not all islands have high schools.

Cartí Sugdup is the main island in the archipelago, because it's close to the mainland and the beginning of the road to Panama City. Before the road was built, it was a 12-hour trek to reach the Pan-American Highway. Most goods from the mainland are delivered here by boat to Cartí Sugdup.

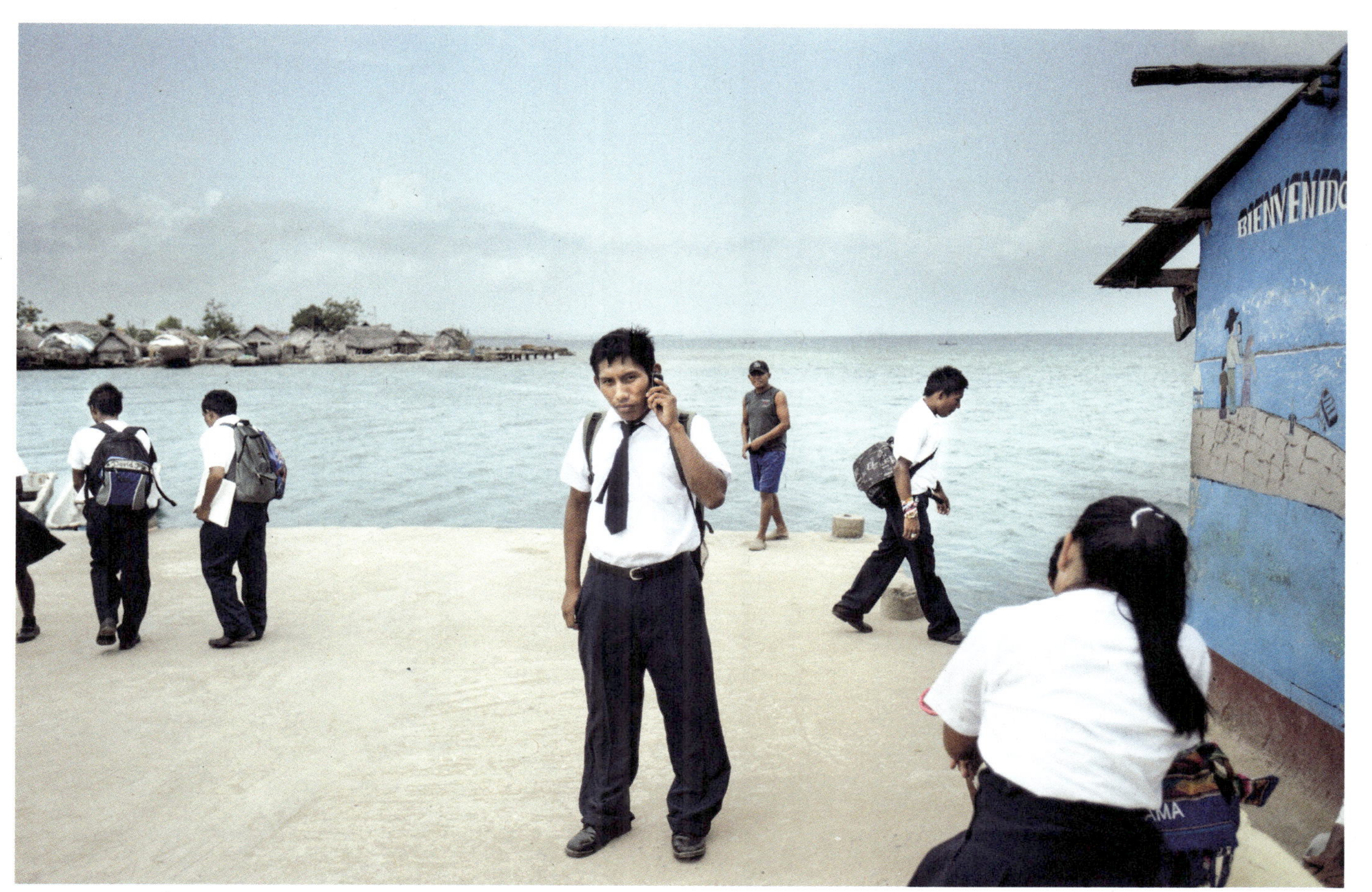

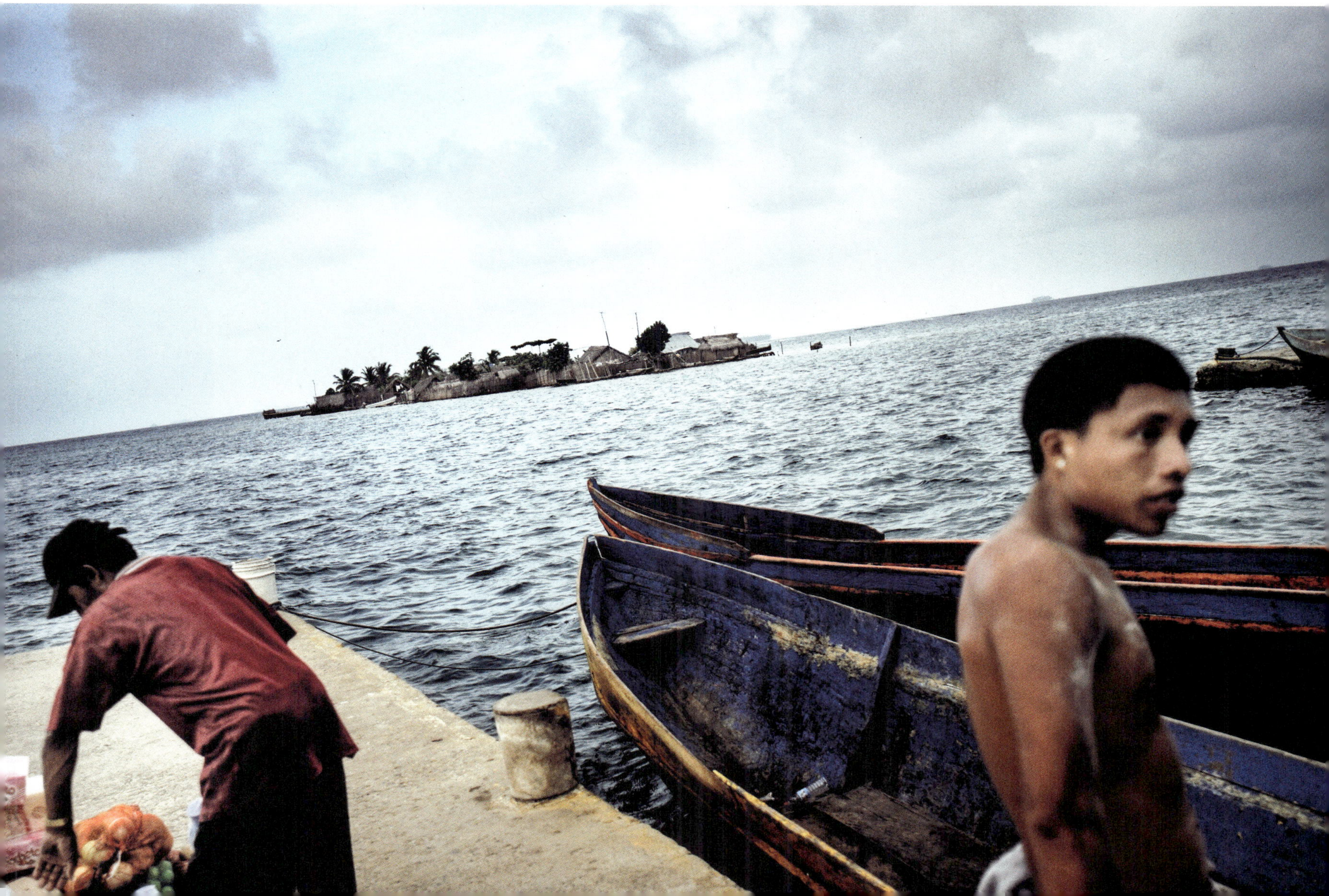

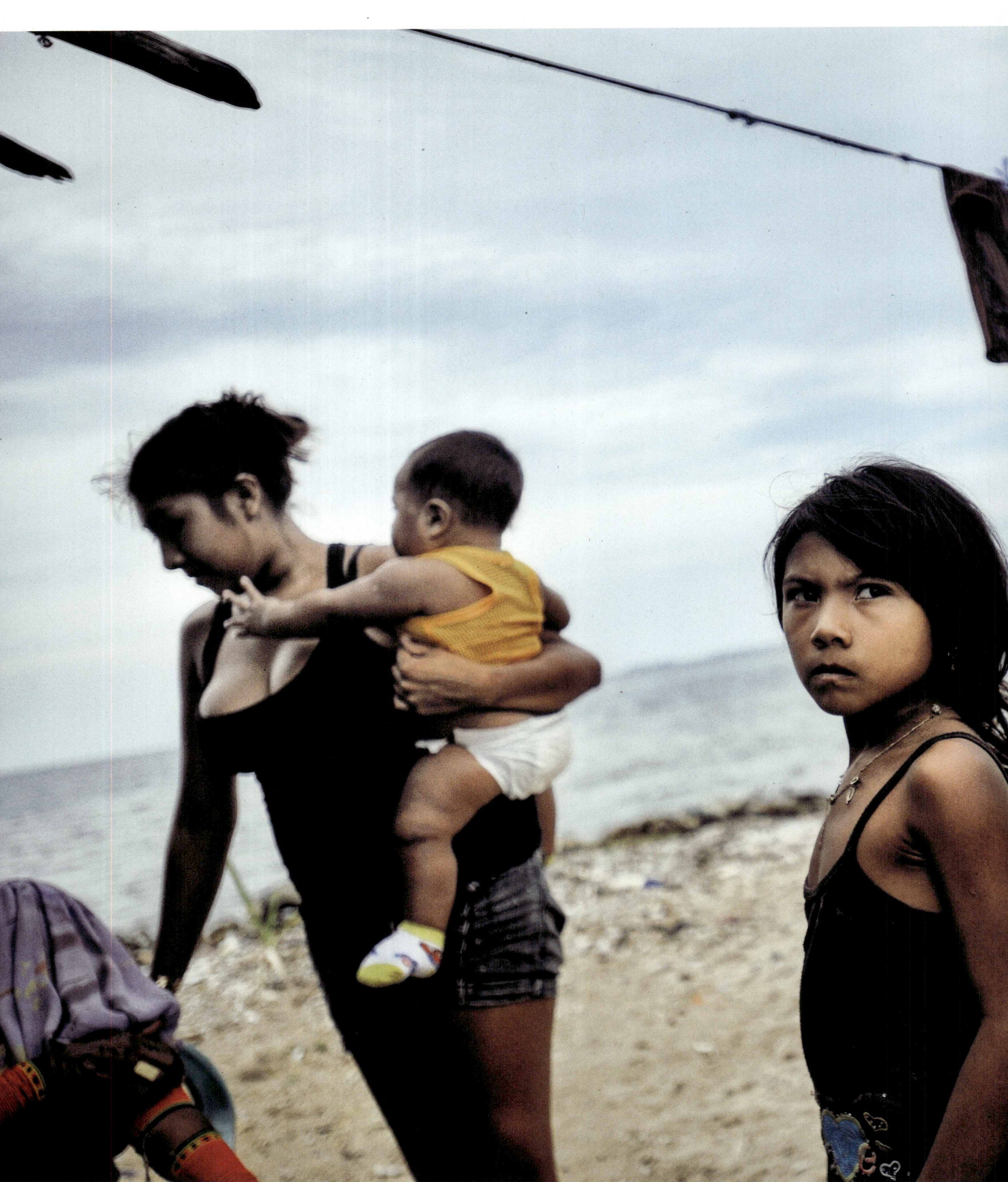

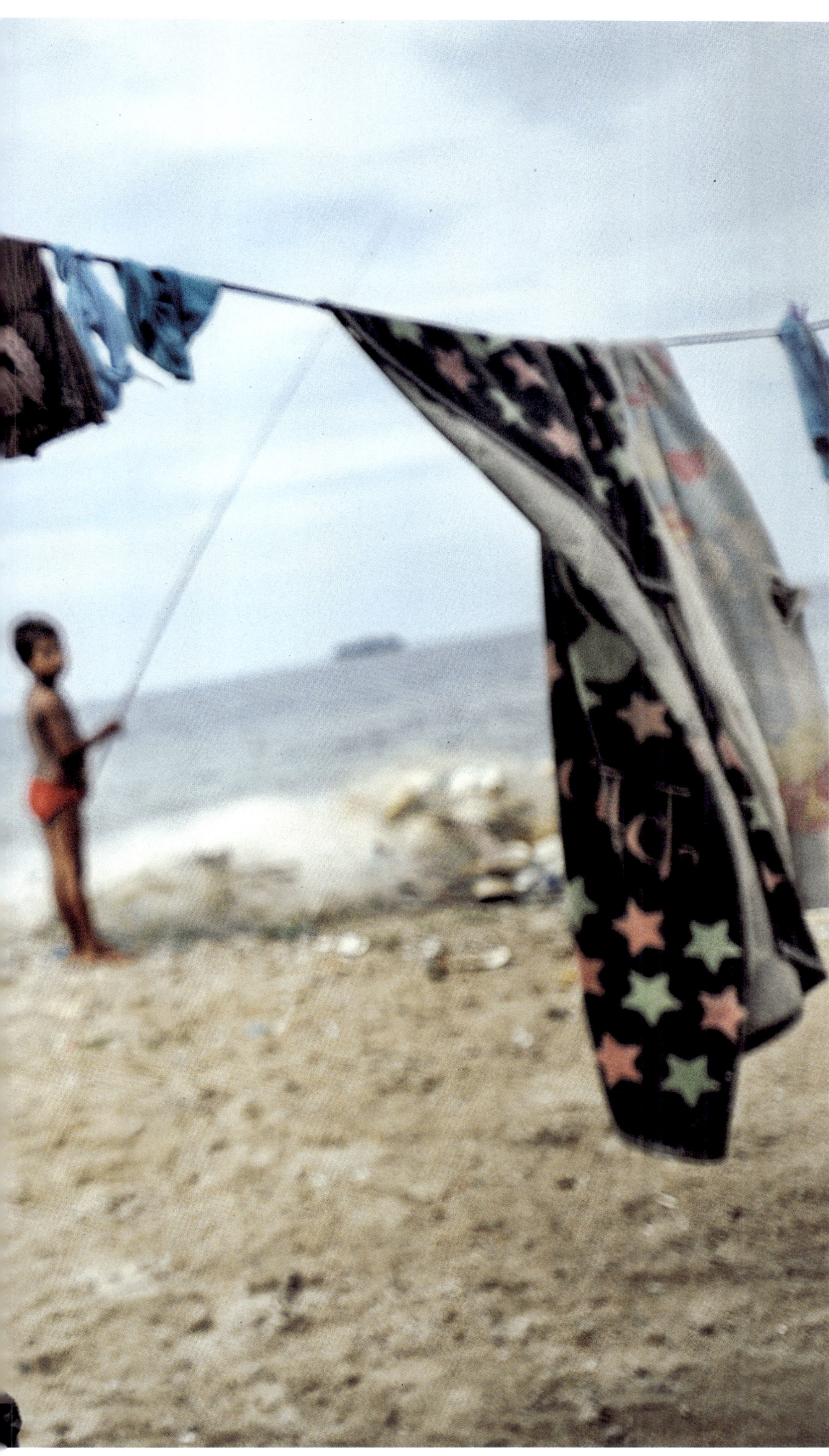

Norberto Hernandez (52) and his wife Olga have been exiled to the island of Sucunguadup, which they heightened themselves using coral. They live on the island with their nine grandchildren and one of their three children.

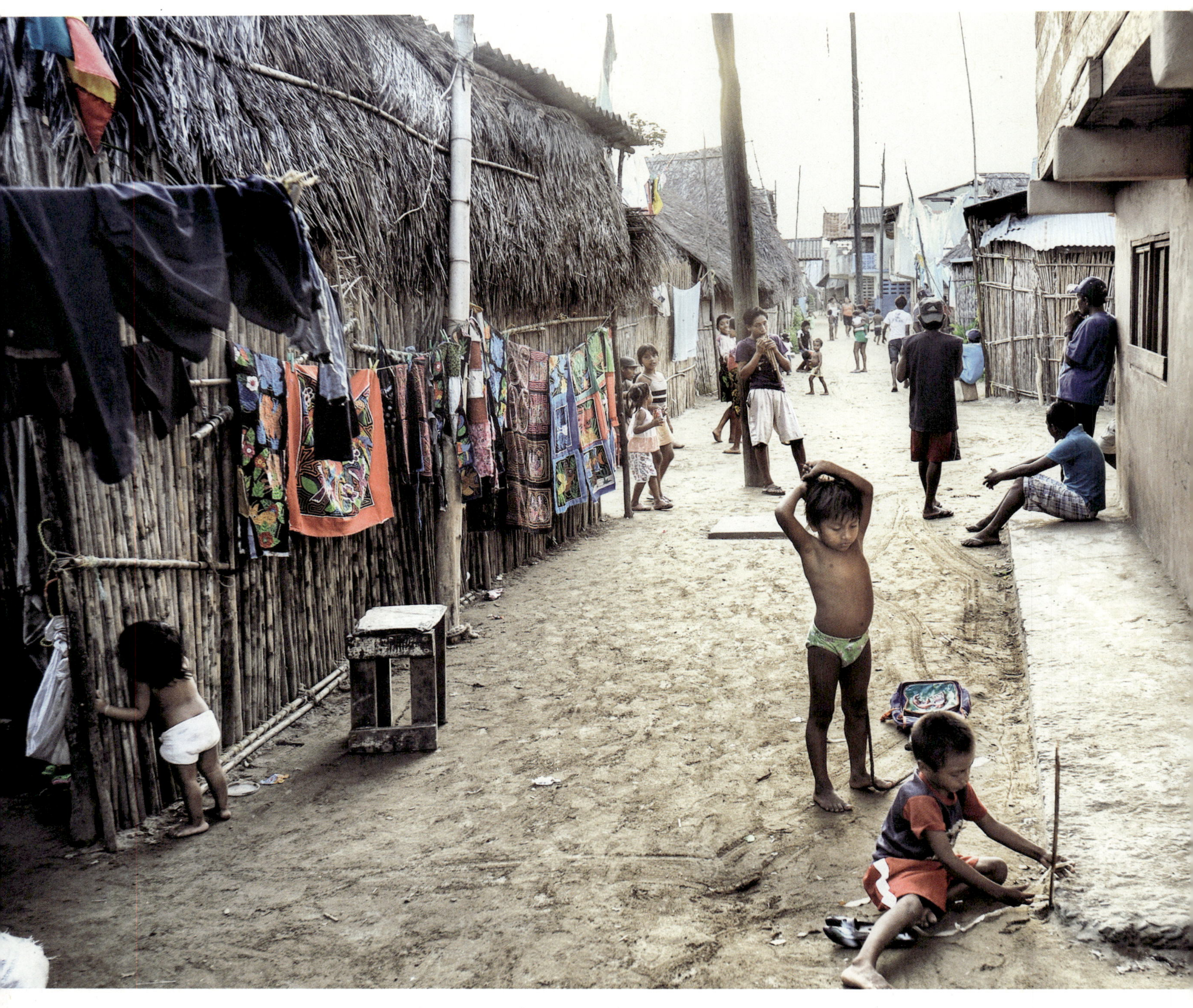

The main street in Cartí Sugdup. This
is the main island in the archipelago,
because it's close to the mainland and
the beginning of the road to Panama
City.

A fisher comes home after a day of
fishing.

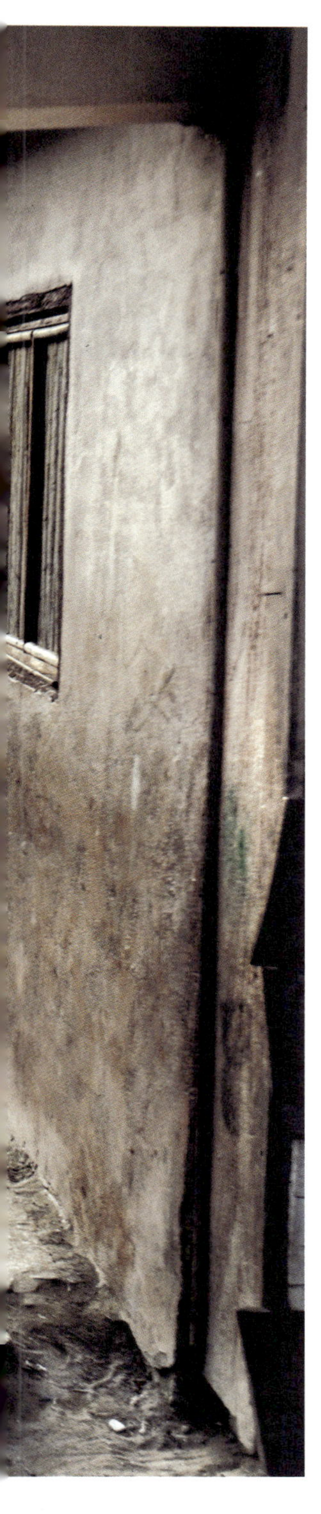

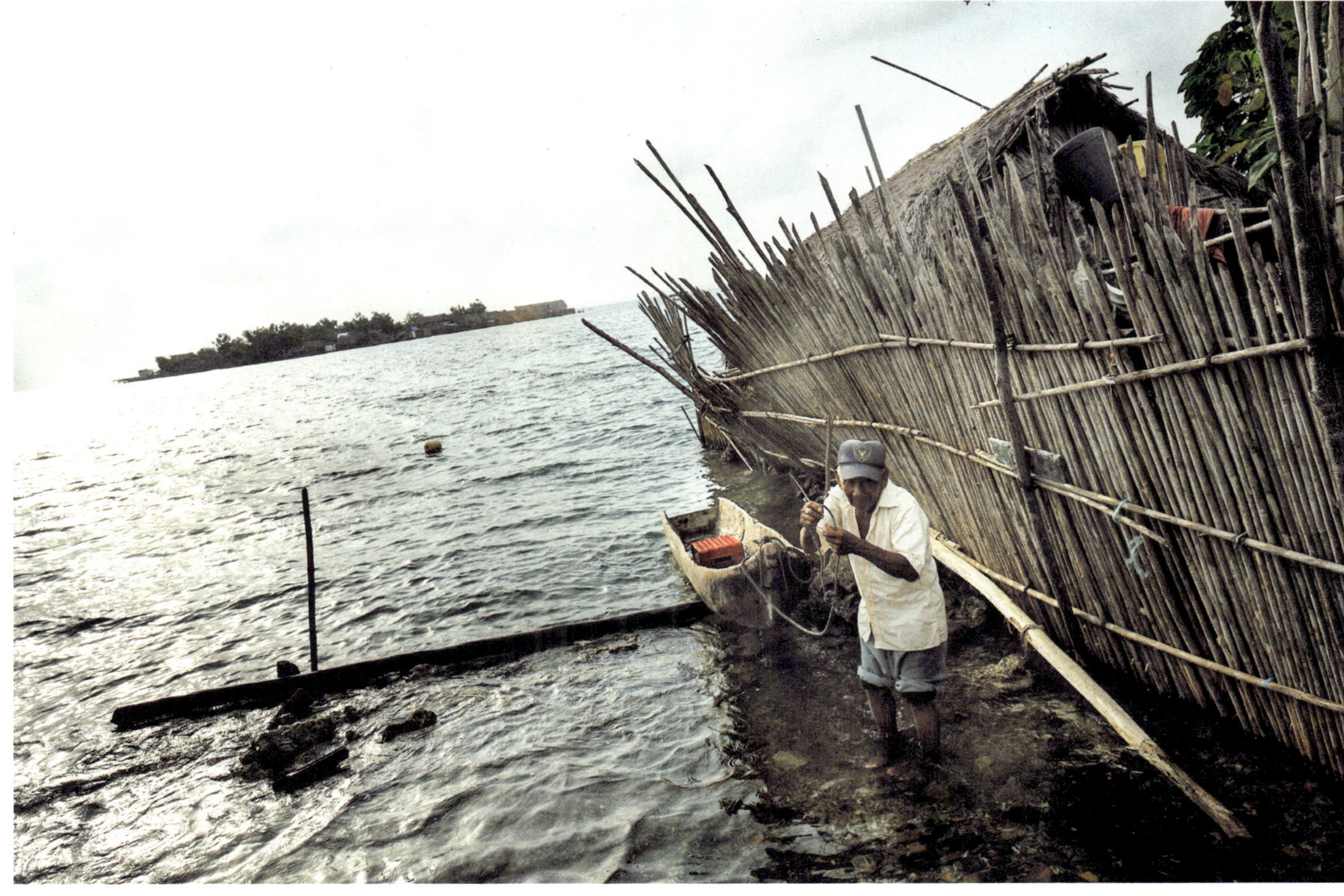

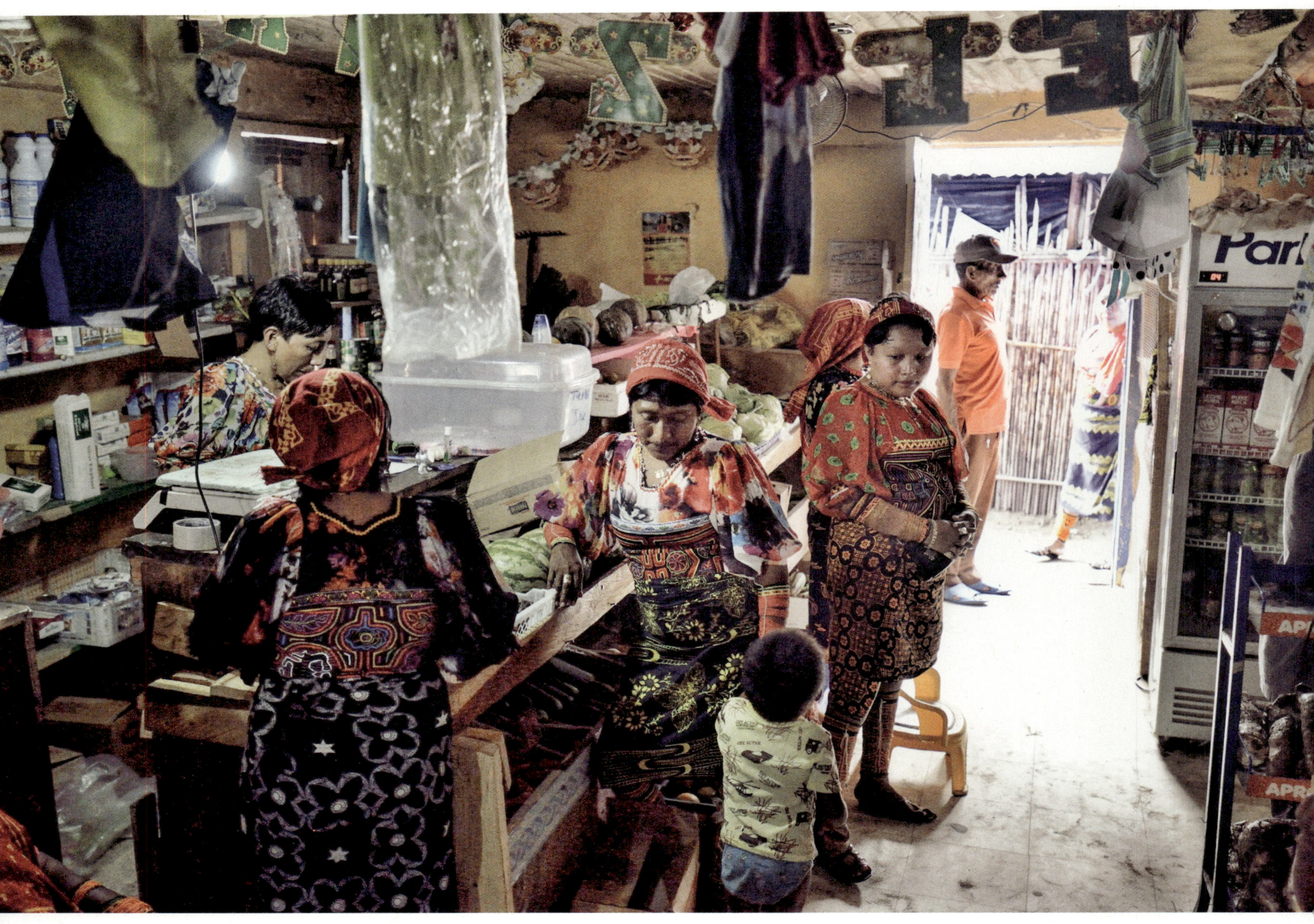

A store on Cartí Sugdup.

A café in the harbour of Cartí Sugdup.

The ferry pier at Cartí Sugdup.

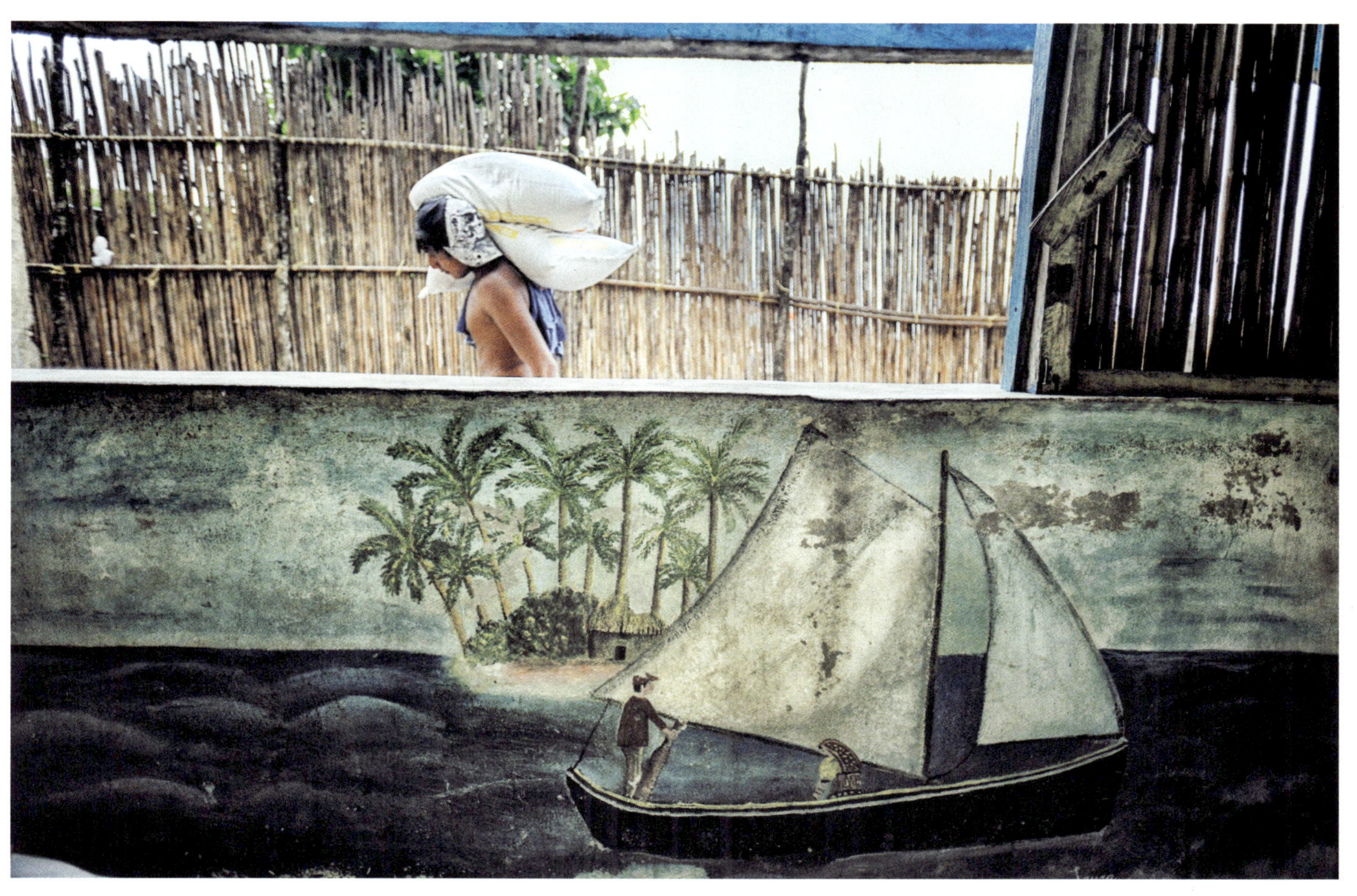

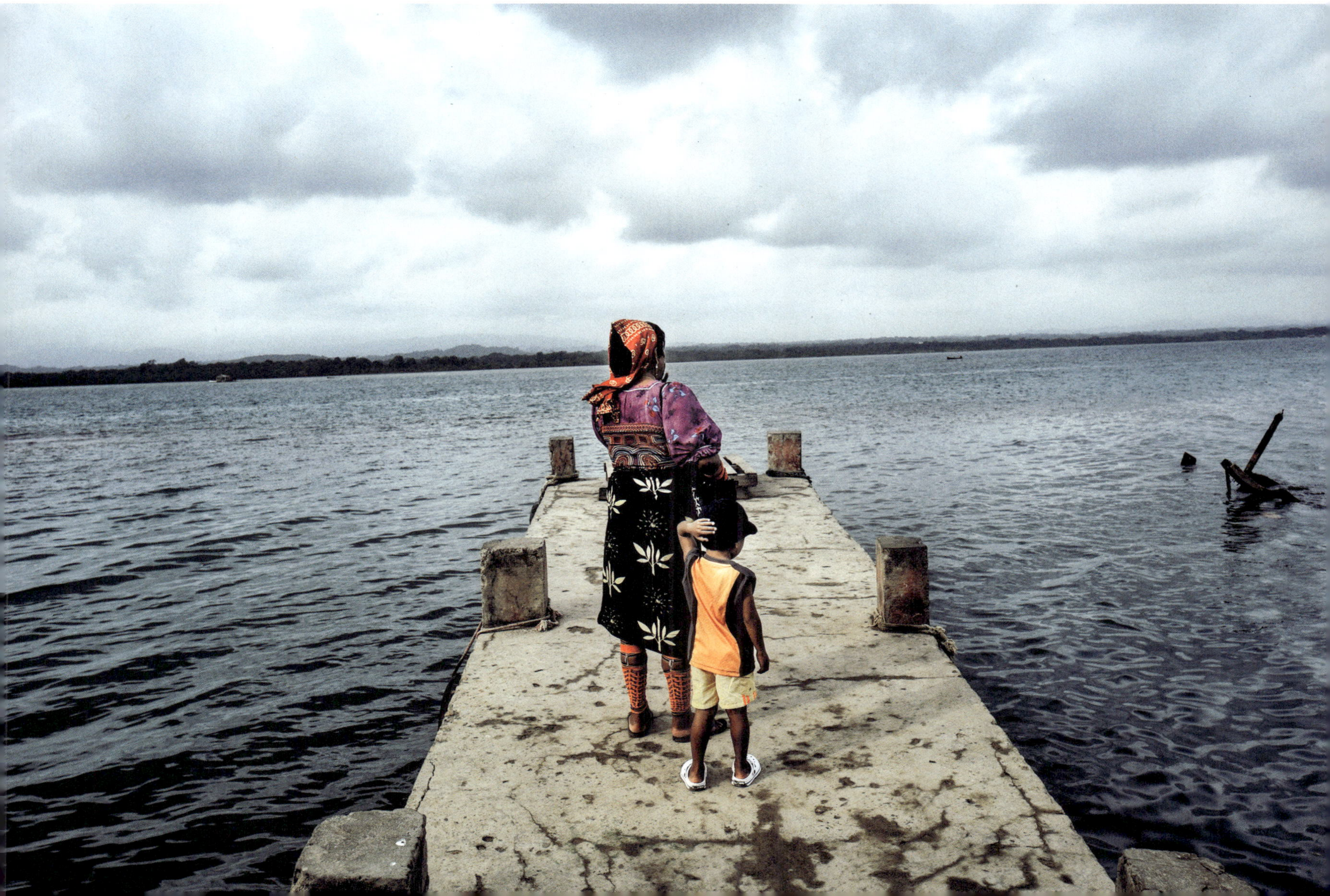

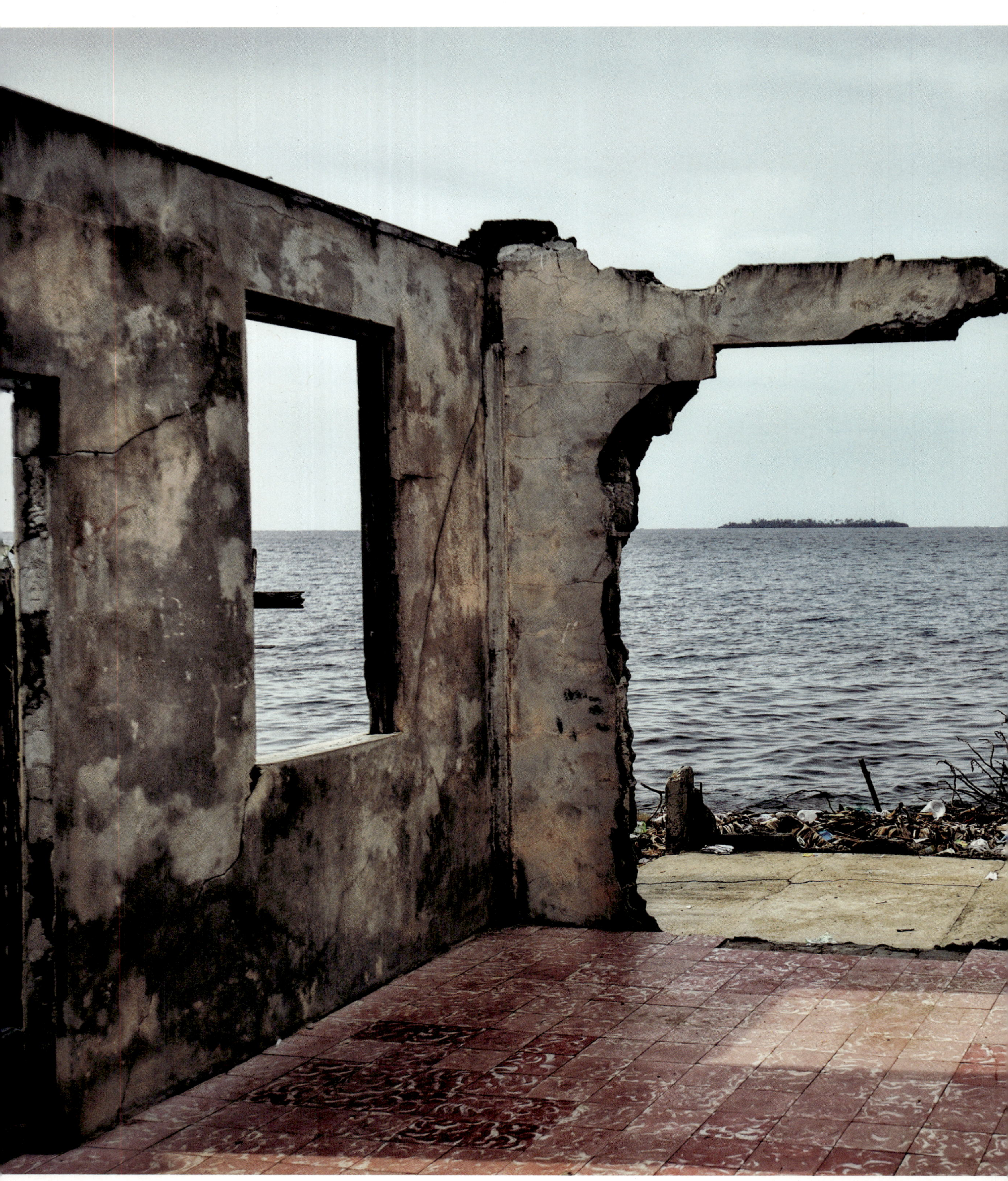

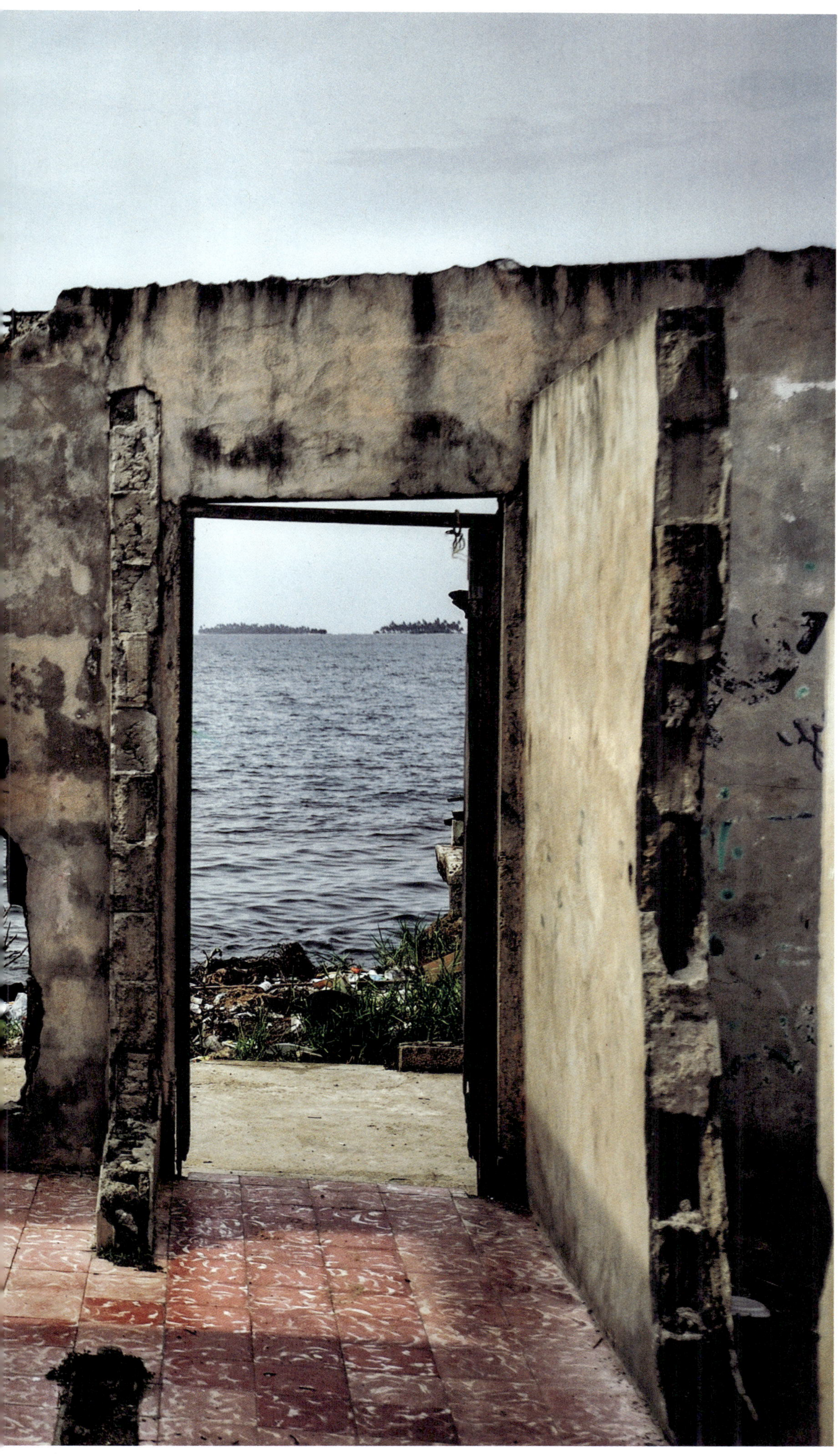

The shoreline of Corazón de Jesús is eroding quickly, and some buildings have already been destroyed by the sea. Inhabitants try to protect the shoreline using dead coral.

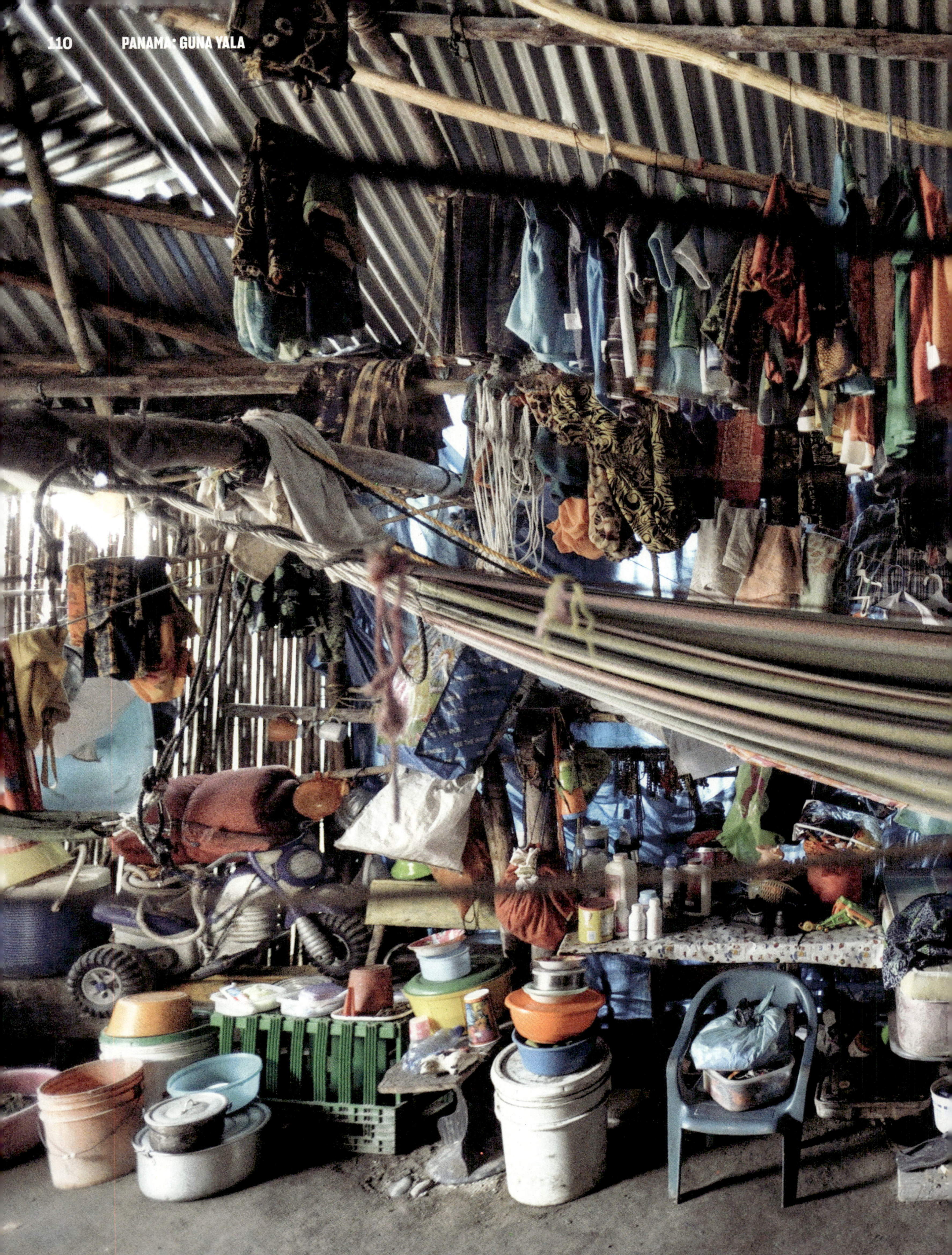

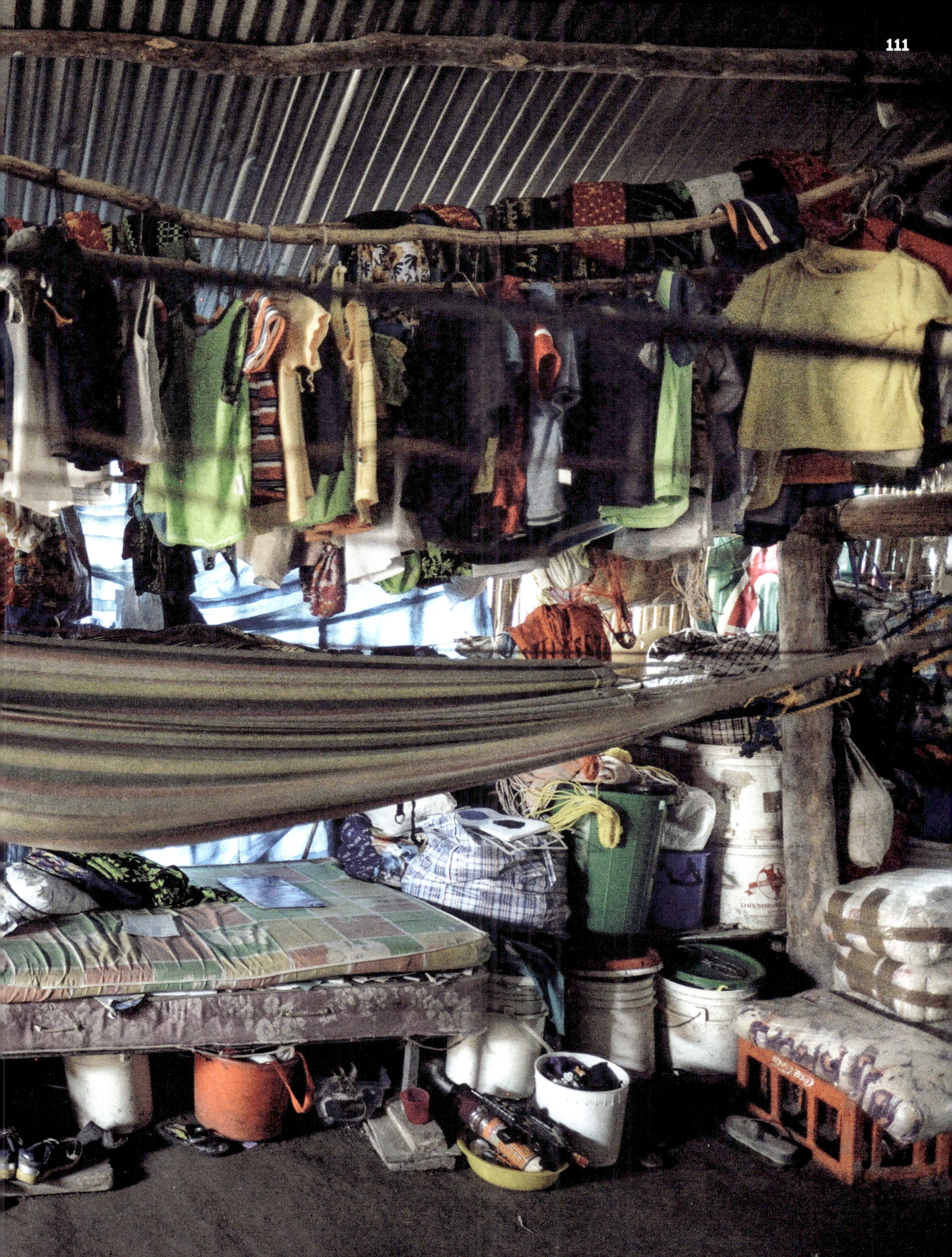

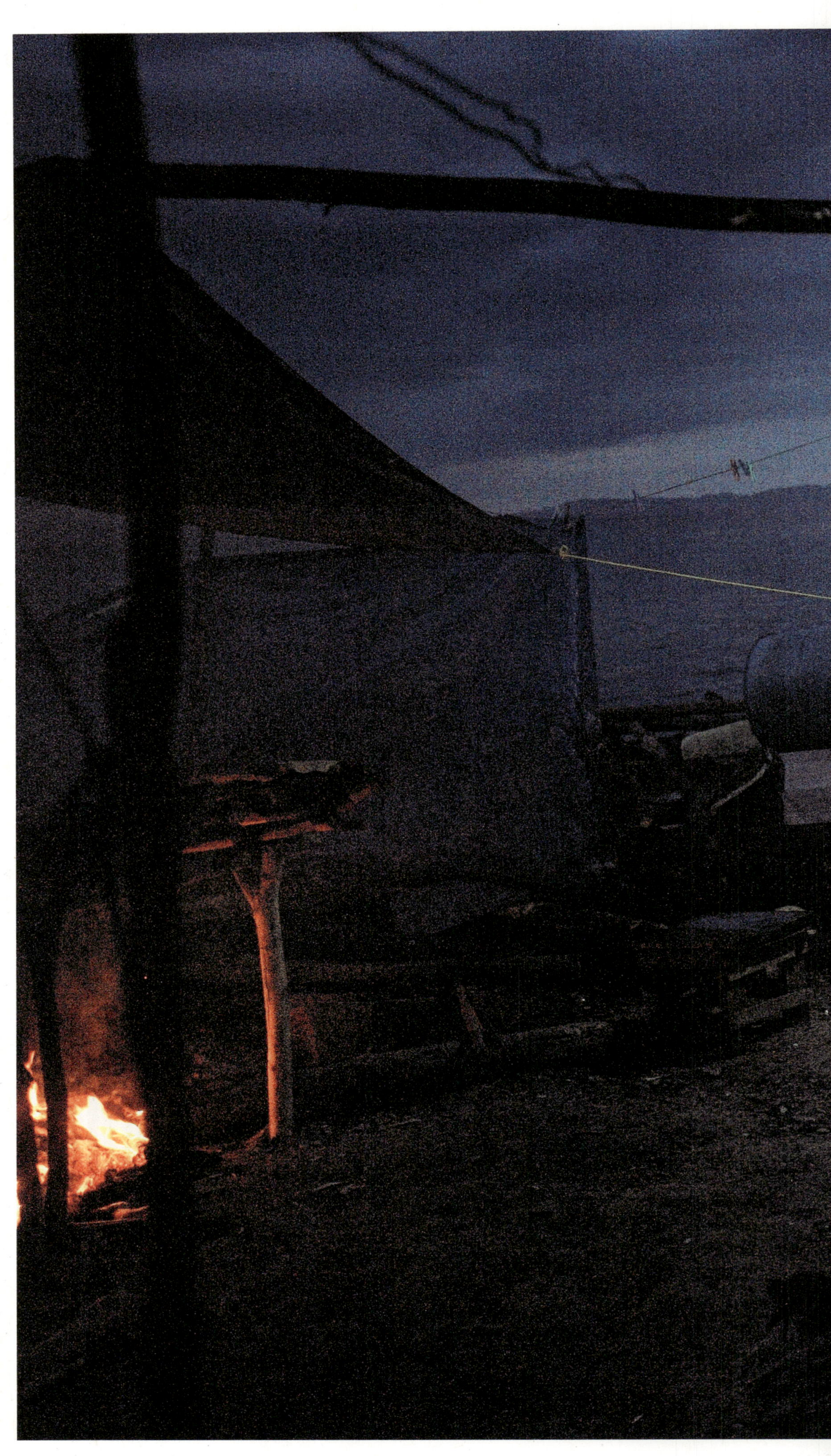

< Pages 110-111: Venilda Arcia lives on Cartí Sugdup.

Venilda Arcia (70) lives with her daugher and son-in-law on Cartí Sugdup. In 2007, part of their house was washed away in a storm. They are rebuilding it now. They live with 15 other family members.

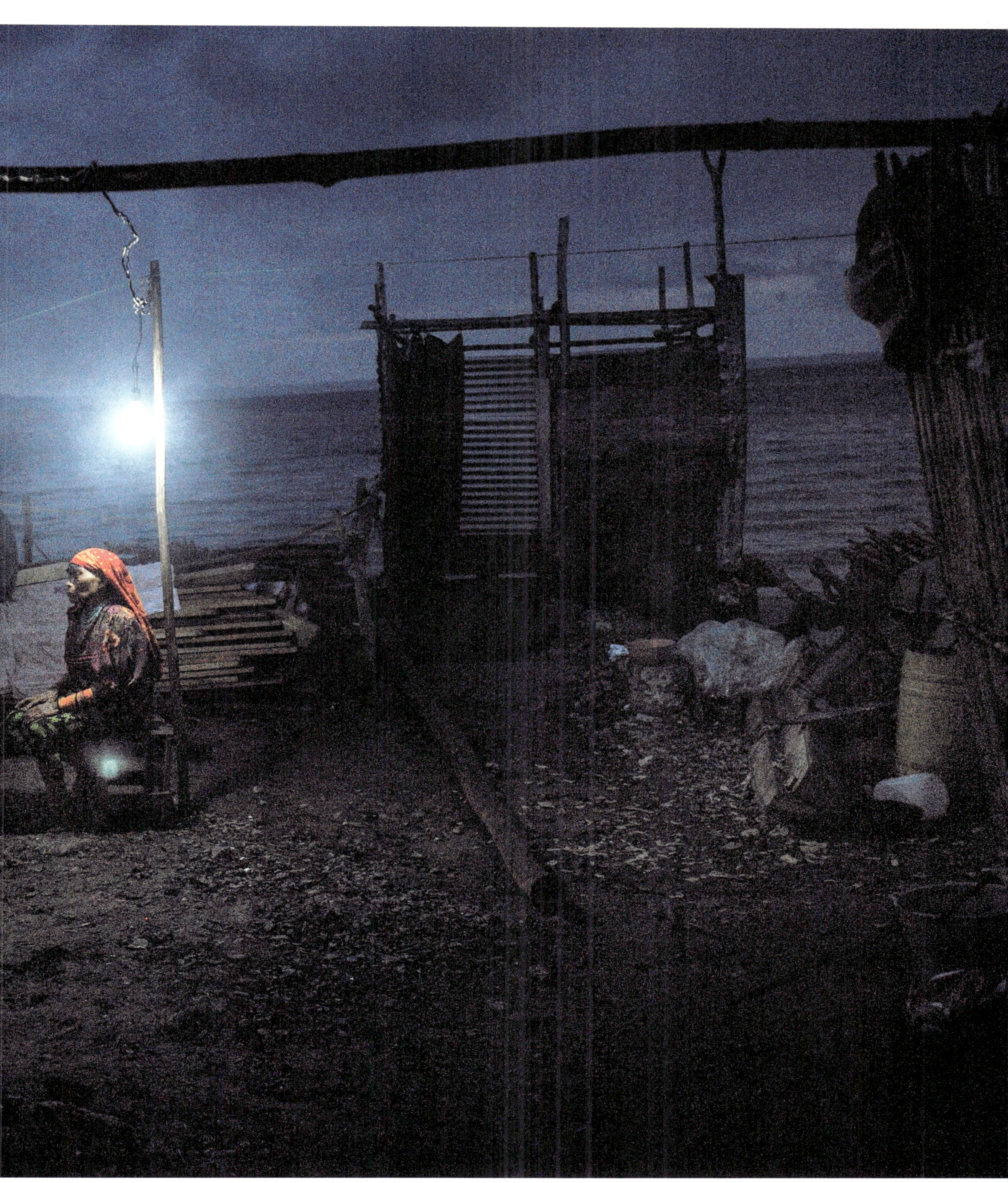

Dialis Morris (61) at the construction site on the mainland where the people of the islands will be relocated to. Currently, Dialis lives on Cartí Sugdup, but she would like to move as soon as possible: 'There are too many people on the islands now, and my house is too small for my large family'.

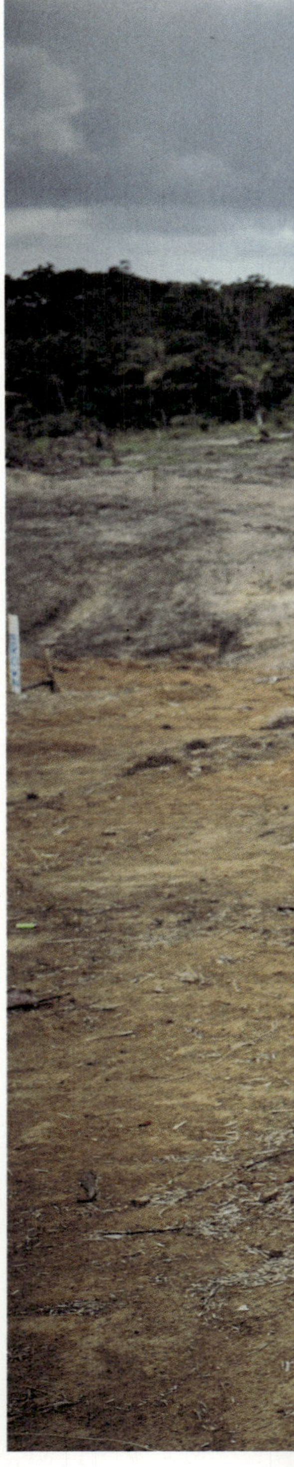

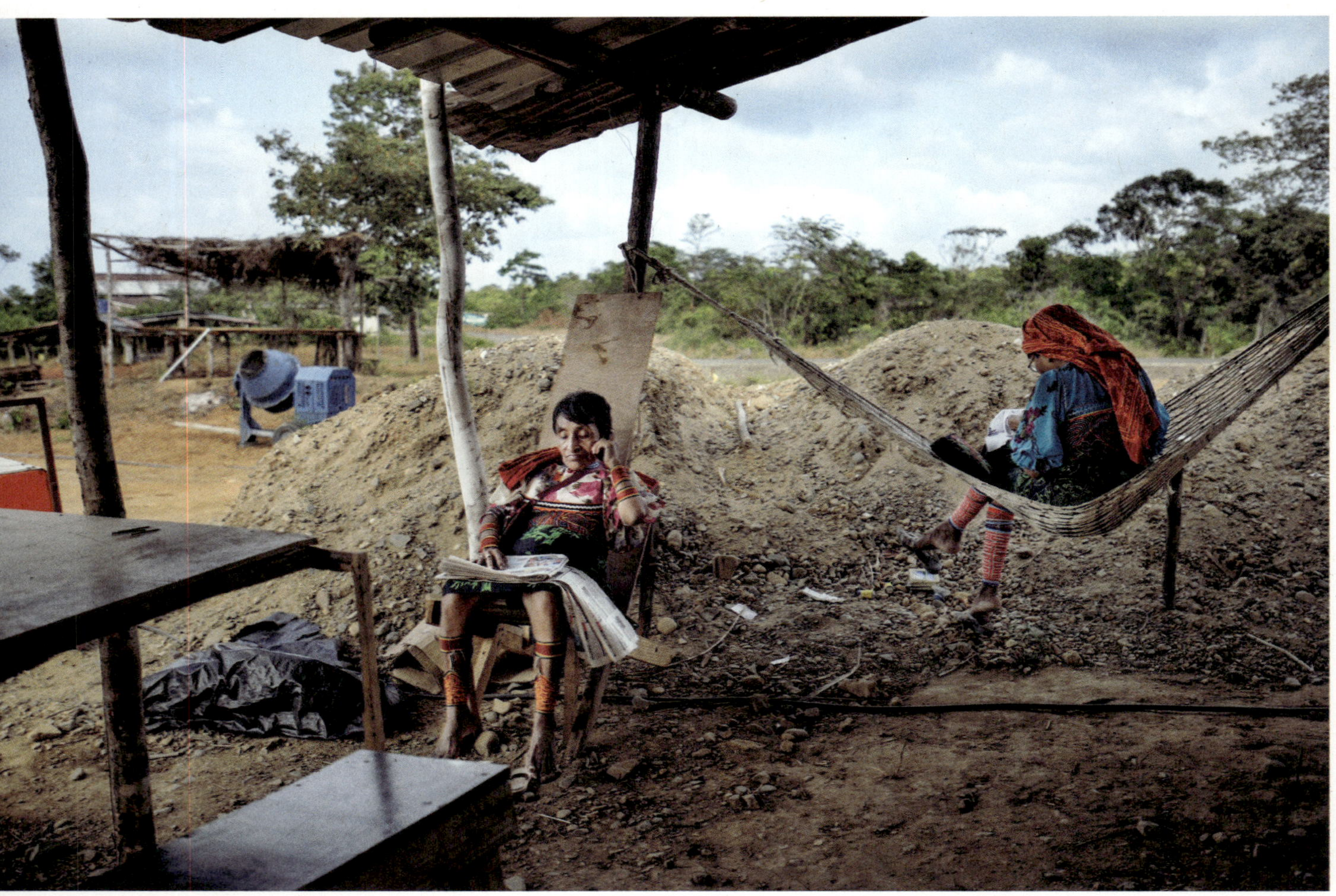

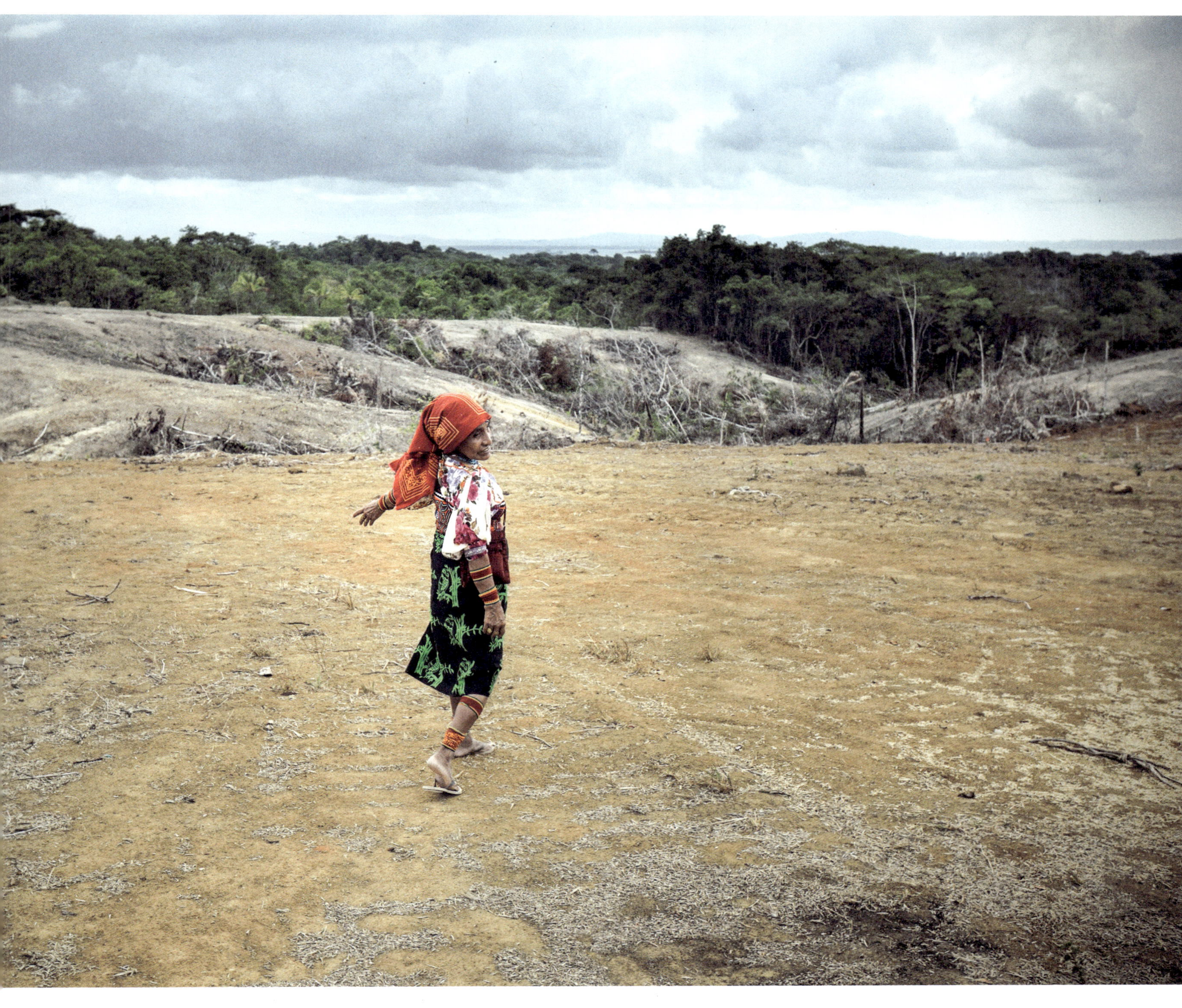

> Pages 116-117: The mainland of
Guna Yala, as seen from the sea.

INDONESIA: JAKARTA
Amalinda Savirani

Due to groundwater extraction, the city of Jakarta is sinking, approximately 25 cm a year in the northern part, where millions of people live. At the same time, the sea level is rising. If nothing is done, large parts of Jakarta will be submerged by 2050. Therefore, the government decided to build a new capital elsewhere. It remains unclear how many people will move there.

Indonesia has experienced many natural disasters, due to a combination of the climate crisis and human activity. The country has recorded an increasing occurrence of extreme weather events such as heavy rainfalls, prompting scientists to warn the public that extreme weather will become a 'new normal' (Kahfi, 2020). They warn that people need to brace for more heavy rains, frequent floods, seawater rise and long droughts. In the first eight months of 2020, almost 2,000 disasters have occurred, according to data from the National Agency for Disaster Management ('Indonesia witnesses 1,928 disasters in 2020 so far', 2020). Between 1866 and 2010, the annual temperature increase in Jakarta was 1.4 times the global average (Holliani Cahya, 2020). Average daily rainfall in Greater Jakarta has doubled in 10 years, from 10 mm to 20 mm. The impact of the climate crisis can be felt in other parts of Indonesia as well.

Indonesia is an archipelago of more than 13,000 islands, and water and the sea have always been a part of life for Indonesians. Historically, coastal areas have played a major role in trading and economic activities. At the same time, water and the sea have threatened the lives of the 35 million Indonesians living in coastal areas, due to the climate crisis and the absence of sound policies on water management. This essay explores how Indonesia has dealt with these issues so far, what solutions it has adopted and to what extent it has prioritized human life. The focus will be on Jakarta, as the fastest-sinking city in the world (Lin & Hidayat, 2018).

JAKARTA AS A SINKING CITY

The climate crisis has affected Jakarta in the form of sea level rise. However, Jakarta is also facing another problem: Severe land subsidence is making the city sink faster than the sea is rising. In fact, Jakarta is the fastest-sinking city in the world, at 15 cm to 25 cm per year, according to a report by the World Bank Group (2016). Geographically, Jakarta is a flood-prone area because it is situated on a deltaic plain, with 13 major rivers flowing through it, from the southern area to Jakarta Bay in the north.

Jakarta, therefore, faces a double threat of flood: first, from the rivers from the upstream areas south of the capital, which have seen deforestation and shrinking water catchment areas, and second, from the rising seawater and land subsidence in the north. Severe flooding from heavy rains in Jakarta and upstream areas like Bogor has also been caused by unchecked urban development that has involved violations of Jakarta's spatial planning. In 2007, the capital was submerged with water, both from the rivers flowing from the south and from seawater coming from the north. Scientists

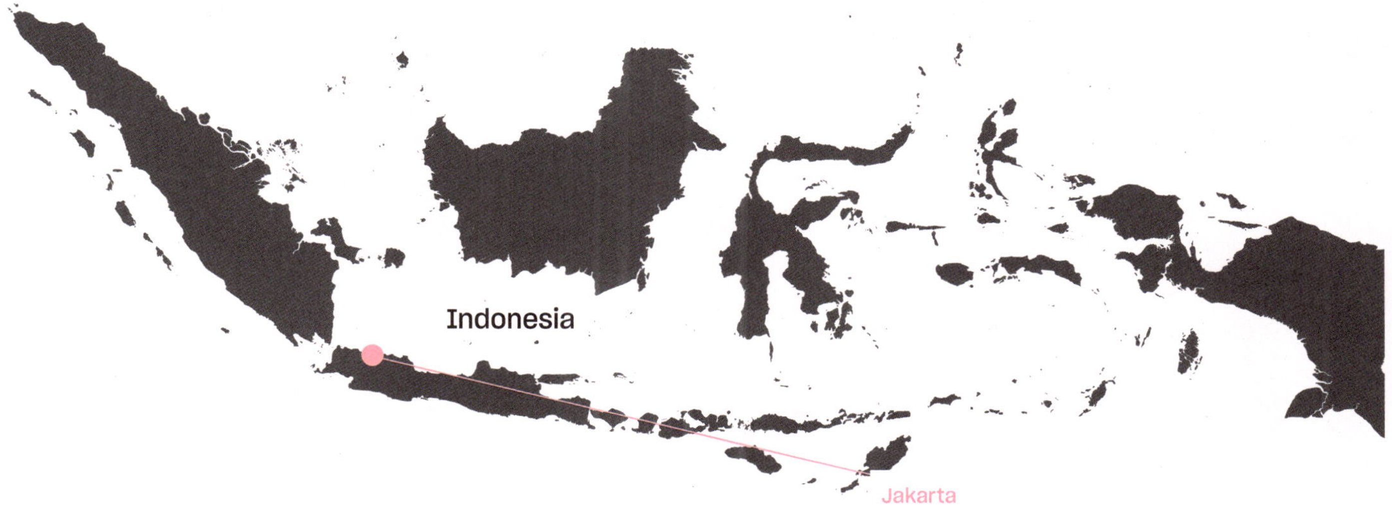

Jakarta, the capital of Indonesia, has grown immensely over the last two decades; over 35 million people currently live in the larger metropolitan area, which is situated in a delta.

Due to rapid growth, more and more groundwater is being extracted, which causes the land to subside and the city to sink. The densely populated north part of the city sinks approximately 25 cm per year. At the same time, the sea level is rising, which makes Jakarta one of the most threatened cities in the world. If nothing is done, large parts of the city will be submerged by 2050.

To protect the city, an ambitious plan was developed to build a large sea wall closing off the bay from the sea and to reclaim land to create 17 islands, where a large part of Jakarta would move to. Due to the extremely high cost, the government put the plan on hold and announced recently that Indonesia will build a new capital elsewhere.

Where to move, who will move and how quickly to move are so far not clear, but rapid action is needed. Jakarta is running out of time.

Population of Indonesia: 273,523,000
Population of Jakarta: 10,770,000
Population of Jakarta metropolitan area: 35,000,000

The Jakarta metropolitan area includes the capital, Jakarta, as well as five satellite cities and four core districts. Jakarta is one of the world's cities that is most at risk from extreme land subsidence and, on top of that, sea level rise. If nothing is done, large parts of the city will be submerged by 2050.

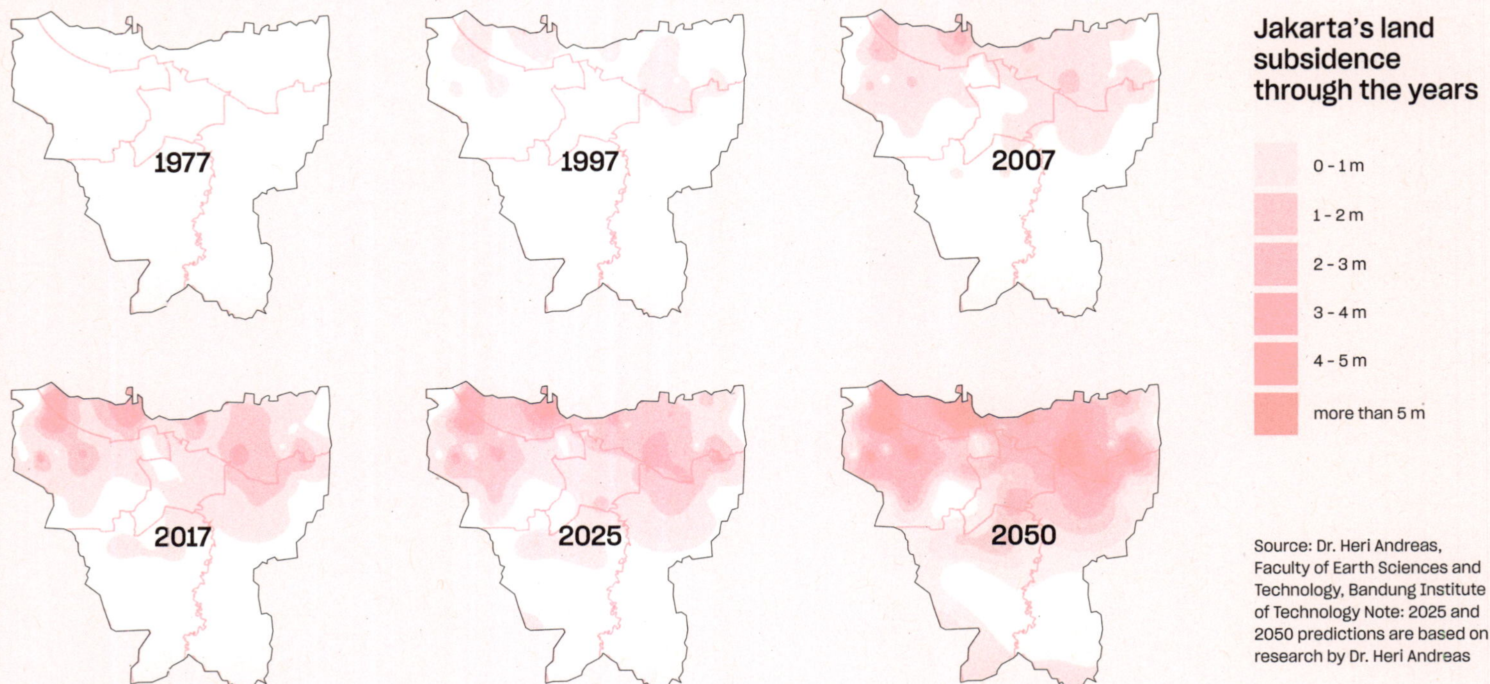

Source: Dr. Heri Andreas, Faculty of Earth Sciences and Technology, Bandung Institute of Technology Note: 2025 and 2050 predictions are based on research by Dr. Heri Andreas

have predicted a grim future pertaining to land subsidence: If nothing is done about it, Jakarta will be under water by 2050.

The following is a detailed explanation of the factors that are causing Jakarta to sink. First, there has been excessive groundwater extraction. A large portion of Jakarta's population, including inhabitants of commercial buildings and upper-income households, depends on groundwater for clean water. The obvious solution is to stop using groundwater, as has been adopted in Tokyo and major coastal cities in Italy. But Jakarta faces difficulties in doing that, as there are no alternative solutions for supplying clean water other than tap water, which has seen a stagnant expansion of service - 60% - despite water privatization in 1997.

Second, unchecked urban development has been going on for decades. Such development has devoured green areas, water catchment areas and urban forests since the 1980s, to be converted into business centres and residential areas. Big developers' notorious proximity to political elites has eased lobbying to speed up the conversion process by changing or even violating spatial planning. For example, in the 1980s, the government designated 70% of Kemang, South Jakarta, as green areas. Forty years later, green areas accounted for less than 5% of the region, replaced by upscale residential areas and commercial buildings. In total, less than 10% of Jakarta is green areas, the city administration often claims.

Third, there has been an absence of sound policies on water management and the climate crisis. Such policies involve several bureaucratic sectors; thus, policymaking needs to go beyond administrative and sectoral boundaries. Integrated policy framing to find solutions is imperative. So far, issues have been dealt with in an isolated manner, rather than being spatially integrated, involving various agencies, levels of government (national and local) and sectors. With all of these human-induced problems unsolved, Jakartans living in the northern part of the city are worried as they witness the protection wall getting higher to keep them safe from the seawater, and some parts of the wall cracking due to the constant splash of seawater.

Hundreds of kilometres to the east of Jakarta, in Central Java's capital, Semarang, another coastal city, severe land subsidence has affected thousands more people. In one research area, hundreds of residents need to relocate to a safer place, as they can see only the roofs of their houses during flooding, and they constantly need to add layers of floor in a race against the sinking land that has made seawater flood their homes (Batubara et al., 2020). Seawater has drowned their sense of security.

INFRASTRUCTURE AS A SOLUTION?

So far, the solution offered by the government is costly physical infrastructure projects to protect the population from the seawater. In Jakarta specifically, the Indonesian government has vowed to continue a project called National Capital Integrated Coastal Development (NCICD), which involves a giant sea wall that will close off Jakarta Bay. An earlier rendition of the wall was shaped like the national symbol of a 'garuda' (eagle); hence the nickname 'Great Garuda'. But after a recalculation, the government decided the giant garuda-shaped wall would be too costly and changed the plan to a simpler, albeit still giant, design.

NCICD is also linked to the controversial construction of 17 artificial islands in Jakarta Bay, known as a 'reclamation project'. Several officials claimed that the developers who received permits to build the islands would have to pitch in some money to pay for NCICD. Later, NCICD announced that the two projects were separate, but they confirmed the connection between them. Many foreign countries have invested in various infrastructure projects in Indonesia, including the Netherlands and South Korea.

The NCICD project in Jakarta, formerly called the 'Jakarta Coastal Defence Strategy', consists of three main phases, with the first phase, heightening and strengthening the existing onshore embankment of Jakarta Bay, completed in 2014. The second and third phases are the construction of a western outer sea wall and artificial islands, to be completed in 2025, and construction of an eastern outer sea wall, to be completed in 2040 (Bakker et al., 2017). The Dutch government will provide the investment, while engineering firm Witteveen+Bos will coordinate the project. The South Korean government joined the project as the third investment partner in 2015. The total cost is predicted to reach US$26.5 billion.

AN INTEGRATED APPROACH

Is infrastructure the only solution? Deltares, an independent Dutch consulting company in the water sector, published a report on 'sinking cities', a study of many delta cities around the world, including Jakarta. The report offered an integrated approach to addressing the double threat of sea level rise and land subsidence, using an approach called 'DPSIR', an abbreviation for 'driving forces, pressures, state, impacts and responses'. It combines aspects of raising awareness, developing knowledge, addressing governance from various actors, monitoring responses and supporting decision-makers. The report does not specifically mention infrastructure as the solution for these sinking cities, let alone giant sea walls, in the case of Jakarta (Deltares, 2015). JanJaap Brinkman from Deltares, in an interview with *The Jakarta Post* (Mariani, 2016), also said the cheapest solution is to stop groundwater extraction. It is not clear why later, Deltares, along with Witteveen+Bos, Grontmij, Ecorys and KuiperCompagnons opted for the infrastructure solution of the 'Great Garuda' (KuiperCompagnons, n.d.).

The Dutch have also made an investment in water infrastructure and water management to protect Indonesia from seawater in Semarang, which is known as the 'Little Netherlands'. King Willem-Alexander and Queen Máxima visited Indonesia in early March of 2020, as part of the Netherlands Economic Mission to Indonesia, bringing along 180 Dutch businesspeople representing 112 enterprises. The Dutch prime minister, Mark Rutte, had previously visited Semarang, in 2016 and 2019. Dutch partners in these water management projects include: Witteveen+Bos, Wetlands International, Bureau Waardenburg, Royal HaskoningDHV, Drainblock, Deltares, and Wageningen University (Partners for Water, 2020). The Netherlands Water Partnership counted 91 Dutch water projects in Indonesia in December 2015, most of them private sector (ter Braak, 2016).

'RIVER NORMALIZATION'

Along with this support, the Jakarta government has initiated a project called 'river normalization', which takes the form of dredging and installing concrete sheet piles along parts of the Ciliwung River. The World Bank Group supports this solution under a programme called 'Jakarta Urgent Flood Mitigation', with a total investment of US$185.95 million (World Bank Group, 2016). To execute the project, the Jakarta government evicted many poor communities from the river banks between 2015 and 2016. More than 16,000 people were victims of the forced evictions, according to a report by Legal Aid Institute Jakarta (LBH Jakarta, 2017). Despite fierce resistance in many areas subject to eviction, including Kampung Pulo in East Jakarta, Bukit Duri in South Jakarta and Kampung Akuarium in North Jakarta, the Jakarta administration went ahead with the forced evictions, relocating some of the eligible residents to *rusunawa*, low-cost rental apartment buildings that the government has constructed in many areas of the city. Some rusunawa are located far away from residents' earlier homes, forcing them to start their livelihood from scratch in the new place (Savirani & Wilson, 2018).

The Jakarta administration claimed that the 'river normalization' was a success, yet, in 2020, when extremely heavy rainfalls hit Jakarta, the capital was submerged again, including the 'river normalization' areas that have been much-touted as being 'free from flood' areas. The latest disaster shows the ineffectiveness of a top-down physical infrastructure solution for flooding in Jakarta.

EFFECTS IN OTHER AREAS

While problems have emerged in the reclamation project in Jakarta, other parts of the country also have ongoing land reclamation projects, including Makassar, Bali, Kendari, Palu, Balikpapan, Lombok, Bitung and Aceh. All are under the umbrella of projects called the 'waterfront city'. Makassar has plans for a waterfront city called Citraland City Losari, part of a 1,000 ha project called the 'Centre

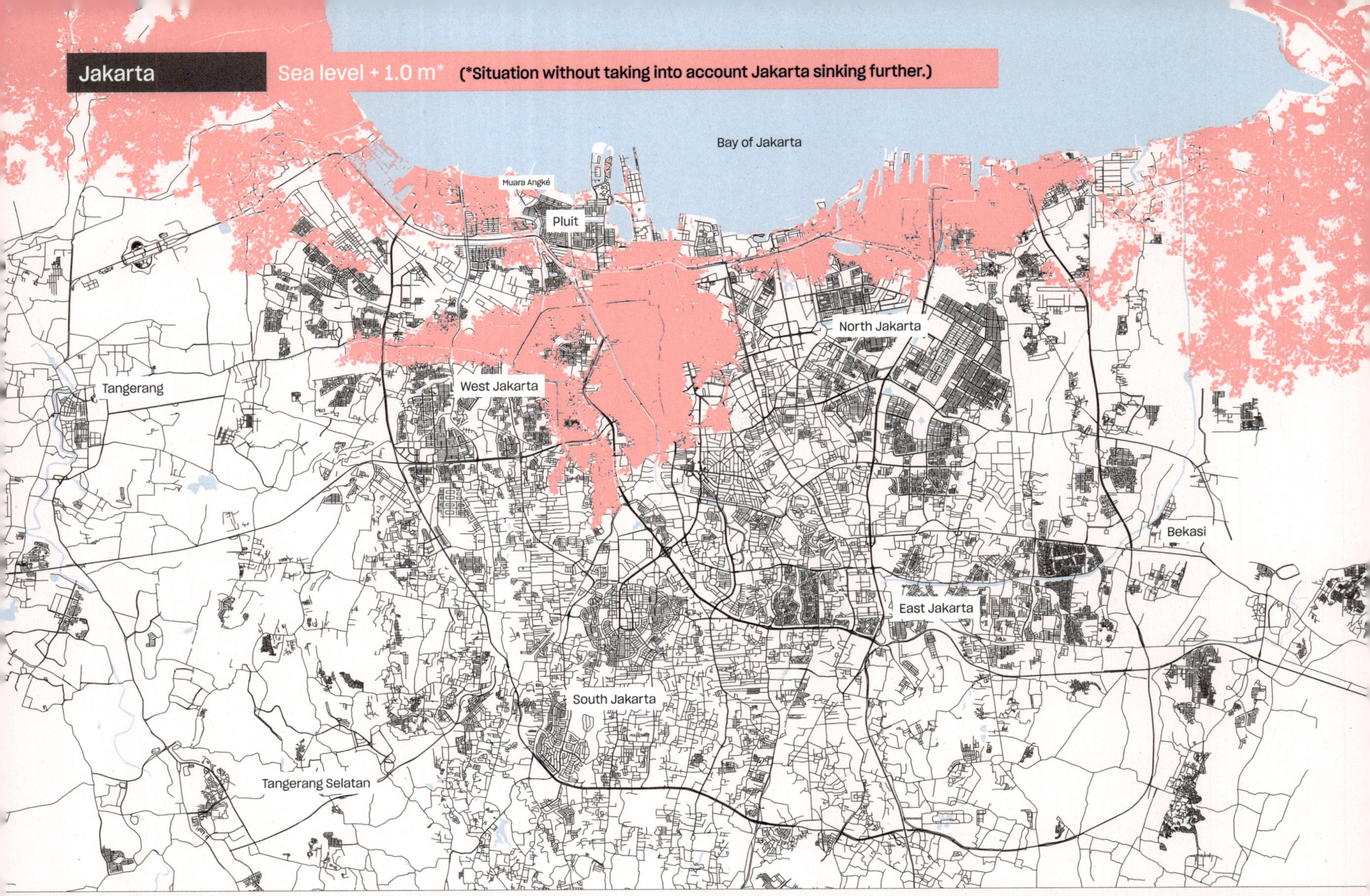

Jakarta
Sea level + 1.0 m* (*Situation without taking into account Jakarta sinking further.)
Bay of Jakarta
Muara Angké
Pluit
North Jakarta
West Jakarta
Tangerang
Bekasi
East Jakarta
South Jakarta
Tangerang Selatan

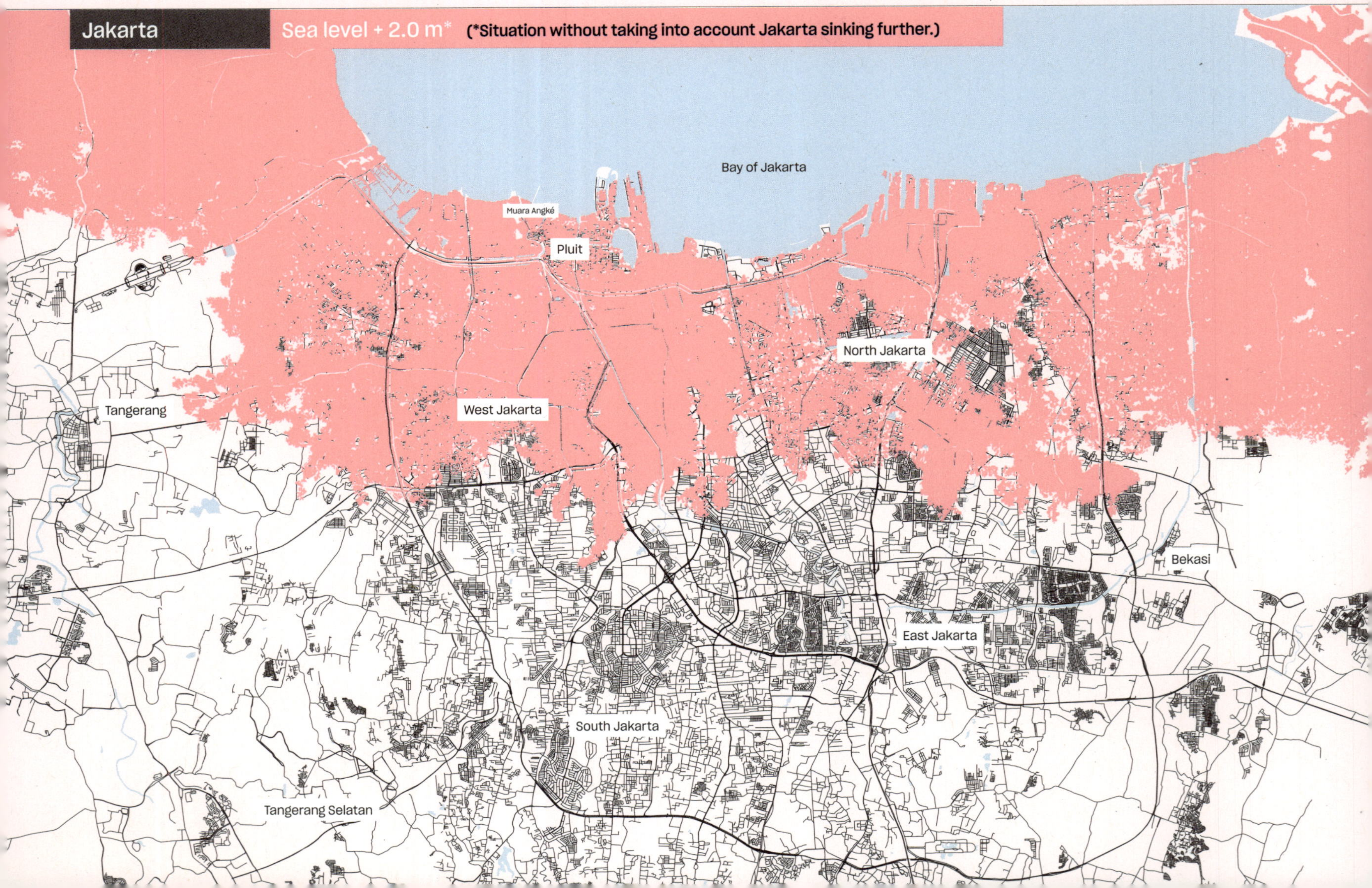

Jakarta
Sea level + 2.0 m* (*Situation without taking into account Jakarta sinking further.)
Bay of Jakarta
Muara Angké
Pluit
North Jakarta
West Jakarta
Tangerang
Bekasi
East Jakarta
South Jakarta
Tangerang Selatan

Point of Indonesia', in South Sulawesi. A Dutch company, Royal Boskalis Westminster N.V., is building five artificial islands in Makasssar, the largest city in eastern Indonesia, at a total cost of €80 million (US$94 million; Dutch Water Sector, 2016). The company has been contracted by Ciputra Yasmin, a high-profile Indonesian developer. Bali started land reclamation in Benoa Bay in 2012 to boost its tourism area, despite resistance from the local *adat* community, the majority of whom are fishers, who lost their livelihood after the reclamation (Burhaini Faizal, 2016).

CAPITAL RELOCATION

Amid this jumble of infrastructure development, President Joko Widodo, also known as Jokowi, announced his plan to relocate the capital city from Jakarta to East Kalimantan, 1,000 km from Jakarta. A total of €27 billion (US$35 billion) will be allocated for this project. The prince of Abu Dhabi, Sheikh Mohammed bin Zayed Al Nahyan, will lead the project. Tony Blair, former British prime minister, and Masayoshi Son, Japanese billionaire and the founder of SoftBank Group Corp., will be members of a team that will supervise the project. Son plans to invest US$40 billion in the project. The planned location is Penajam-North Paser Regency, 76 km from Balikpapan, an affluent oil-and-gas city in East Kalimantan. The new capital city will be 40,000 ha in size, four times larger than Jakarta. The project is expected to be completed in 2024. The COVID-19 pandemic has halted the project temporarily, as the government needs to reallocate the budget and prioritize solutions for the pandemic. But still, the rush with which Jokowi pushed the new capital project is another example of how the Indonesian government fails to adopt an integrated approach in finding solutions to water problems. For the government, concrete infrastructure is the panacea of many problems.

HUMANITY AT STAKE AND PEOPLE AS INFRASTRUCTURE

Flood has caused the loss of life and livelihood for many people in Indonesia. On average, flooding costs Jakarta more than US$400 million per year (Stepputat & van Voorst, 2016). The loss of life is also significant. In 2013, more than 83,000 people were displaced due to flooding. In 2015, water covered 75% of Jakarta, halting the daily activities of inhabitants. Water comes to houses regardless of the inhabitants' social status. Rich and middle-class residents get floods too. Yet the urban poor have always been the ones that bear the greatest brunt.

Adopting infrastructure as the sole approach to dealing with environmental problems in Indonesia positions the government as the most important

actor, along with its resources, including from foreign sources. What would happen if the government left or lost its capacity? We can predict that disaster will continue, as no person or institution can stop the force of nature. This is one of the limitations of using infrastructure as a primary approach to solving environmental problems.

Another limitation is that an infrastructure-oriented solution also creates victims. Critical views on this type of solution were expressed by various NGOs during the construction of the artificial island project and the Jakarta Bay reclamation project in North Jakarta. More than 100 fishers that fish in Jakarta Bay have lost their livelihood and become scavengers (Pramita, 2015). Before the land reclamation, they could find fish within 5 miles (8 km) of sailing in their small boats. After the reclamation, they have to sail further to catch fish, with the risk of being caught by the sea patrol, which allows only certain ships to sail past this point. Previously, they could earn 300,000 IDR to 500,000 IDR (US$20 to US$33) in a day; now they are grateful if they earn one-third of that. The fishers have also reported pollution during the construction of the artificial island, which involves drilling the sea bed. In Makassar, in August of 2020, three fishers were arrested for protesting land reclamation projects there.

One policy alternative is to strengthen citizens in a collaborative way. Here we encounter what Simone (2004) refers to as 'people as infrastructure', which emphasizes the modality and capacity of ordinary citizens in coping with their daily struggle, including in economic activities. The climate crisis and water management problems are insurmountable, and relying solely on government policy will be insufficient. The government needs to revive local knowledge and stories about how people have dealt with water over generations, and include them in school curricula to raise awareness. The government can facilitate this knowledge accumulation and reproduction. In the meantime, NGOs and actors in the grassroots movement need to keep talking, organising and making more noise to alert the government. Without this stubbornness, floodwater will fill our homes, and the homes of generations to come. To be sure, this approach is less efficient and requires longer time to execute. Yet, it is another option to consider. ●

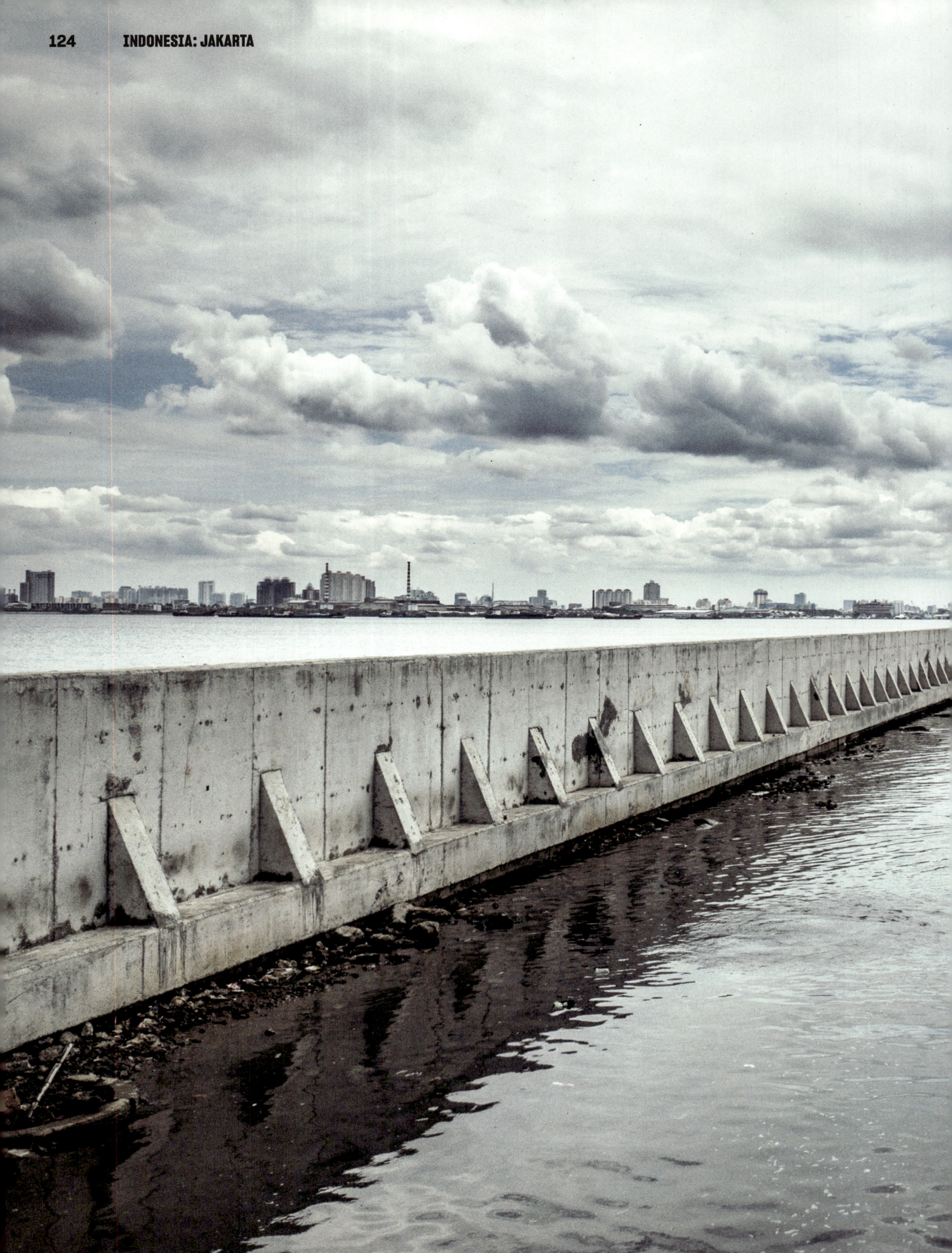

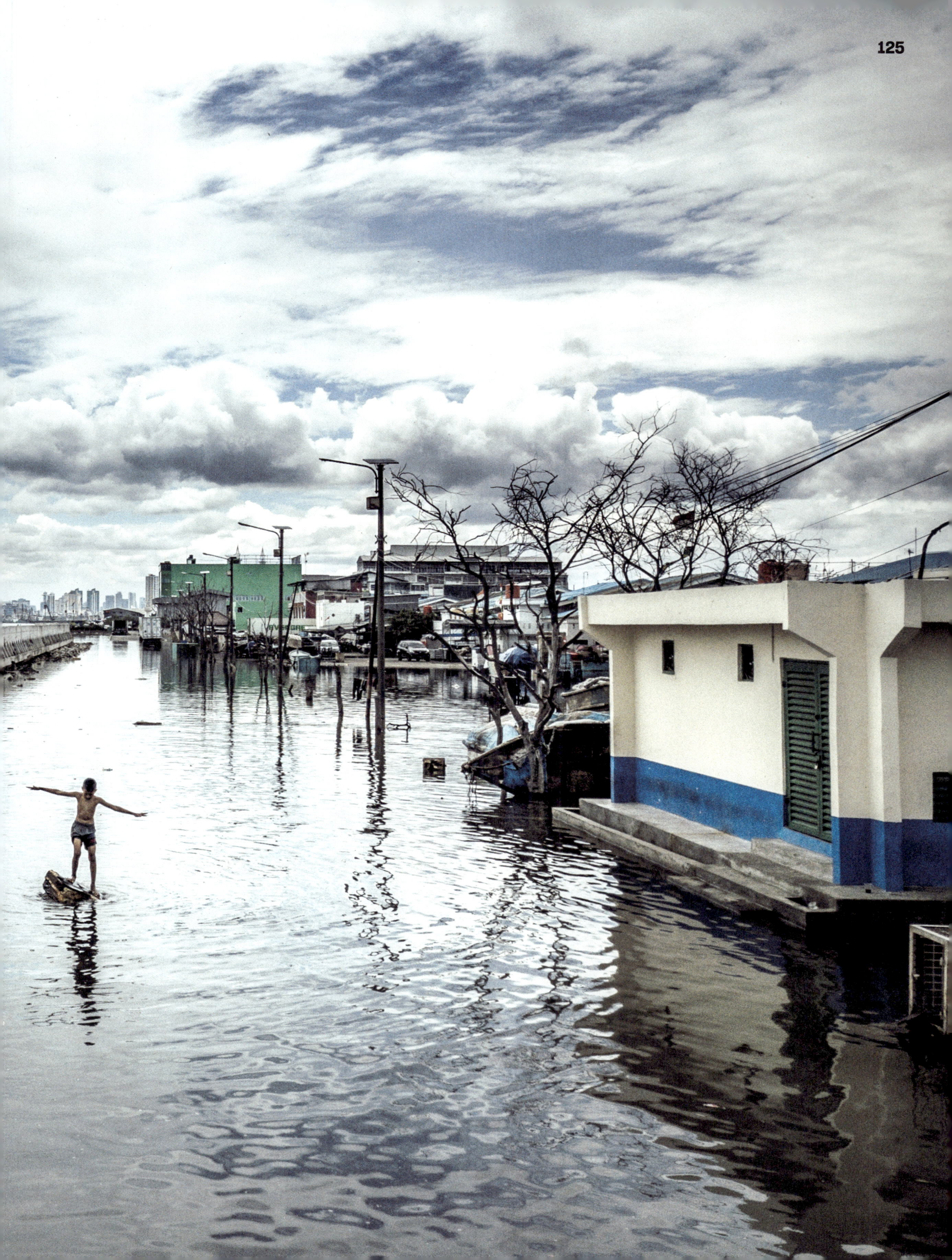

< Pages 124-125: A newly built sea wall
is not very effective.

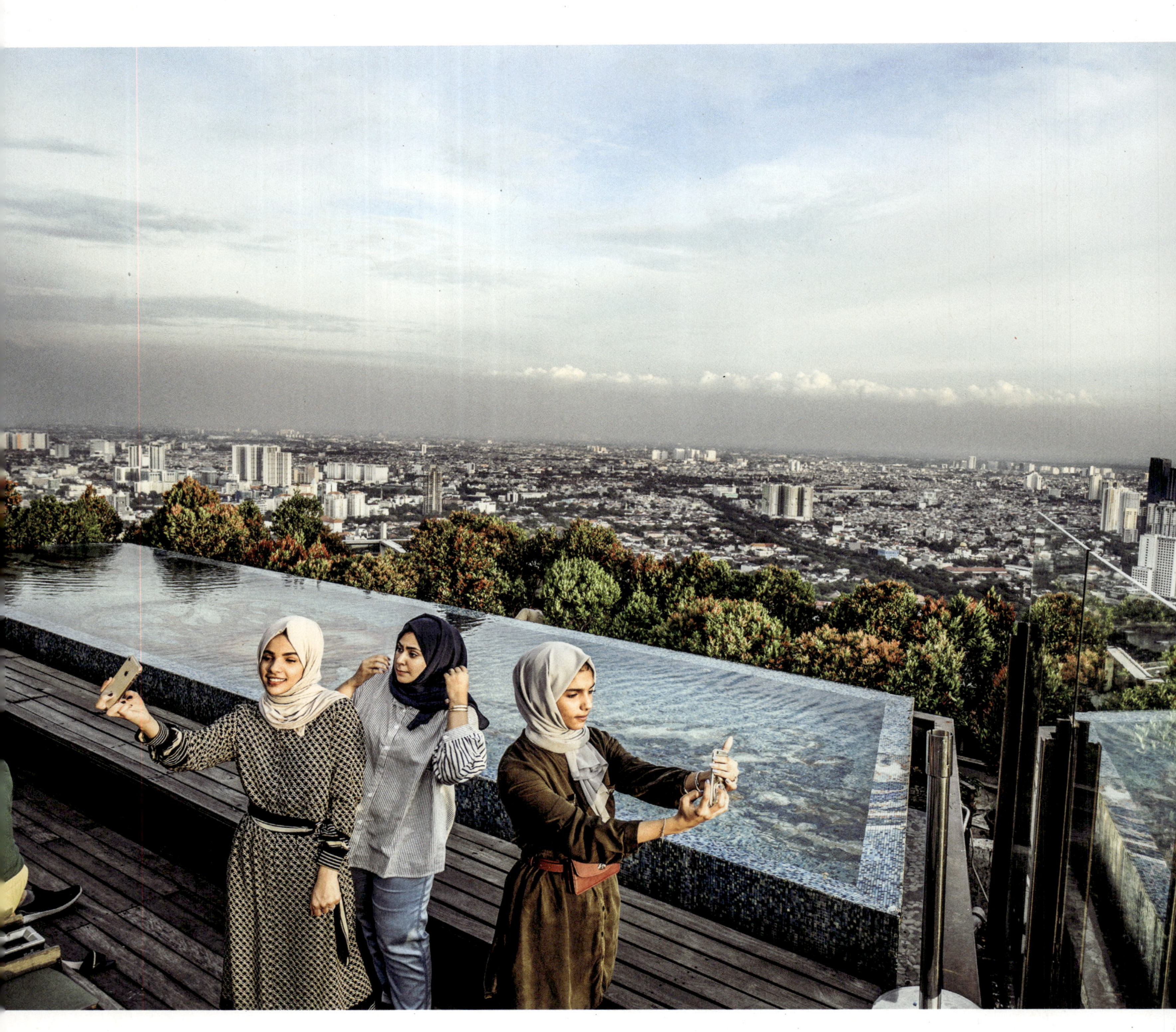

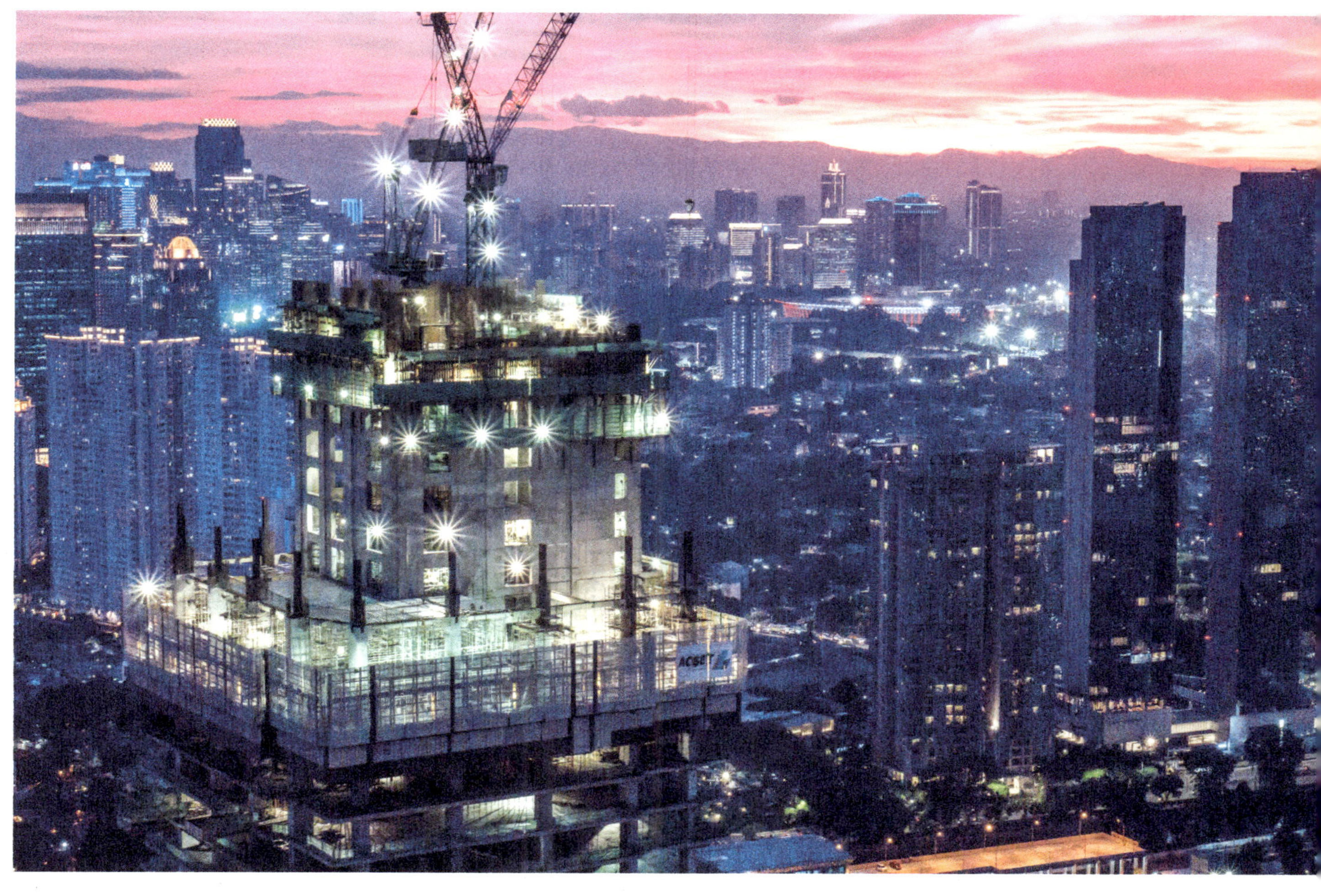

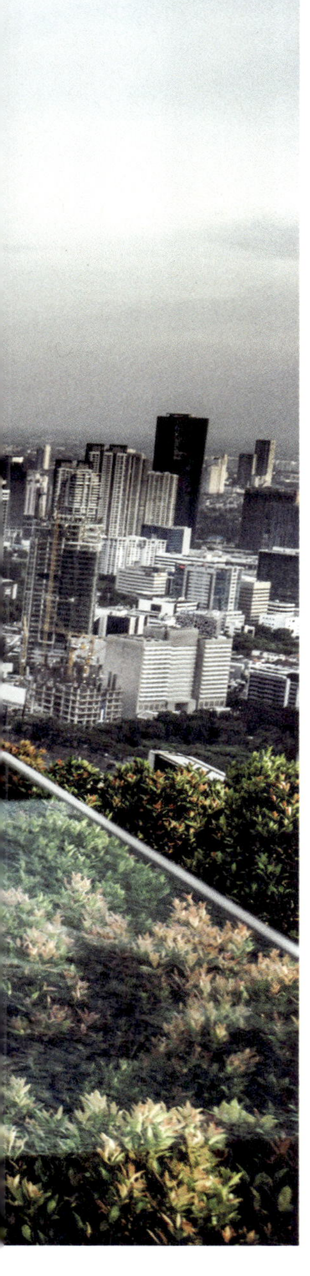

A view of central Jakarta.

20 years ago Jakarta for the greatest part was above sea level. Now 70% is below, due to land subsidence. Northern Jakarta is mostly affected. This part of the city sinks 25 to 30 cm per year.

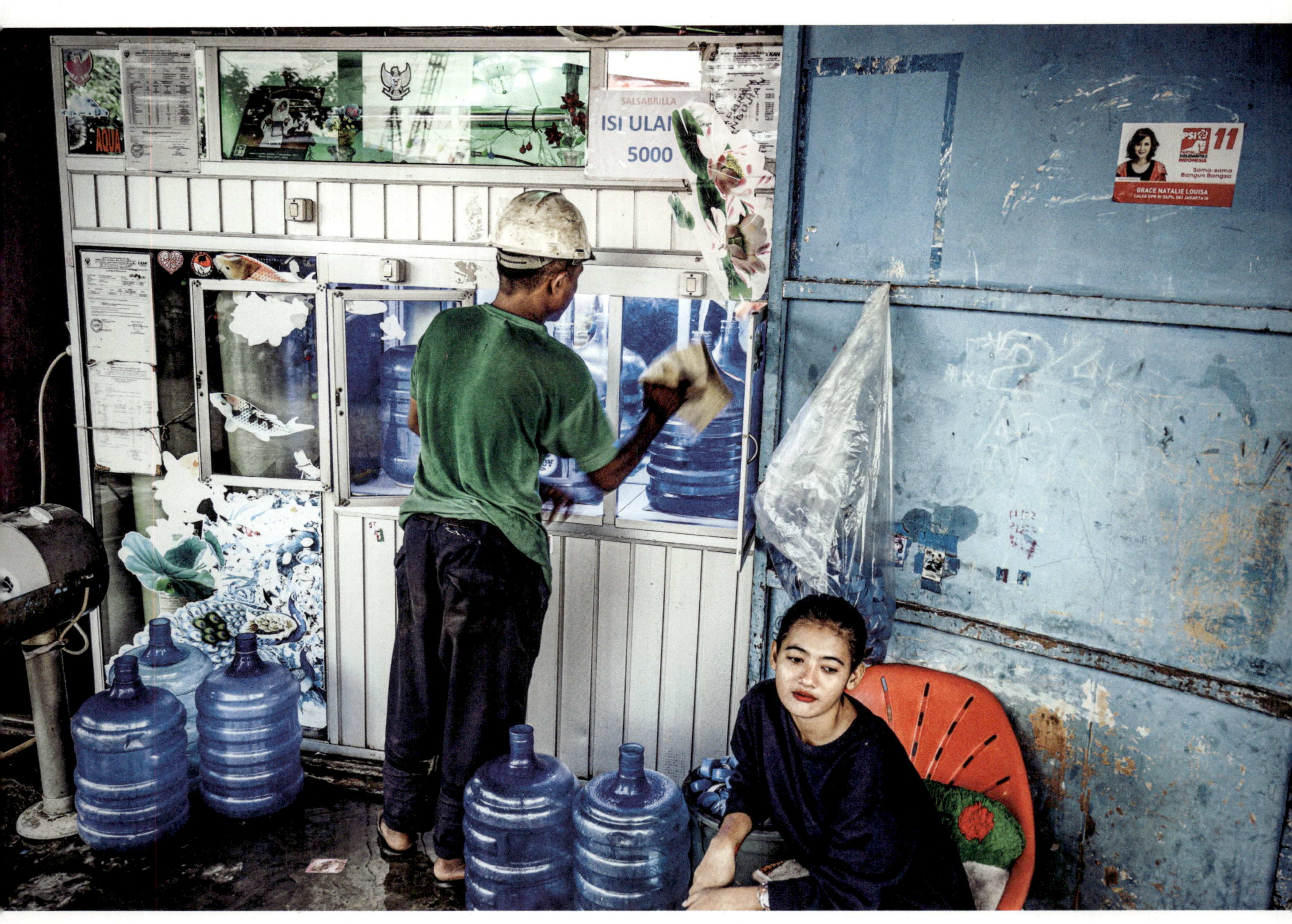

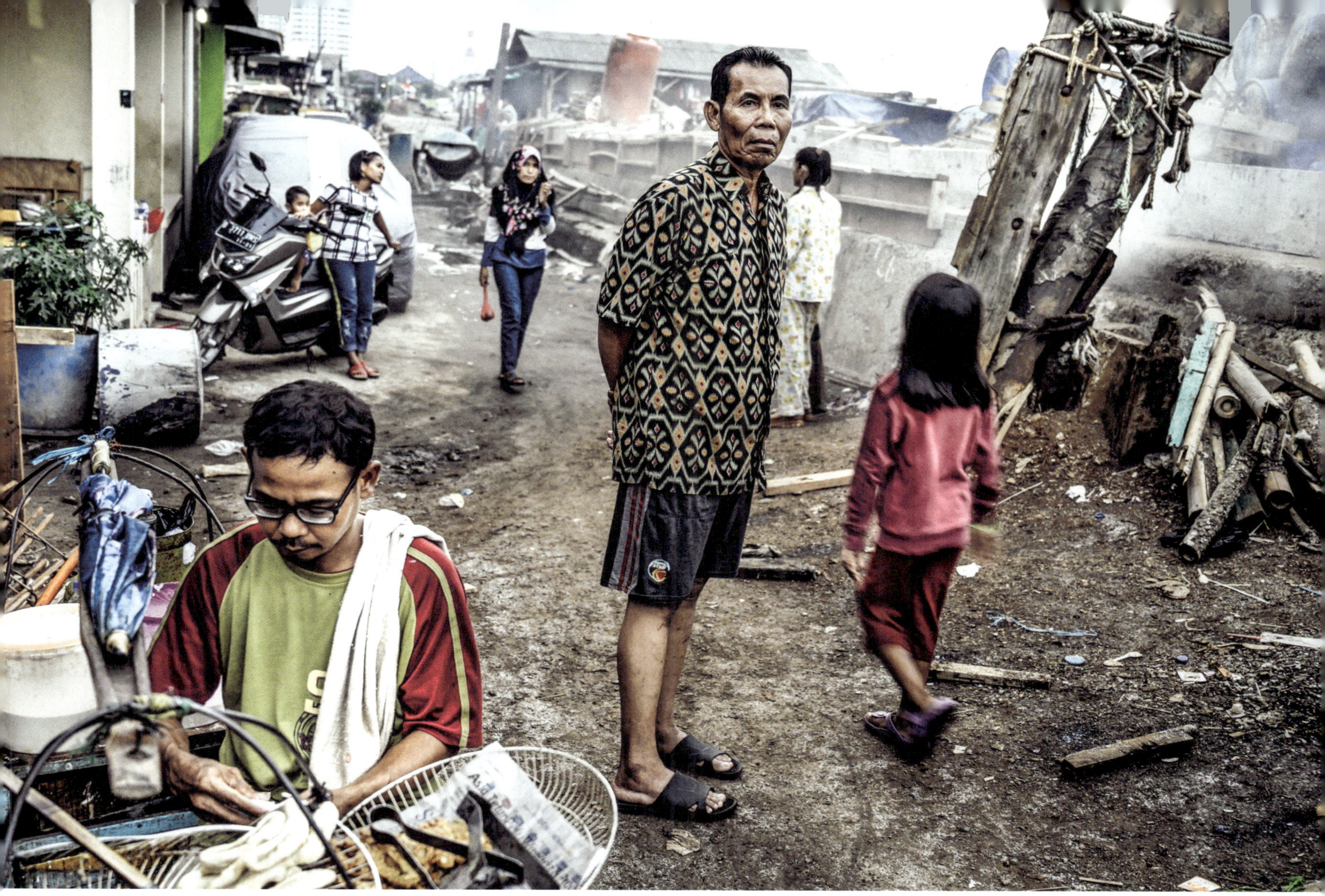

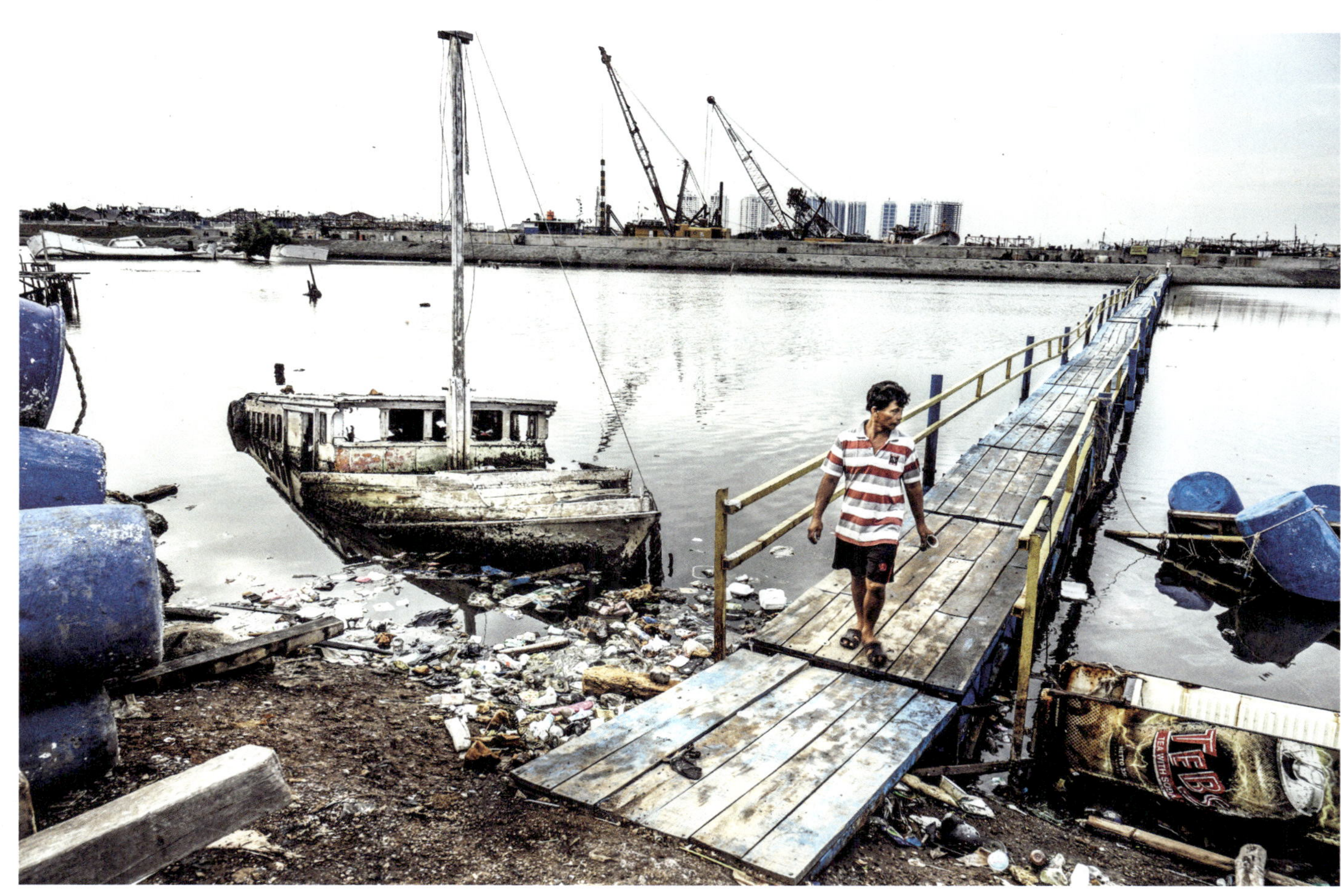

After a heavy rainfall, large parts of Jakarta flooded, partly due to the canals being clogged with garbage, but also due to land subsidence; too much groundwater is being extracted.

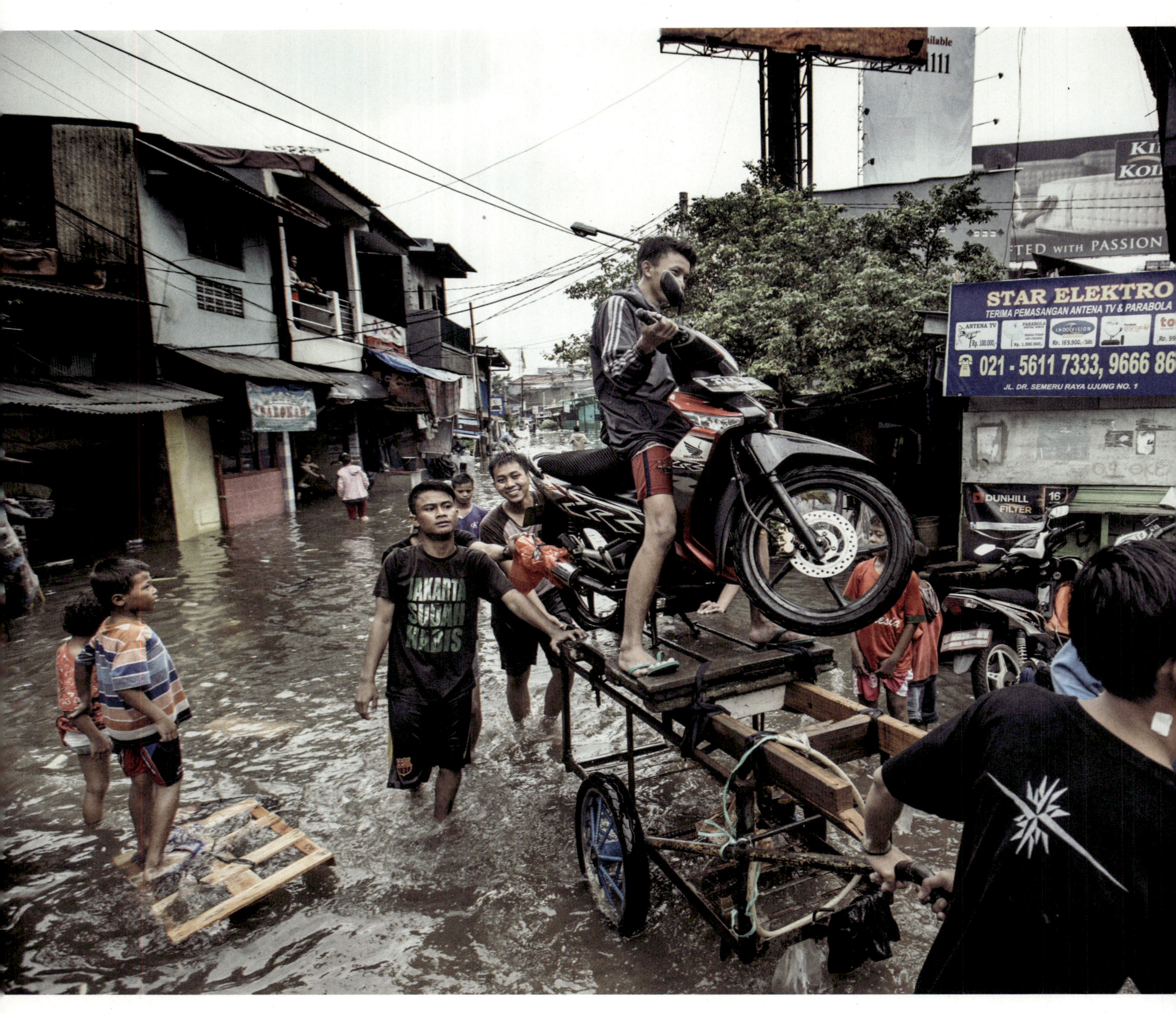

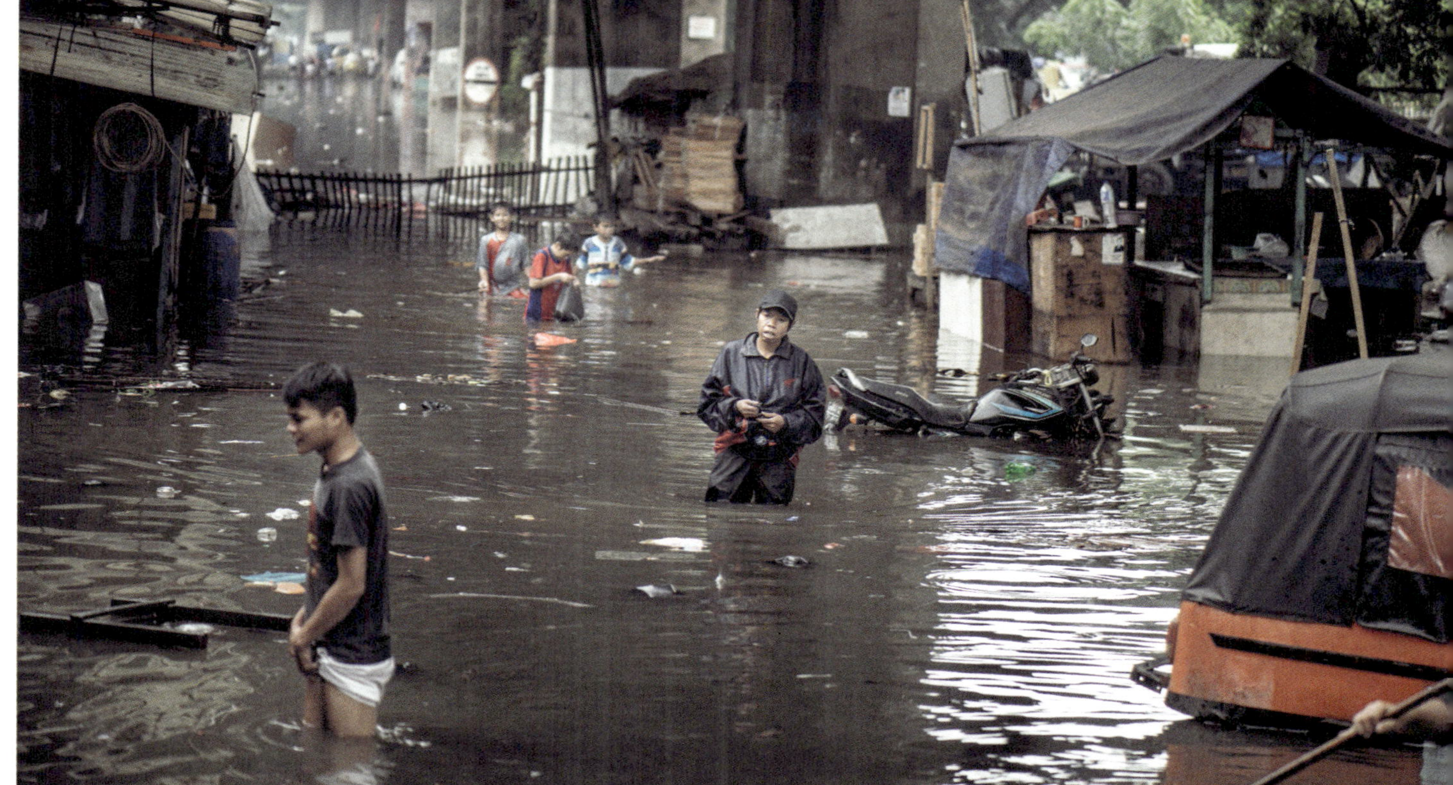

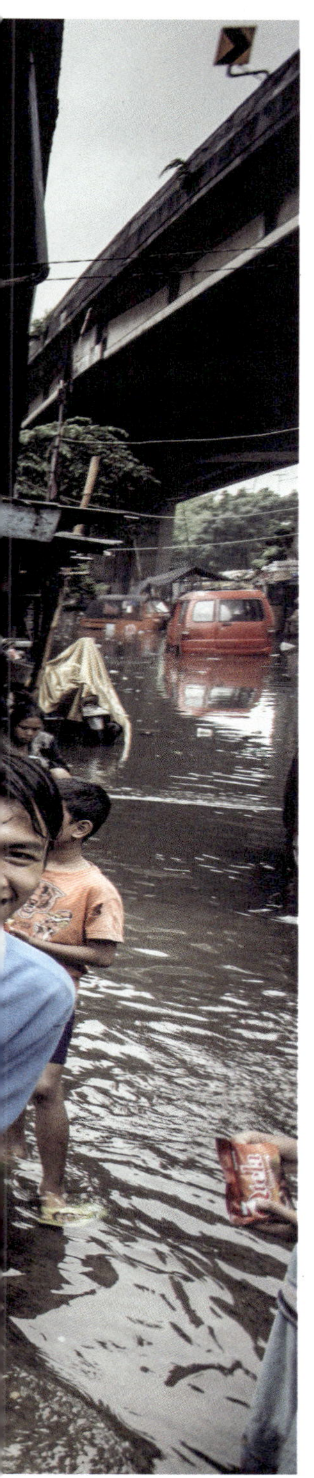

> Pages 132-133: Muara Angke during high tide. This is an area in the north of Jakarta that is particularly vulnerable to sea level rise and land subsidence. The area floods at high tide.

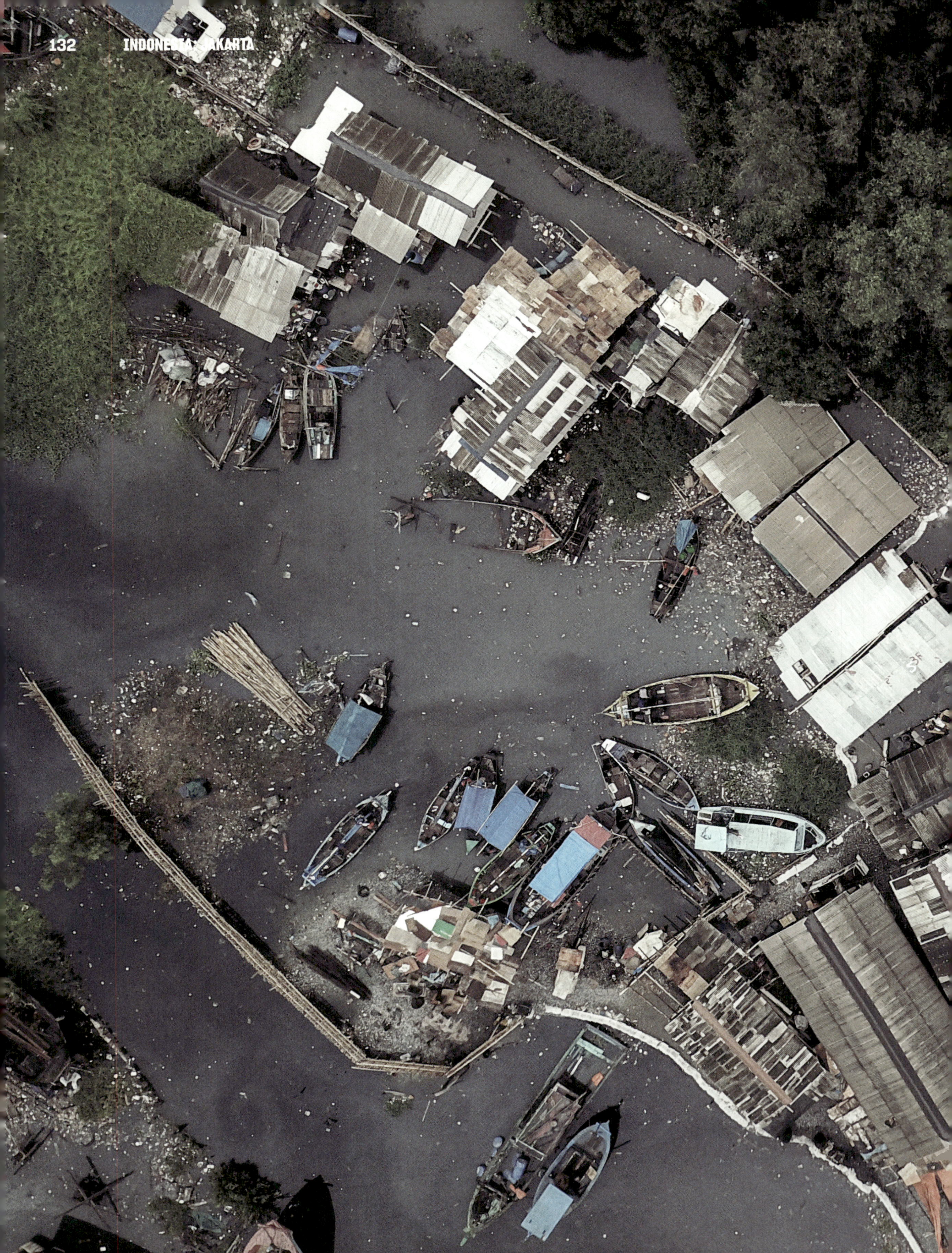

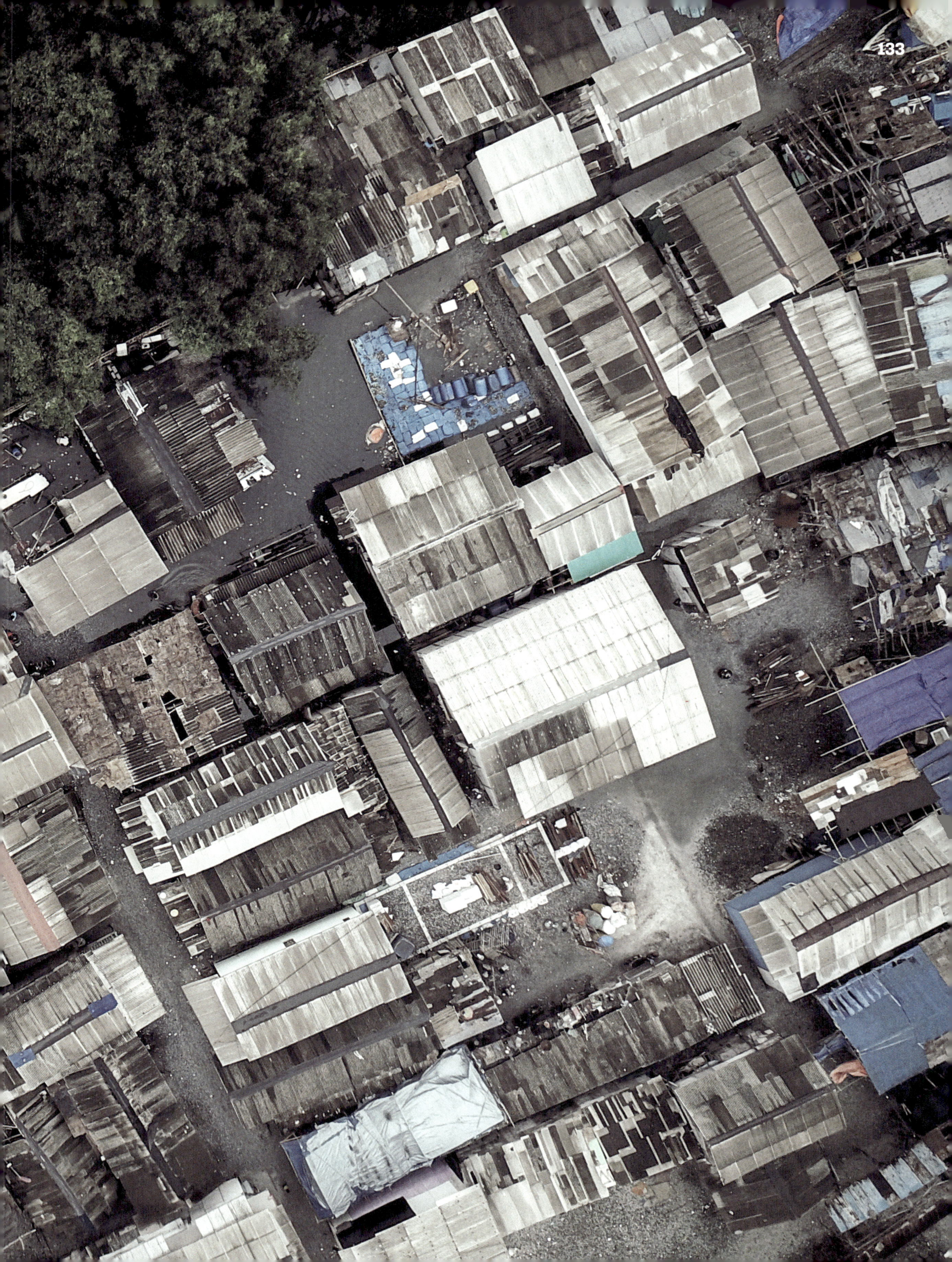

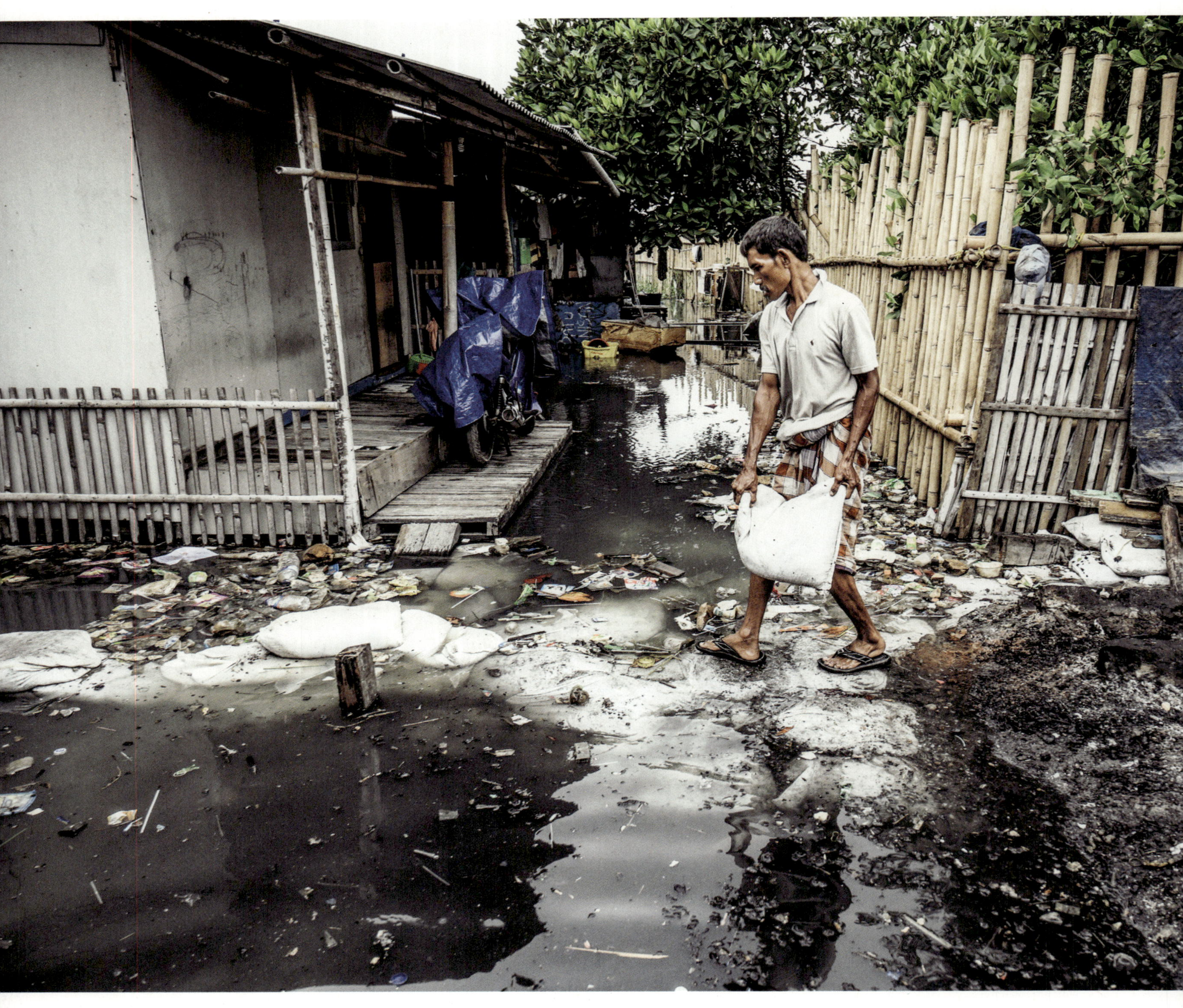

Pak Khalil, a resident of Muara Angke,
tries to protect his house from flooding
with sandbags.

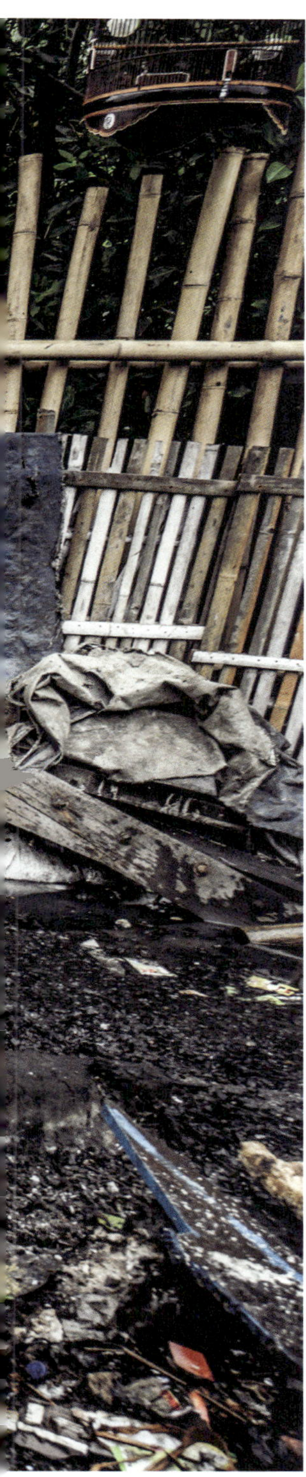

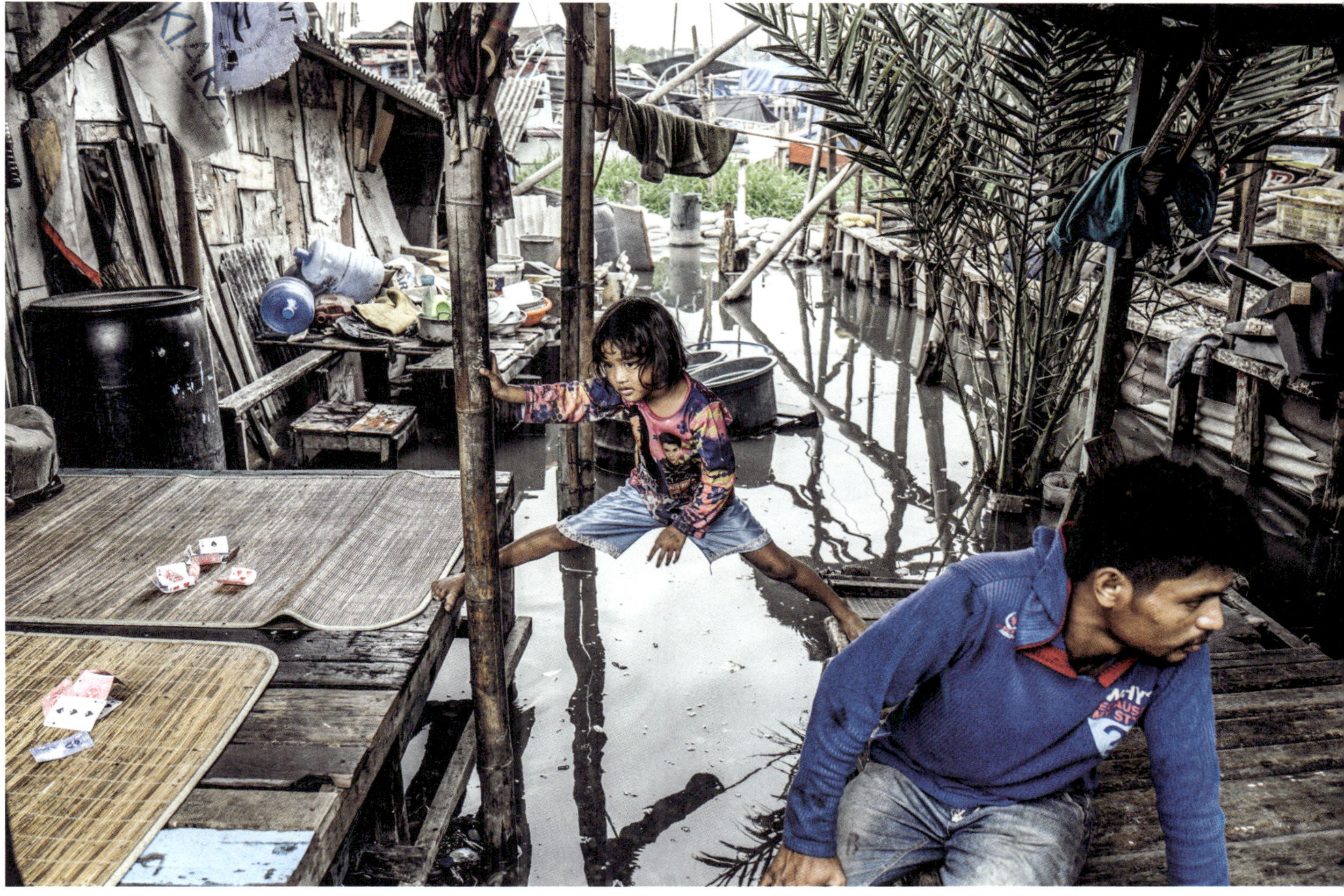

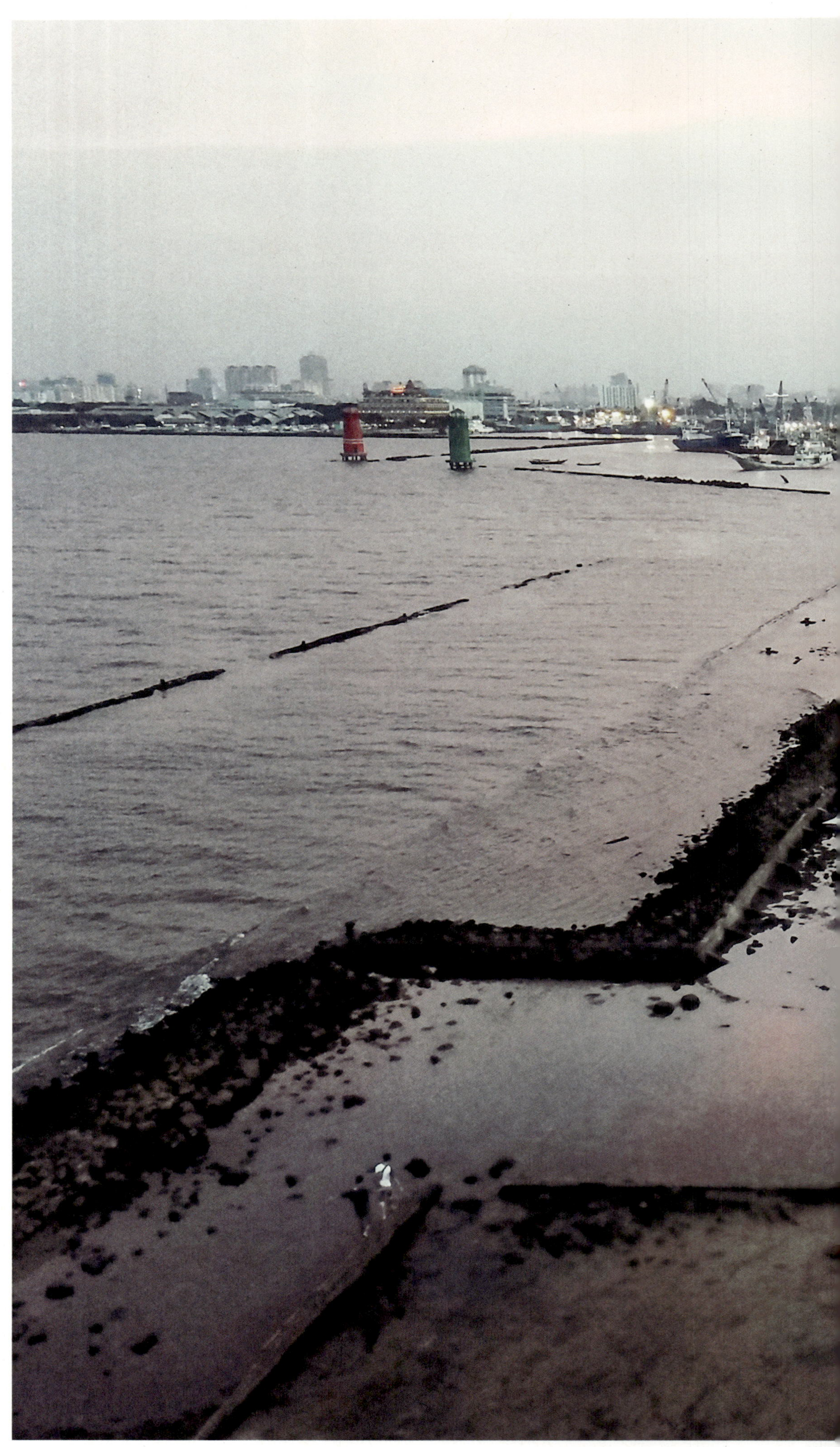

A newly built sea wall in Pluit. The mosque has been abandoned.

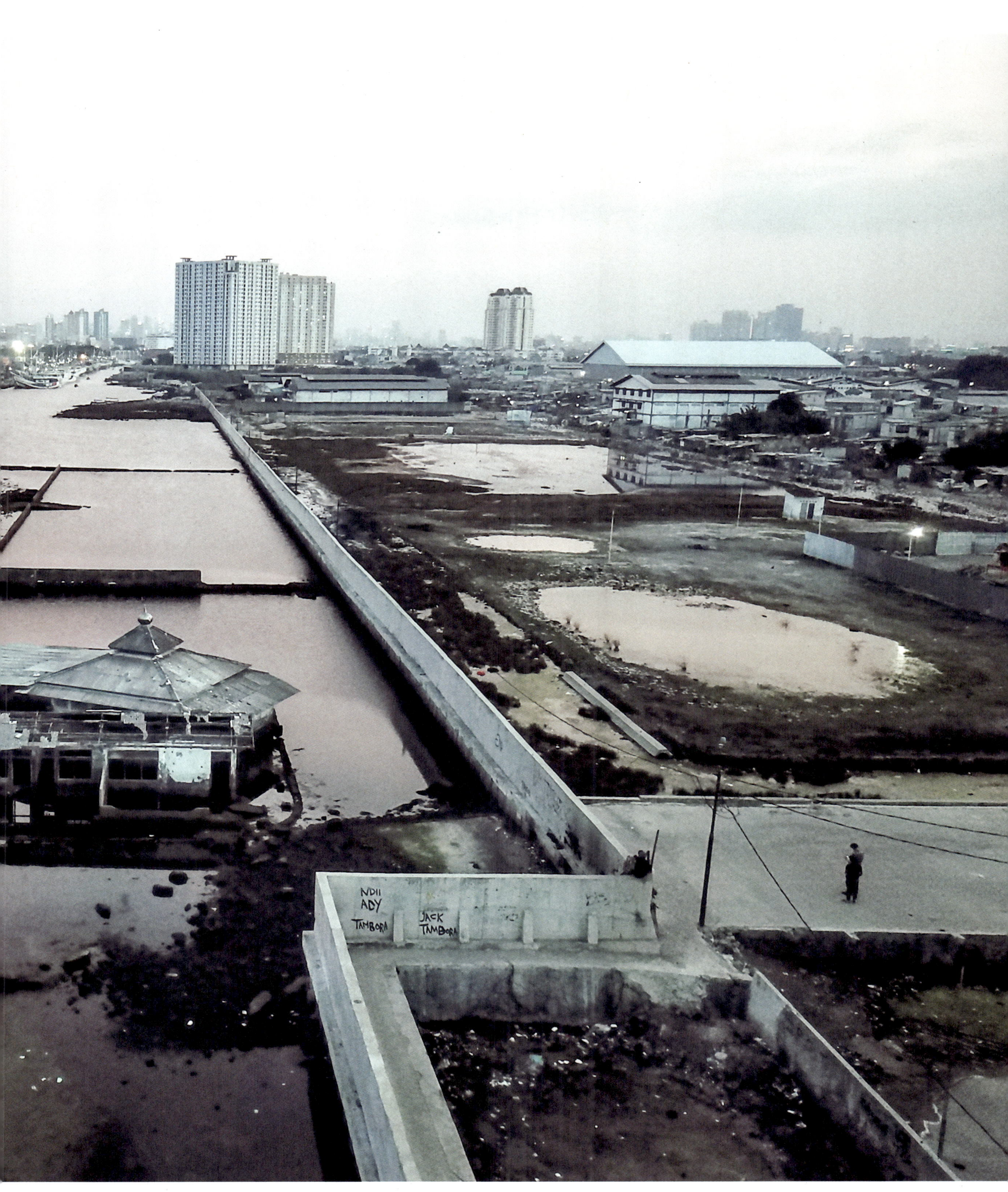

NDII
ADY
TAMBORA
JACK
TAMBORA

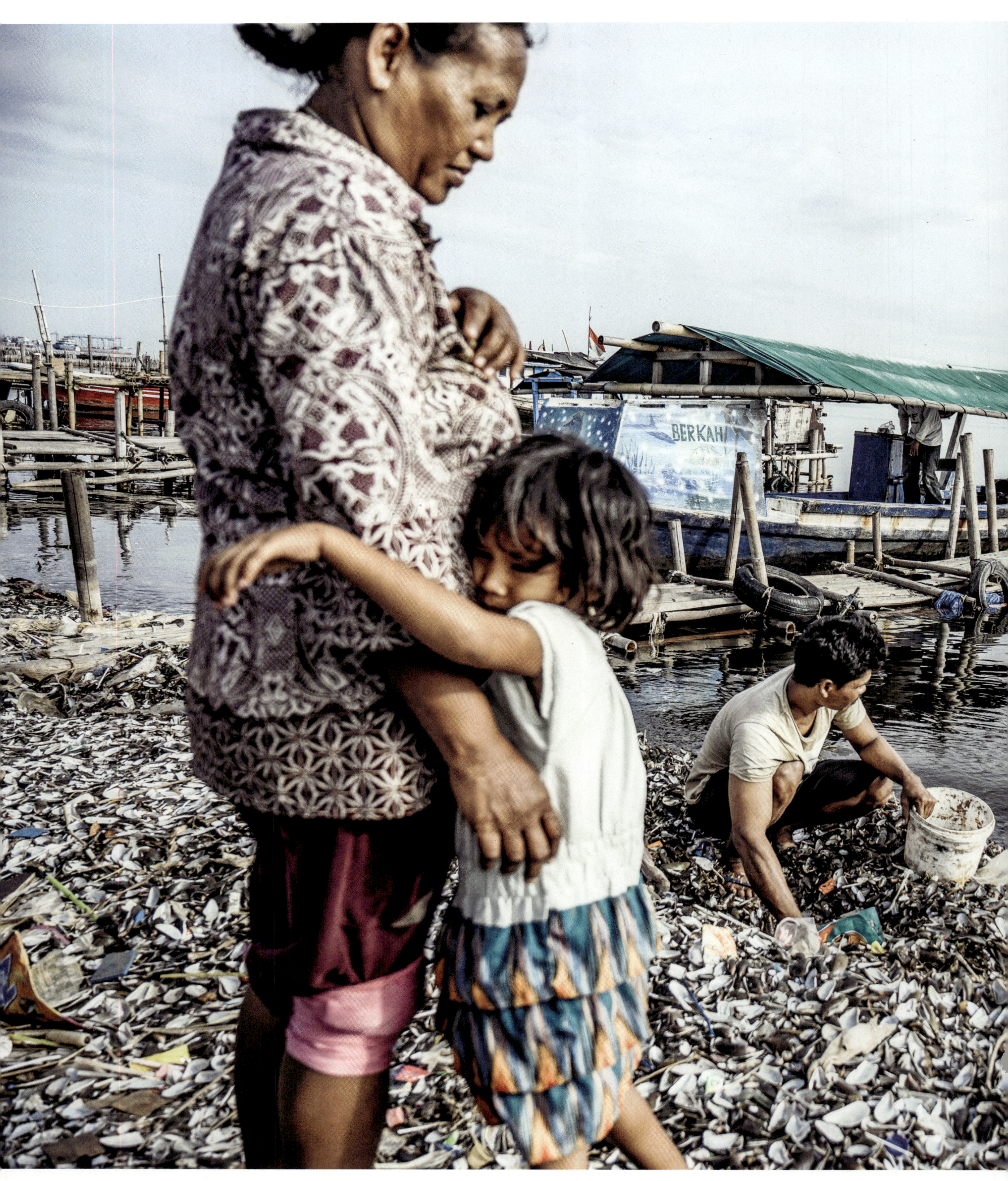
BERKAH

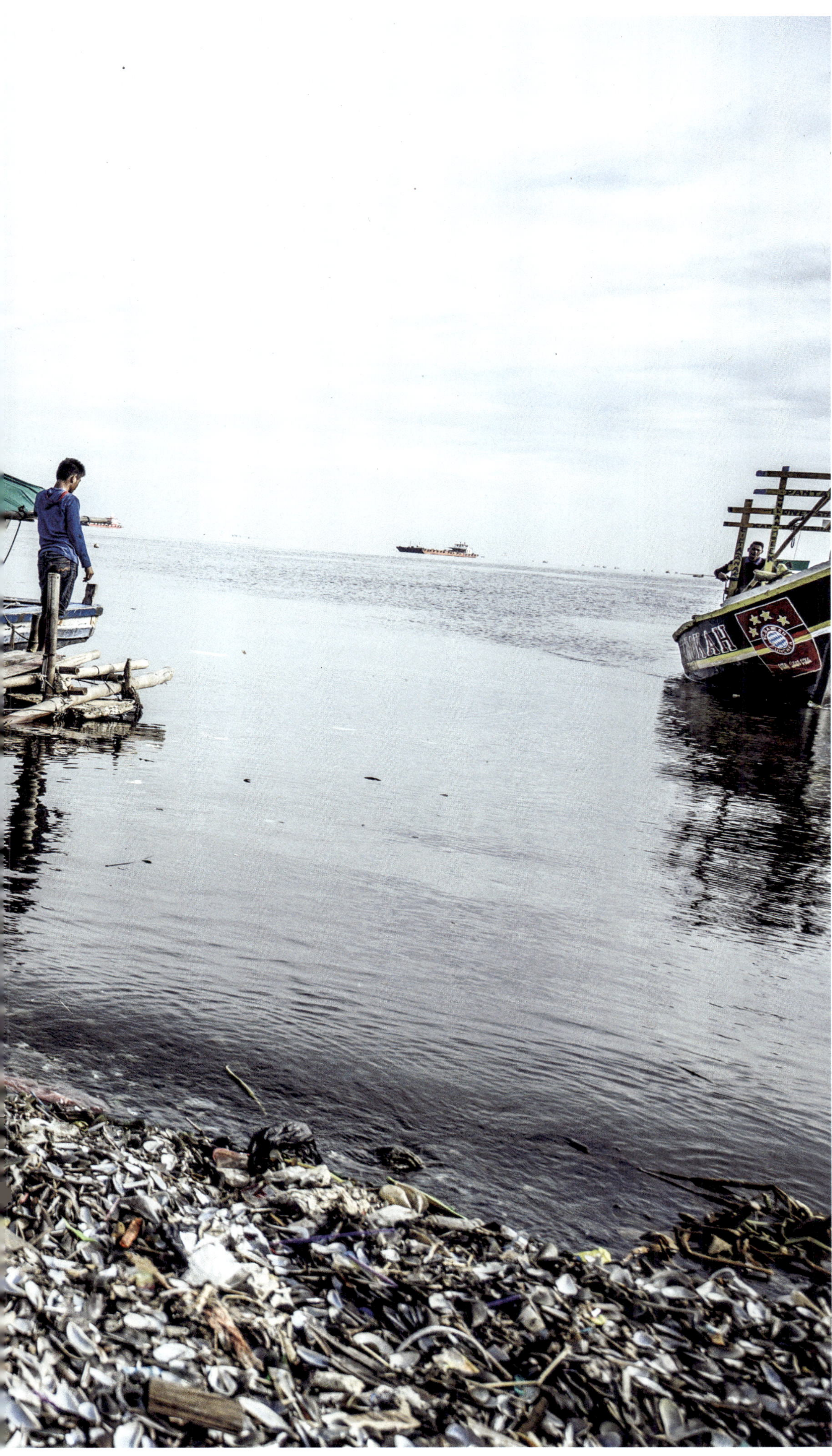

The beach of Muara Angke.

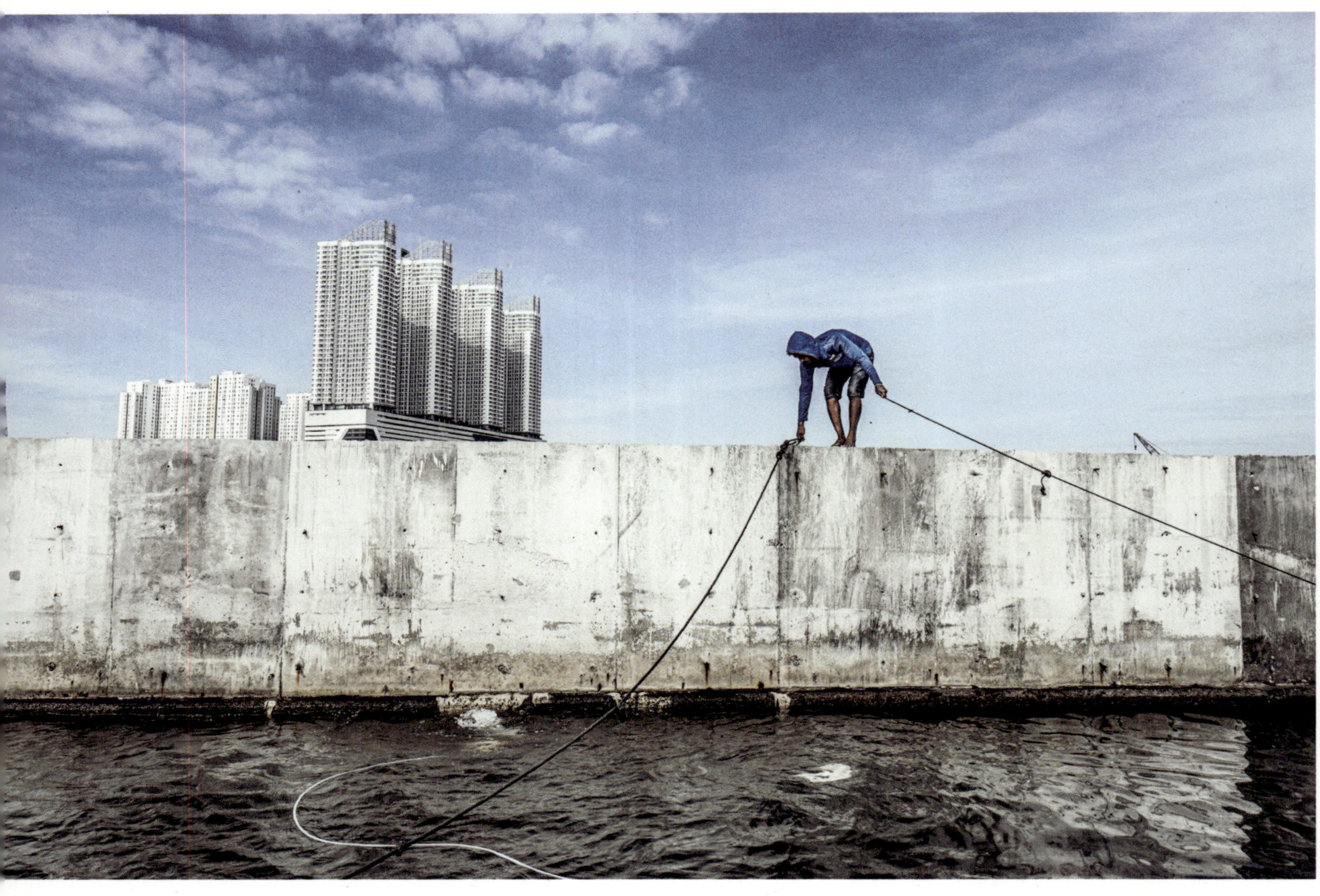

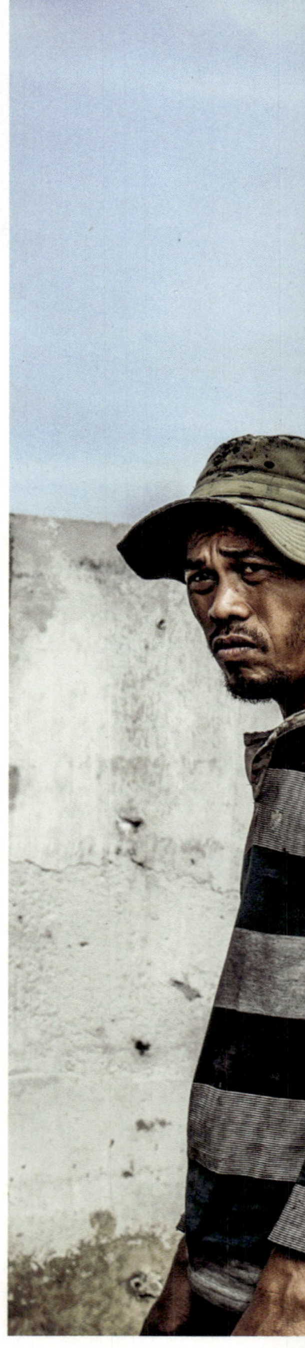

Fishers off the coast of Muara Angke resist the closing of the bay to protect Jakarta from rising sea levels, because they will lose their fishing grounds.

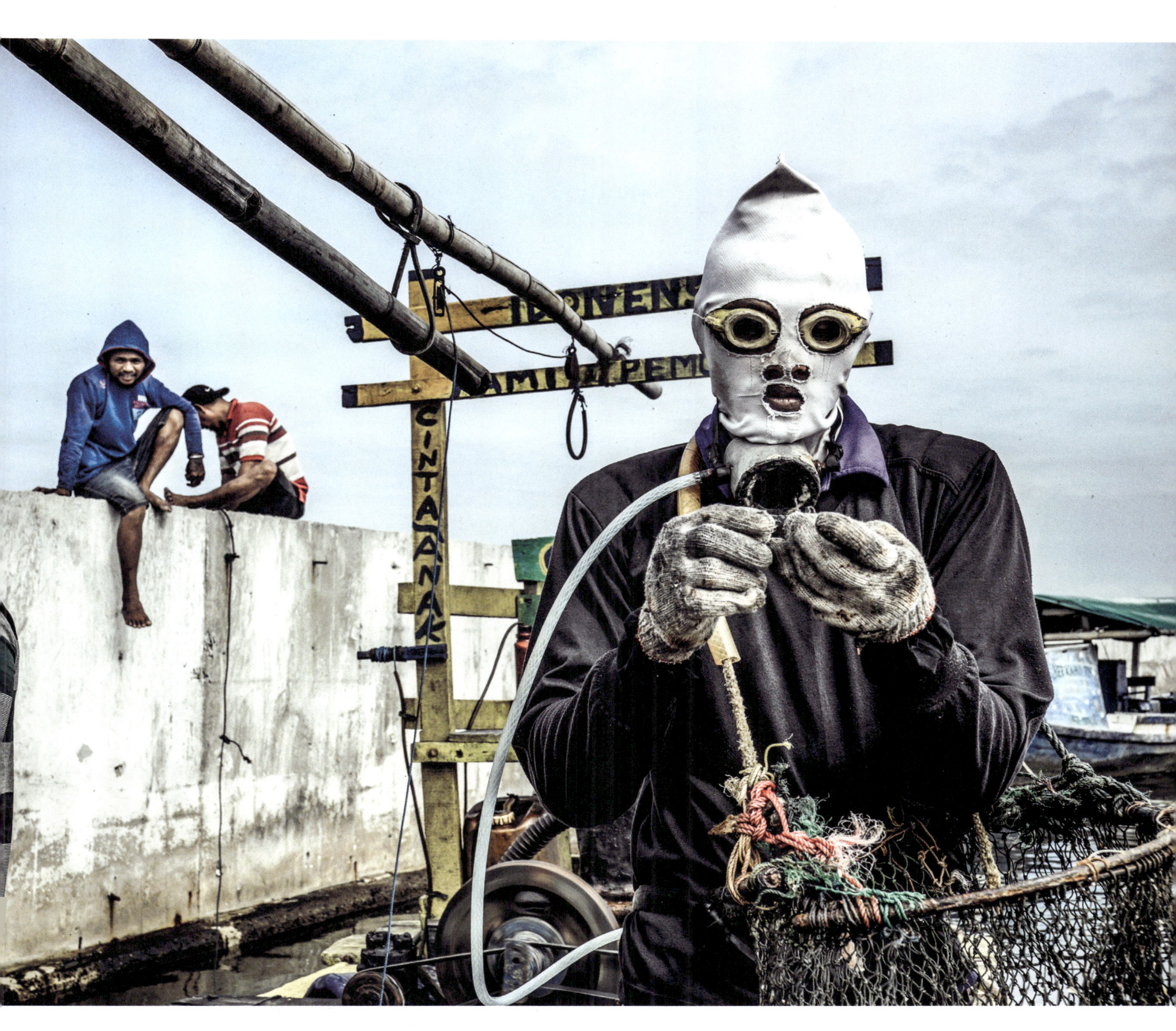

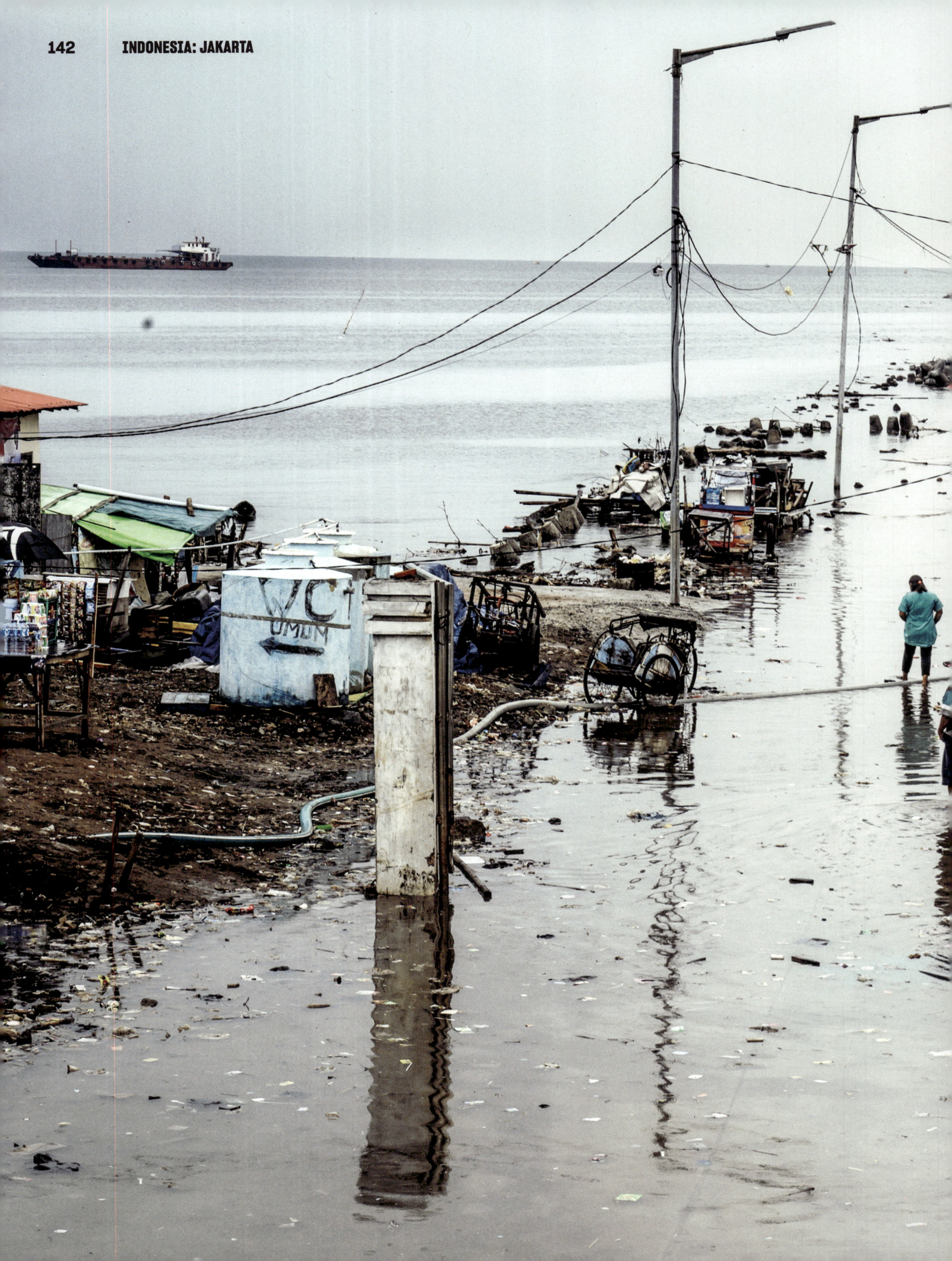
WC
UMUM

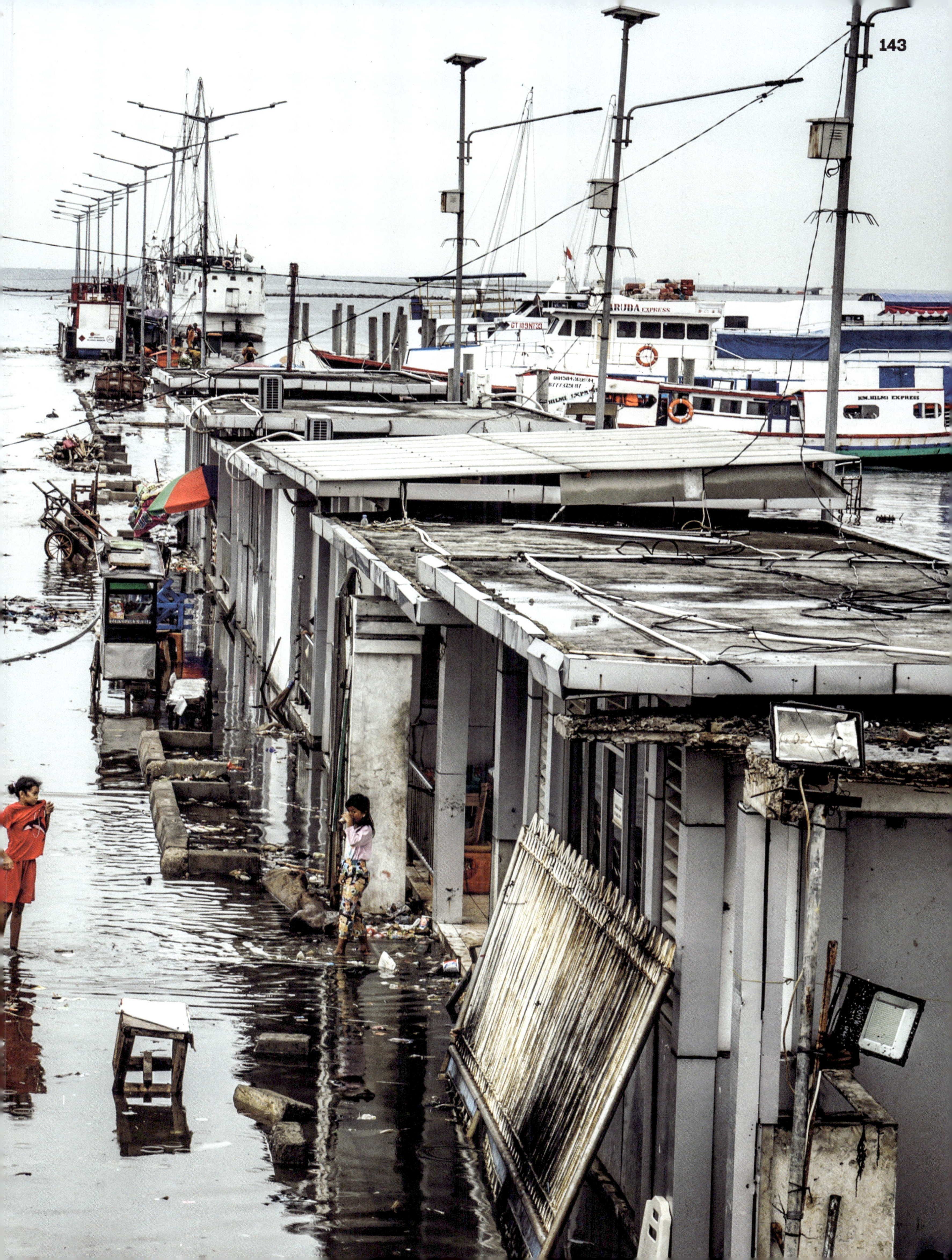

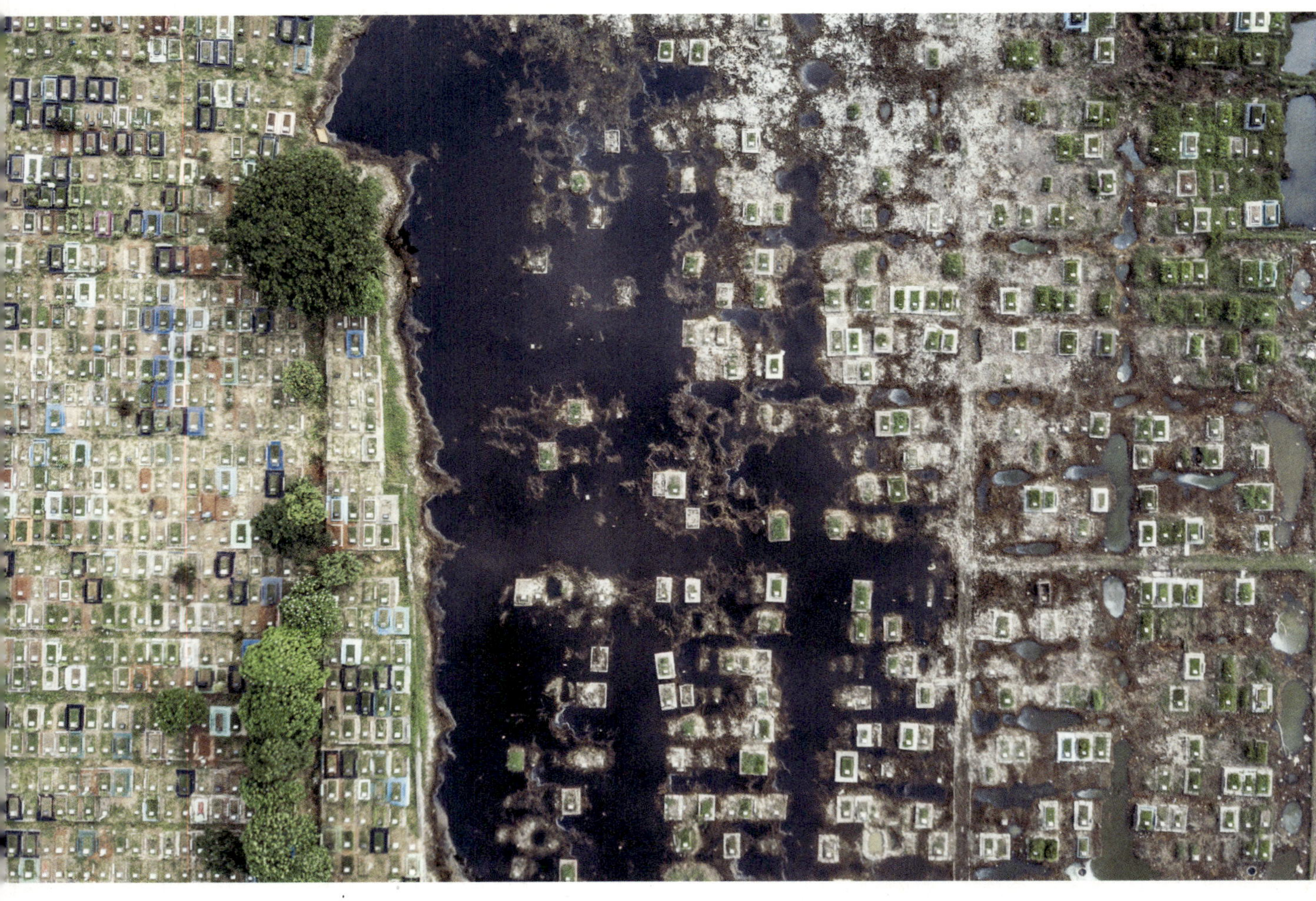

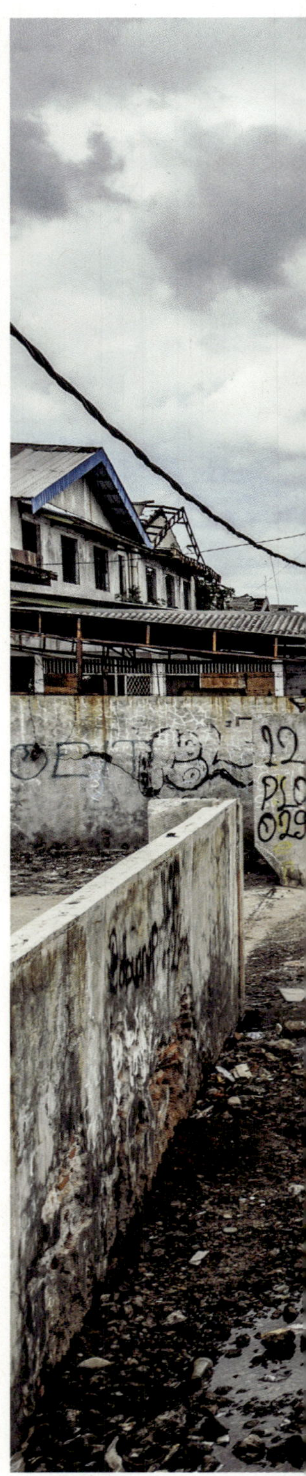

A cemetery in Northern Jakarta is
partly flooded due to the sinking of
the city. Sea level rise will only make
it worse.

< Pages 142-143: The ferry pier off
of Muara Angke during high tide.

A newly built sea wall in Pluit is not very effective. Pluit is an area in the north of Jakarta that is particularly vulnerable to sea level rise and sinking. The city is fortifying the area with a new sea wall. Until now, millions of people have been protected by only a small wall.

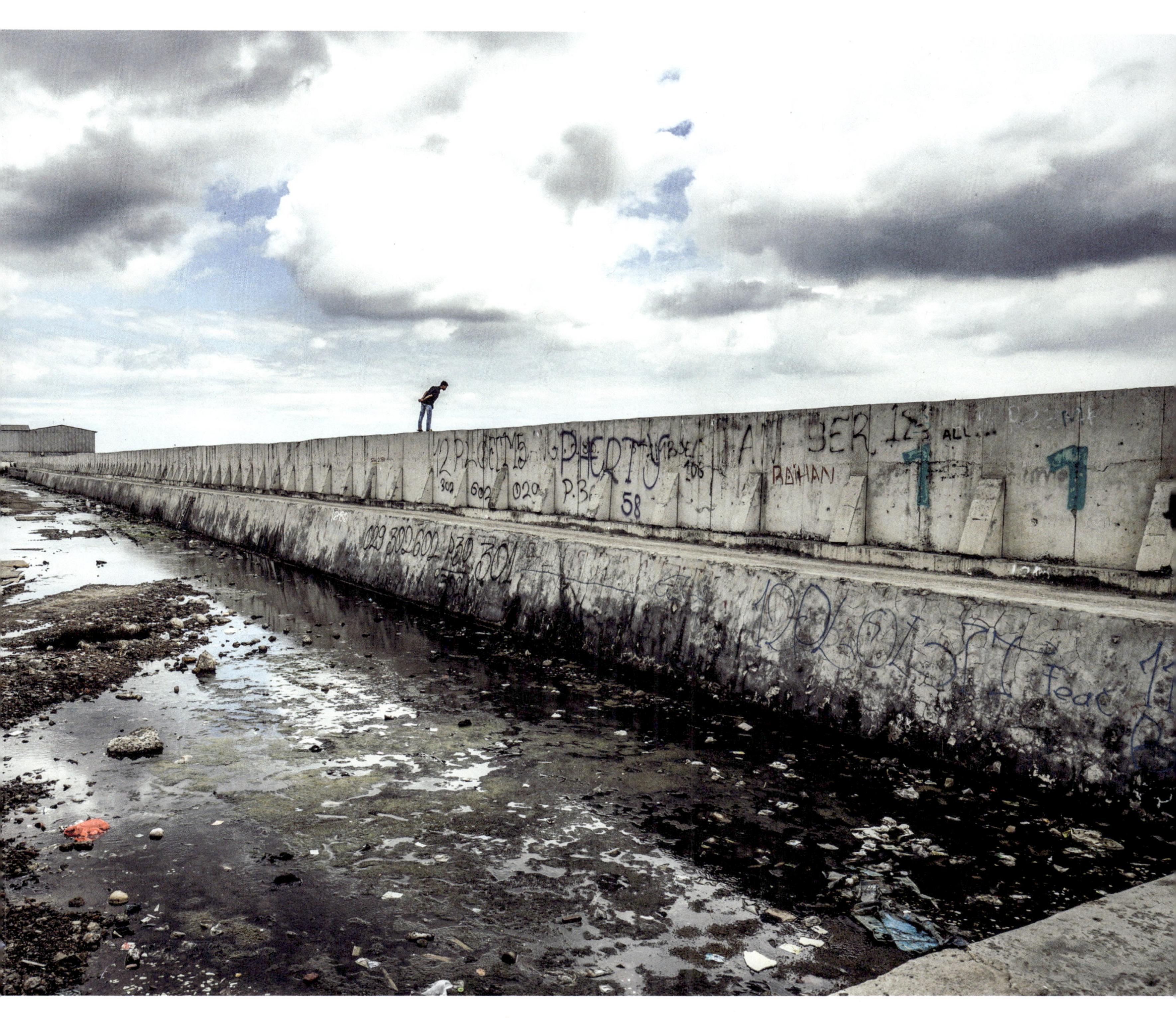

PACIFIC: KIRIBATI
Anote Tong

Kiribati is on the front line of the climate crisis. The impact of storm surges and coastal erosion is already very visible across the land, and the sea level is rising. Much of the land is not more than 1.5 m above sea level. Other islands in the Pacific, among them the Carteret Islands (Papua New Guinea) and the Marshall Islands, seem to await the same fate. A few Pacific nations and others will cease to exist, and where they will move to is unknown.

MIGRATION WITH DIGNITY

On coming into office as president of Kiribati in the middle of 2003, my first task was to consolidate my domestic political agenda, and I therefore decided not to attend the United Nations General Assembly (UNGA) meeting in September of that year. Later, as I turned my attention to foreign policy, I noted that much of the international focus, as reflected in the UNGA leaders' statements, was on terrorism, and of course the usual developmental issues that feature in every UNGA debate.

GAINING AWARENESS

In the course of my briefings, I had also become aware of the ongoing discussions on the United Nations Framework Convention on Climate Change (UNFCCC) and, previously, the Kyoto Protocol. However, what immediately caught my attention were the successive reports of the Intergovernmental Panel on Climate Change (IPCC). I noted that the scenarios the IPCC was predicting would have very serious implications for low-lying atoll island countries. I was aware of the concerns the former prime minister of Tuvalu, The Right Honourable Bikenibeu Paeniu, had expressed on this issue when speaking on behalf of the smaller Pacific Island nations at the Fourth session of the Conference of the Parties (COP 4) to the UNFCCC in November 1998 in Buenos Aires, Argentina. Unfortunately, he did not stay in office long enough to maintain the momentum of his campaign. I was also well aware of the ongoing controversy regarding the validity of the science on climate change, which we now know was orchestrated by the energy corporations, who saw this emerging concern as a threat to their interests. I firmly believed, however, that even if there was no consensus on the science of climate change, the threat posed by the impacts being predicted, even if only a remote possibility, were far too serious for low-lying atoll island countries like Kiribati, Tuvalu, and the Marshall Islands to be ignored and far more relevant to us than the threat of terrorism.

TAKING ACTION

In my first statement at the UNGA in 2004, I made my first attempt at drawing international attention to the human dimension of the climate crisis, what it will mean for those people and countries on the front line of the impacts. Until then, the focus had been on the fascination with the science and the impacts on the environment, and of course what it would mean for the polar bears. By my recollection, there was hardly ever any mention at the highest international political level of the likely implications for those people whose home islands would be destroyed. Ever since then, I have taken every opportunity at every international forum to emphasize the existential threat posed by the climate crisis to countries on the front line. No doubt also realizing their own vulnerabilities, other Pacific Island countries and the Maldives in the Indian Ocean soon picked up the campaign and added momentum to the call for action on the climate crisis.

IMPACTS IN KIRIBATI

What has been our experience of the climate crisis in Kiribati? For anyone to truly understand why we should be concerned, one needs to have lived on an atoll island. An atoll island is simply a solid coral formation (usually below the high-water mark) and an accumulation of coral sand (gathered by constant wave action) atop a submerged seamount. The narrow strips of land surround an almost enclosed lagoon and are on average about 2 m above sea level. Kiribati straddles both the equator and the International Date Line, and, being in the doldrums, is not prone to cyclonic storms.

Throughout my own lifetime, I have seen storms (gale-force winds of around 30 km/hour), the occasional flood, waves overtopping the land and erosion of the coastal areas, sometimes resulting in the destruction of homes and food crops. These events are not new experiences to those living on such fragile atoll islands. I am often asked what I consider to be a somewhat silly question: 'Have you seen any rise in sea level?' I don't believe anyone can see a rise in sea level as one may see a rise in a cup or a tank. Perhaps a more relevant question would have been 'Have you experienced any changes in the severity of these impacts?' And the answer to that is definitely yes.

GOVERNMENT ASSISTANCE

During my almost 13 years in office, as a government, we were inundated by requests for assistance from communities seeking protection against erosion that was threatening their homes, their community infrastructure and the source of their livelihoods. Some village communities also needed assistance relocating from their once-thriving community. As a government, we also had to dedicate huge amounts of resources building sea walls to protect public infrastructure. Due to the limited resources available for sea defences, the government did not provide assistance for protecting private property.

Within the last two decades, there have been occasions when a state of emergency had to be declared and assistance provided for communities on islands which had experienced severe floods resulting in unprecedented damage to homes, food crops and

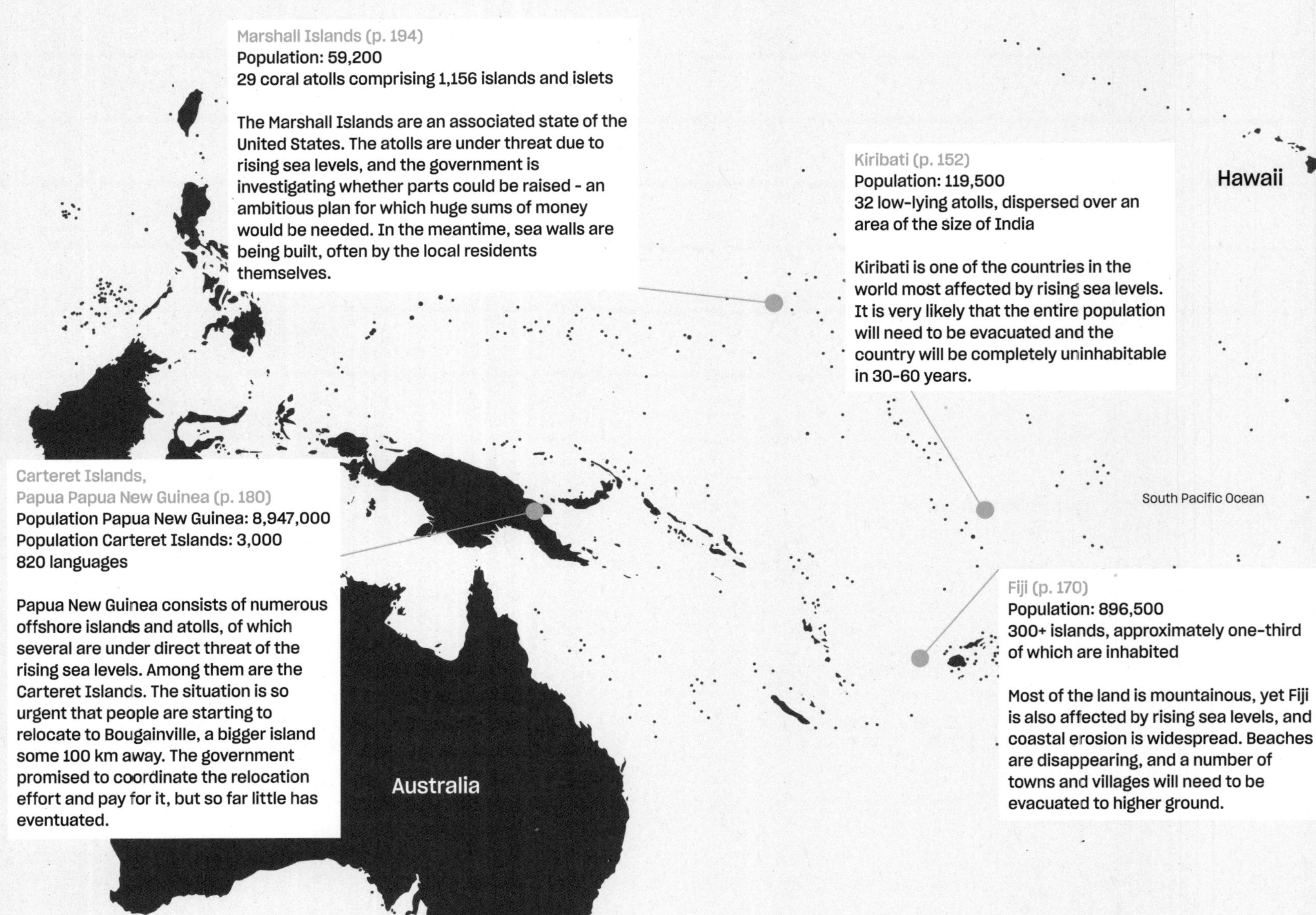

the ever-so-precious underground aquifer, our tradi-tional source of potable water.

In 2015, Cyclone Pam devastated the islands of Vanu-atu, then veered northwards and seriously flooded all of the islands of Tuvalu, together with our southern-most islands, and, to a lesser but still unprecedented extent, the rest of the Gilbert Islands group, damag-ing decades-old food trees. This incident reflects a serious departure from the normal weather pattern, a trend which may in fact pose a more immediate threat than the more gradual rise in sea level. Of course, both in combination would be devastating.

THE GLOBAL COMMUNITY

As we were experiencing these events in Kiribati, other events were also unfolding in other parts of the world. Hurricane Katrina hit the southeastern United States in August 2005, resulting in destruction on a scale which even the most developed country on the planet had difficulties dealing with. But of course, this was at a time when the United States did not believe in climate change. The chronic summertime bushfires kept happening in the United States and Australia, but so what? Heatwaves of unprecedented magnitude started sweeping across Europe, Japan, India and other parts of the world, resulting in fatalities. More recently, there have been un-precedented bushfires in California; 13 million acres of land burnt and 2,439 homes destroyed in one of the worst bushfire seasons in history in Australia, another country whose government consistently denies climate change; unprecedented bushfires in Siberia and the Amazon; floods in Central Europe; Tropical Cyclone Idai, the most devastating storm ever to hit Central Africa, in 2019; the unprecedented melting of the Greenland ice sheet and of the Arctic region itself, which is unlikely to be reversed; and the calving of huge masses of ice in the Antarctic.

All of these and more are taking place, initially regarded as part of the normal cycle of annual events, but now that they are happening on a totally different scale, they are beginning to sound alarm bells. The developed world, which once regarded climate change as not only irrelevant but inconvenient to its future and way of life, is now beginning to realize that climate change is a threat not only to those of us in low-lying island countries and that the impacts of climate change are not confined to a rise in sea level.

The special report issued by the IPCC in October 2018 predicted cataclysmic events if the global community does not take urgent and radical measures to stop the ongoing rise in emissions. Since then, we have been seeing a daily release of reports and studies on various aspects of climate change in different parts of the world, clearly indicating that an apocalyptic future awaits us all, perhaps even if, as a global community, we are able to cut emissions in the foreseeable future. Just over the last few years, I have been reading reports suggesting that human civilization as we know it is under threat. The surge in momentum on climate action, especially the active participation of young people, reflects a new and welcome dimension in the movement. The developed world has now begun to realize the existential threat posed by the climate crisis, which is not confined to those of us on low-lying atoll islands. The young are demanding more accountability by those responsible for causing the climate crisis and from those with the authority to do something about it.

CONTEMPLATING THE FUTURE

Considering all of the above - our experiences in Kiribati, Tuvalu, the Marshall Islands and other low-lying atoll island countries in the Pacific; events unfolding in other parts of the world; successive IPCC reports and the scientific information coming out on a daily basis with increasingly dire projections - what is the future scenario likely to be for us? The defiant cry often shouted during climate protest marches is 'We are not sinking; we are fighting!' It is a cry of frustration and of anger at the gross injustice of what is happening, but more so at the casual dismissal of the plight of those of us on the front line of the climate crisis by the very privileged few who uncaringly continue doing and benefitting from what science now confirms to be an attack on the future of our people and indeed all of humanity. It is a cry of anger and frustration against the complicit acquiescence of those governments and administrations which have the capacity to remedy the situation but have chosen instead to become willing victims in their regulatory capture. It is a self-flagellating declaration of our commitment and desire to ensure that our children and their children continue to live on our home islands.

A MULTIPRONGED APPROACH

In the face of these challenges, I have always subscribed to the idea of retaining the integrity of our home islands and to doing whatever is necessary to ensure that our future generations are able to continue to reside on these islands. But in order to adapt and to build the necessary resilience to survive these projected impacts, extremely radical adaptation measures, which would require a huge amount of resources, would be needed. However, based on my two decades of personal experience observing the lack of international response to effectively cut emissions, I am resigned to the reality that the sacrifice of providing us with the resources needed to implement the required adaptation measures is unlikely to be forthcoming from the international community, especially now that developed countries are also facing climate impacts themselves. It was against this background that I had proposed a multipronged approach, such that whilst we should give priority to building climate resilience in order to remain on our home islands, we should also concurrently plan for the relocation of those who choose to migrate. I have been very conscious of the strong opposition by other Pacific Island leaders to the notion of relocation, seeing it as compromising our negotiating position in assigning responsibility for compensation to those responsible for causing our problem. I do not believe that the above options are mutually exclusive, and it is my view that we have to acknowledge the brutal reality that some of our people will choose to relocate, and that if we are unable to build the necessary resilience, most, if not all, will one day have to relocate, whether we like it or not.

Over the years, many of our people have been migrating as a matter of choice, mostly to Fiji, Australia, New Zealand and even the United States, through the Marshall Islands. That is a constitutional right which they cannot be denied. More recently, under the Pacific Access Category (PAC) scheme, there has been a steady outflow of people migrating from Pacific Island countries to New Zealand, taking up their annual country-allocated quota. In Kiribati, we have our share of internal migration, mostly in search of better economic opportunities, resulting in over half of the country's almost 120,000 residents residing in the urban centre under extremely congested conditions. In the 1960s, under the British colonial administration, people from drought-affected islands in the southern Gilbert Islands group were resettled in the Phoenix Islands group and later in the Solomon Islands, also due to severe drought conditions. There has also been a general drift of people from the islands which are more vulnerable to drought to other islands within Kiribati. Relocation is not a new phenomenon for Kiribati or indeed for the rest of the Pacific Islands.

MIGRATION WITH DIGNITY

Therefore, in acknowledging that climate-induced migration is inevitable, from the very first time I raised the issue at my inaugural statement at the 2004 UNGA, I proposed that if and when our people relocate due to the climate crisis, they must do so with dignity and not as climate refugees; hence the genesis of the concept of 'migration with dignity'. It is simply a proactive response to an inevitable eventuality, involving the upskilling of our people through education and training to gain qualifications acceptable in their countries of destination, so that they can be gainfully employed and not become second-class citizens in their newly adopted communities. Without the proper and necessary preparations, those who migrate will indeed become climate refugees and will most likely be residing in the slum areas of the cities they move to. Migration with dignity, as I defined it, must also involve the preparation of the communities into which our people move in order to ensure smooth integration and avoid the instabilities often experienced in communities with sizeable migrant populations.

RETAINING SOVEREIGNTY

The issue of sovereignty is also an actively debated aspect of the uncertain future facing the small island countries on the front line of the climate crisis. Various formulations of how and what will happen have been hypothesized. Can we relocate as an entire nation and establish a sovereign nation in another country? Unlikely. What is to become of our extensive exclusive economic zones (EEZs), with their abundant resources? A number of formulations have been suggested, including being allowed free access into countries like Australia in exchange for control of the EEZs. Another suggests the formation of compact-type arrangements along the lines of existing arrangements between the island countries in Micronesia and the United States, and between some Polynesian island countries and New Zealand in the South Pacific. Needless to say, the reaction from the island countries has been a clear rejection of the possibility of losing their sovereignty and an unequivocal desire to retain the integrity of their home islands so that even those who choose to migrate will always have their homeland to return to.

I believe that for Kiribati it is possible to generate resources to build the resilience needed to maintain our sovereignty, perhaps for the next century. I do not see the likelihood of being able to do so for all of the islands in Kiribati, but it is critical that we do what is necessary to maintain our EEZs. The resources needed to build climate resilience must be sourced from our abundant fisheries resources and deep seabed minerals. In order to do so, we must strive to obtain a higher rate of return from our multi-billion-dollar fishing industry and our extensive fisheries, potentially by restructuring existing partnerships. With respect to the ongoing debate over mining of the deep seabed, Kiribati, like other small island nations with abundant deep seabed minerals, will have to make a decision between protecting the living organisms on the deep seabed or protecting the future of their people. I have also suggested that, if island countries facing the existential threat from the impacts of the climate crisis are to be dissuaded or prevented from mining the deep seabed, then they must be duly compensated for not doing so.

LEADING BY EXAMPLE

In the course of the almost 13 years of my presidency, in addition to my passionate advocacy on the climate crisis, I was also involved in two somewhat radical ventures, over which I continue to be questioned to this day. The first was the designation of the Phoenix Islands Protected Area (PIPA) in 2008, which was inscribed as a World Heritage Site by UNESCO in 2010. At the time, it was the largest marine protected area in the world, comprising more than 400,000 km^2 of the most fertile fishing ground in the centre of the Pacific Ocean, where no extractive industries are permitted. That was our sacrifice to humanity and our climate crisis challenge to the global community to say if even a small country like Kiribati can make a contribution to the conservation of the planet, then why can't you? The second was the purchase of a more than 6,000 acre plot of land in the island of Viti Levu in Fiji in 2012. Immediately, I was swamped with media enquiries asking if my intention was to relocate our people due to the climate crisis. My honest answer was that I had no plans for the relocation of our people there. I said it was a long-term investment, which was true enough, as the price we paid was around A$1,500 per acre, compared to what one would pay in Tarawa, around five to 10 times that amount. But the real objective was to draw the attention of the global community to the reality of the existential threat we face from the climate crisis and challenge them to step up to their responsibility. Even if they will not assist us, we have to try to resolve our own challenges, as the alternative is unacceptable. I also believed that if and when it became necessary for our people to relocate, it would be a standby safety net for our people should they become displaced in the future. And the most gratifying response of all has been the publicly announced offer by Fiji to accommodate our people and those from Tuvalu if ever they should become displaced due to the climate crisis. So far, Fiji has been the only country to have risen to the greatest moral challenge facing humanity. •

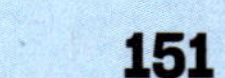

Kiribati: Tarawa — Sea level + 1.0 m

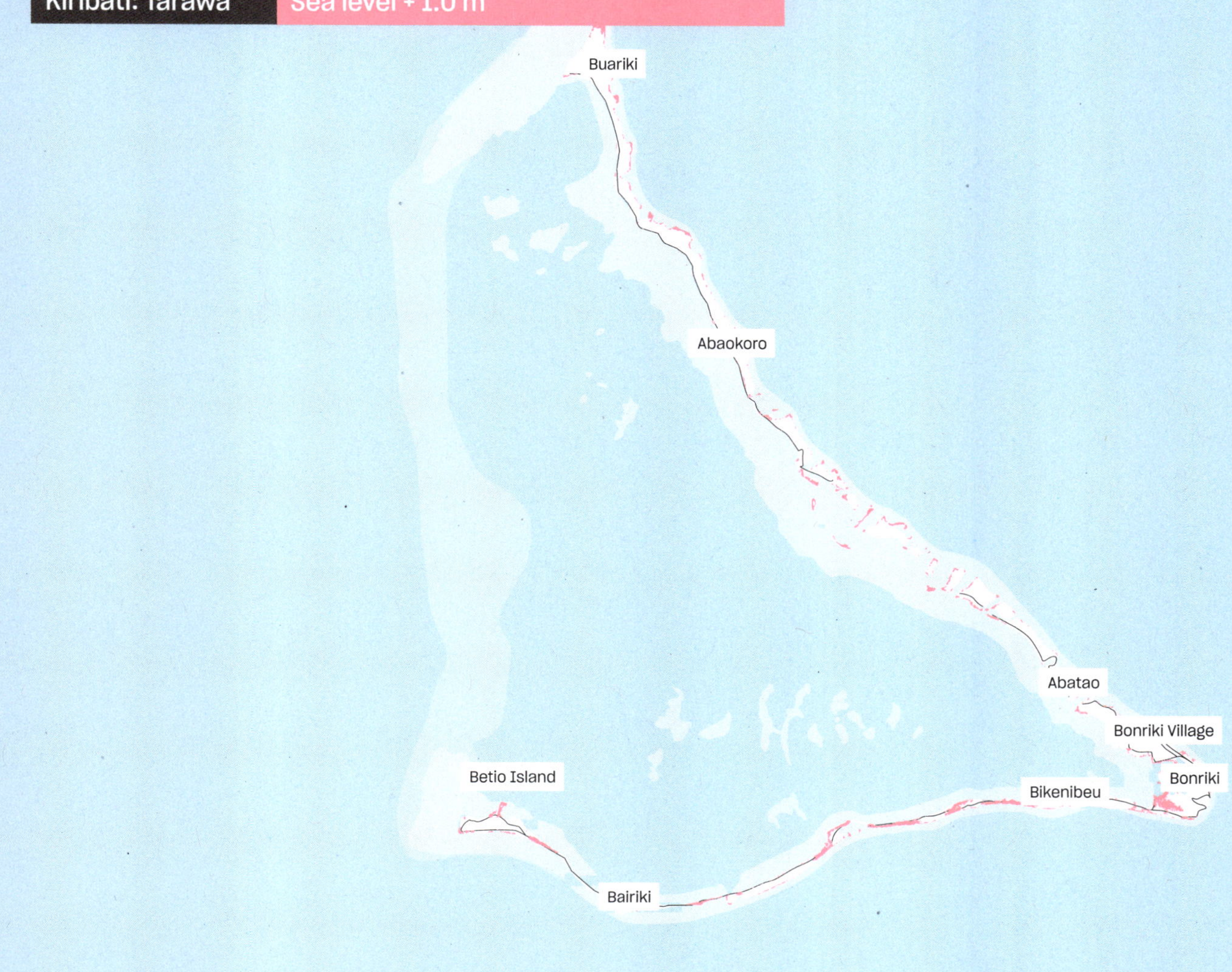

Kiribati: Tarawa — Sea level + 2 m

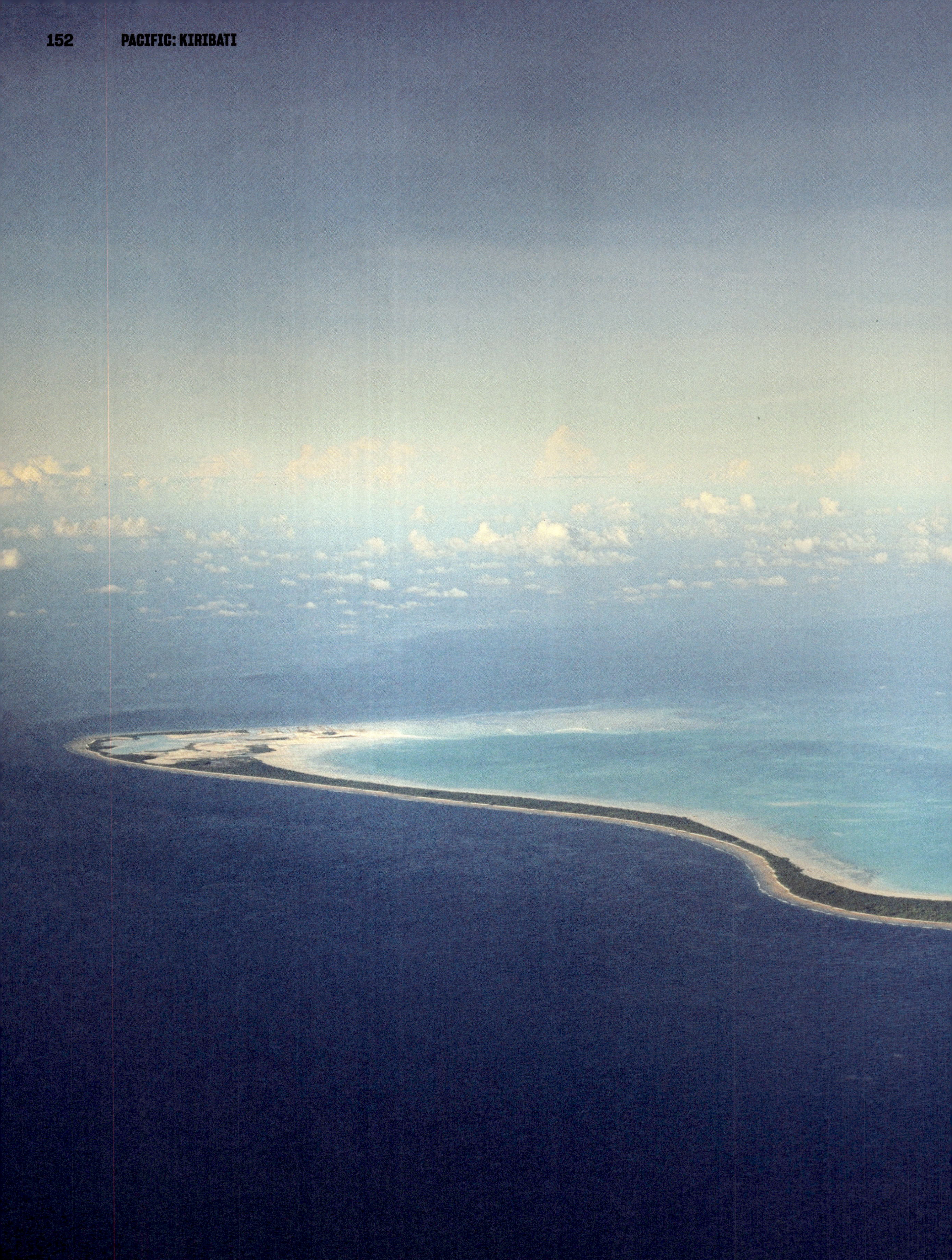

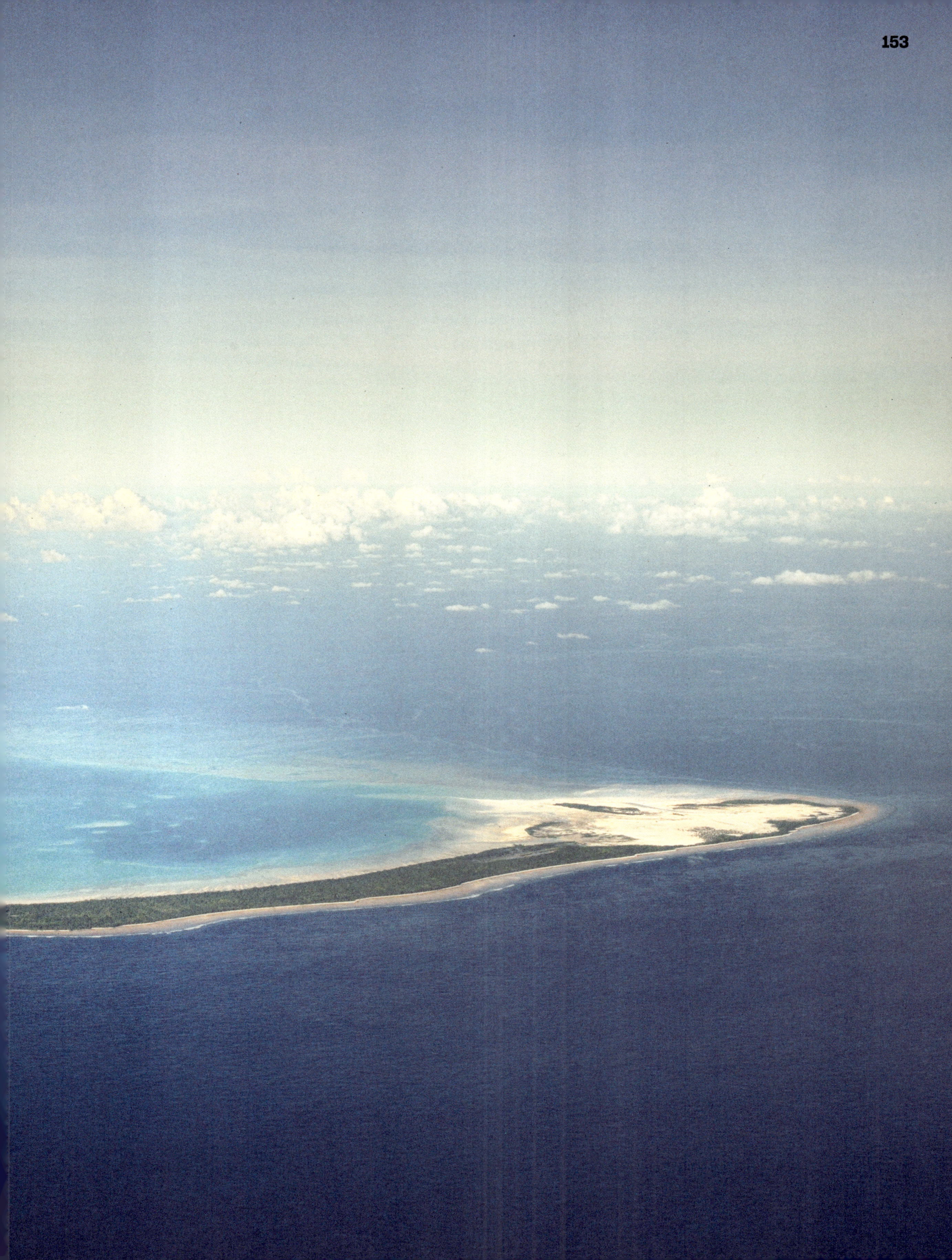

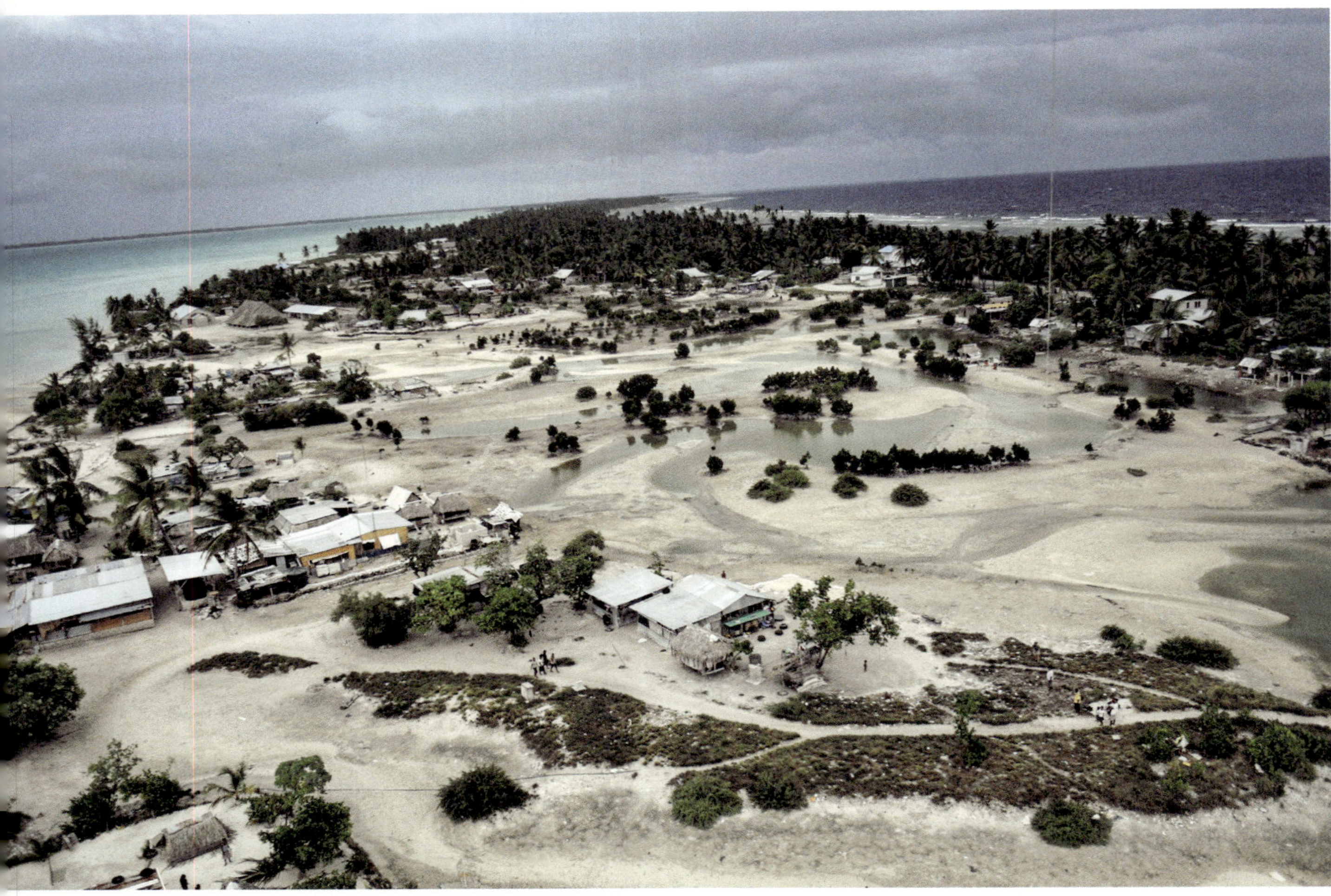

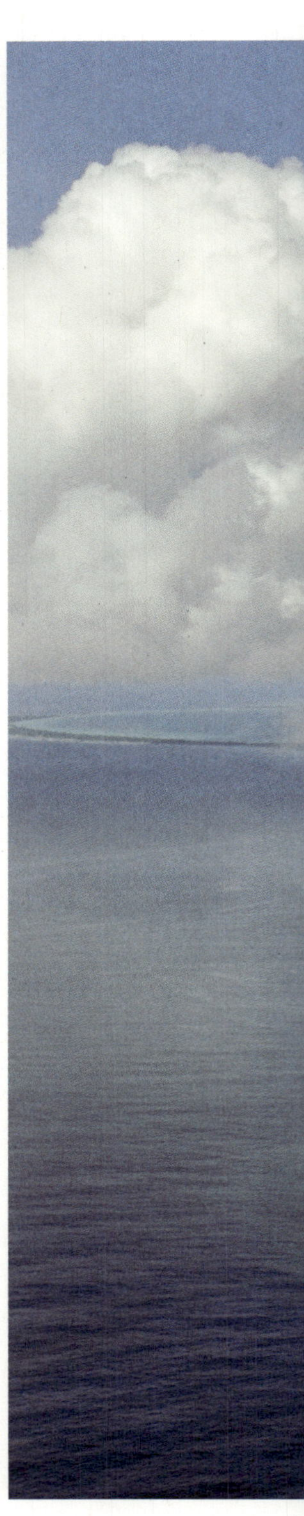

Tebike Nikoora ('Golden Beach') is a residential area in the densely populated part of South Tarawa. People live here under very vulnerable conditions.

< Pages 152-153: Maiana, just south of Tarawa Island. Tarawa is where the government resides and almost 60,000 people live, out of a total population of 119,500.

The south of Tarawa Island; on the right is the international airport.

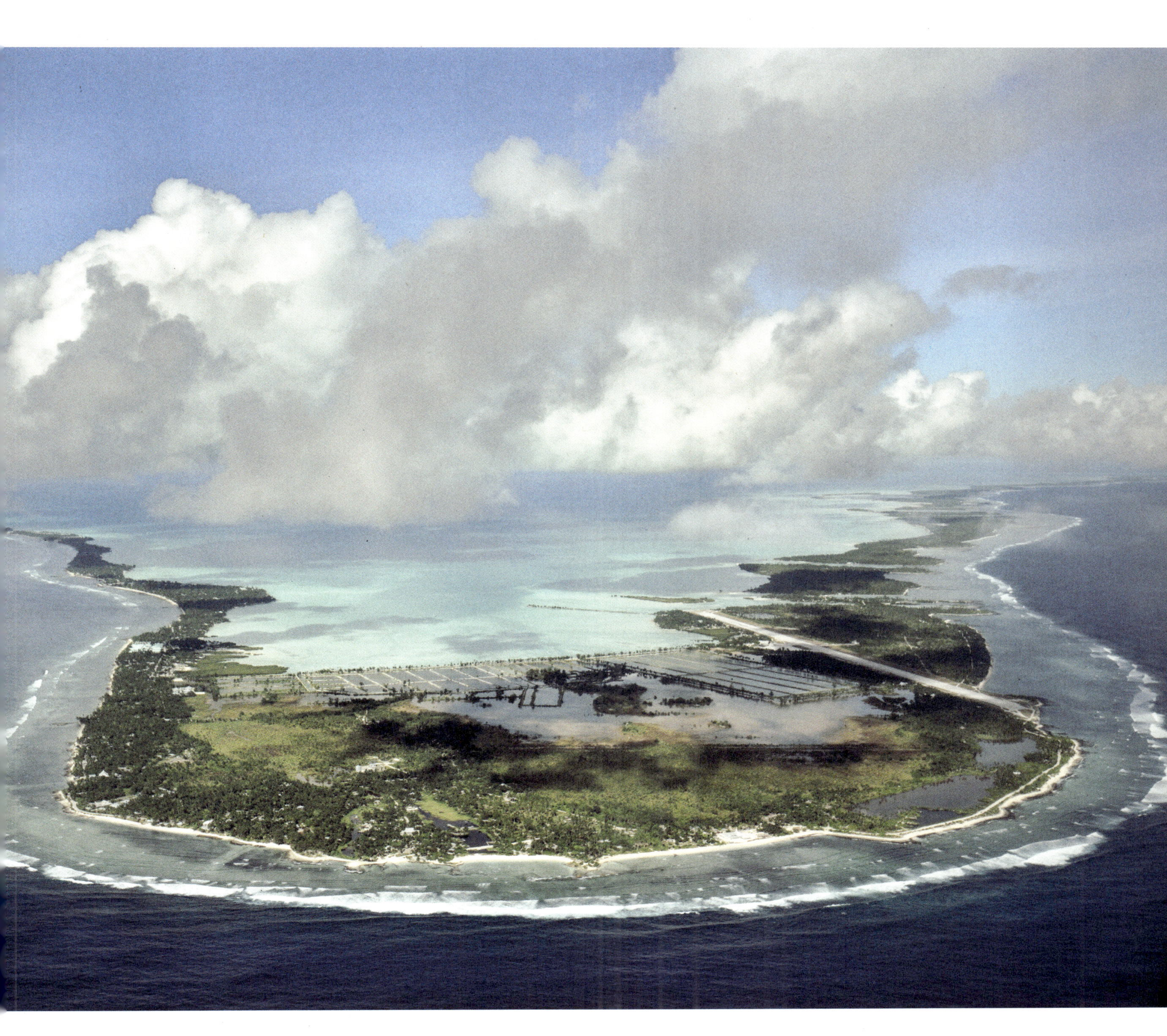

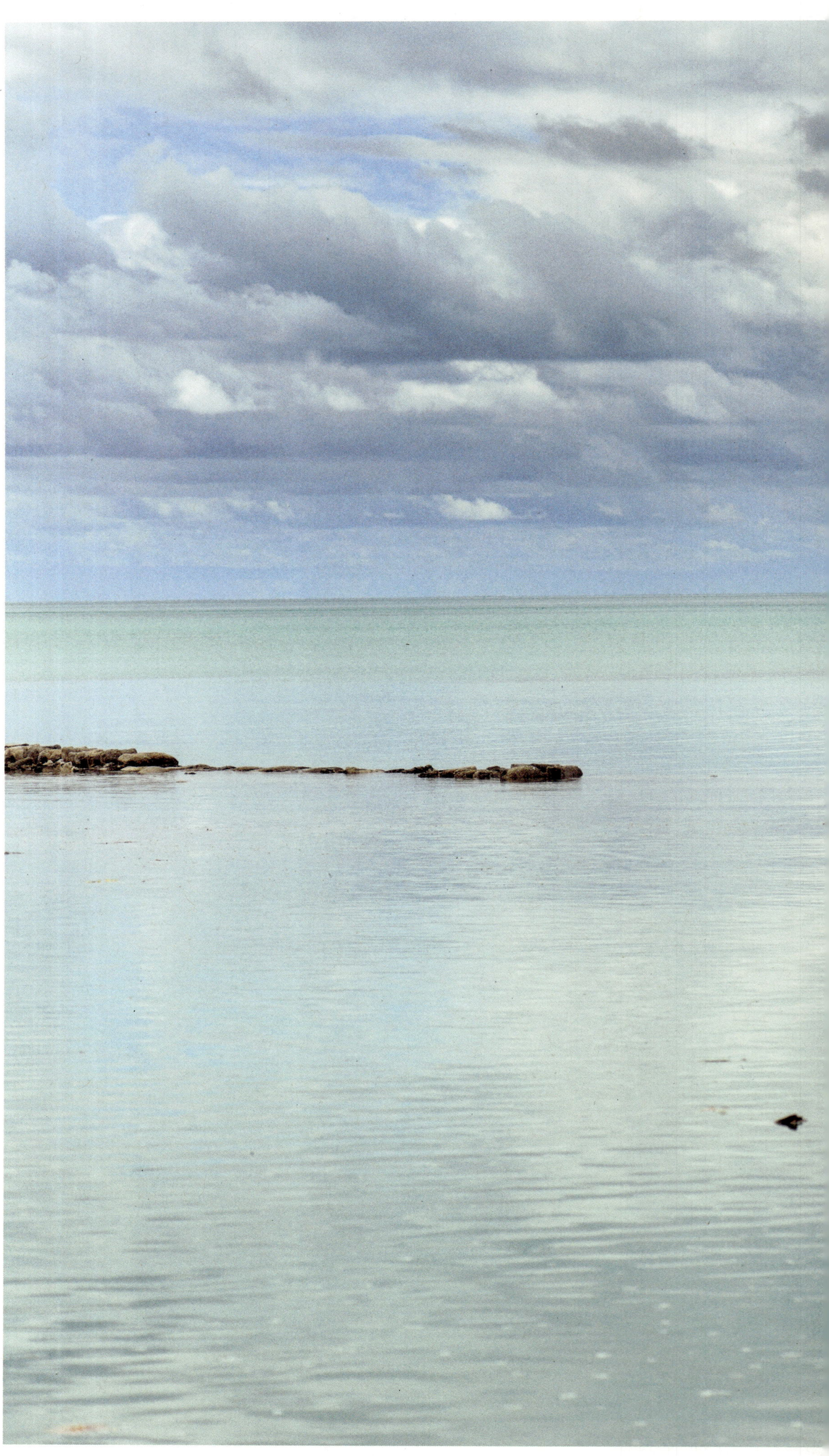

A failed attempt to reclaim land in
the lagoon of Tarawa.

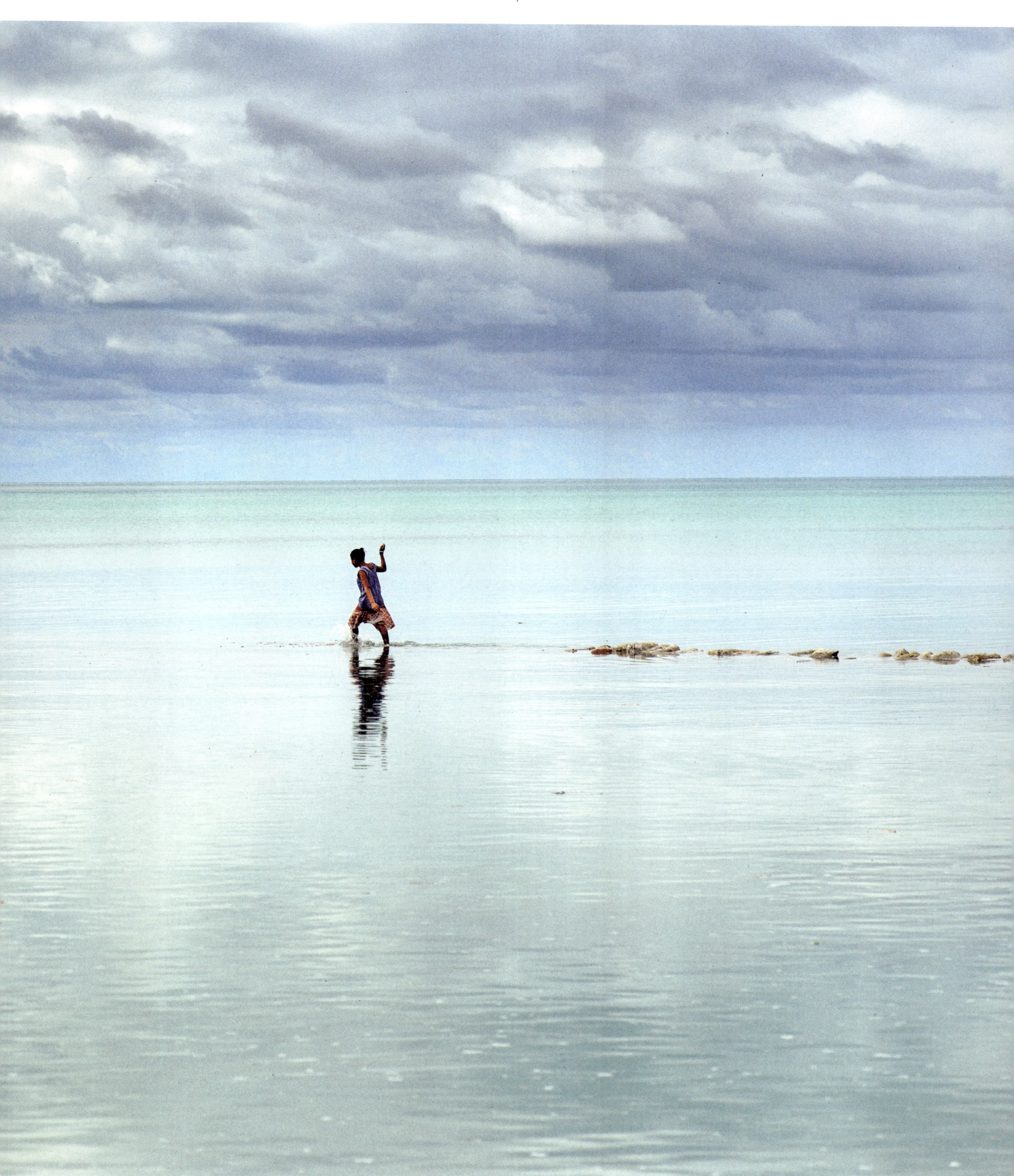

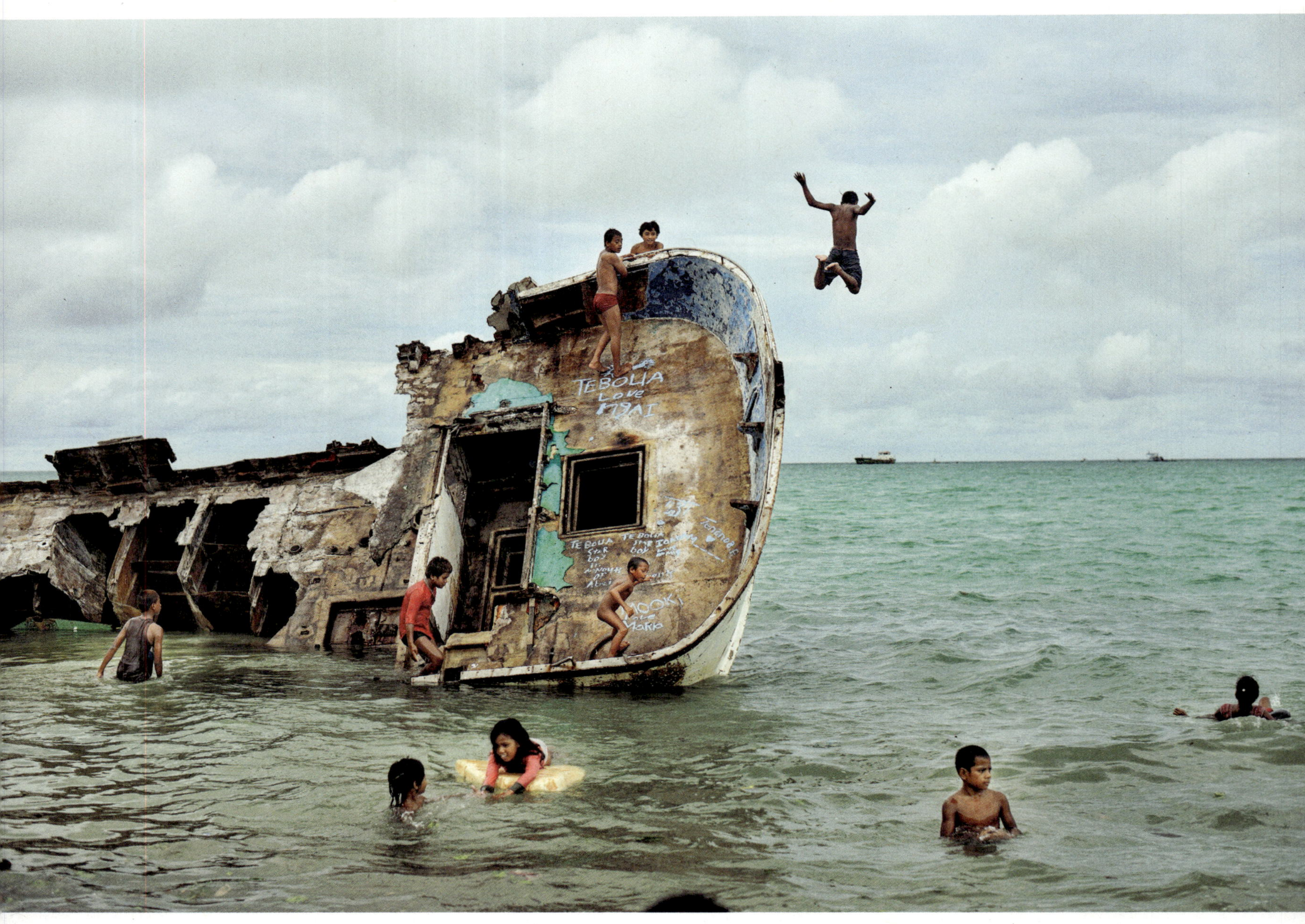

A Taiwanese fishing trawler was swept onto the beach, after it was previously stranded on the reef. Due to rising sea levels and a higher frequency of storms, the reef is not the natural protector that it used to be.

Residents protect their homes from the rising sea levels with sandbags on Betio Island, the most populated part of Tarawa. Due to seawater intrusion and the high population, there is a serious lack of drinking water and agricultural land.

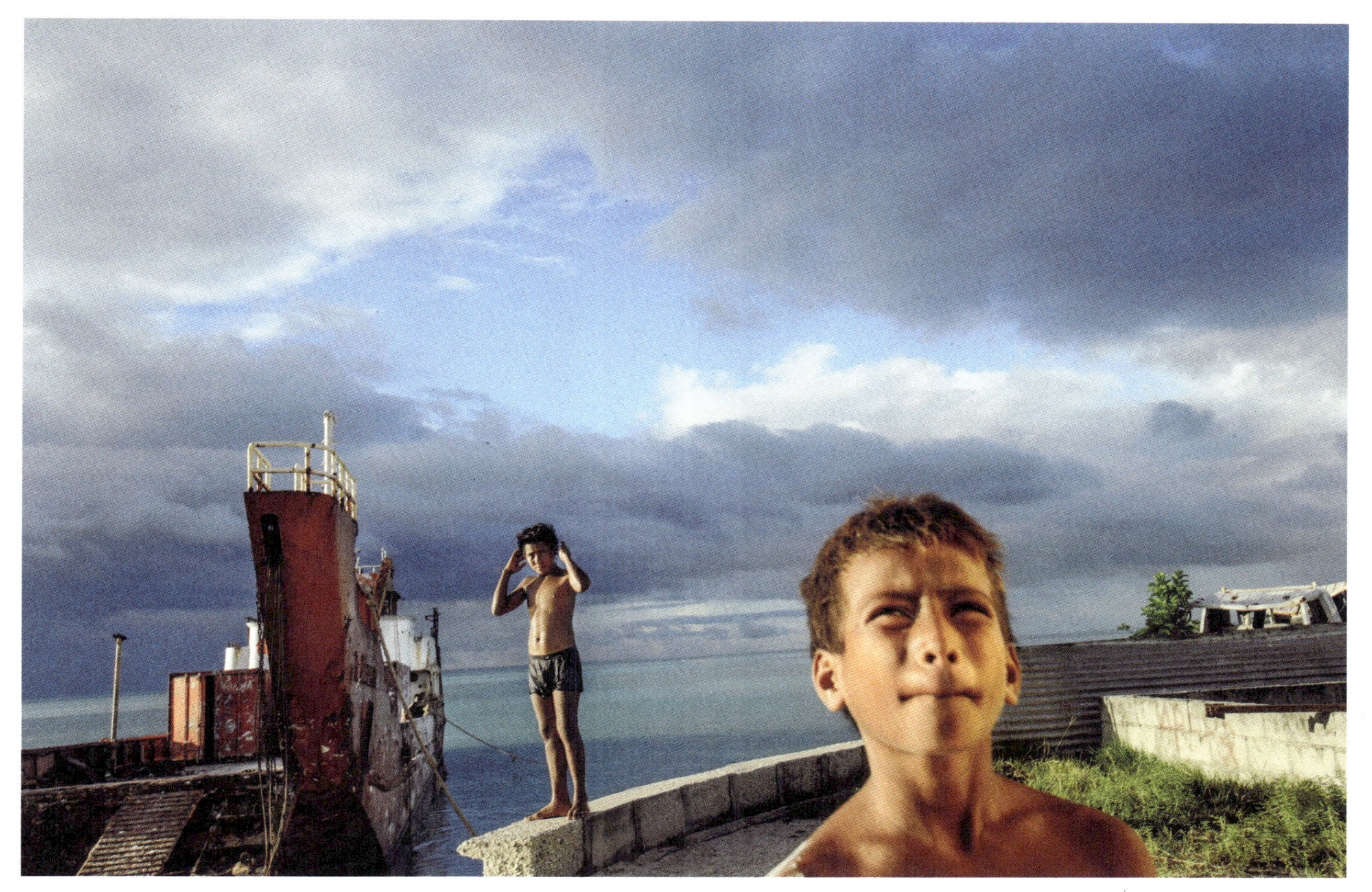

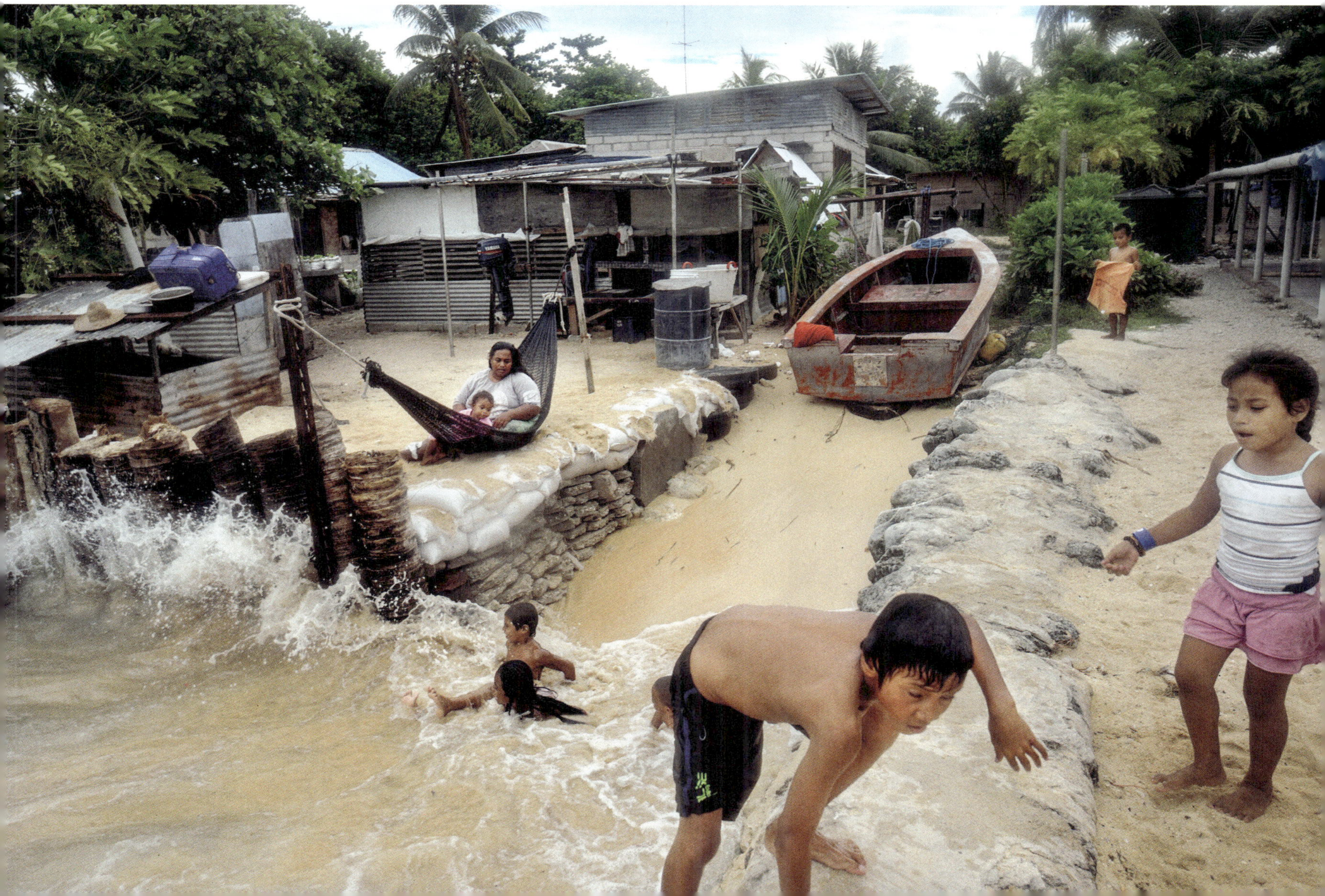

A wedding at the ferrycrossing
to Abatao Island.

Tebike Nikoora ('Golden Beach')
is a residential area in the densely
populated part of South Tarawa.

Kids play in Bonriki at low tide. The
sea wall protects the airport runway.

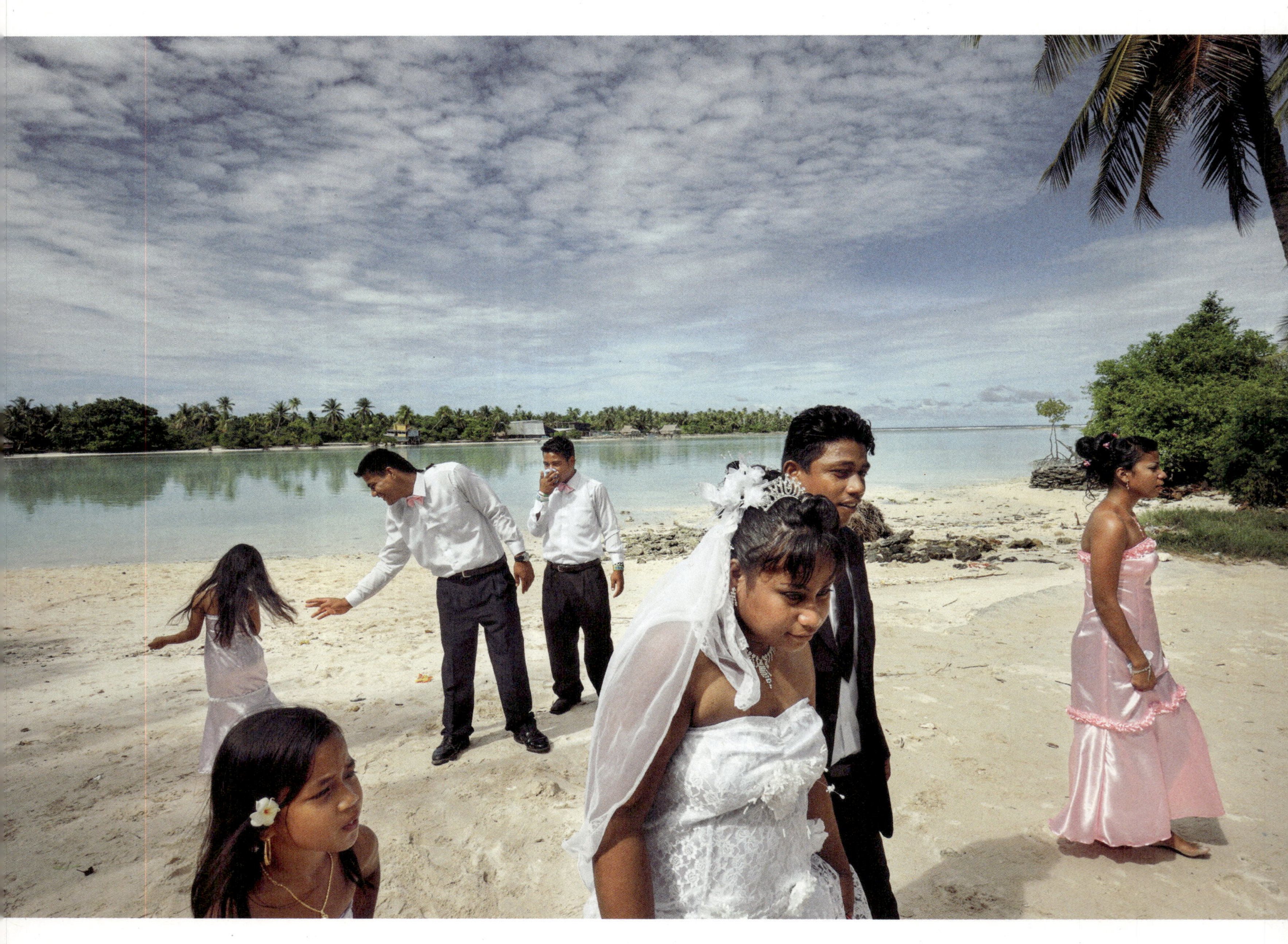

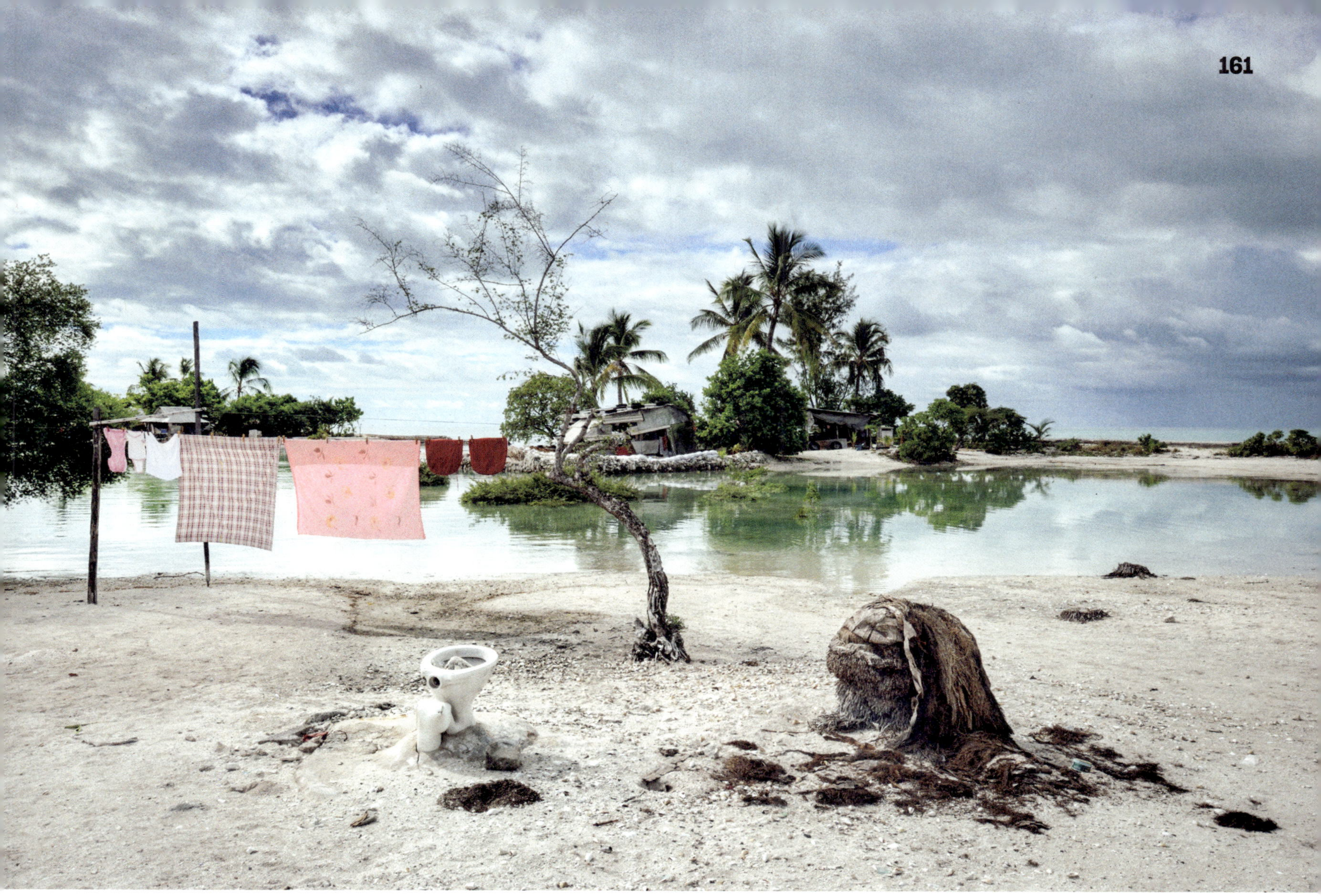

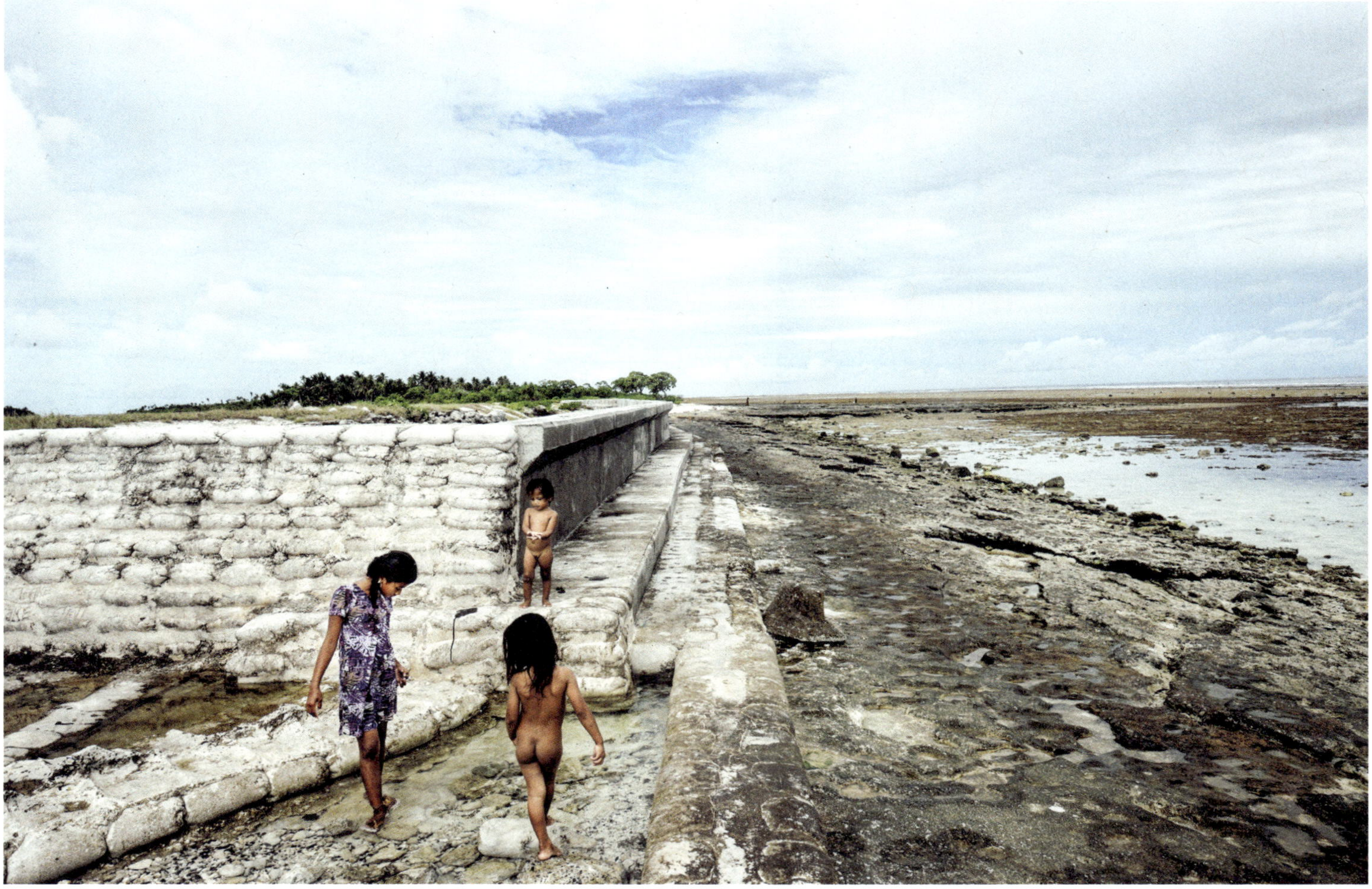

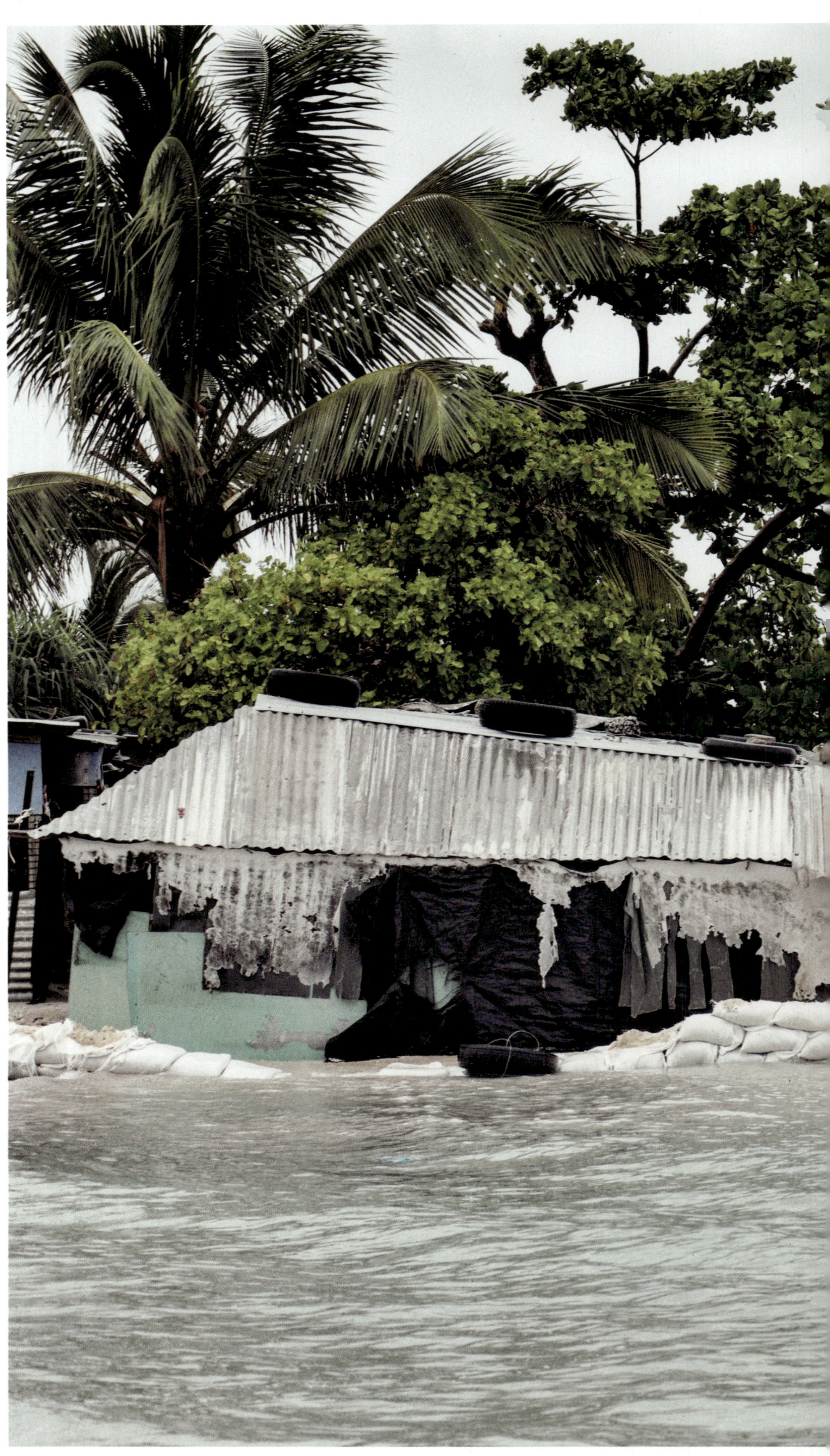

Residents in Tebike Nikoora fishing during high tide. Tebike Nikoora ('Golden Beach') is a residential area in the densely populated part of South Tarawa.

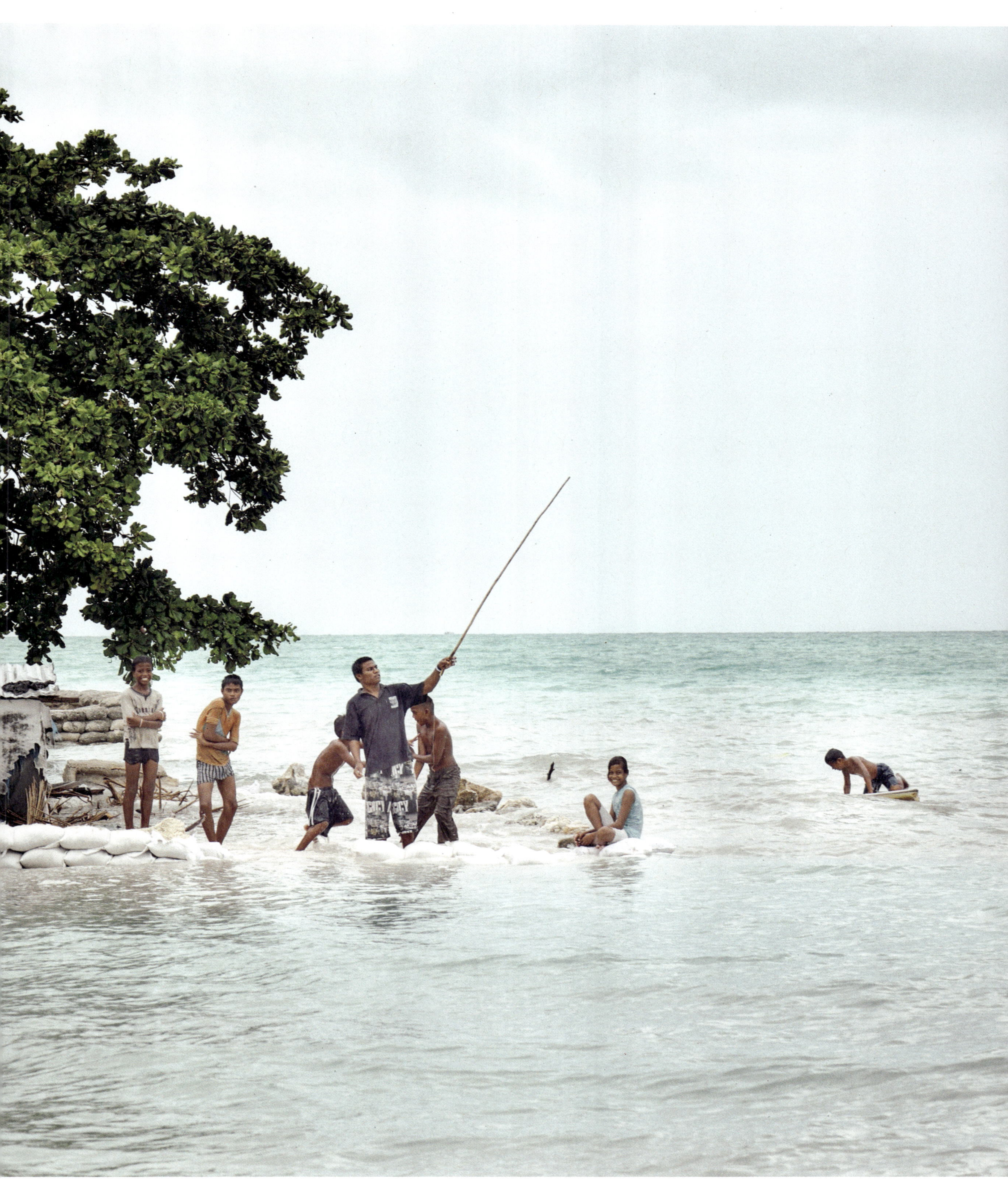

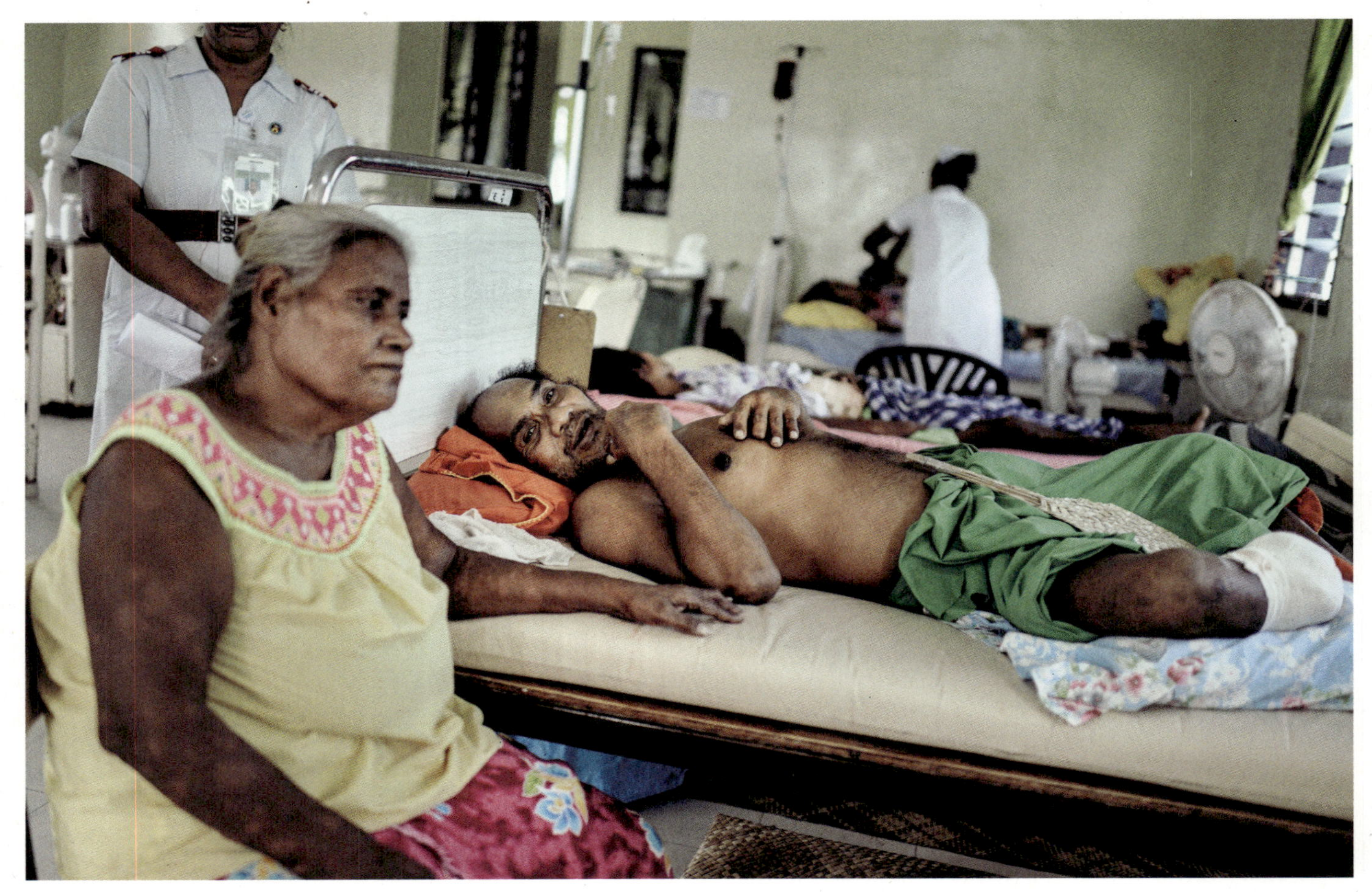

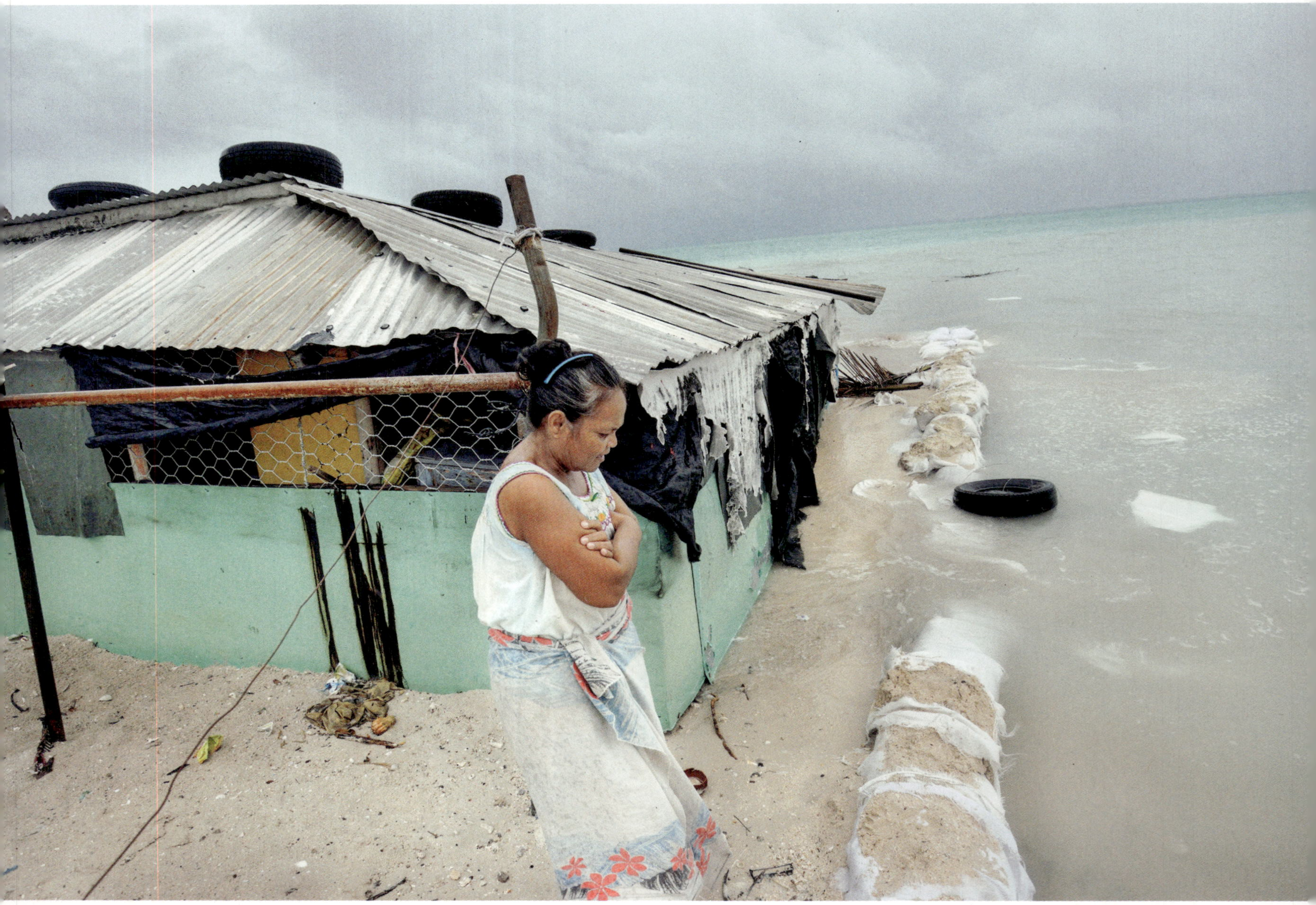

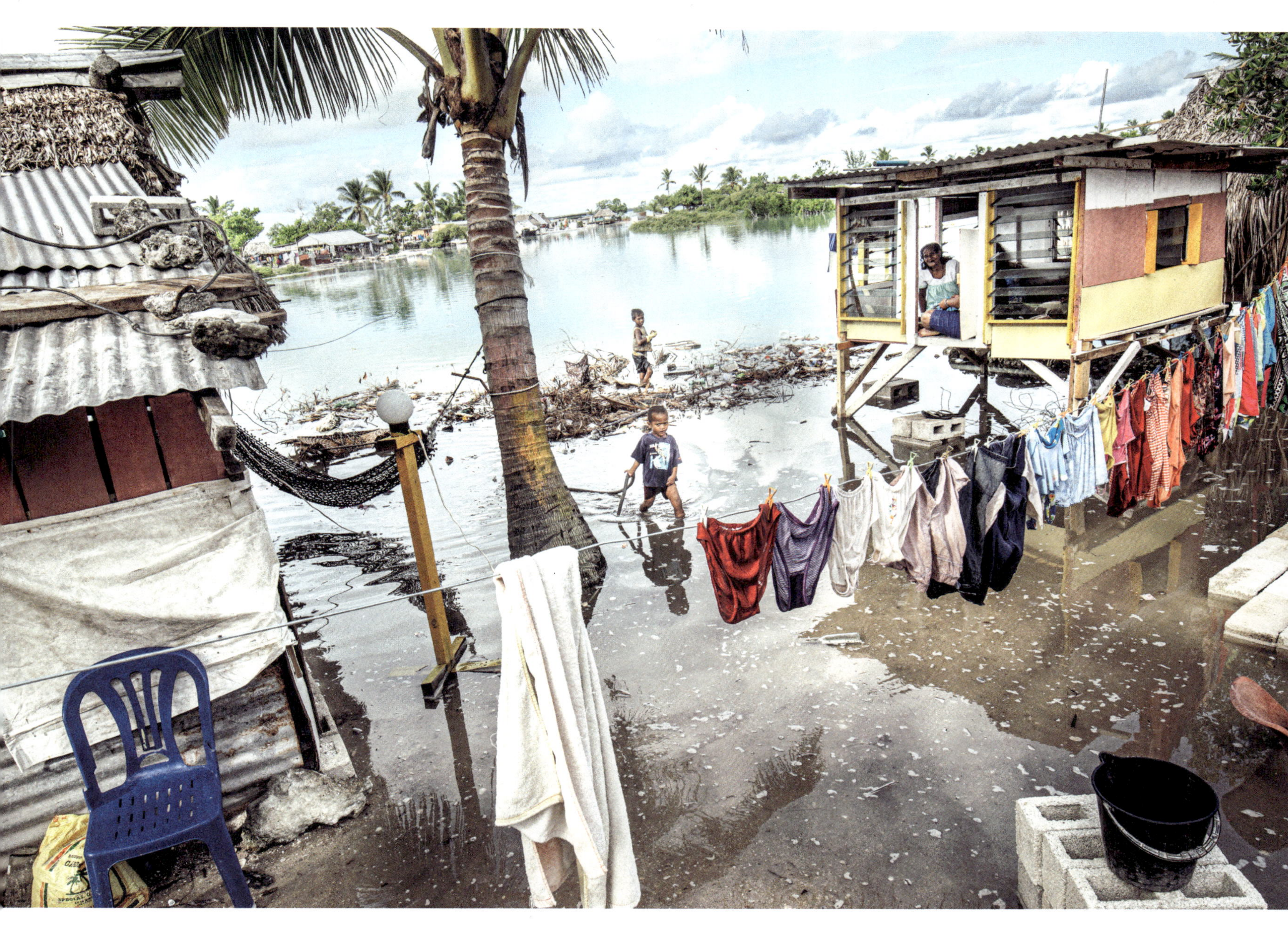

The Nawerewere Hospital in Bairiki has many patients with diabetes. A very high percentage of the population suffers from the disease due to a lack of exercise and an imbalanced diet, as very few crops can be grown due to saltwater intrusion. In the worst-case scenario, a limb needs to be amputated.

A resident in Tebike Nikoora.

Residents in Tebike Nikoora watch the high tide coming into her house. Tebike Nikoora ('Golden Beach') is a residential area in the densely populated part of South Tarawa.

Mangroves are being planted to protect the land from the sea. They are a better alternative than sandbags, which get washed away. The problem is that mangroves grow very slowly.

Tebike Nikoora ('Golden Beach') is a residential area in the densely populated part of South Tarawa. Due to seawater intrusion, many coconut trees have died, and agricultural land has disappeared.

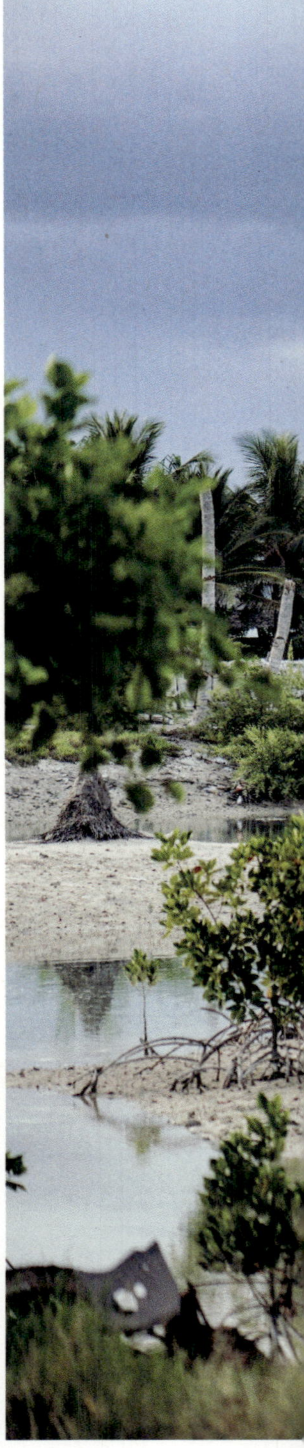

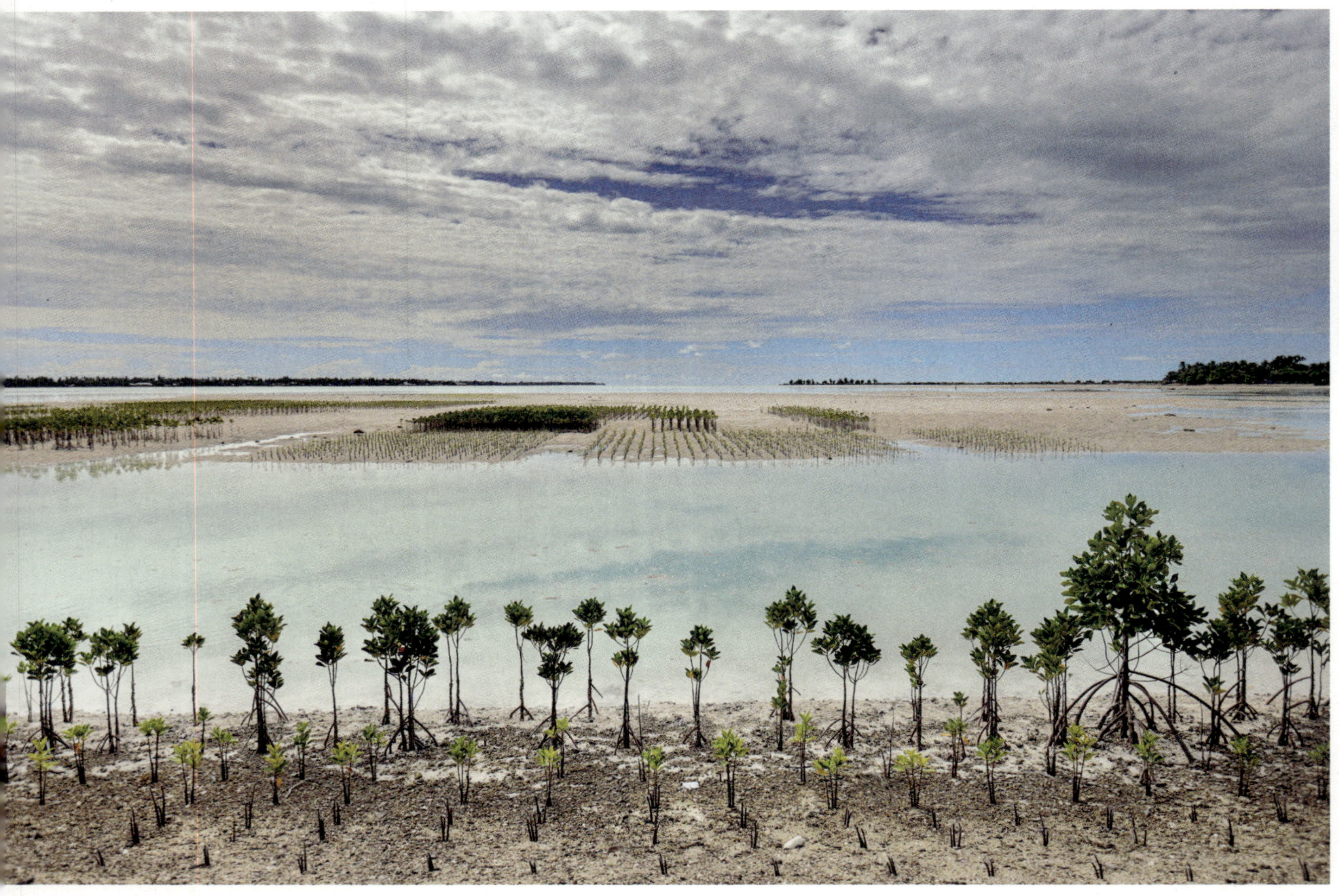

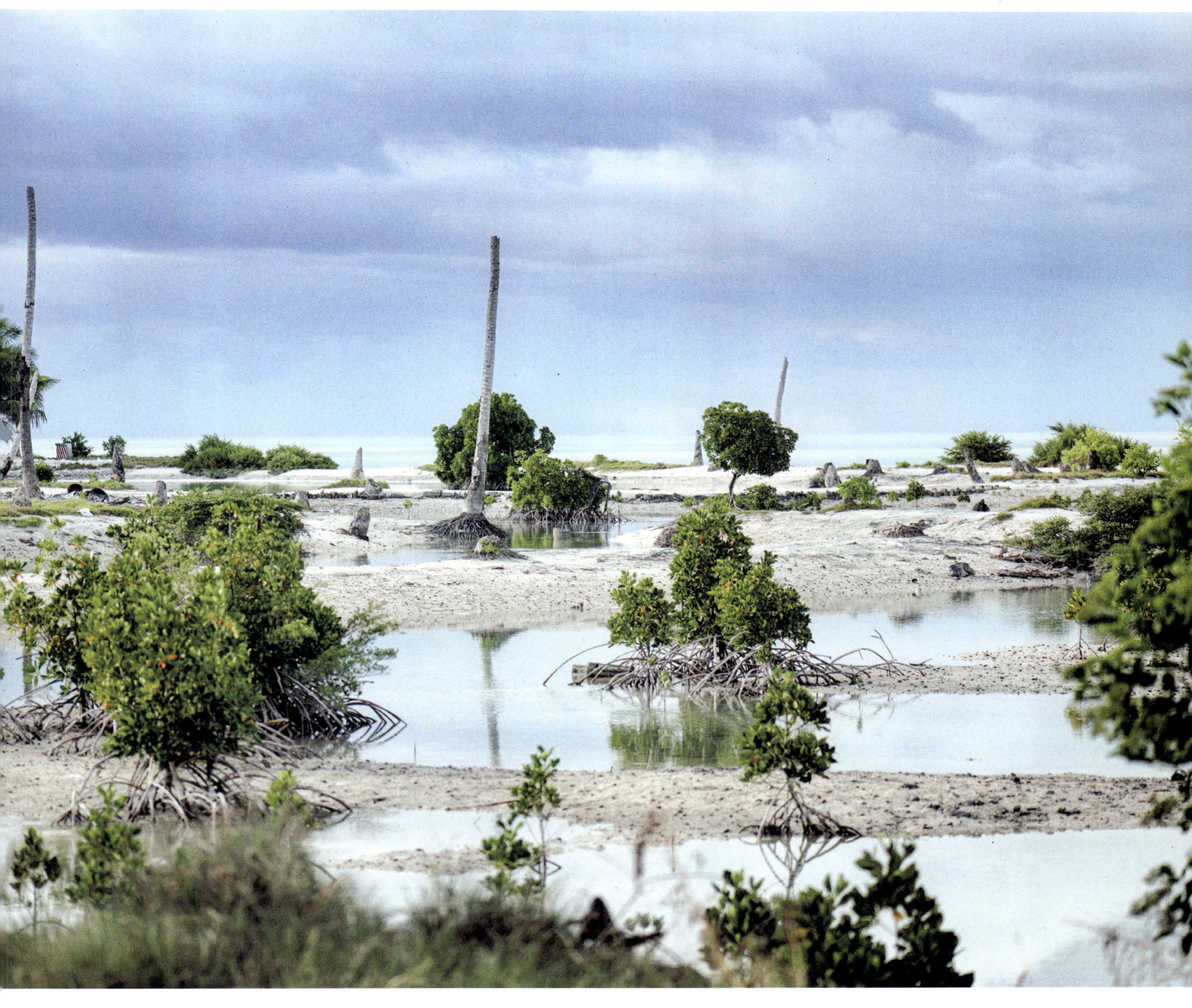

> Pages 168-169: Children play on the beach in Temwaiku, a vulnerable village on South Tarawa. Sandbags have been placed to try to hold back the ocean. In 2012, waves washed away this bulwark and rolled inland, leaving behind flooded homes, salty soil and tainted wells.

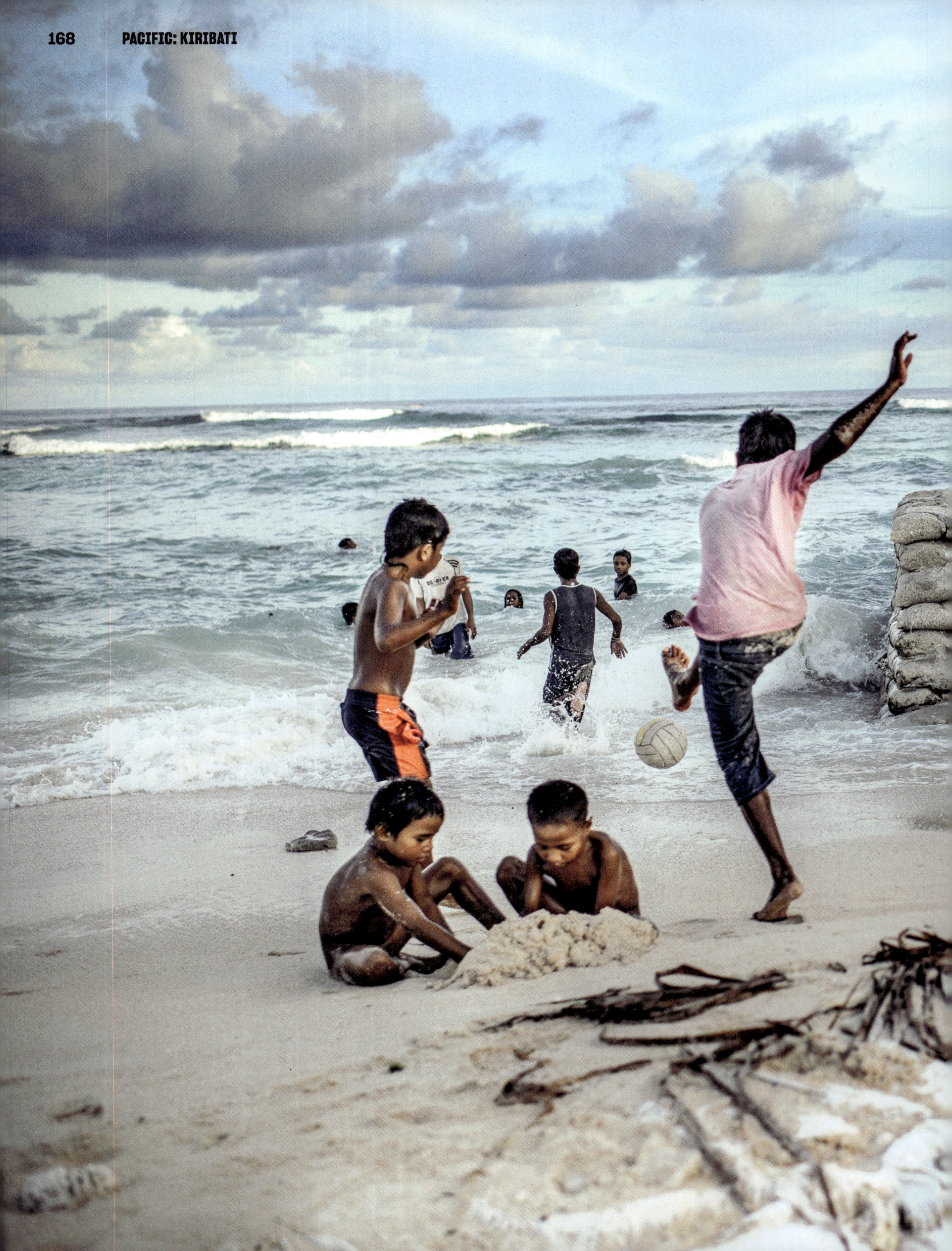

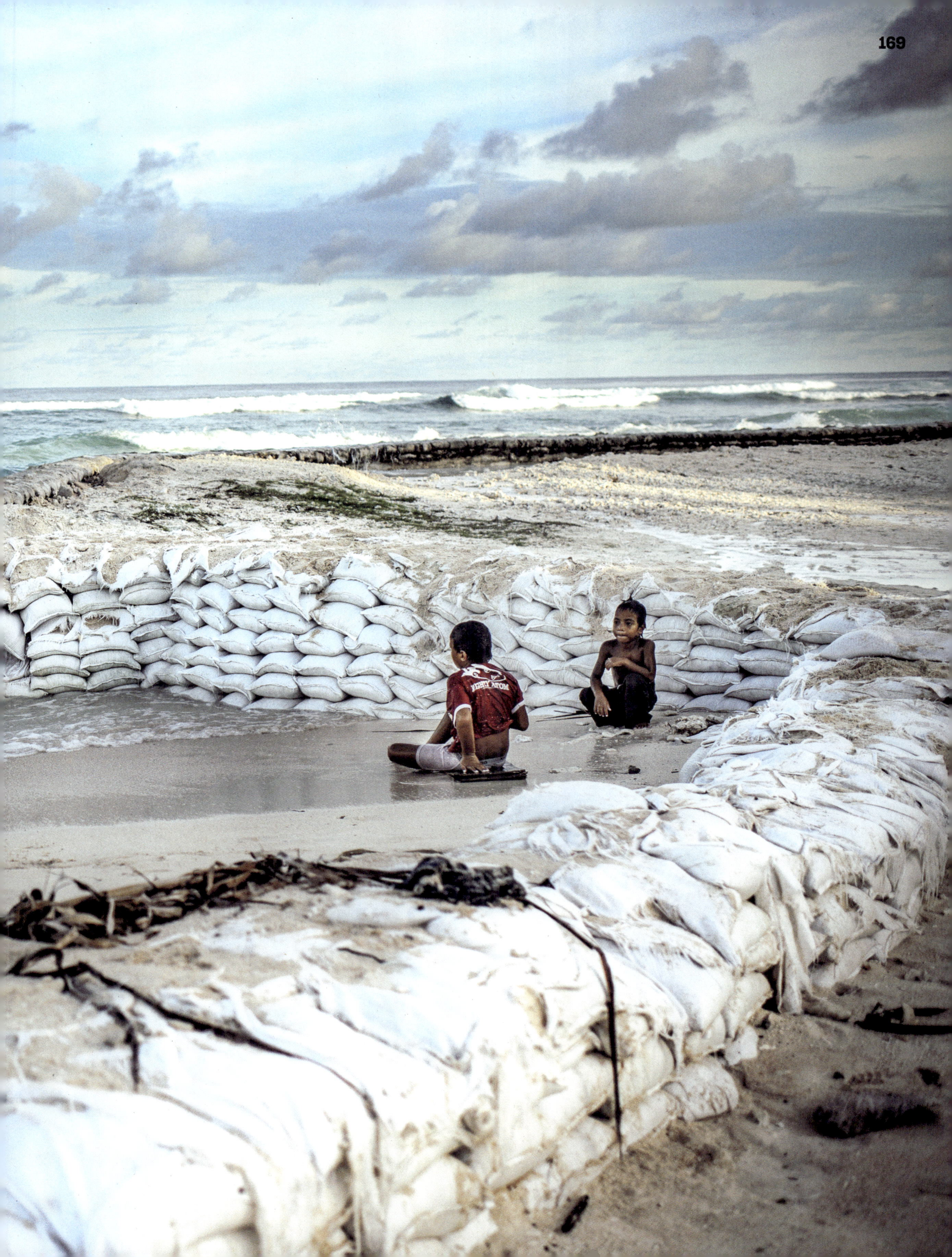

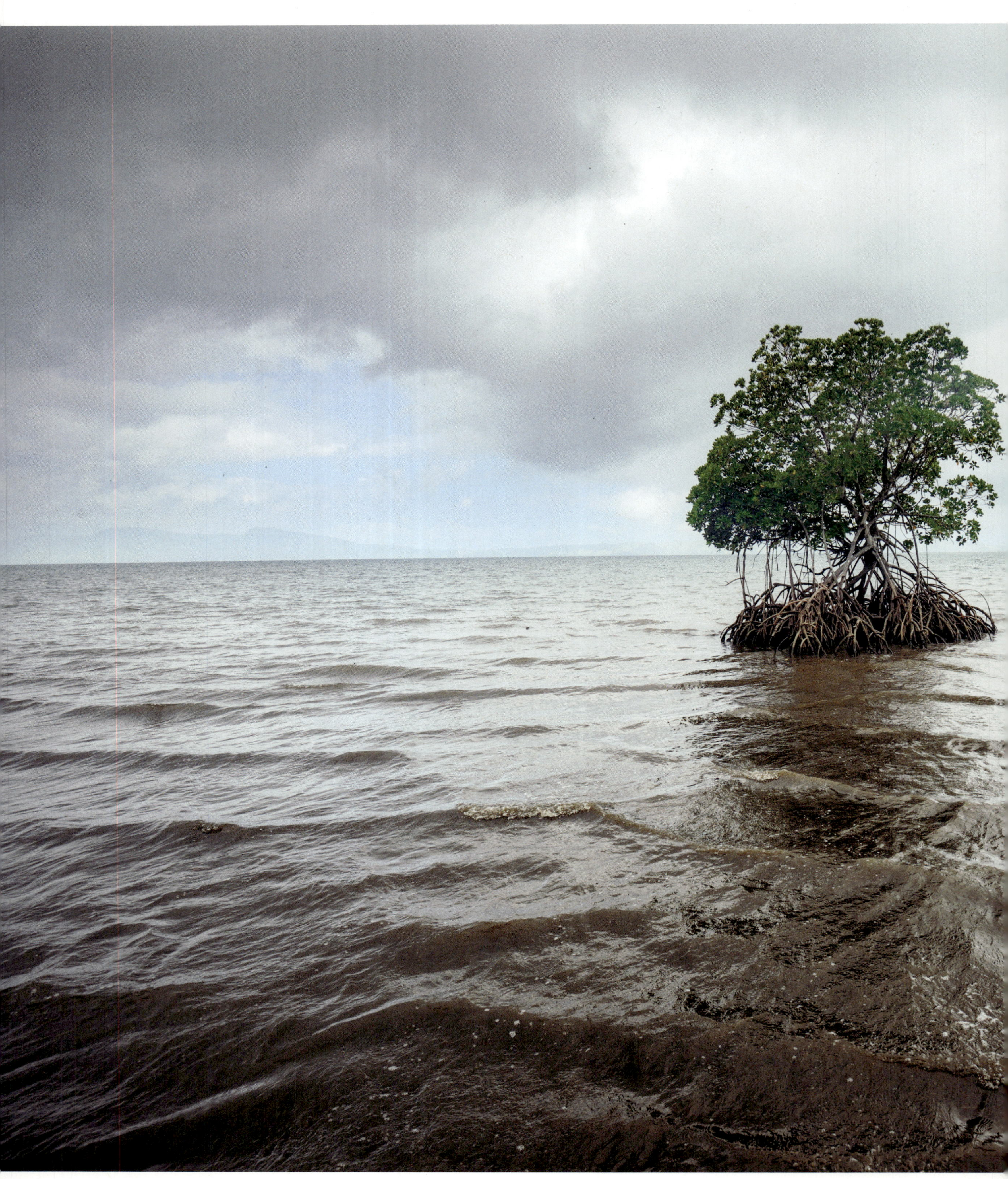

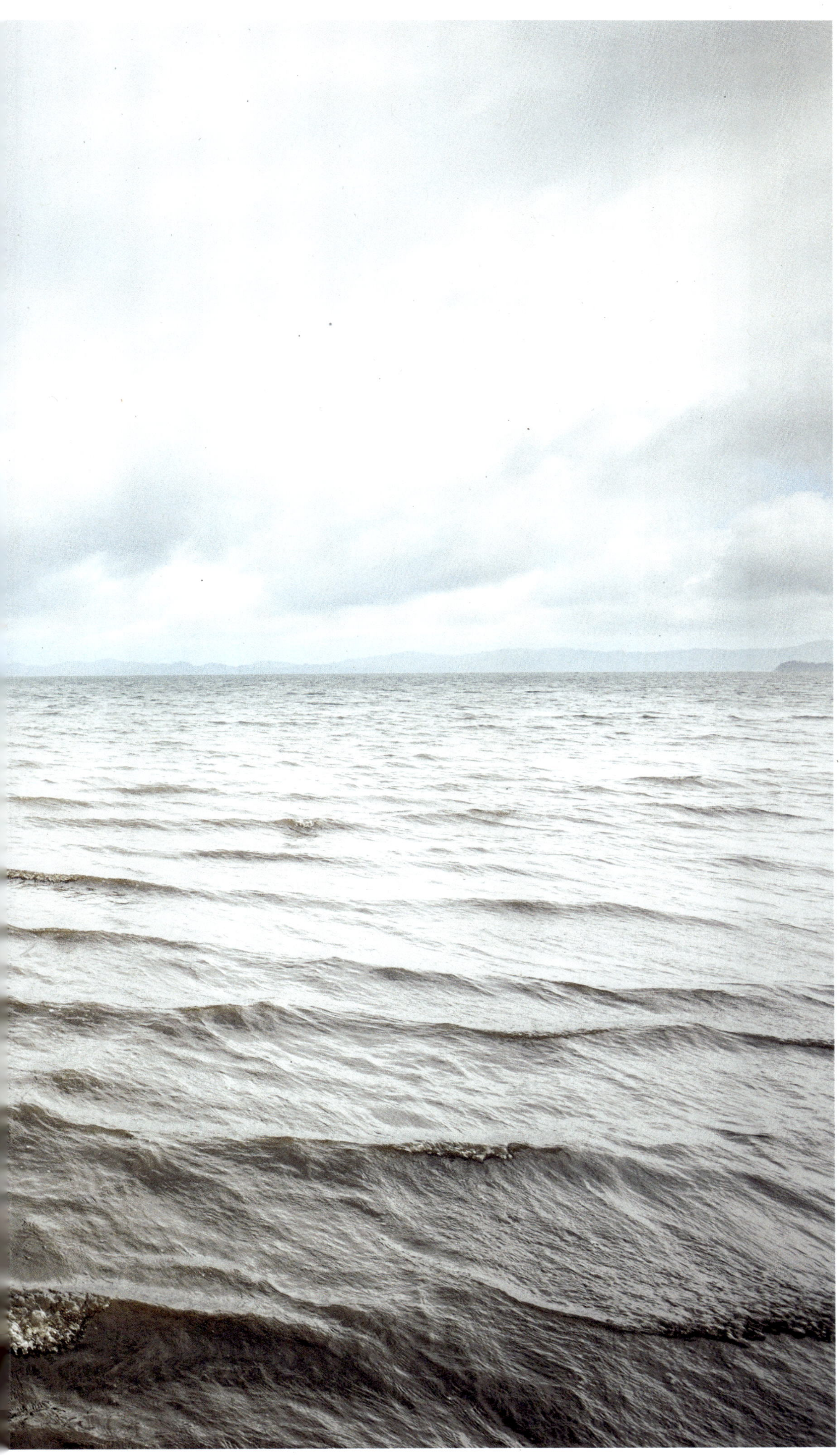

The shoreline of Vunidogoloa is heavily eroded due to the rising waters. Vunidogoloa is situated in Natewa Bay on Viti Levu, Fiji's main island. The situation became so precarious that the government decided to relocate the village. Unfortunately, the new site was poorly designed and was eroding even before anyone had moved there.

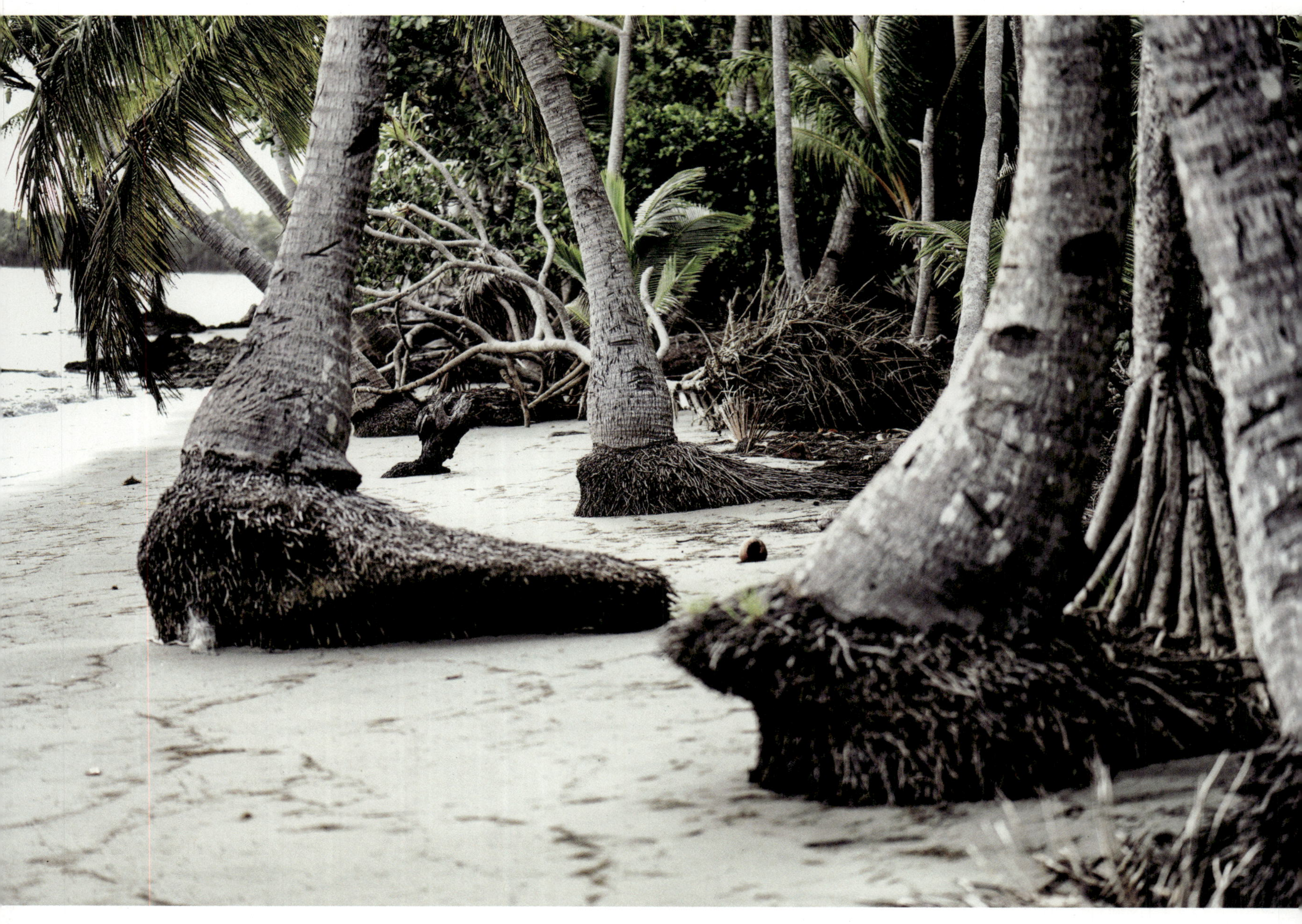

Dying coconut trees at the beach of
Todoru village. Todoru was founded by
James Dunn, an Irish settler, in 1835.
Since then, generations of the Dunn
family have lived here.

Vunidogoloa is situated in Natewa
Bay on Viti Levu, Fiji's main island.
Vunidogoloa has 153 inhabitants and
frequently floods due to the rising
waters.

Sailosi Ramutu (52) stands at the
shoreline of Vunidogoloa. He is the
headman of the village. His family
has been in Vunidogoloa for four
generations.

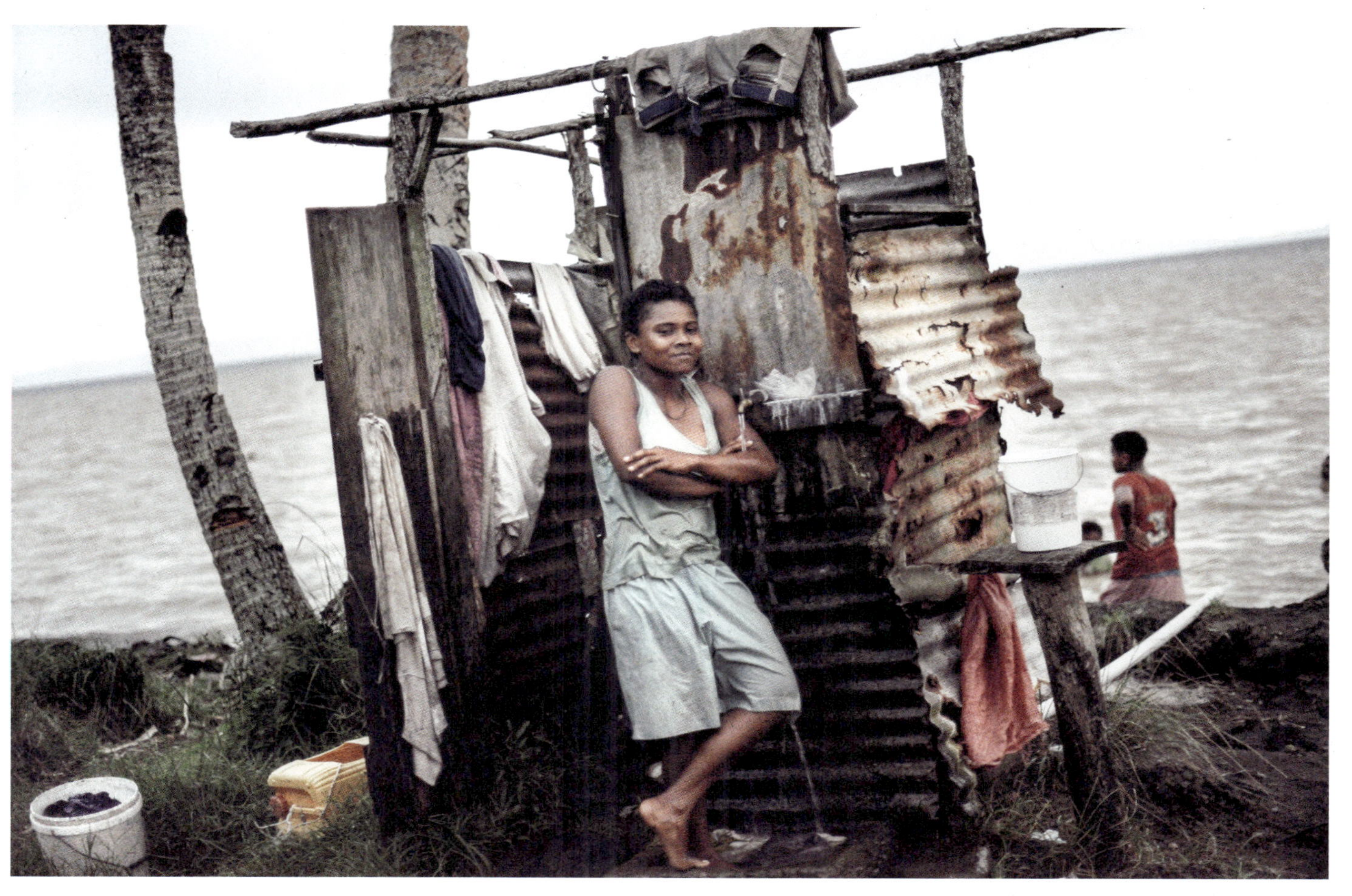

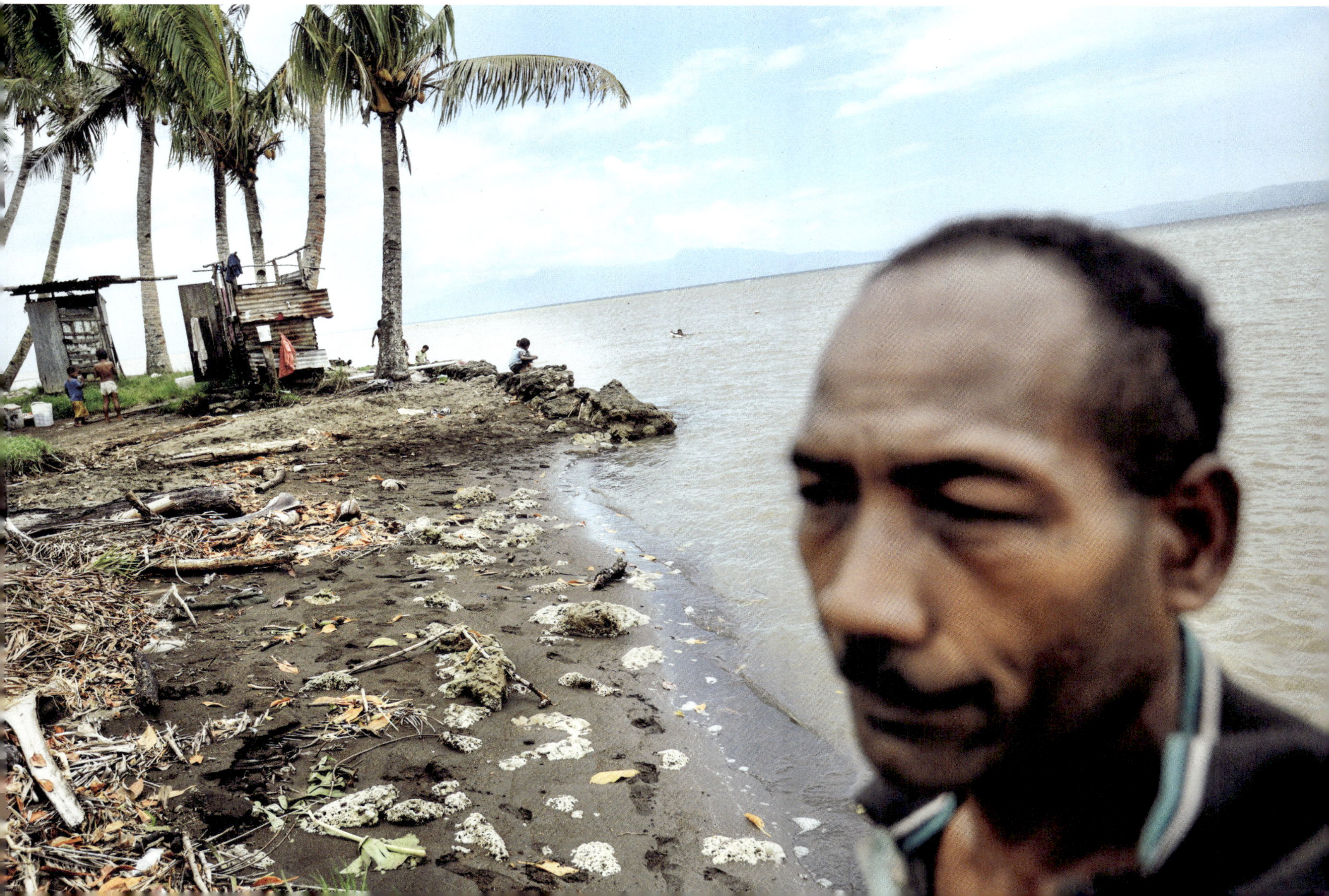

The shore of Kioa is eroding due to rising sea levels. Kioa is an island east of Vanua Levu with a population of c. 400. The island is inhabited by people from Tuvalu. The first settlers came to Kioa in 1947 for economic reasons. Nowadays, Tuvalu is threatened by rising sea levels, and the population will have to evacuate in the future. Whether Kioa is an option for relocation remains an open question.

Elisabeth Kidimoce (72; left) lives with her husband Keresoni (79). Including all of their children, grandchildren and great-grandchildren, 32 people live in the house. Keresoni says, 'I know I have to leave, but I would like to stay'.

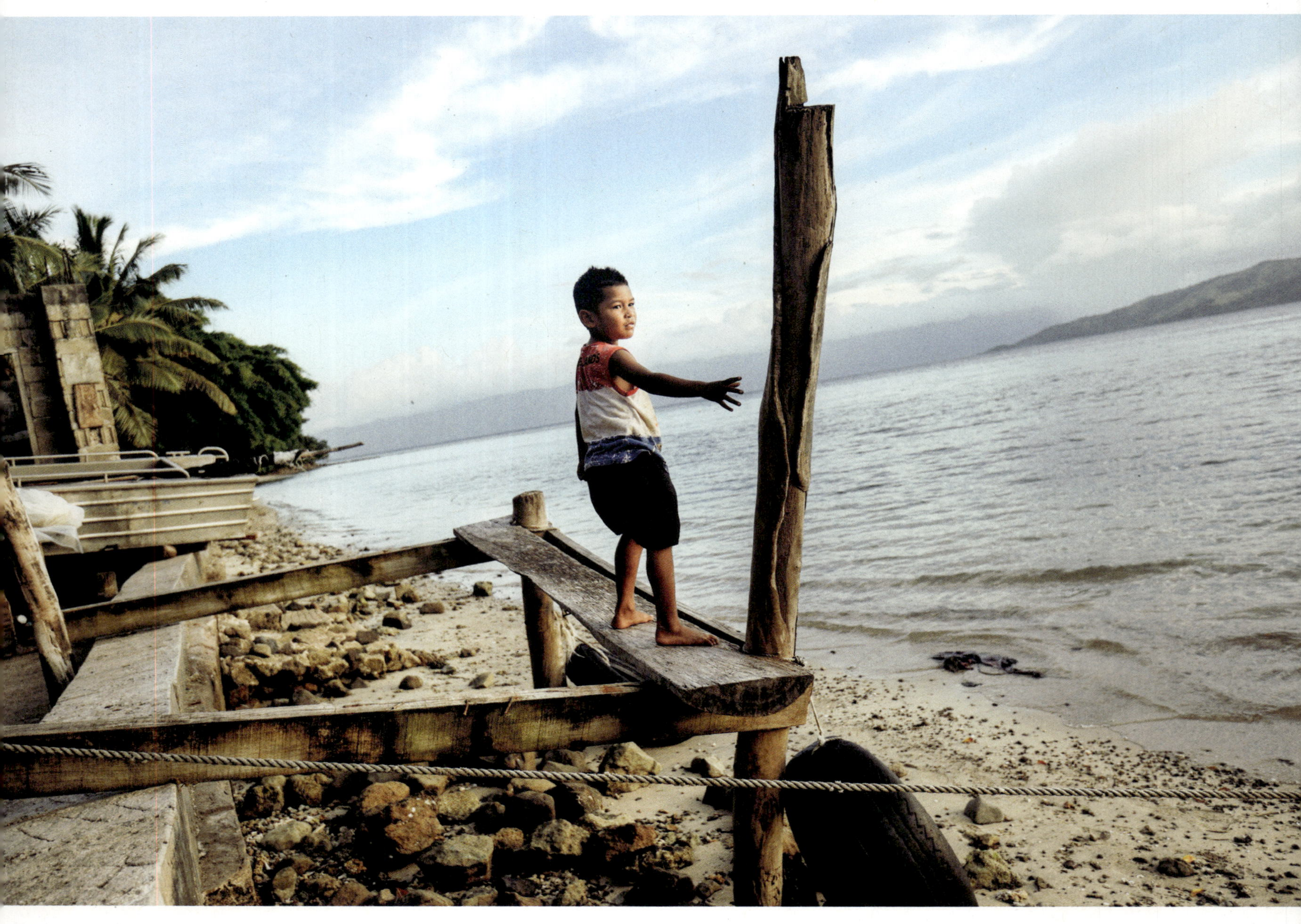

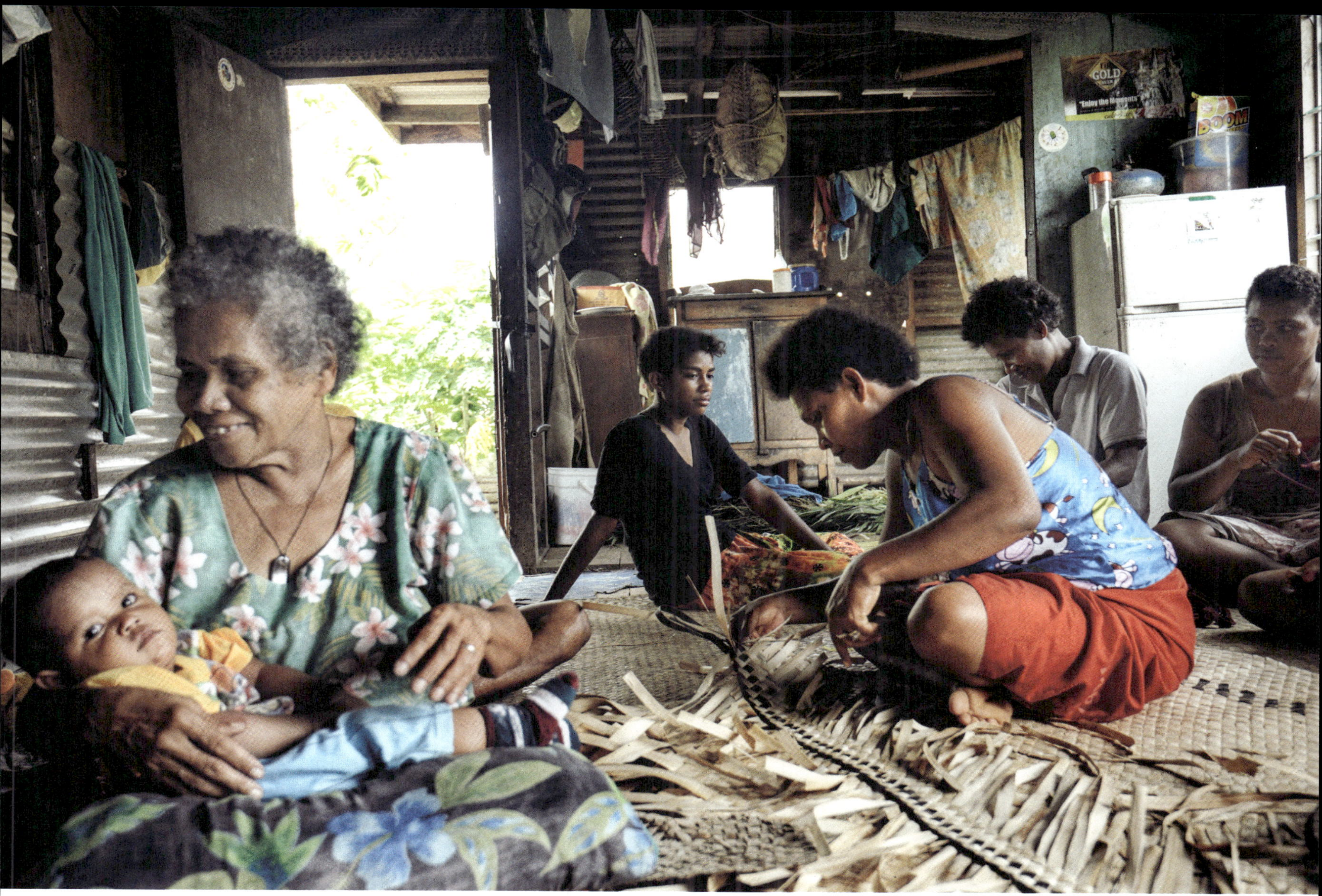

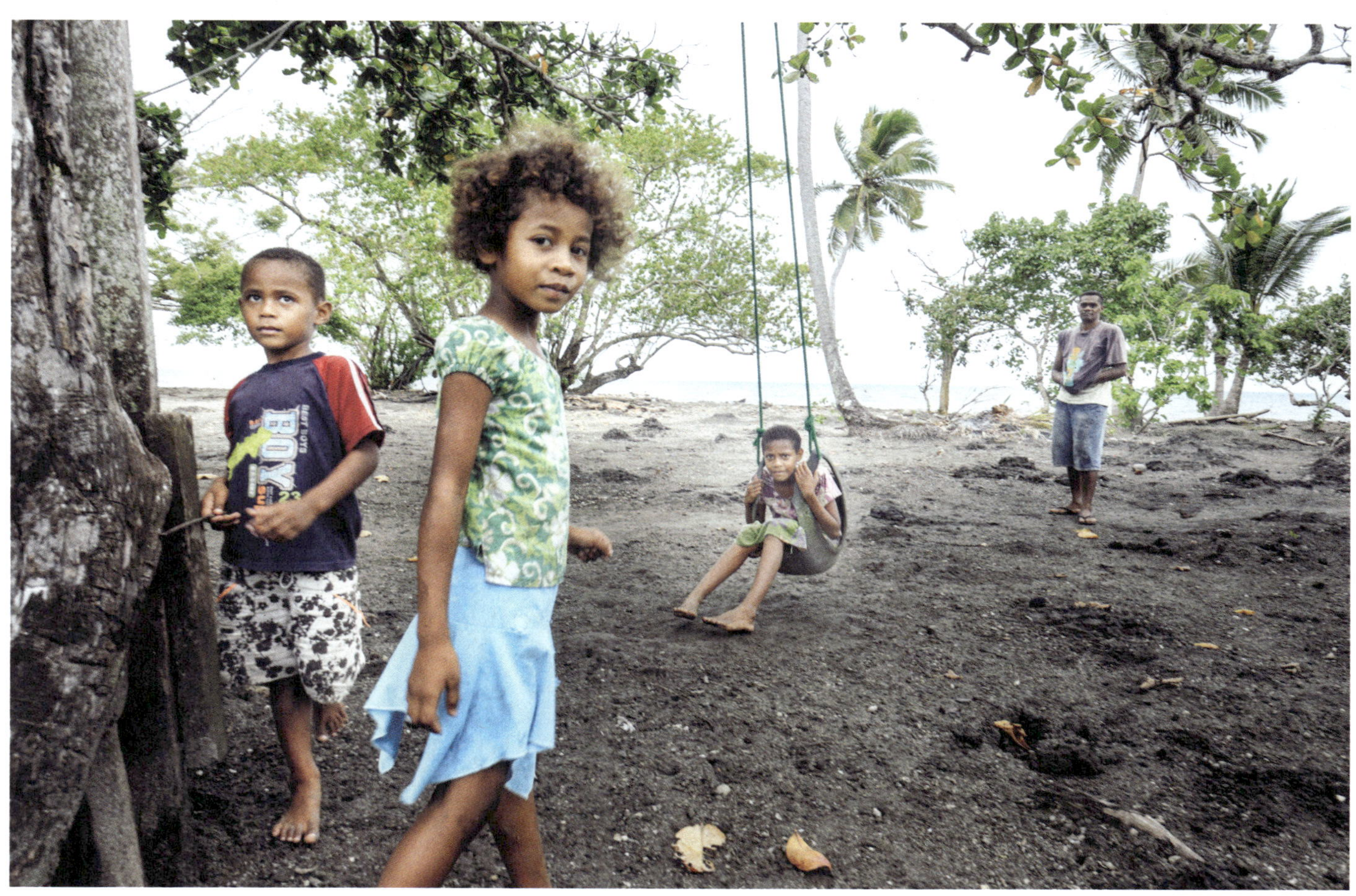

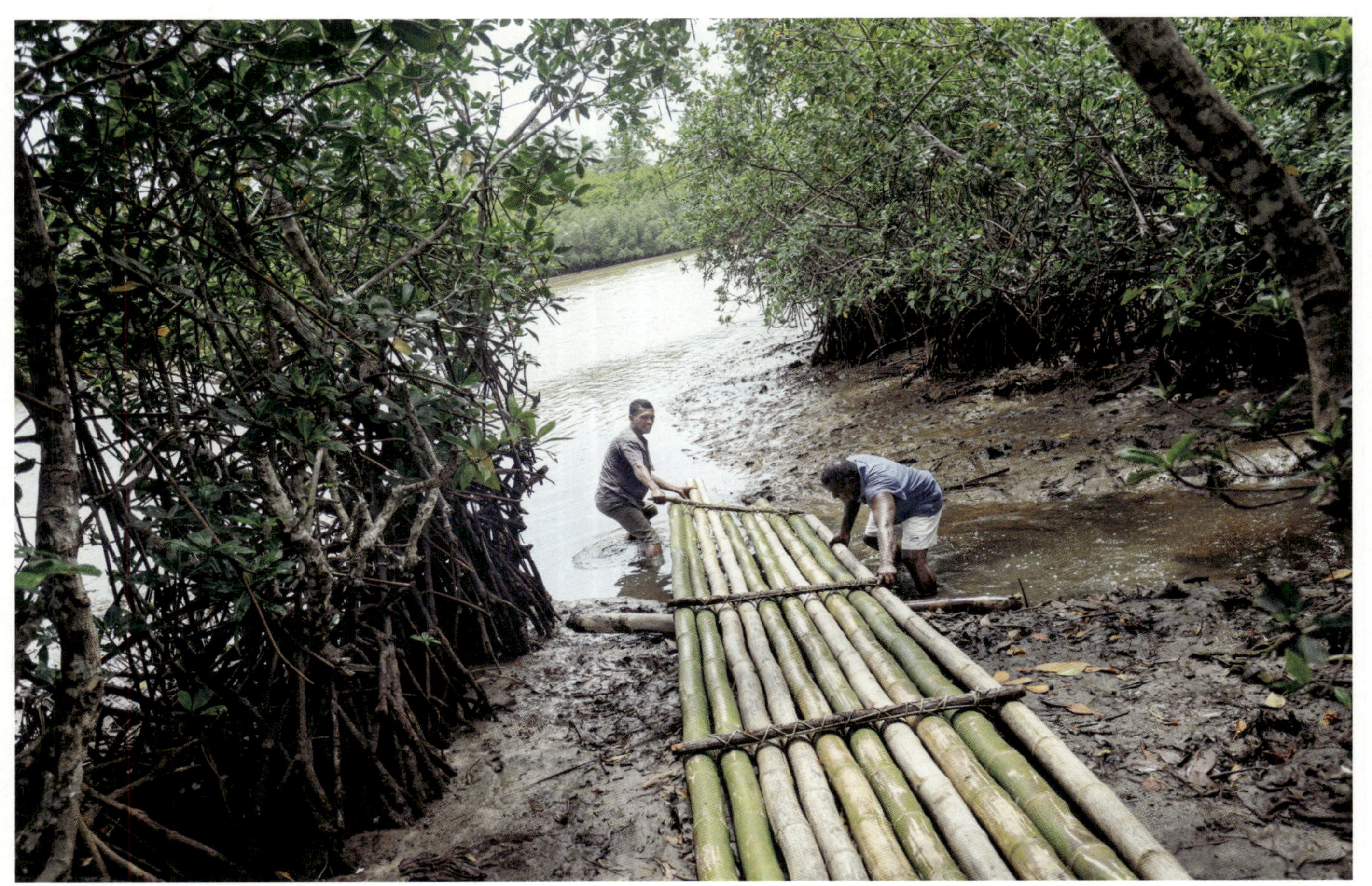

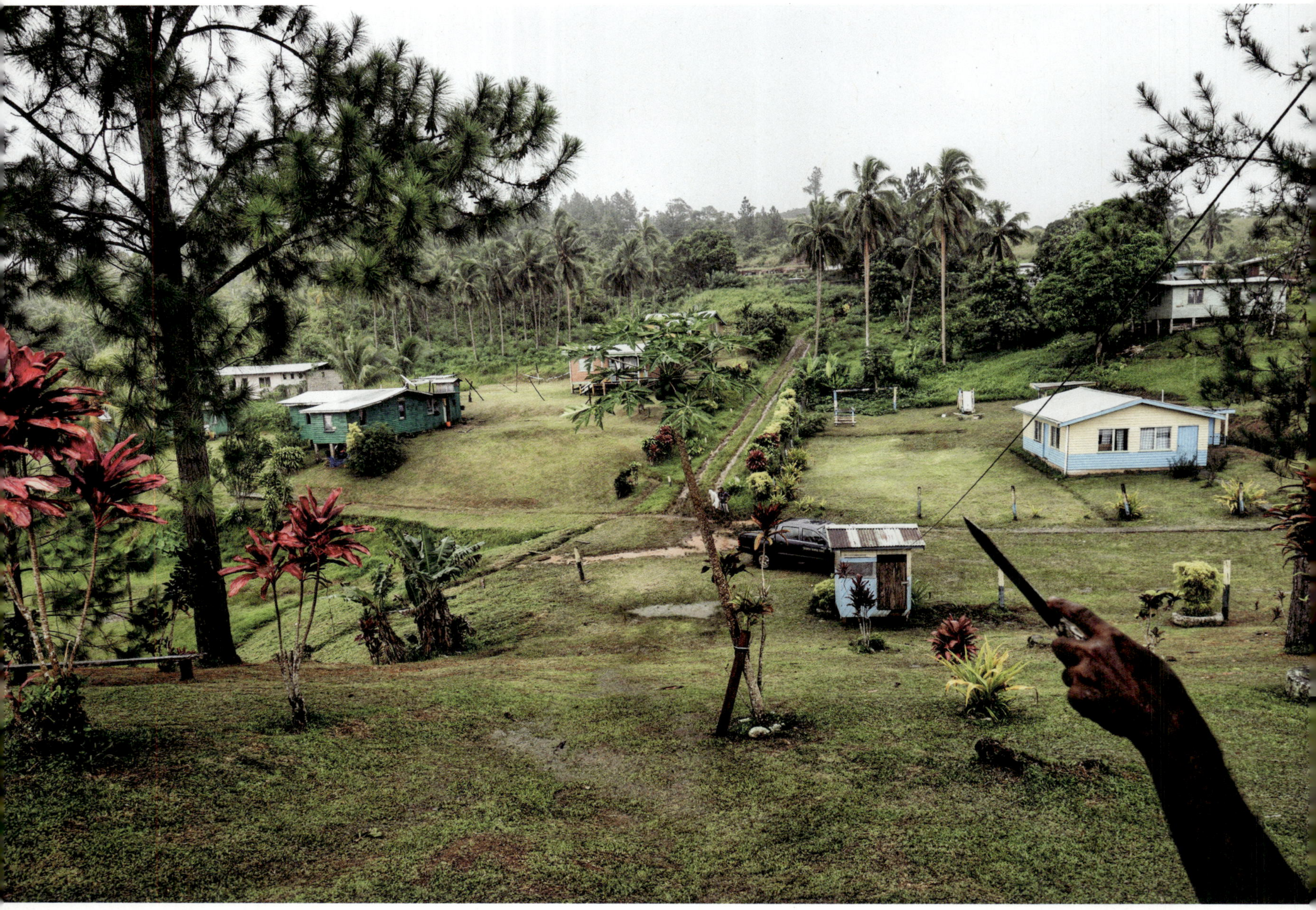

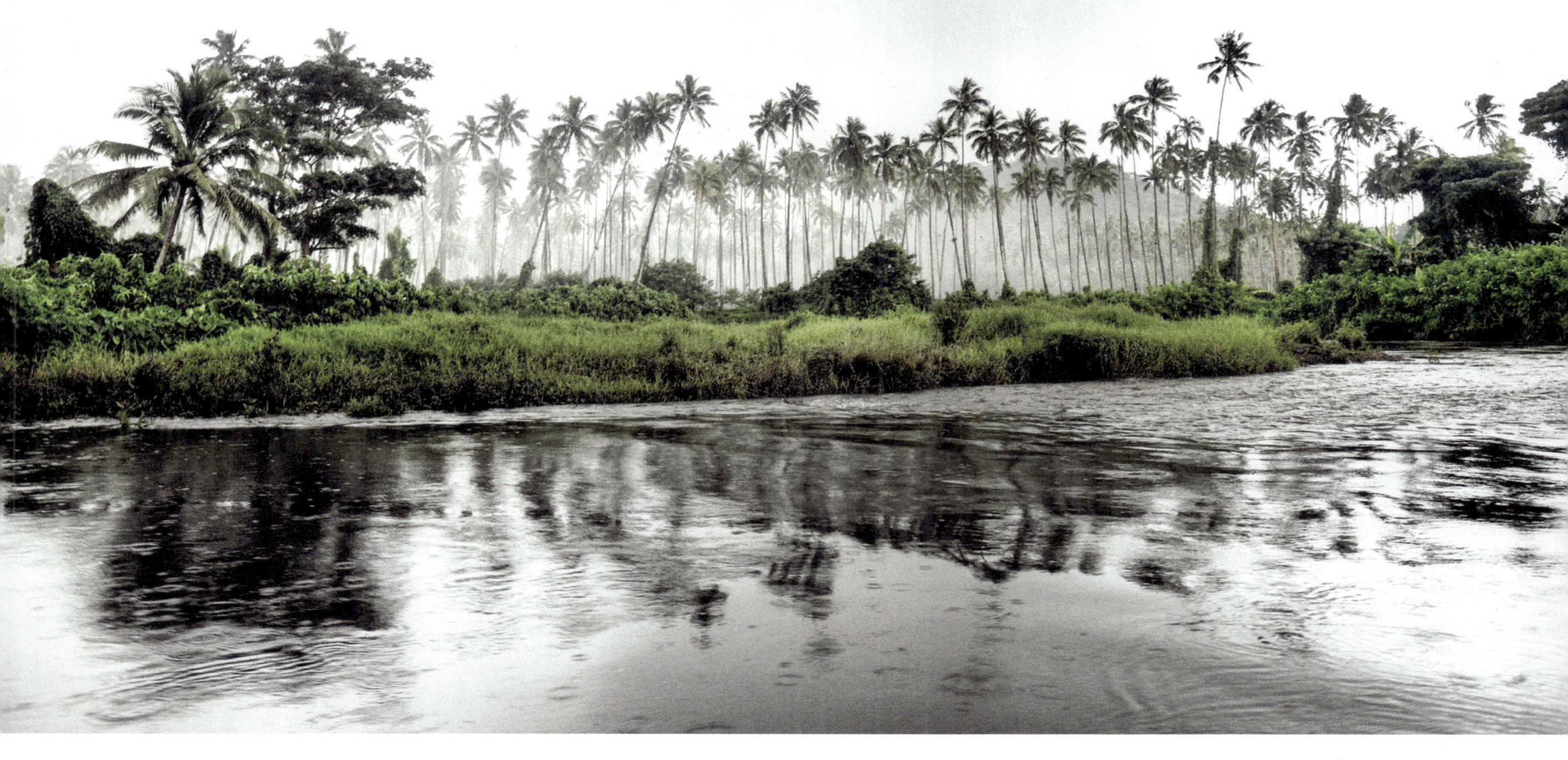

The villagers of Vunidogoloa use rafts for fishing.

The village of Naviavia is at the centre of an area of 6,000 acres that was purchased by the government of Kiribati in 2011 for possible relocation of their people. Naviavia houses 500 people, former slaves from the Solomon Islands. The land is owned by the Anglican Church.

> Pages 178-179: The cemetery of Todoru village is nowadays at high tide under water. Todoru was founded by James Dunn, an Irish settler, in 1835. Since then, generations of the Dunn family have lived here. Barney Dunn (46) is one of them: 'We have built seawalls to protect us, five times, but every time the sea destroys them. The cemetery where our ancestors are buried is in the sea now'.

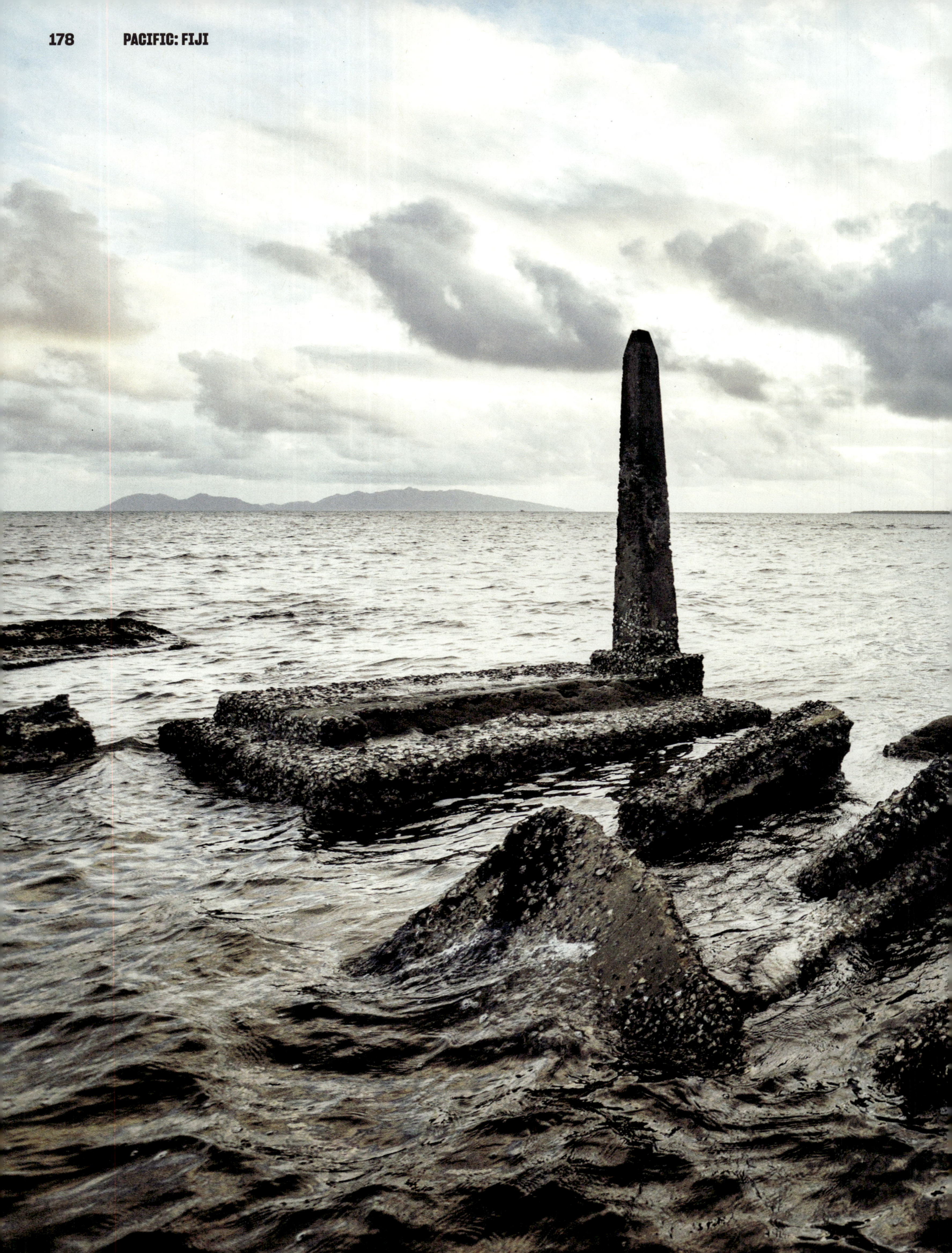

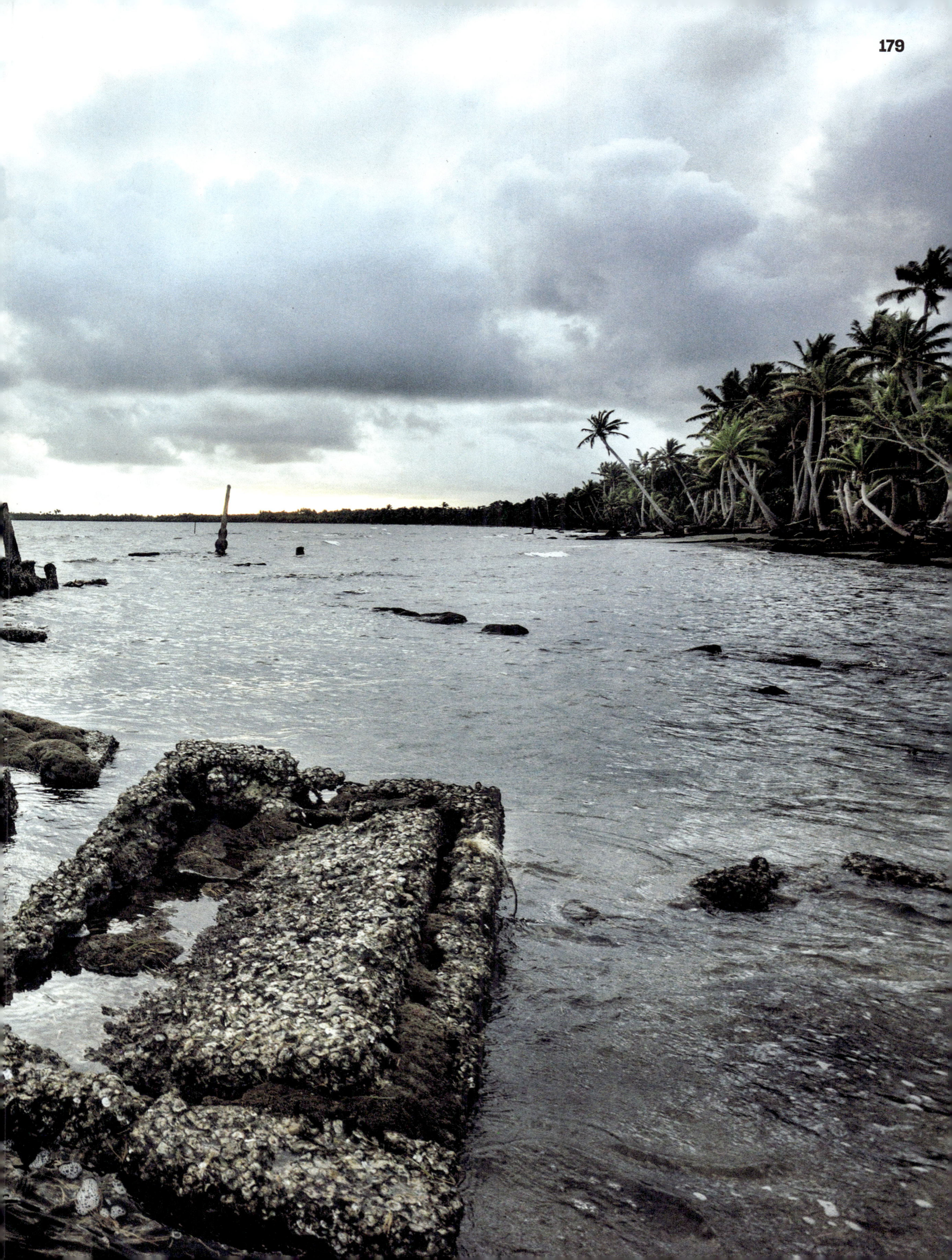

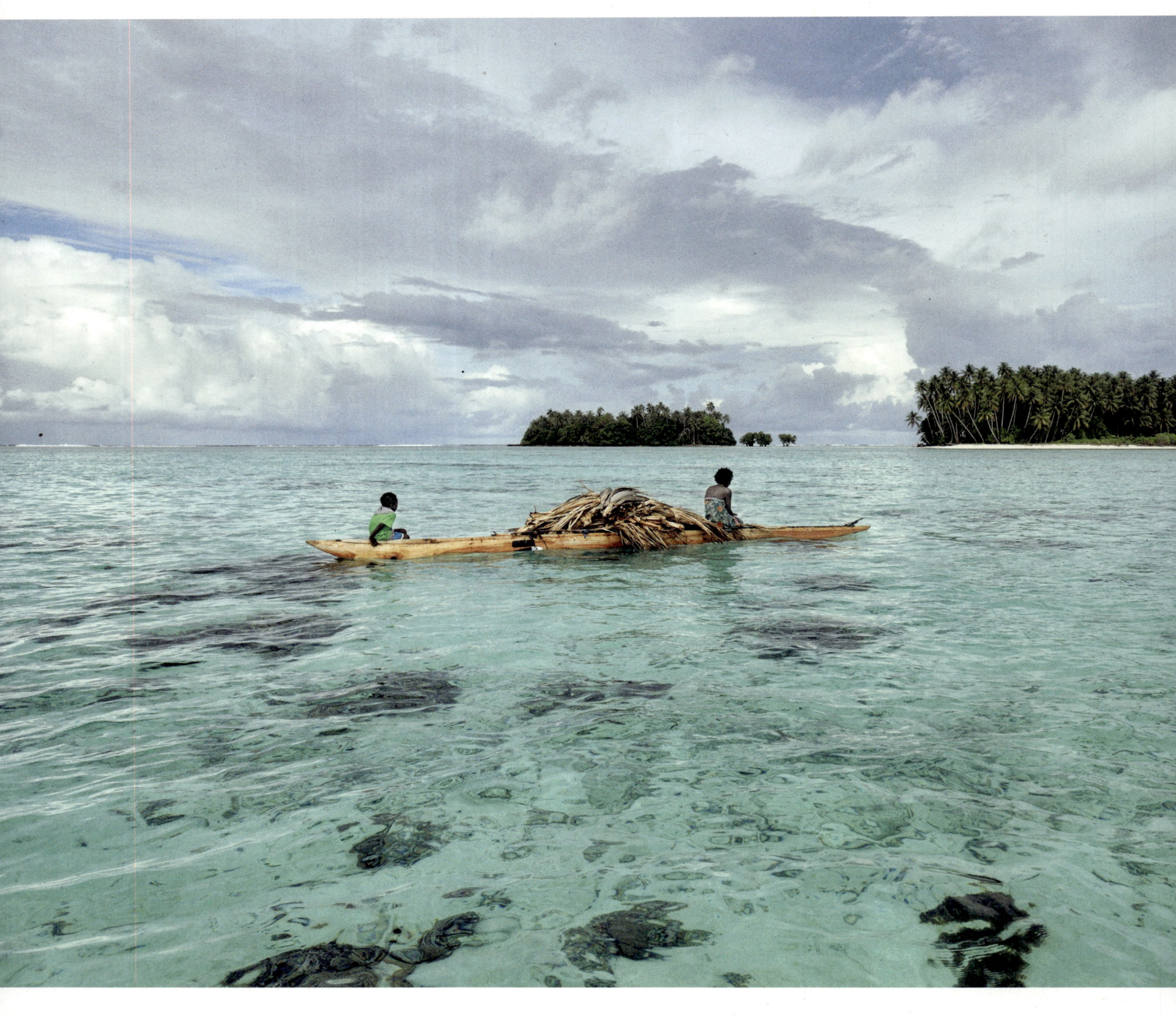

Yolasa Island in the background, the
island that broke in two some years
ago.

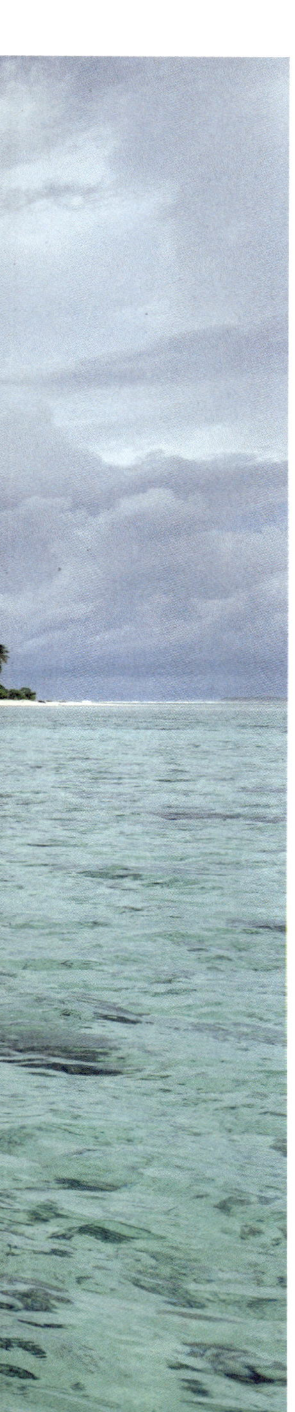

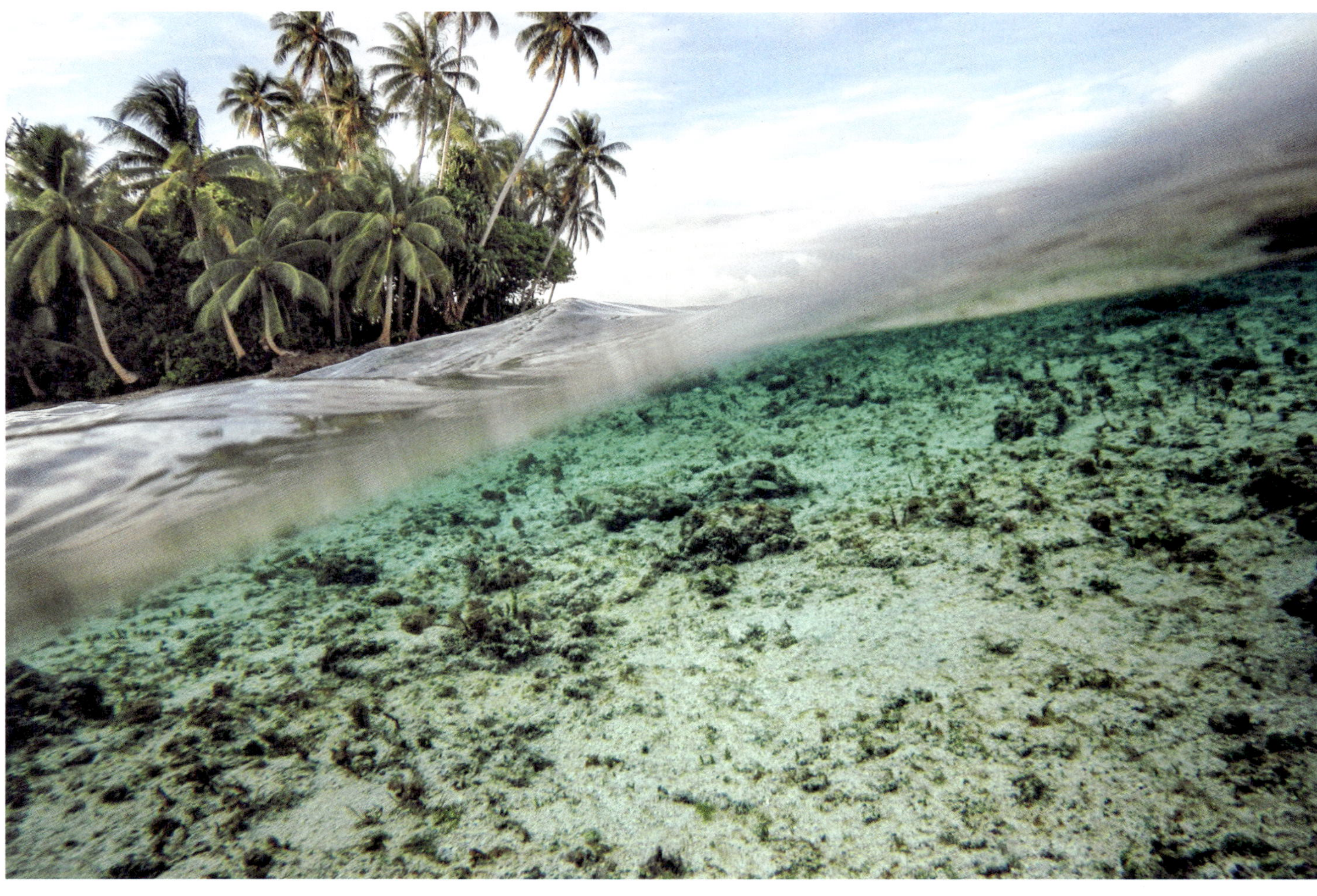

Han Island.

Due to seawater intrusion on the Carteret Islands, people can hardly grow crops, and the drinking water is too salty and not safe to drink. Part of the island has turned into a swamp. As a direct result, mosquitoes have become a serious problem - people walk around with smudge pots to keep them away.

Han Island is eroding away at an alarming rate.

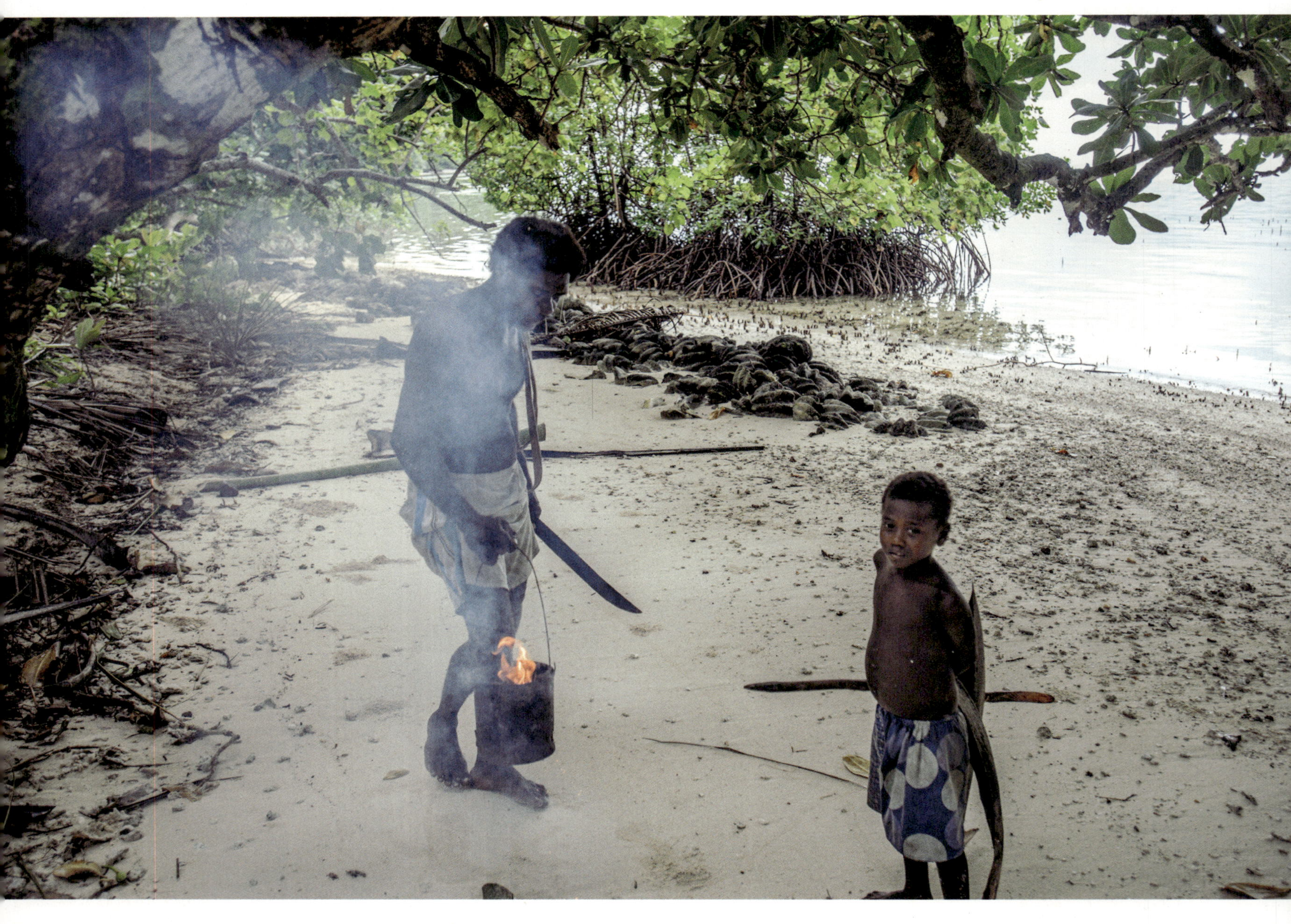

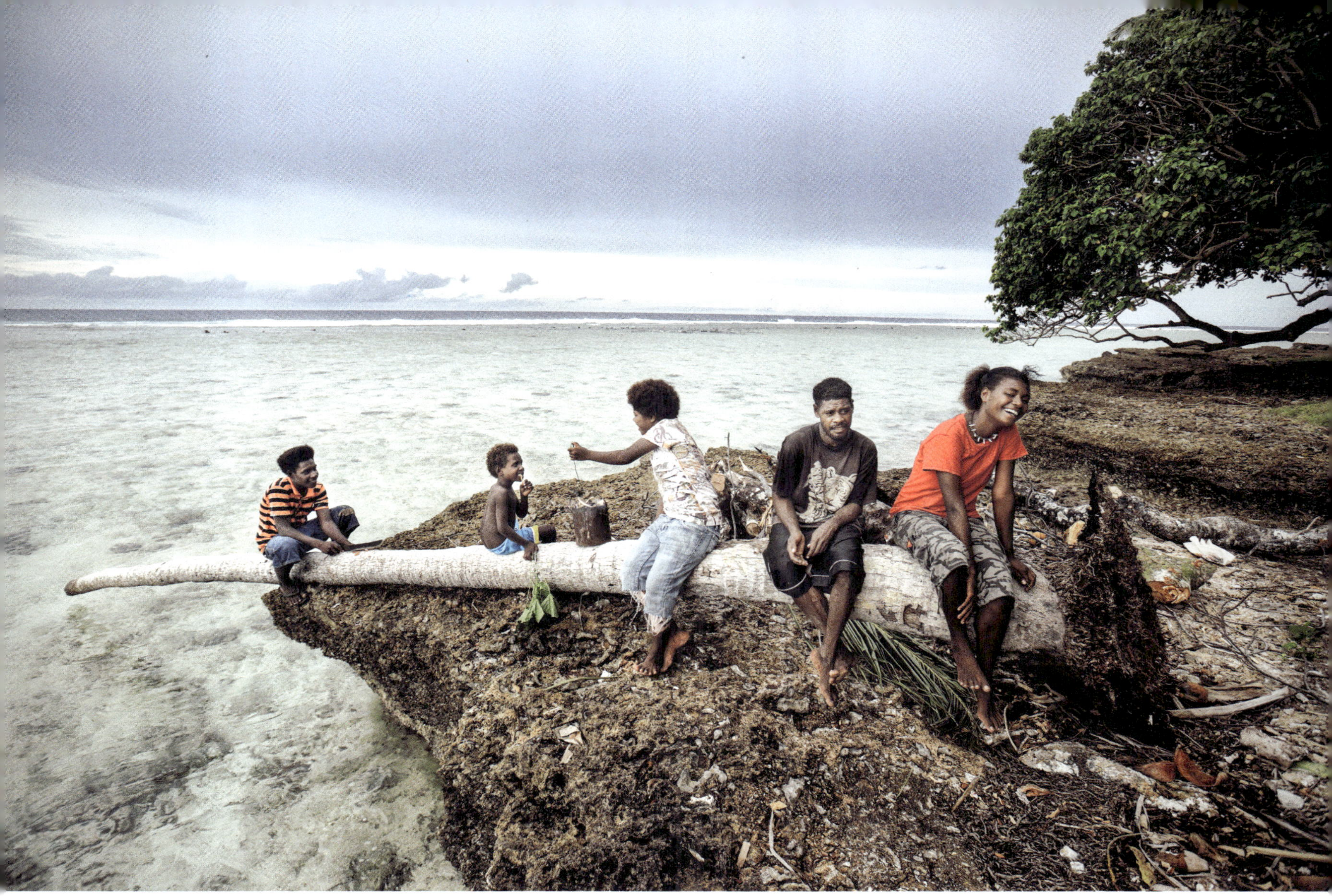

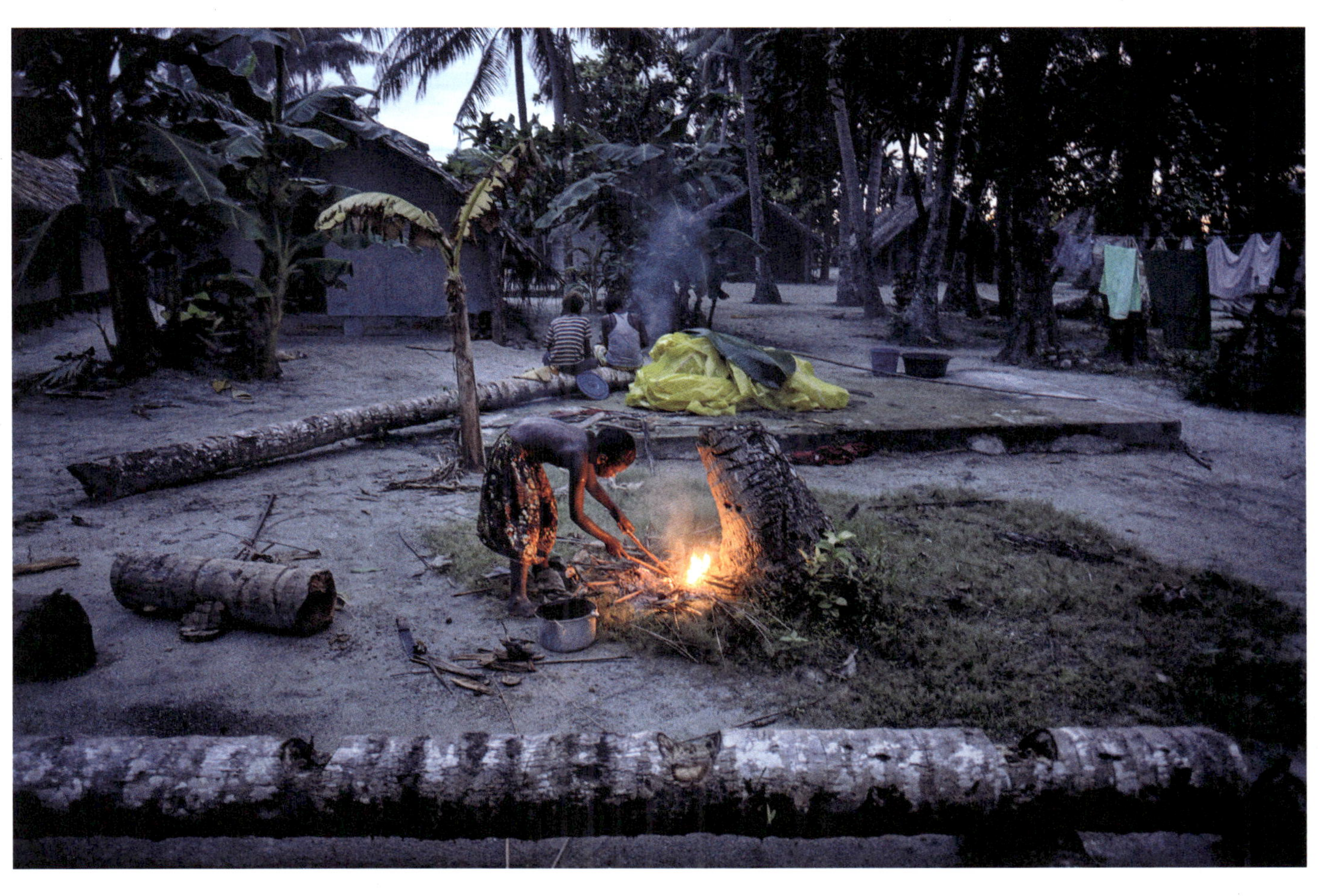

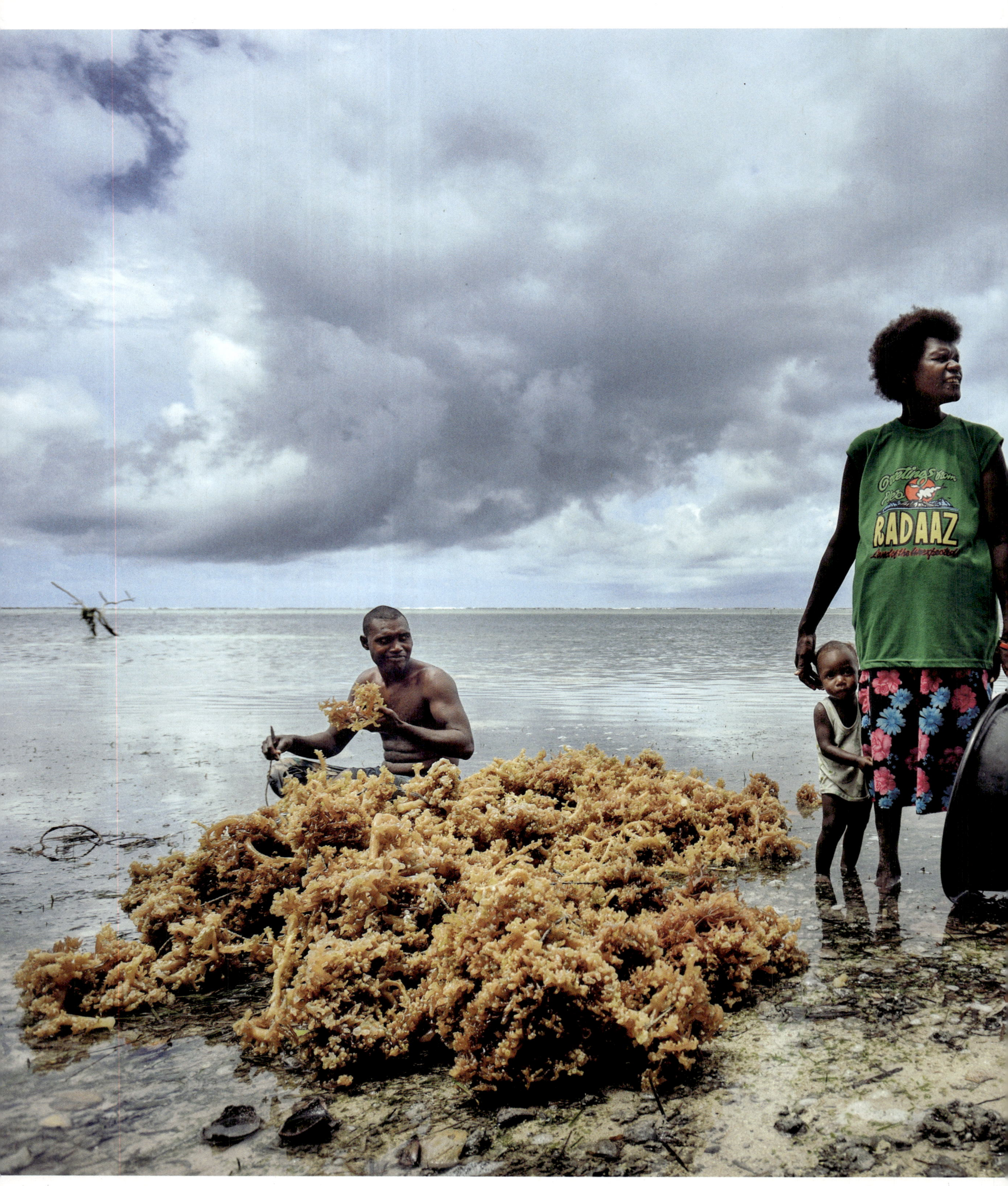

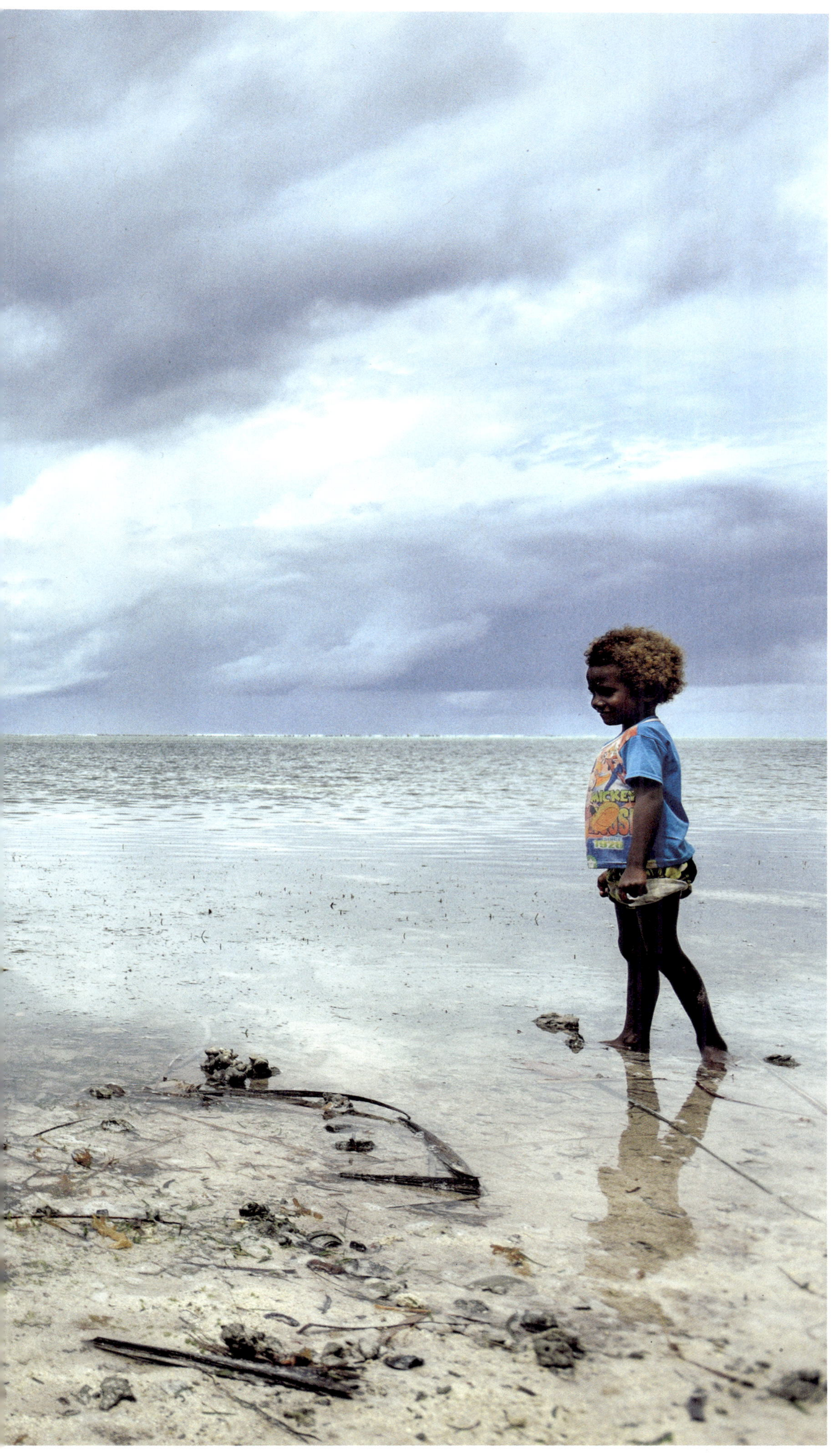

Harvesting seaweed for the cosmetics industry, is a project where the Islanders can earn a little income.

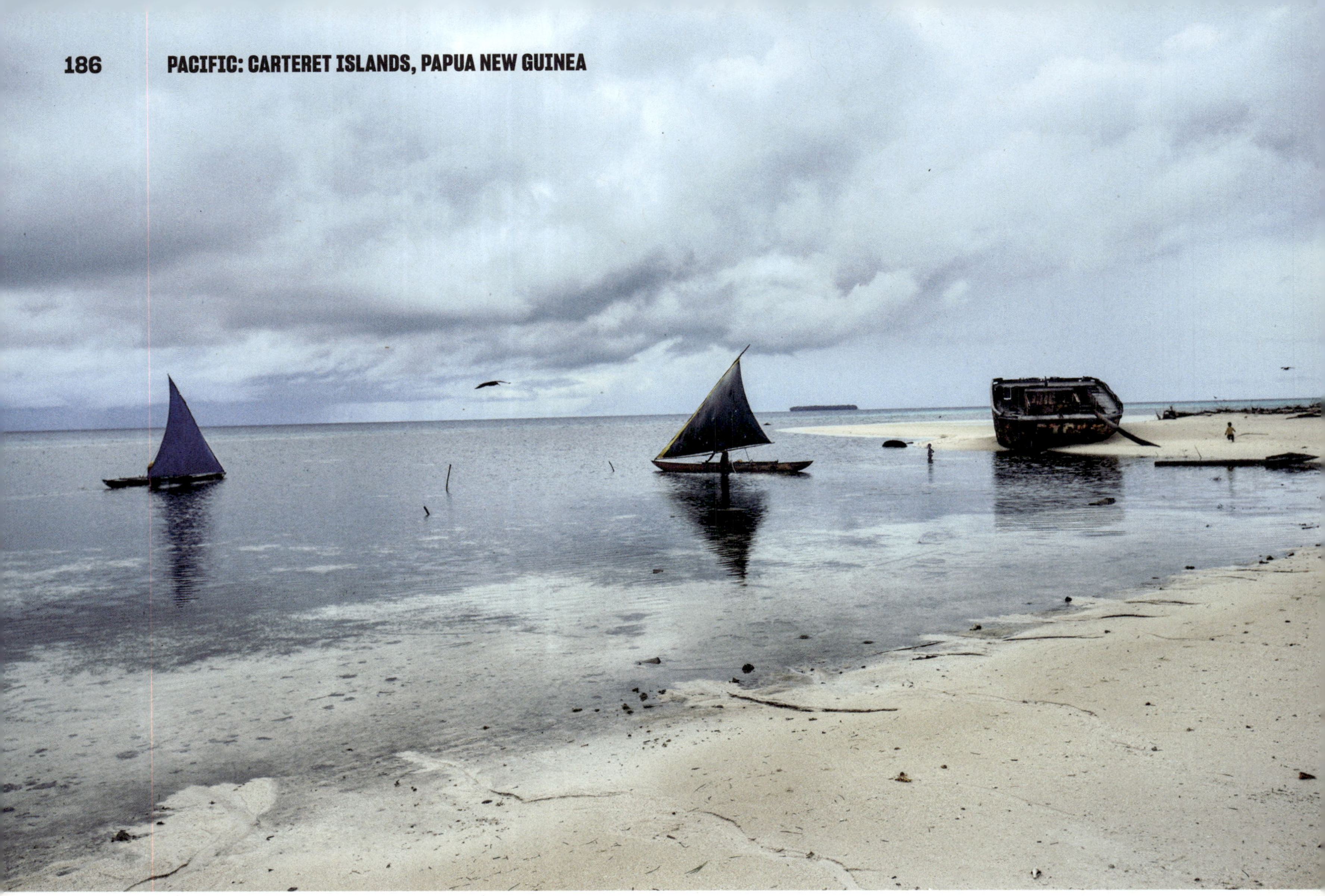

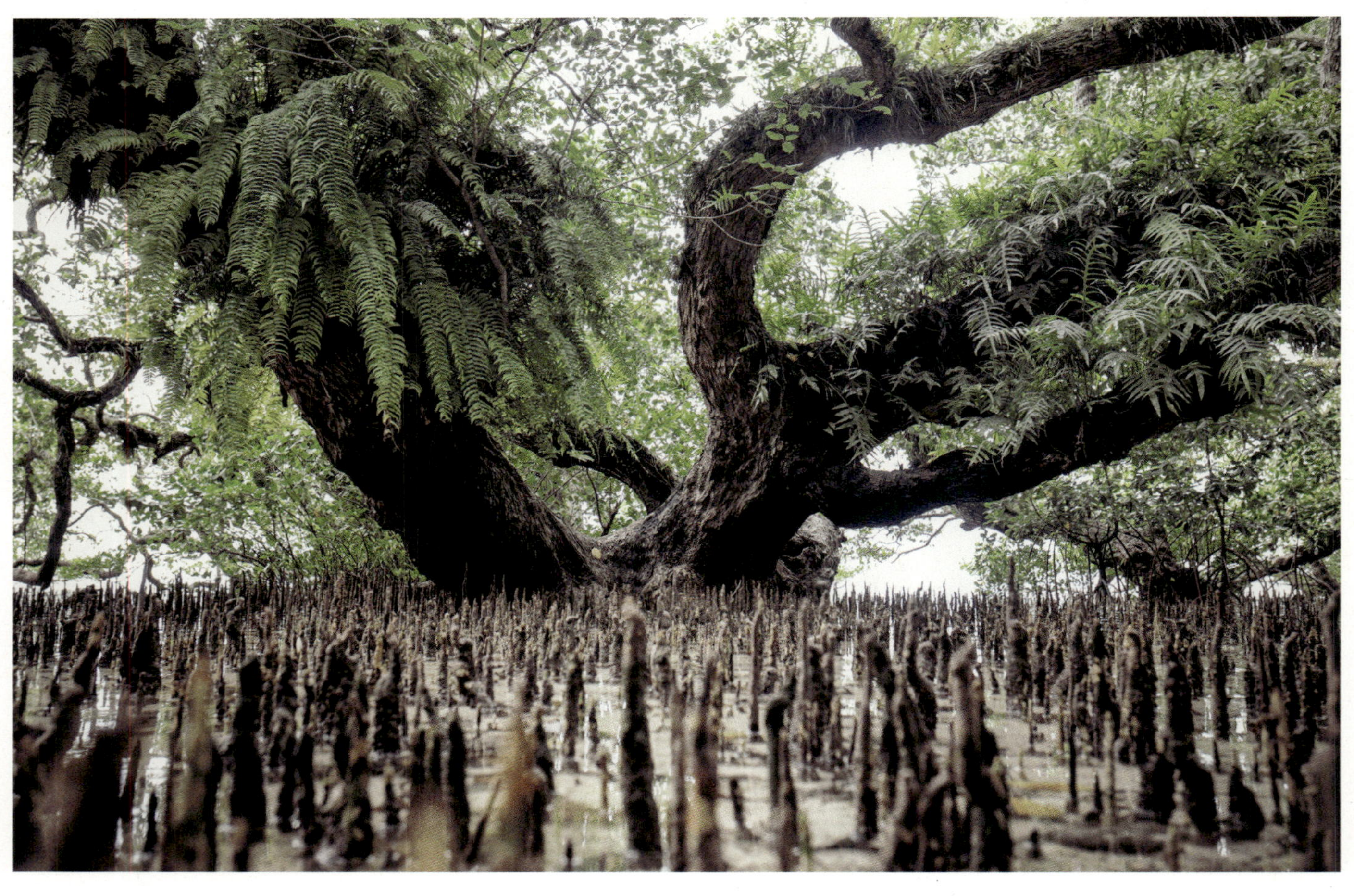

Mangroves are being planted
to prevent coastal erosion.

Inhabitants of Yesila Island face even
more challenges than the people of
Han Island, including very serious food
shortages. Malnutrition is common.

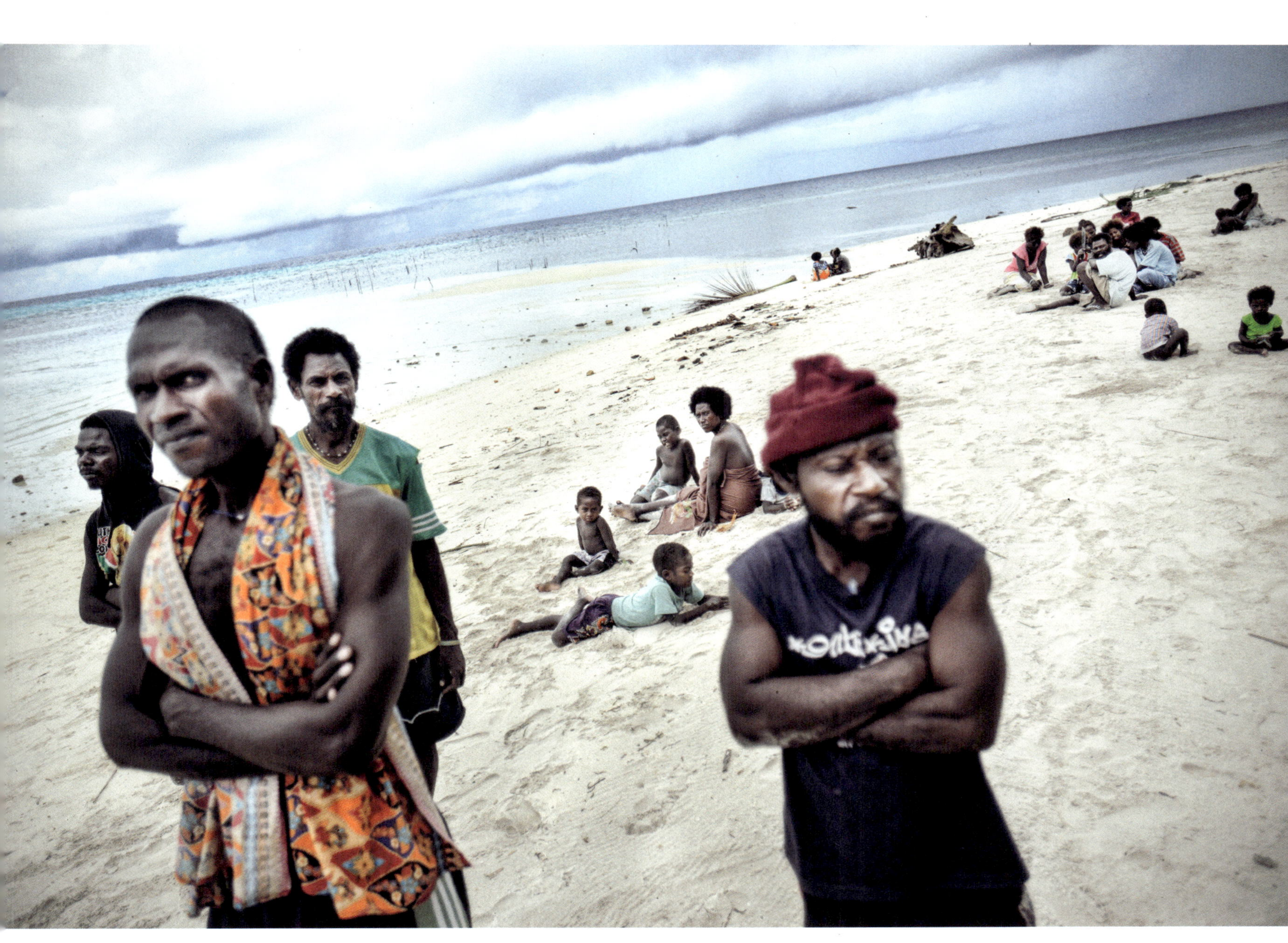

After months of delay, a ferry is about to leave the harbour of Buka on Bougainville Island with relief supplies for the Carteret Islands. Unfortunately, the ferry carries only 600 bales of rice, which will not last for more than two weeks. Flour, oil, medicines and more are also needed.

The stores on Han Island are empty. There is no regular boat connection to Bougainville Island anymore, and promised relief supplies haven't arrived.

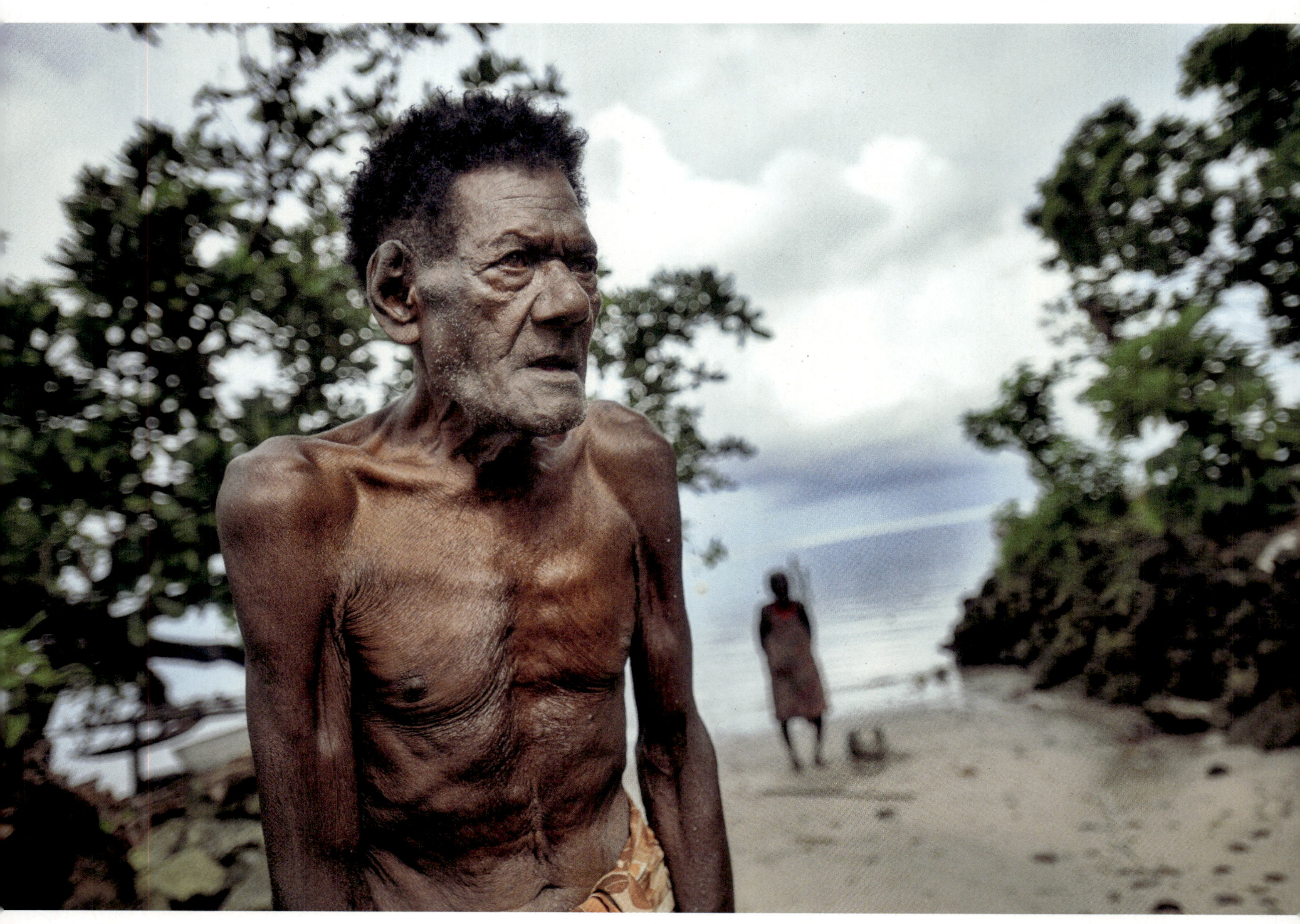

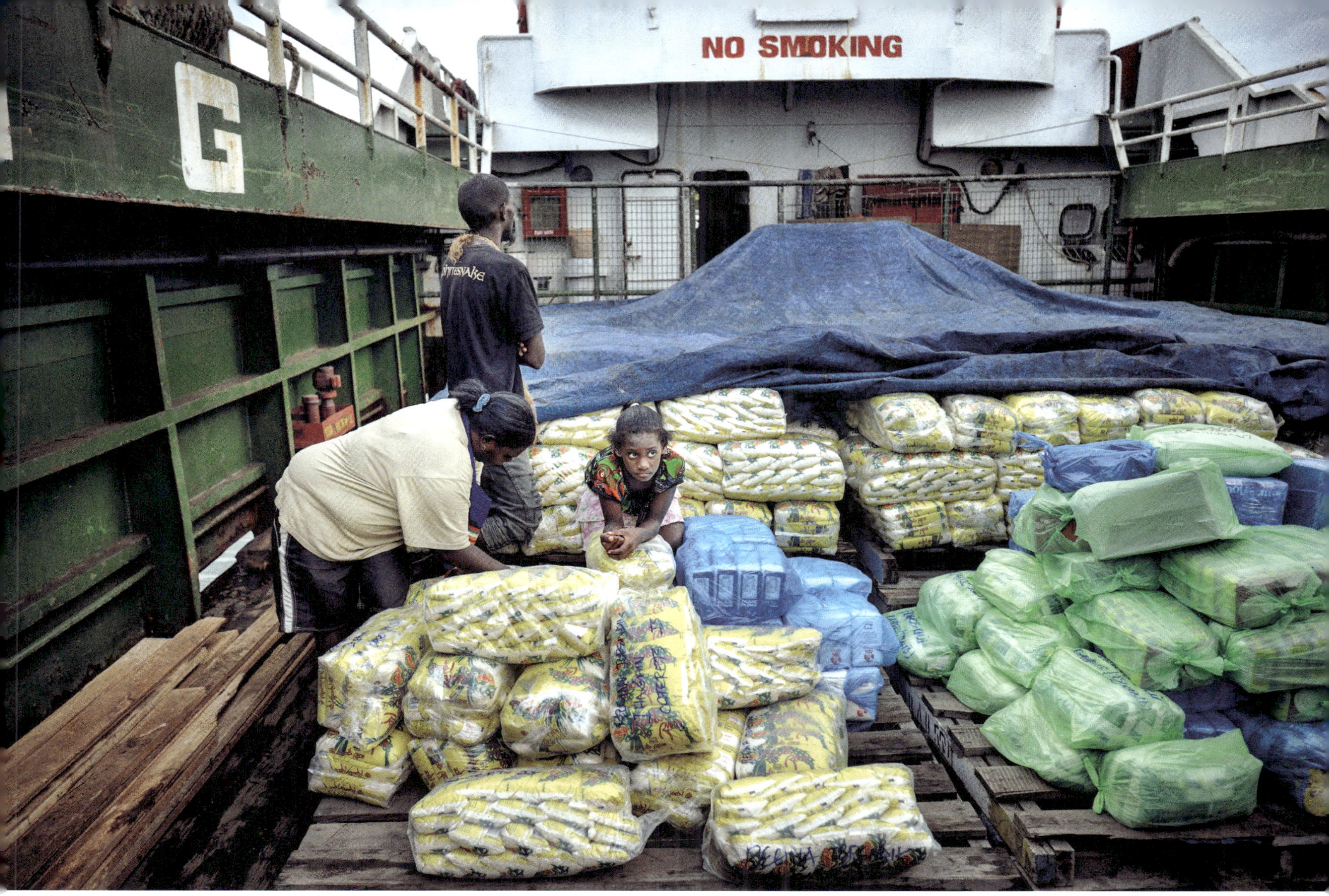

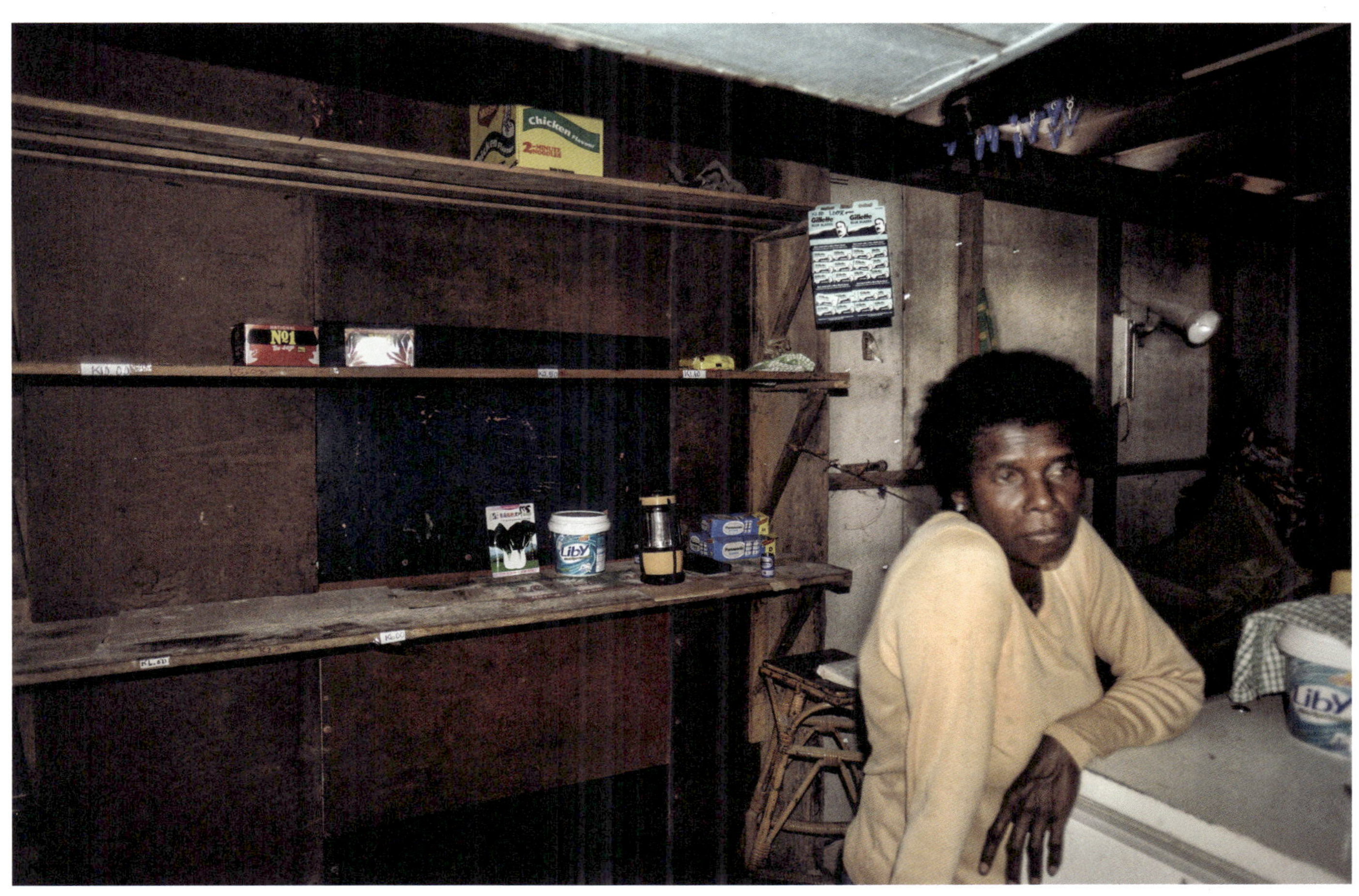

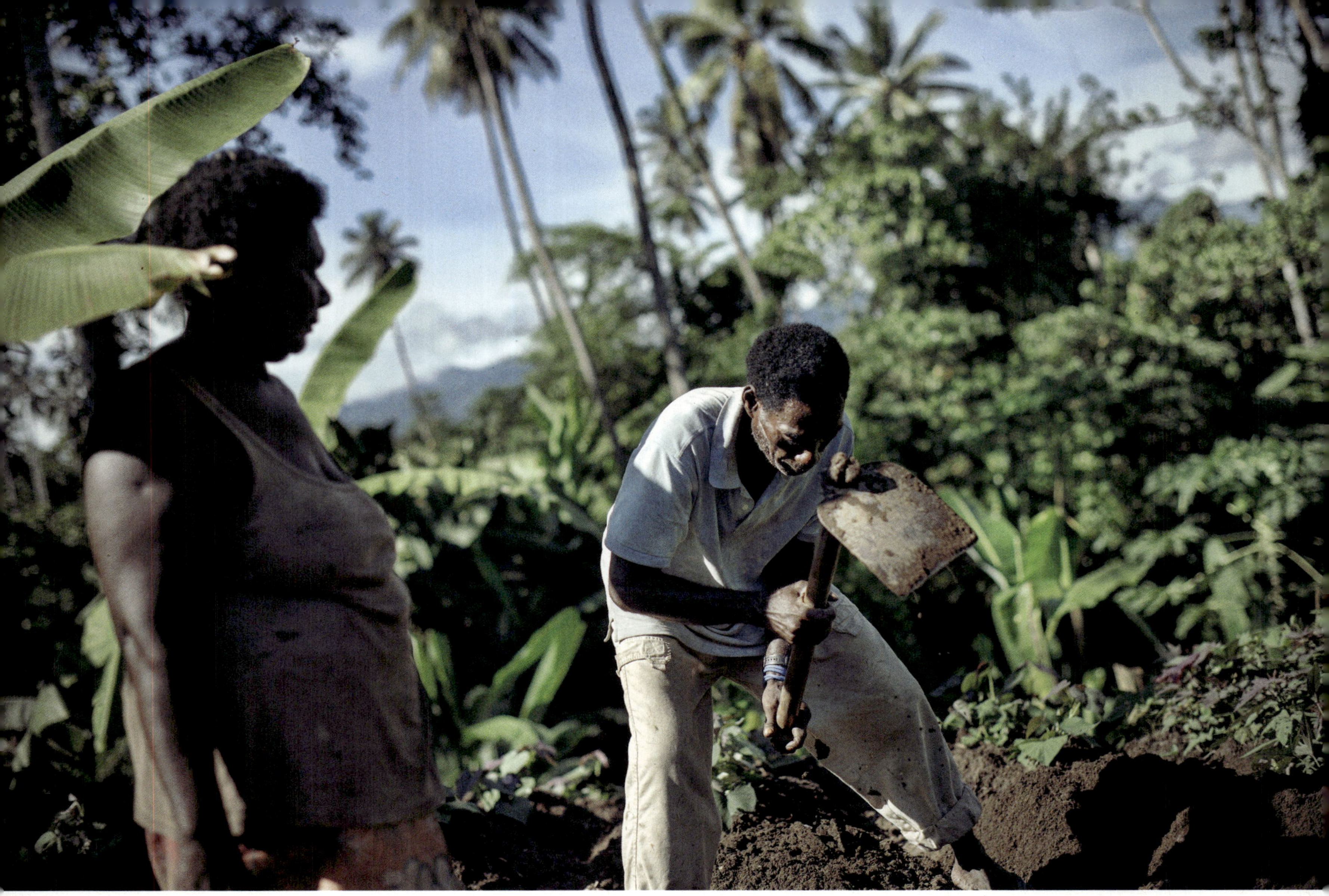

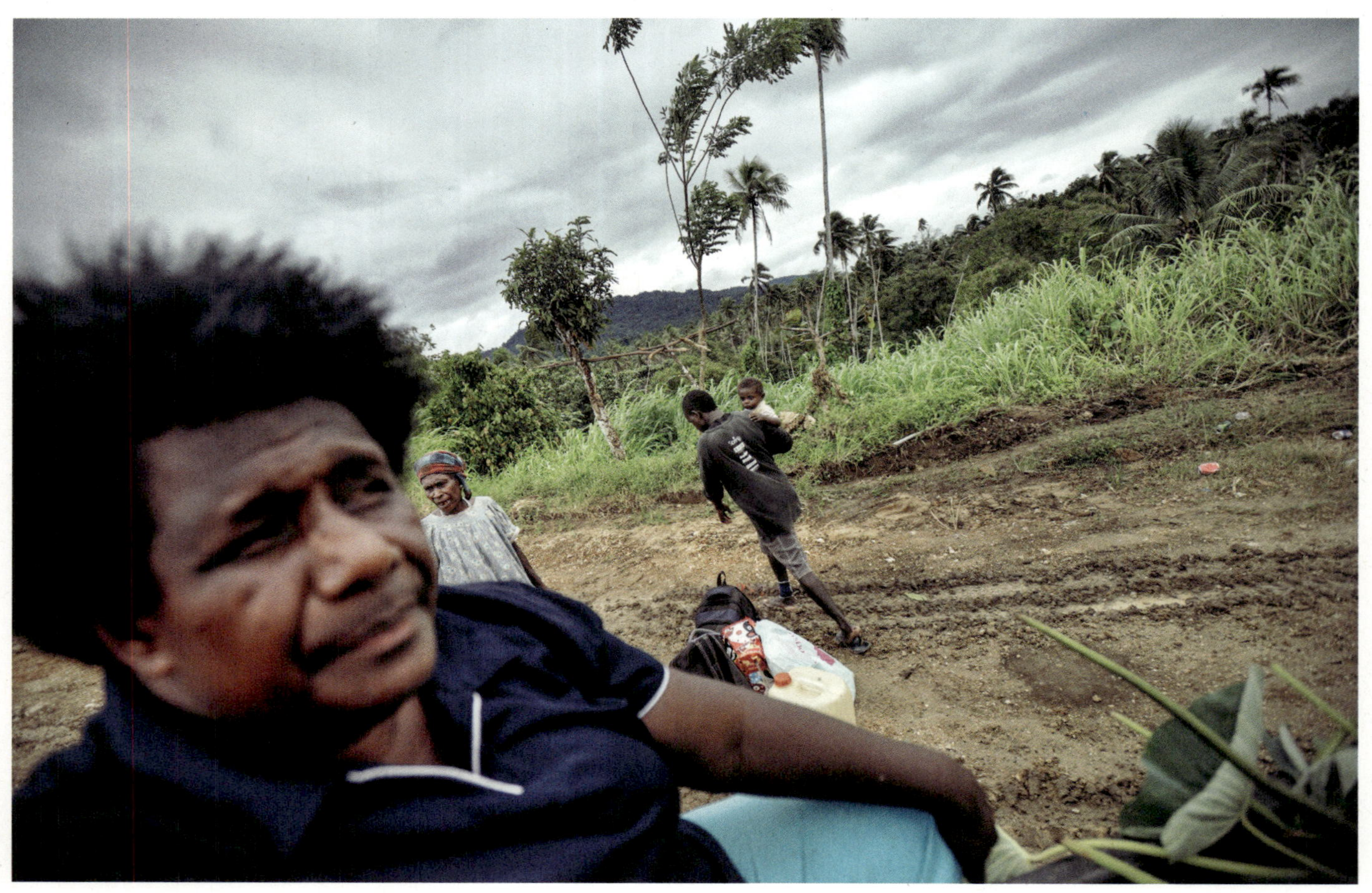

The people from the Carteret Islands have taken the initiative to relocate to Tinputz, on Bougainville Island, after the government's planned relocation failed.

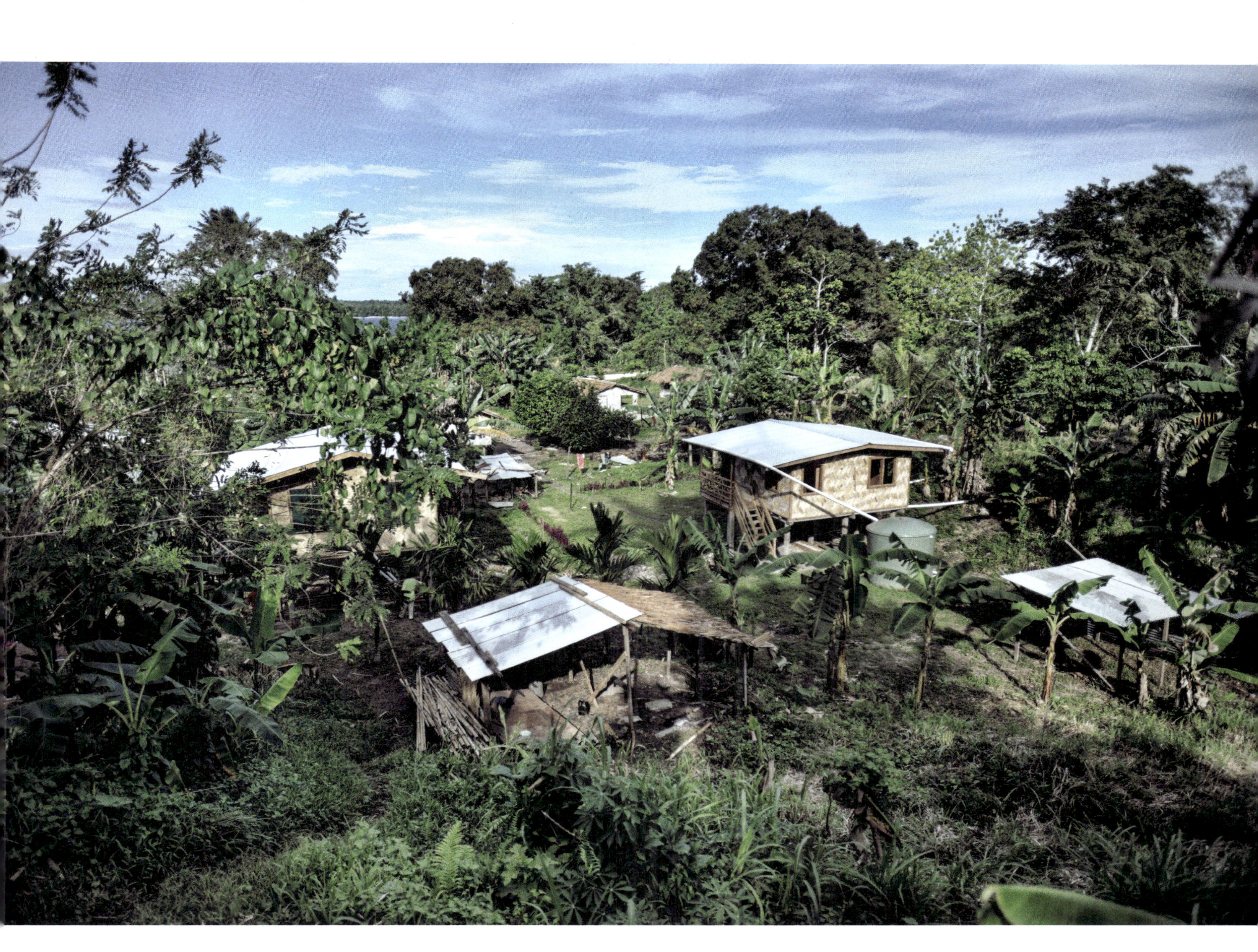

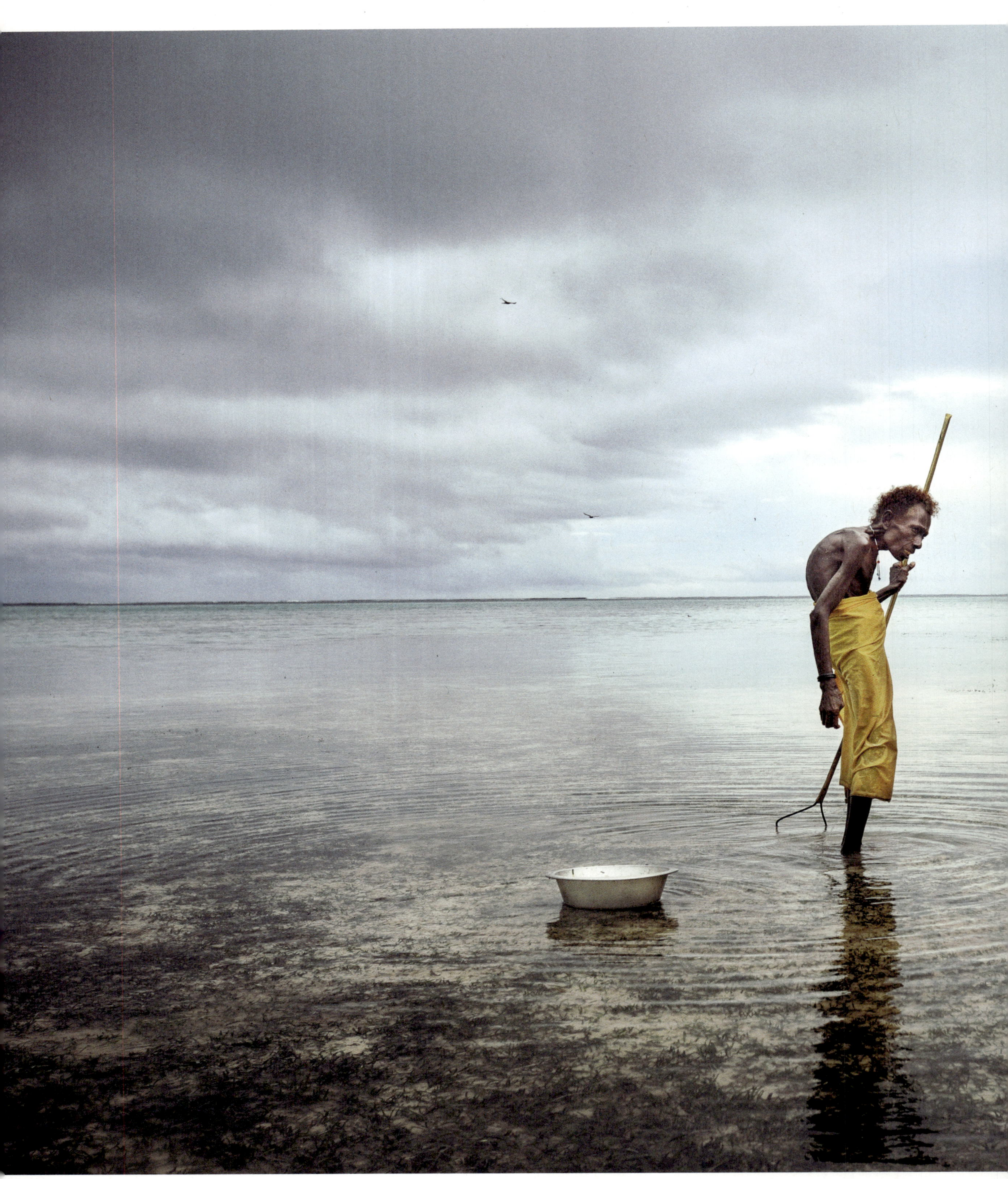

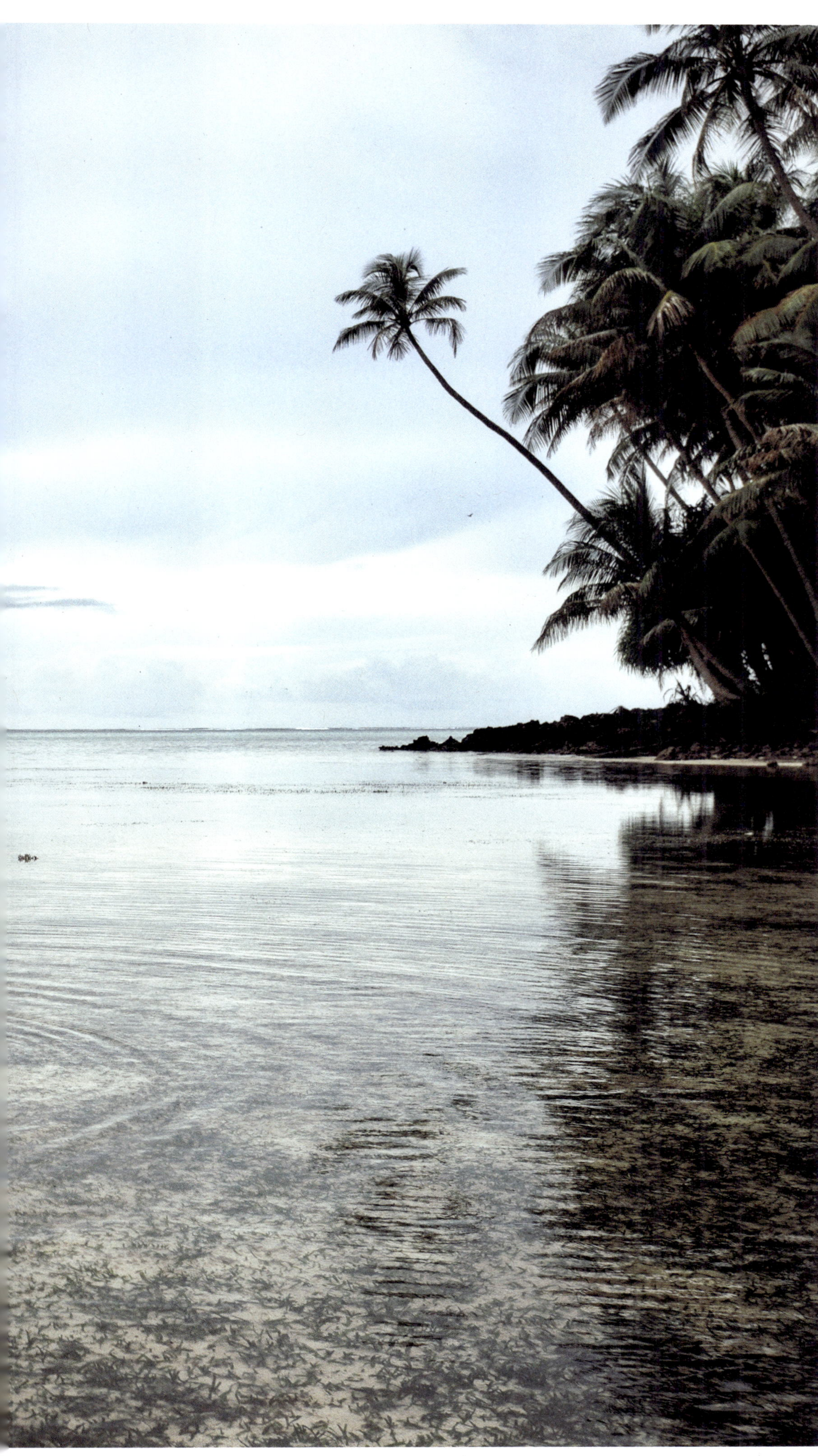

A woman 'harvests' the seeds
of water plants to cook to have
something to eat.

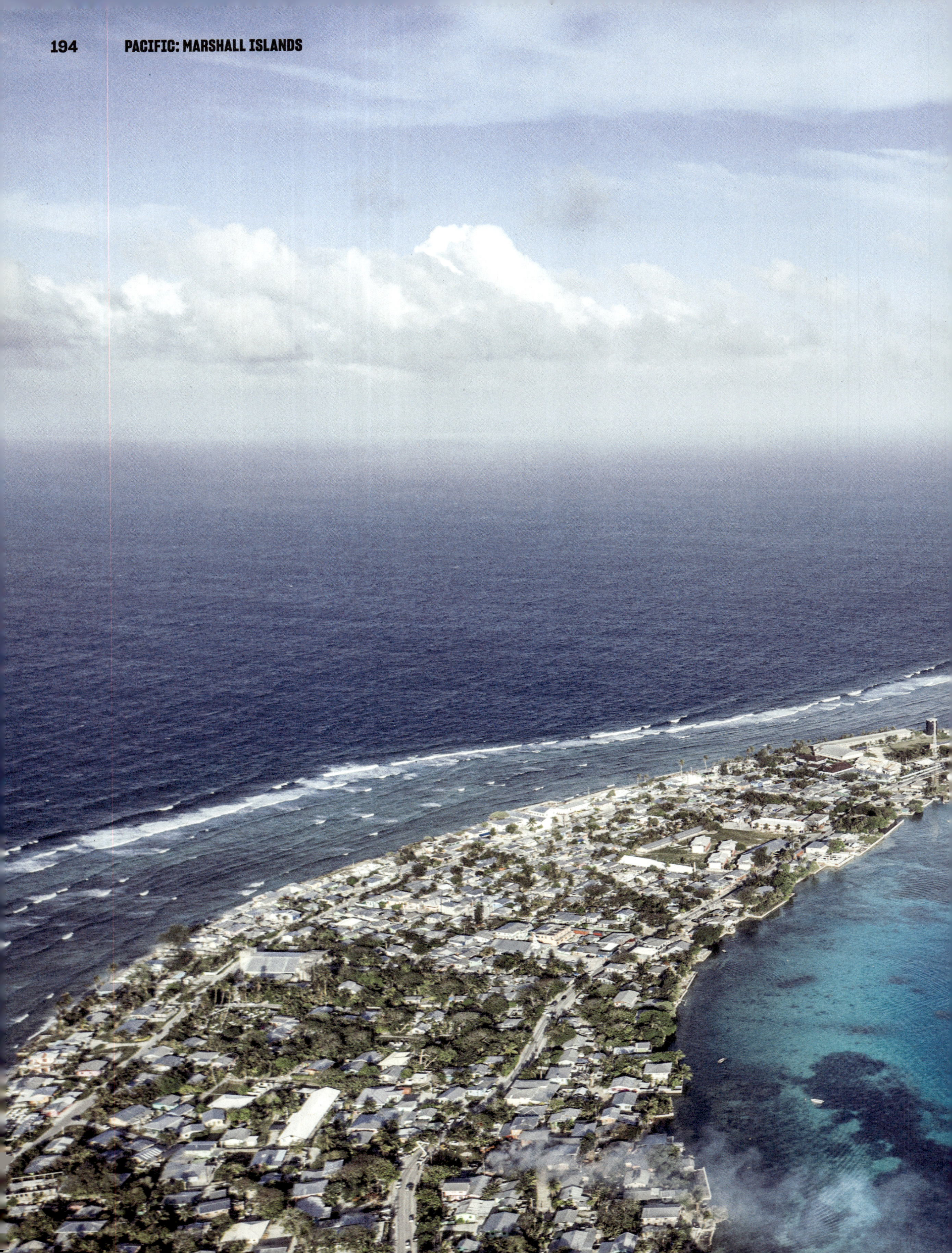

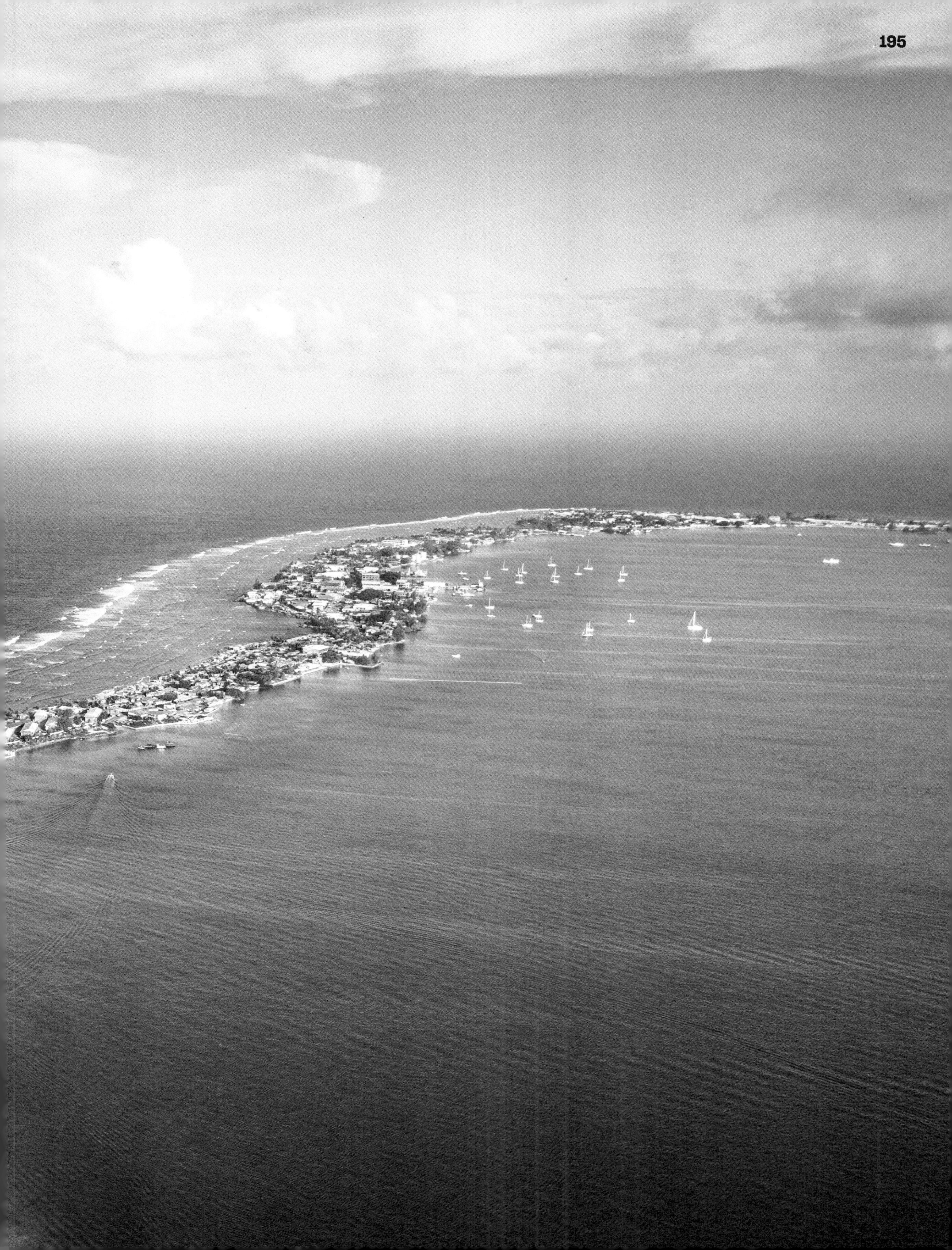

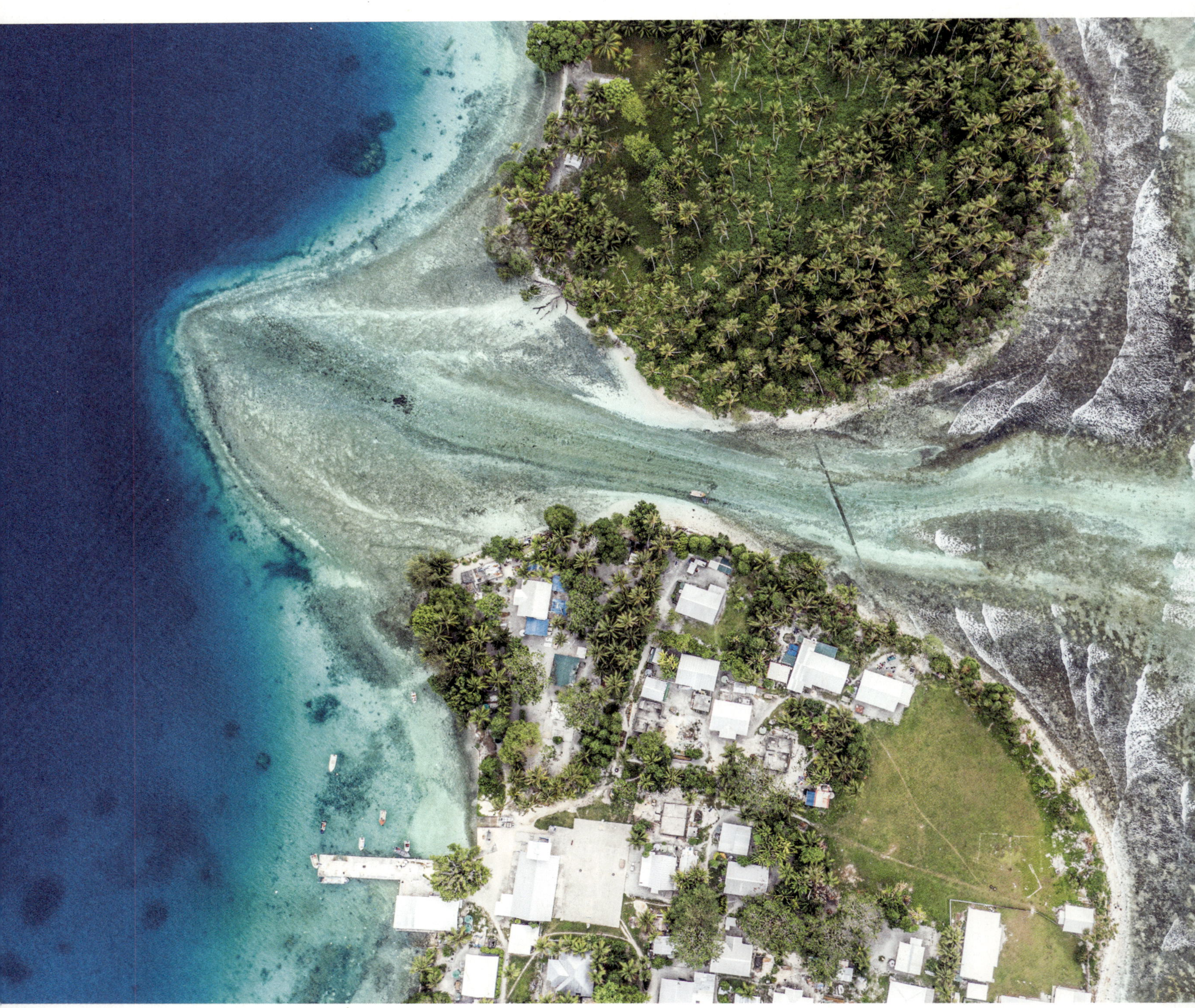

< Pages 194-195: Majuro is the capital of the Marshall Islands and is very densely populated. The government of the Marshall Islands wants to raise several of the islands to make sure they don't get submerged.

The Marshall Islands are an independent nation in the Pacific and an associated state of the United States, which means that its people can also get residency in the United States. The country is probably best known for the nuclear tests the United States conducted on Bikini Atoll. The population is close to 60,000, spread over different atolls.

Ejit Island is part of Majuro, the capital of the Marshall Islands. Ejit is inhabited by people from Bikini Atoll who were evacuated because of nuclear testing. The country is very vulnerable to sea level rise, and if nothing is done, it will become submerged.

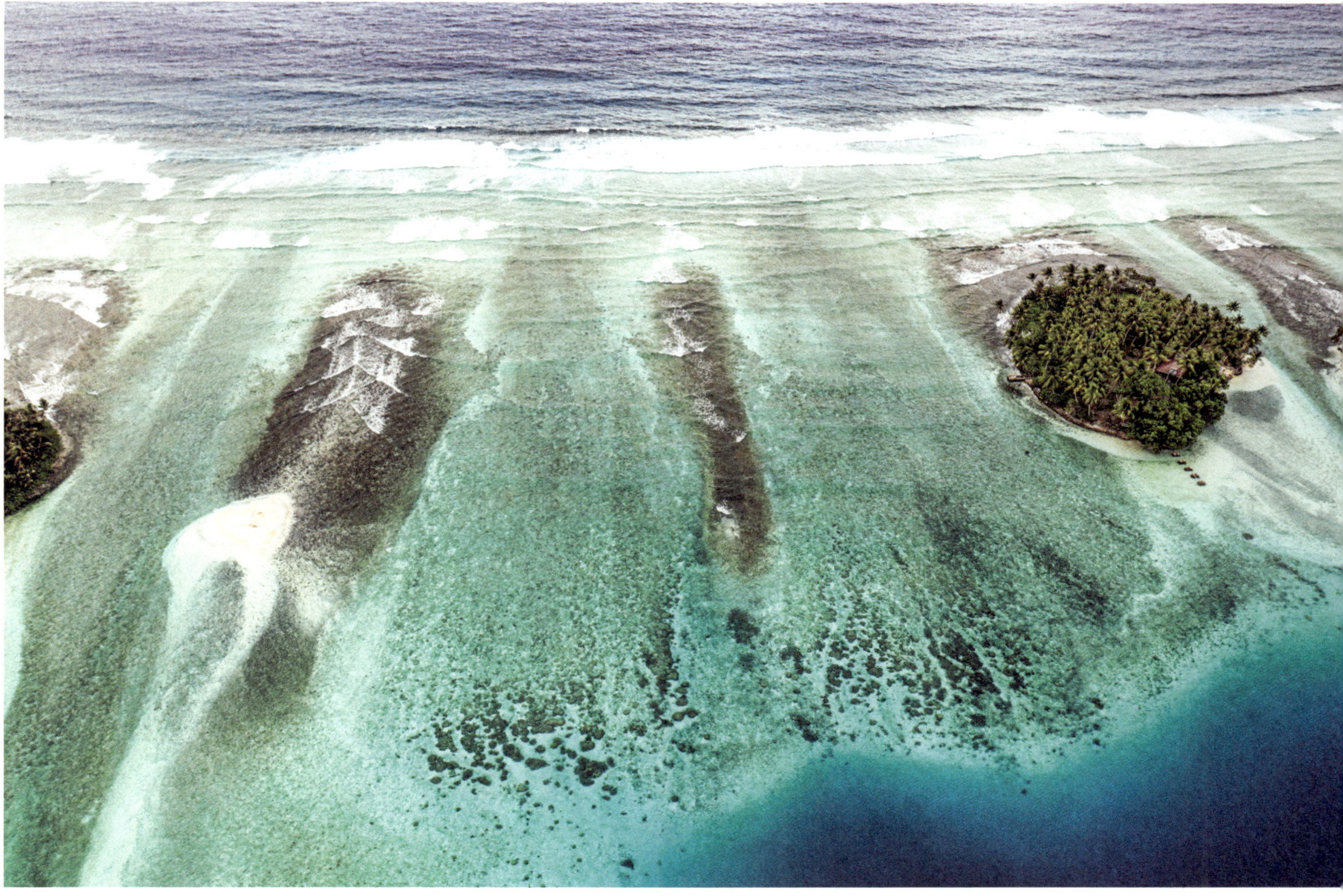

Listoro Ralpho lives with her family
on Majuro.

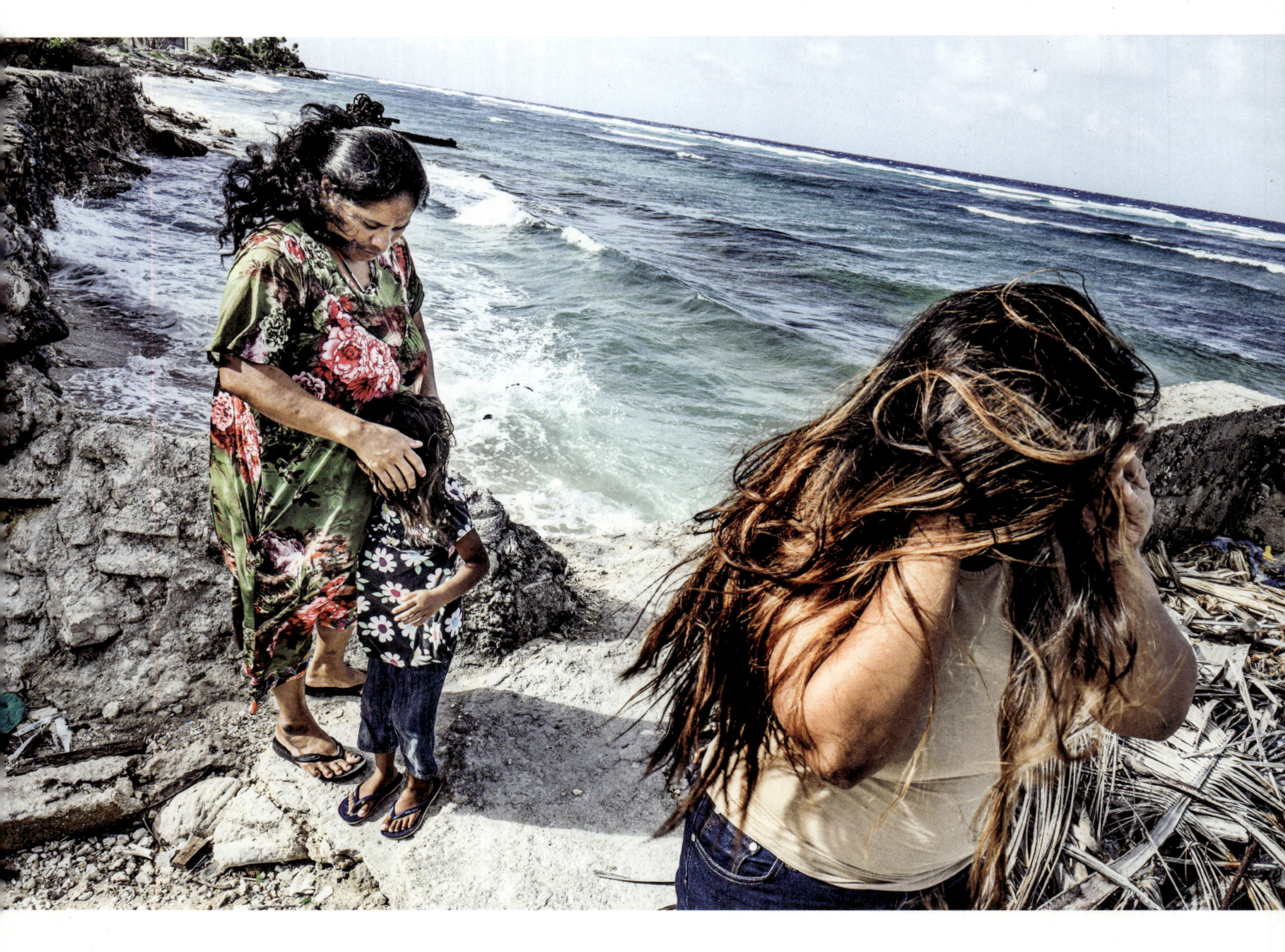

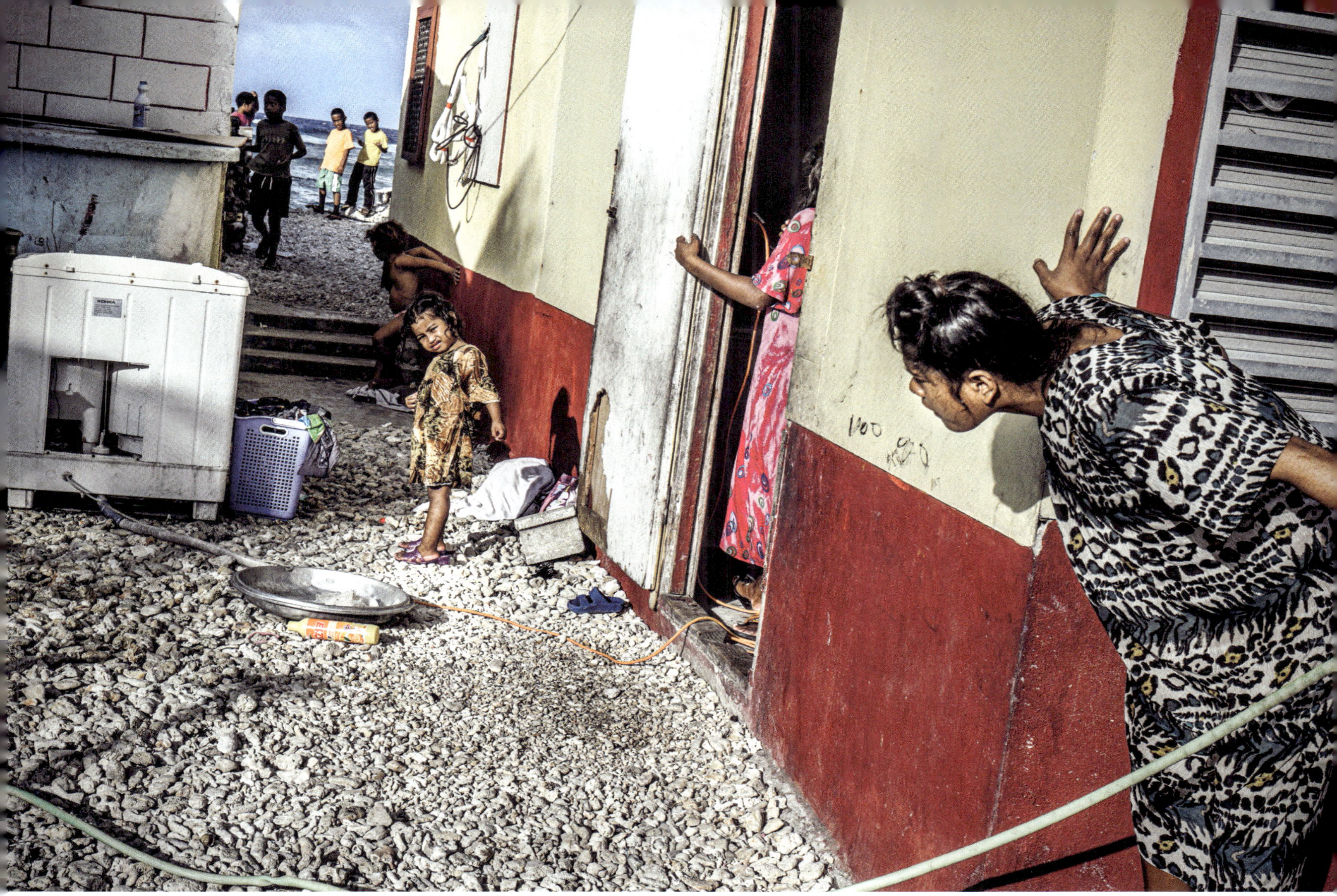

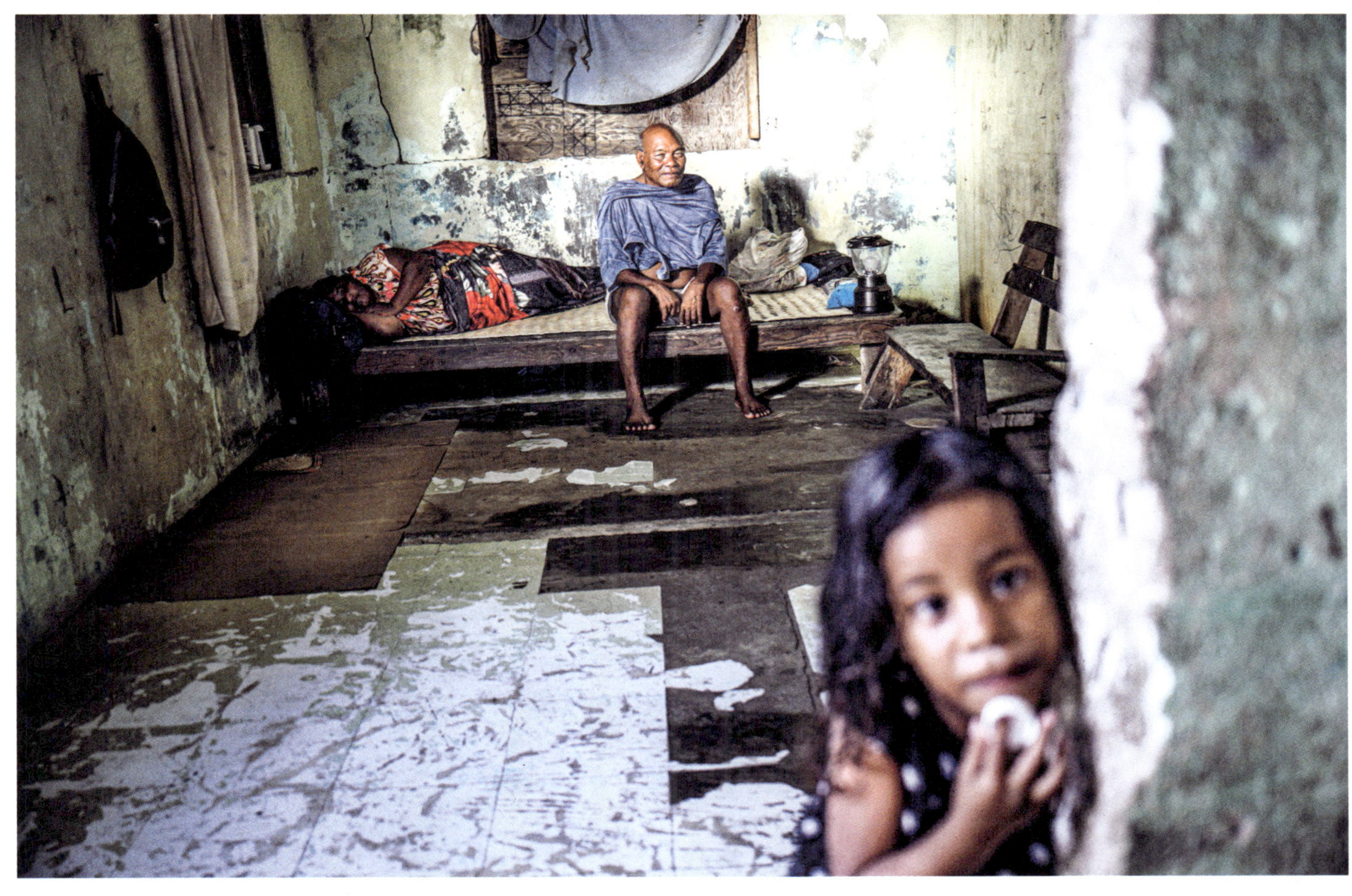

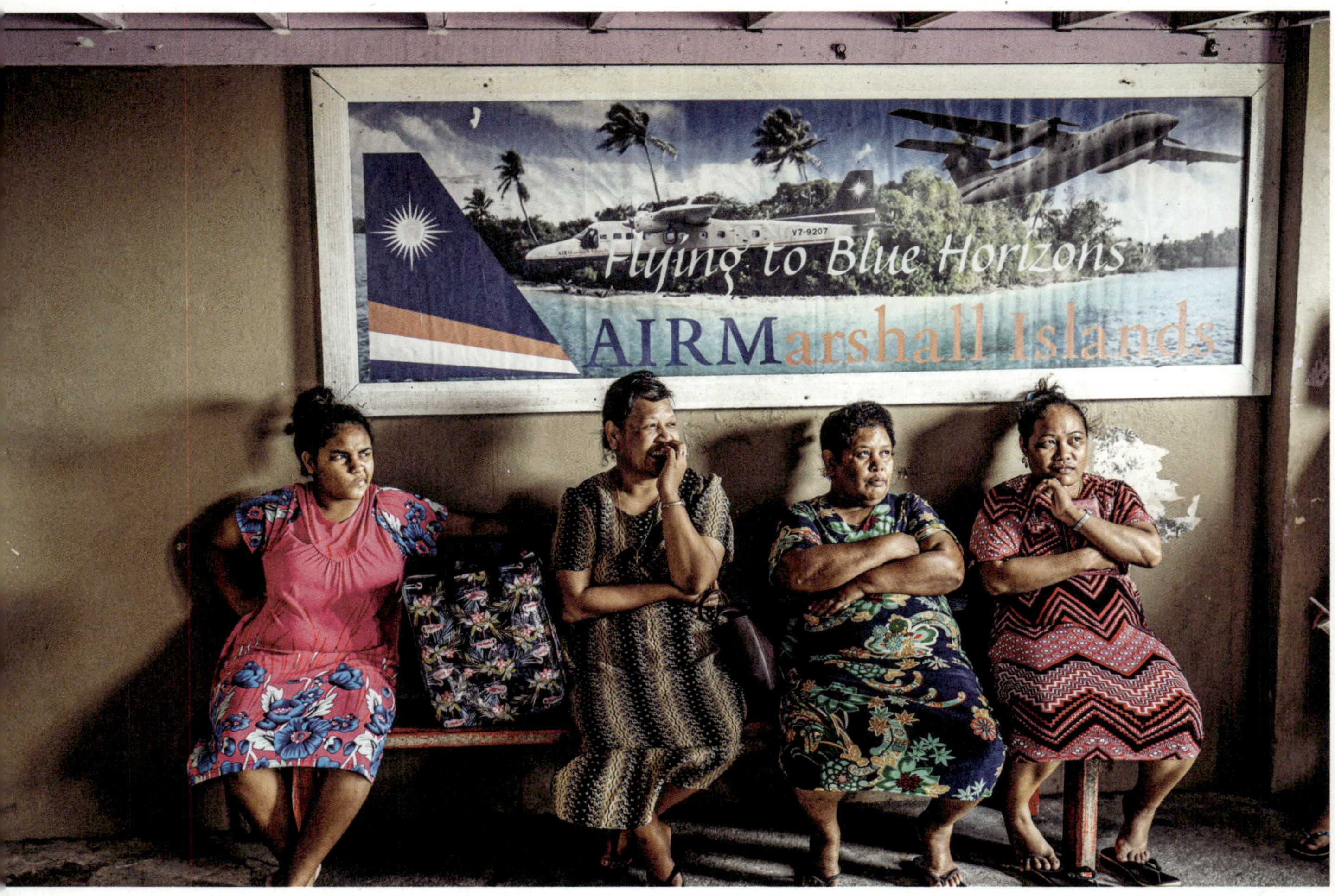

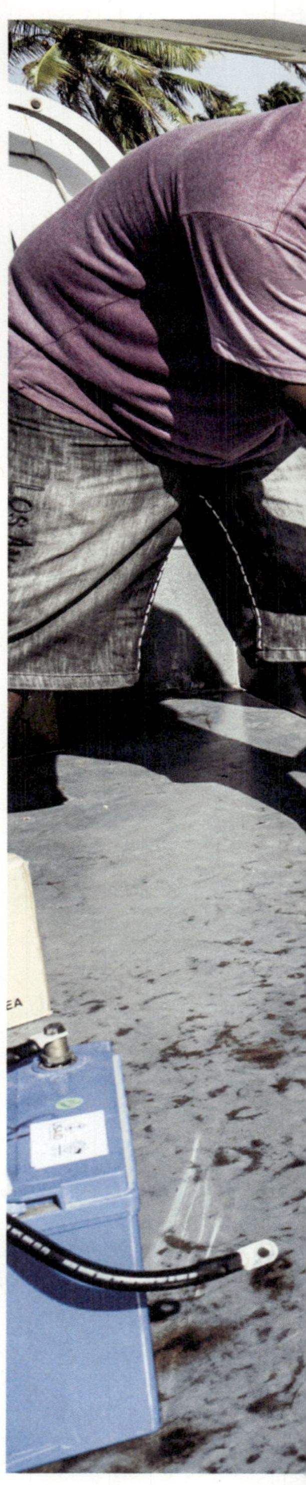

The departure terminal at Marshall
Islands International Airport on
Majuro.

Air transport is often the only means of transport between the atolls. In addition to passengers, planes bring supplies.

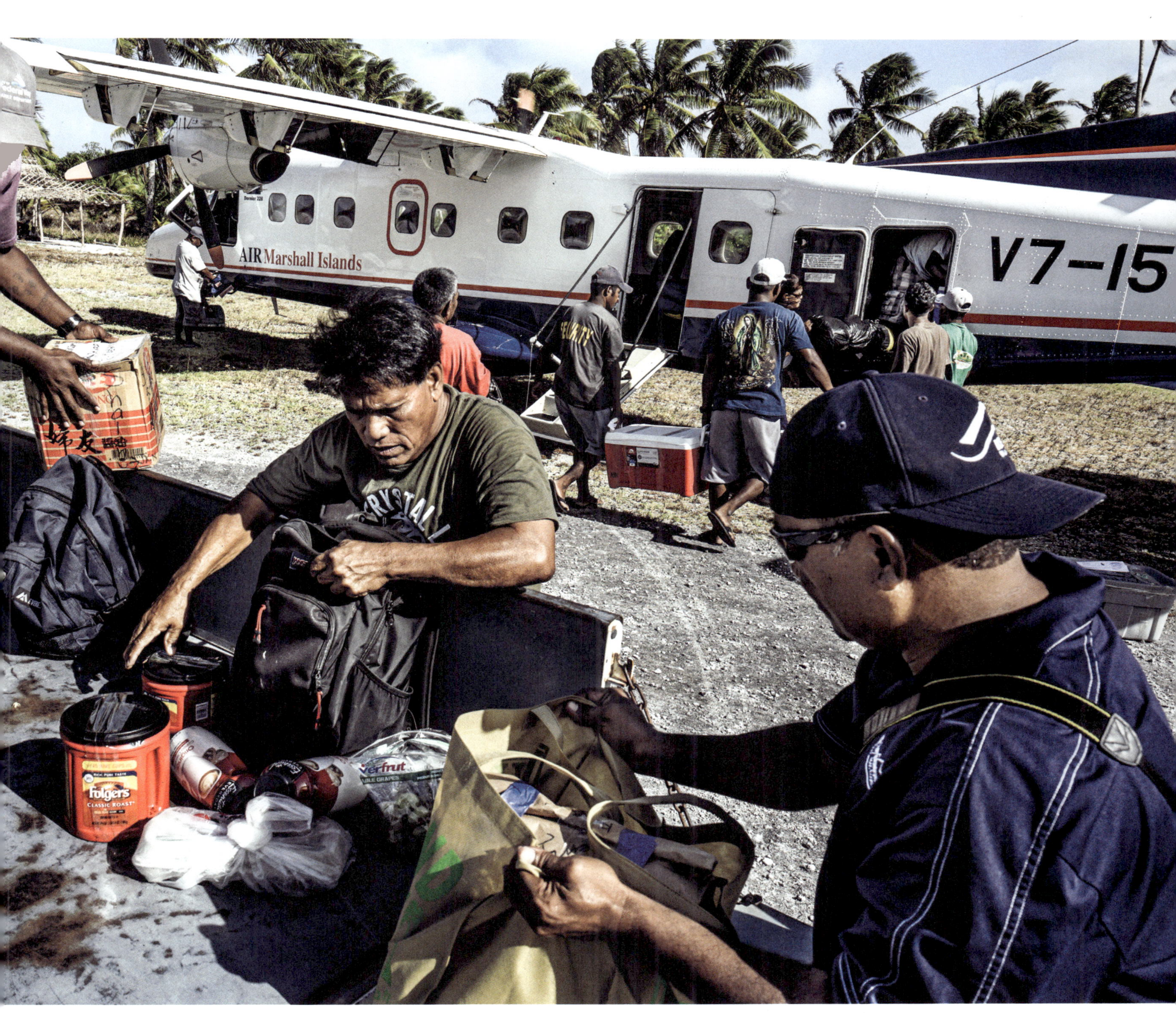

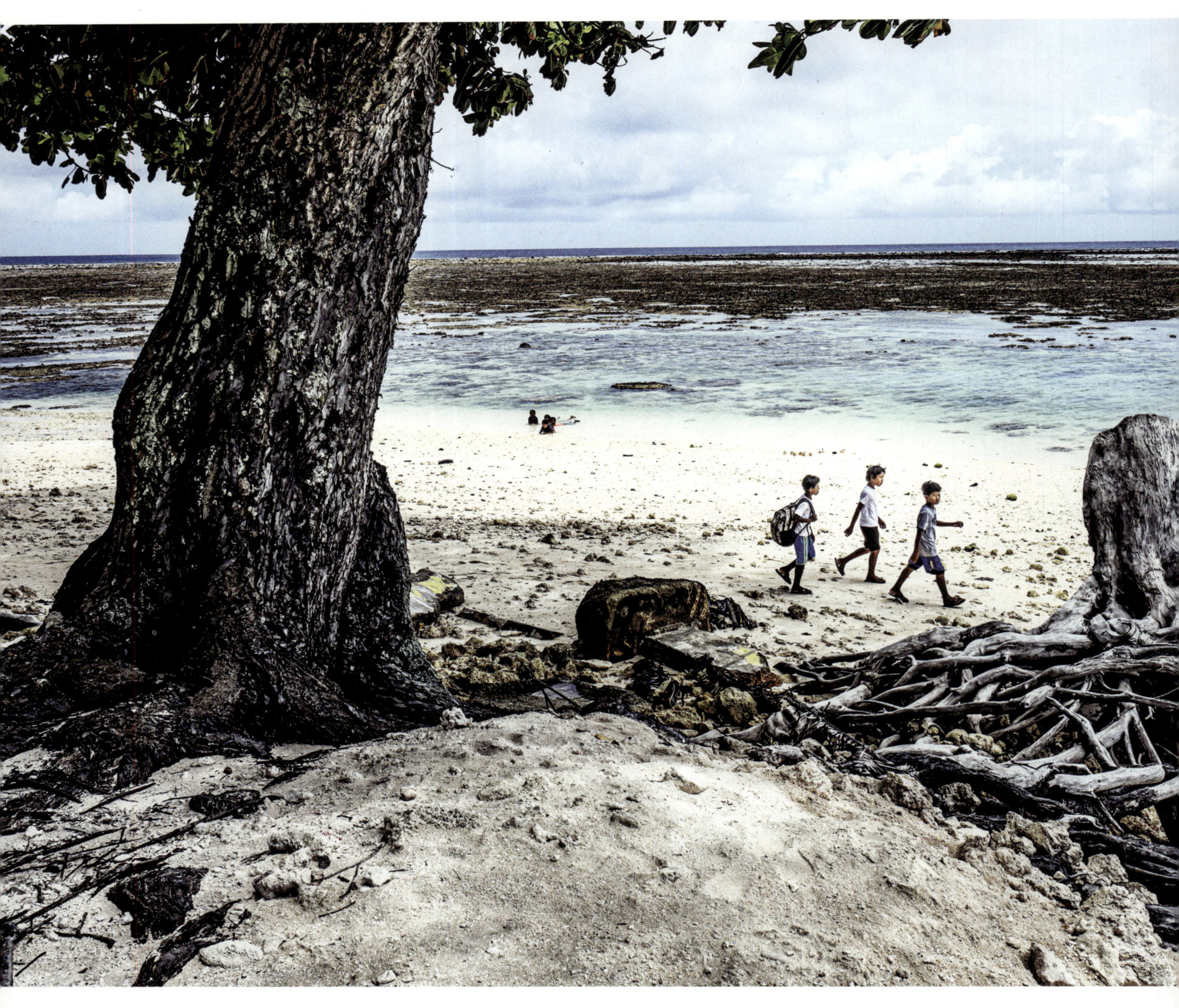

The coast is eroding at a rapid pace
on Majuro. Most beaches have disappeared, and the coast has been
fortified by often self-built sea walls.

A private sea wall is being constructed
in Majuro.

> Pages 204-205: A home on Majuro, which is right on the shore. Most beaches have disappeared, and the coast has been fortified by often self-built sea walls.

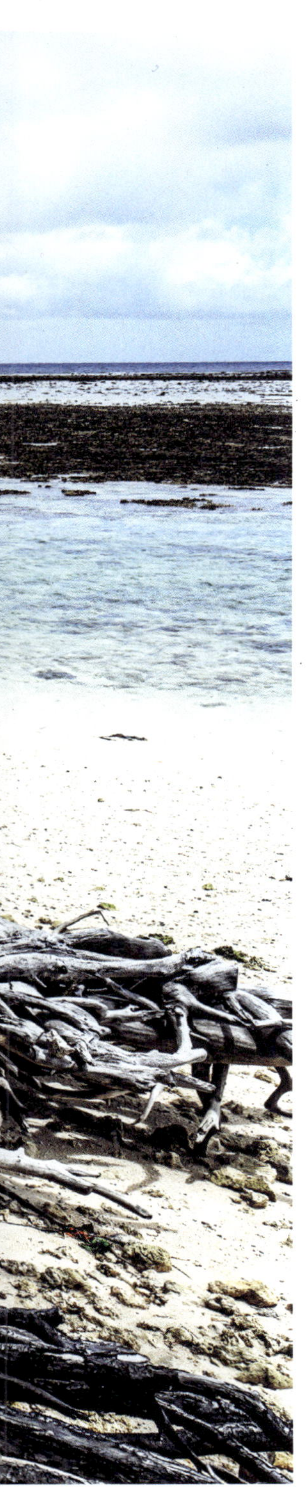

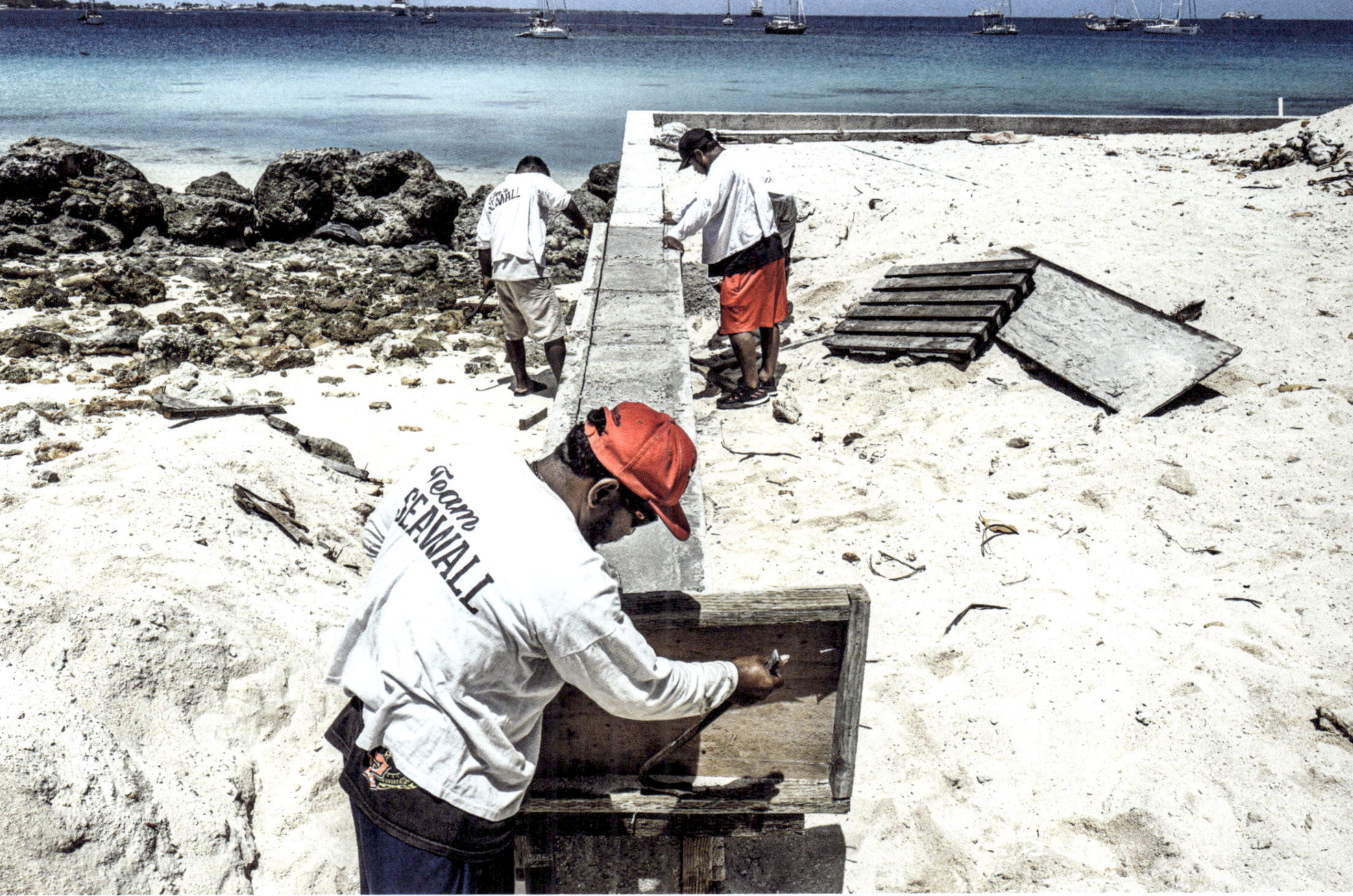

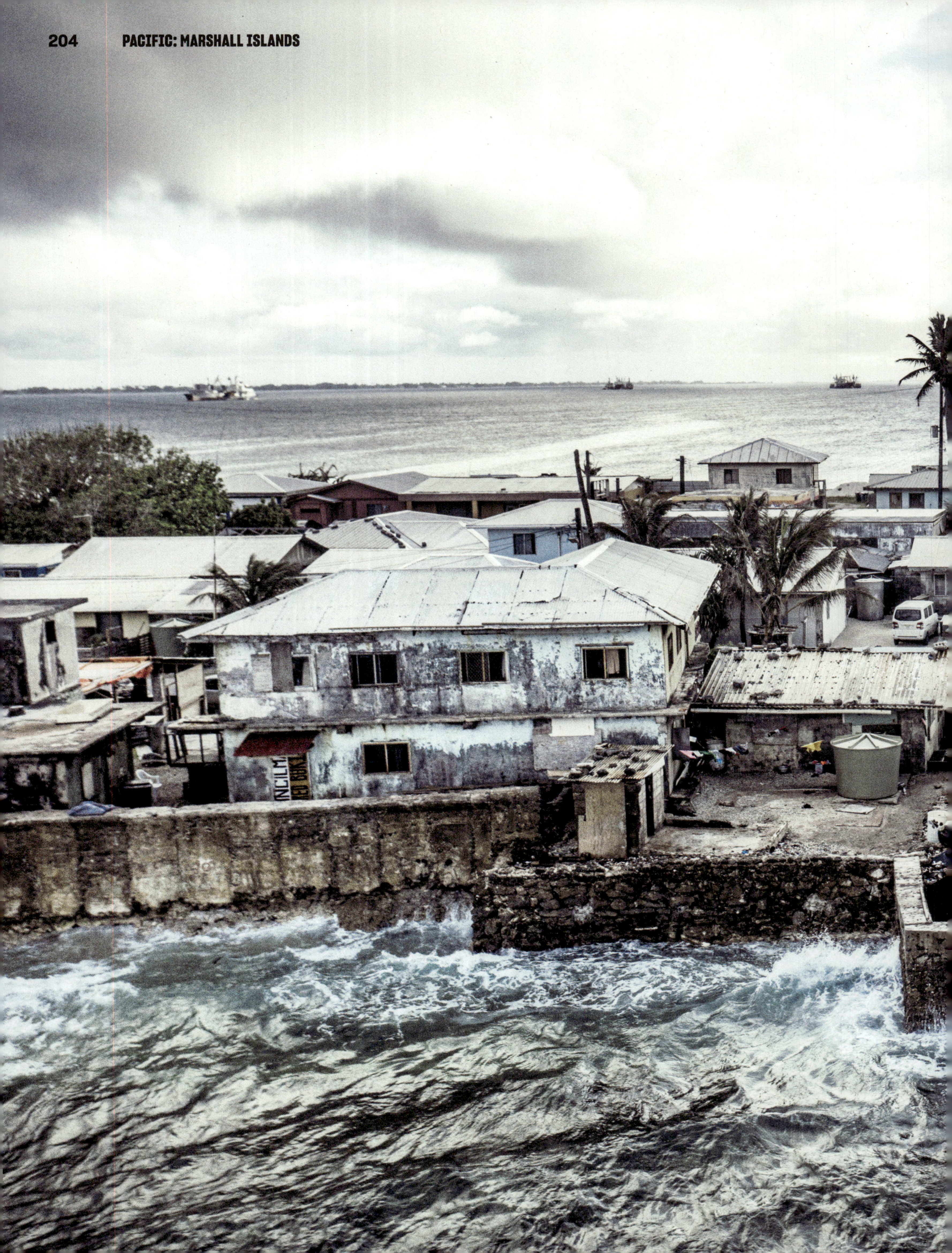

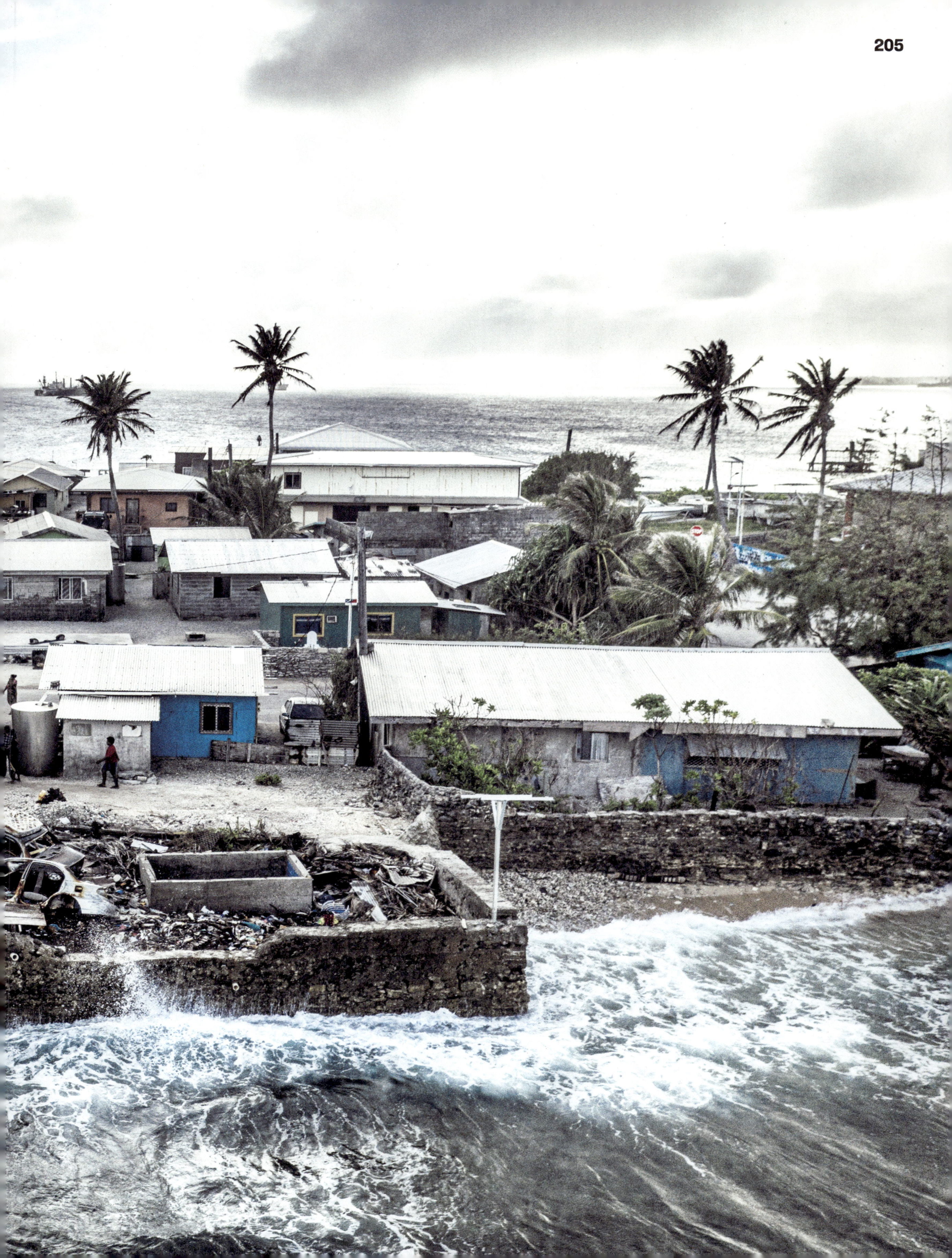

BANGLADESH
Sharif Jamil

Perhaps no country is more affected by rising sea levels than Bangladesh. In the near future, up to 50 million people may need to be relocated from the delta. Many people have fled already due to rising waters, the salinization of farmland and erosion, moving to the poor parts of the larger cities. But Bangladesh is not waiting for future disasters to happen and has adopted a US$40 billion delta plan, which might keep some coastal zones safe a little longer.

If it weren't for the life-shattering impacts of the climate crisis, Manjum, a 34-year-old father, would be living with his parents, siblings and three children. The family shared a cottage in Padma village, where the Baleshwar River flows into the Bay of Bengal. Padma was a beautiful village inhabited by thousands of people, an administrative village in Ward 6 of Patharghata Sadar Union in the Barguna district of Bangladesh. Manjum and his 50-year-old brother were both born in this village. Their childhood was passed in the school and *madrasah* with friends and relatives. They met friends in the fruit garden that wrapped around the big ponds in the village, played together in the paddy fields and swam in the river. Most of the villagers earned a living from fishing or farming. The fertile lands surrounding the village yielded crops three times a year. Fishing was possible year-round since there were no restrictions on the abundant aquatic life in the Baleshwar River.

But on 15 November 2009, everything changed for Manjum and all of the people in Padma. Super cyclone Sidr hit with 215 km/hour winds and a 5 m storm surge that washed away the entire village. People lost their homes, their vehicles and all of their belongings, leaving them destitute in one night. A few kilometres away, the villages of Jintola and Patuakhali also underwent the same fate. Cyclone Sidr killed 3,447 people in Bangladesh and caused 196.25 billion taka (US$2.31 billion) in economic damage. It was one of the deadliest cyclones to ever hit Bangladesh.

COMMUNITY LOSS AND MASS MIGRATION

Eleven years later, people living on the banks of the Pashur River in Bagerhat, the Andharmanik River in Patuakhali and the Meghna River in Noakhali, as well as in other coastal areas of Bangladesh, are experiencing recurrent cyclones, storm surges, flooding and sea level rise. Land where people live and work is being washed away at ever-increasing rates, forcing people to migrate from ancestral homelands in the densely populated country. Presently, the scale of the community loss and mass migration is unprecedented in the coastal areas. In a country having less than 150,000 km² of total land area, there are more than 160 million people striving to better their economic condition. It is significant to note that all of these migrations are from rural areas to urban areas. Recently, Tazkin Ahmed, the mayor of Satkhira Municipality, said, 'Of the 1.5 lakh [150,000] people in Satkhira Municipality, some 50,000 to 60,000 people are climate refugees from Sundarban areas', ('Our cities not ready for climate migrants', 2019). Dhaka, the capital of Bangladesh, is the most alluring tar

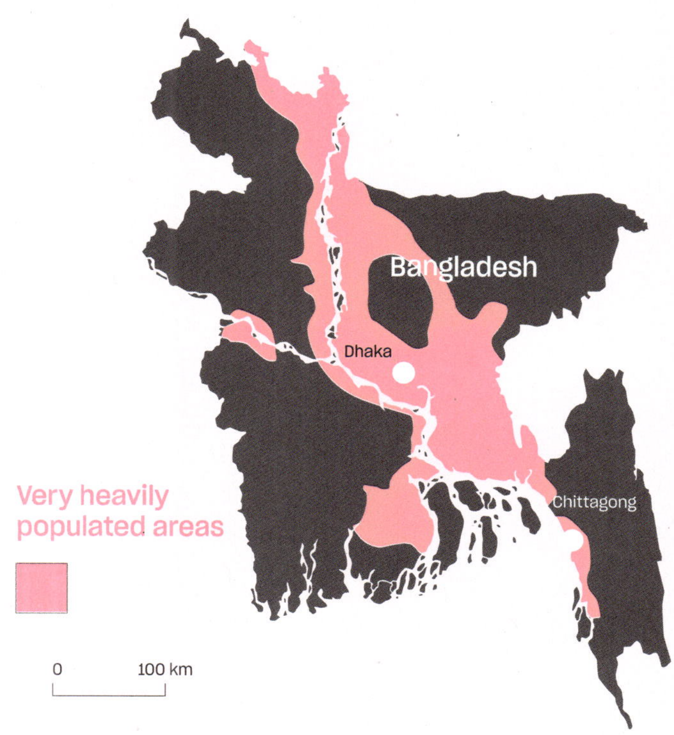

Sources: Dhaka University; Intergovernmental Panel of Climate Change (IPCC).

get to migrants. According to a report by *National Geographic*, on average, 700,000 people were internally displaced each year over the last decade, with up to 400,000 migrants arriving in Dhaka every year (Bodrud-Doza et al., 2020). According to the United Nations Human Settlements Programme, Dhaka is the world's most crowded city (McPherson, 2018).

The projected number of climate-induced migrants by 2050 varies from 13 million (Wernick, 2019) to 18 million (Rekacewicz, P., & GRID-Arendal, as cited in UCAR Center for Science Education, n.d.), with a staggering 17,000 km² to 22,000 km² of land being submerged.

Overall, the number of displaced people by 2050 could be even more than 18 million if all of the various impacts of the climate crisis are considered, making the country number one in the list of internal migration (McDonnell, 2019). Bangladesh's prime minister, Sheikh Hasina, on one occasion mentioned that a 1 m rise in sea level - a plausible scenario this century - would submerge a fifth of the country and turn 30 million people into 'climate migrants' (Darby, 2017). Between 2008 and 2014, an estimated 4.7 million people were displaced due to natural disasters in Bangladesh. Those displaced are the poorest and most marginalized people in the country (CDMP, 2015).

CURRENT EFFECTS AND ADAPTATIONS

The impact of sea level rise has been readily apparent in 2020. During the third week of August,

Bangladesh's frequent flooding and cyclones are widely known and have long featured as leading news when they hit particularly hard. What is different now is that instead of flood waters receding followed by people returning to their homes and lands, in today's Bangladesh, all too often, the water simply doesn't recede.

But Bangladesh is not waiting for future disasters to happen, adopting the US$40 billion *Bangladesh Delta Plan 2100* in September 2018 to protect many of its coastal areas. Whether there will be enough time to implement the plan remains to be seen, but at least Bangladesh is one of the very few countries in the world which is really acting.

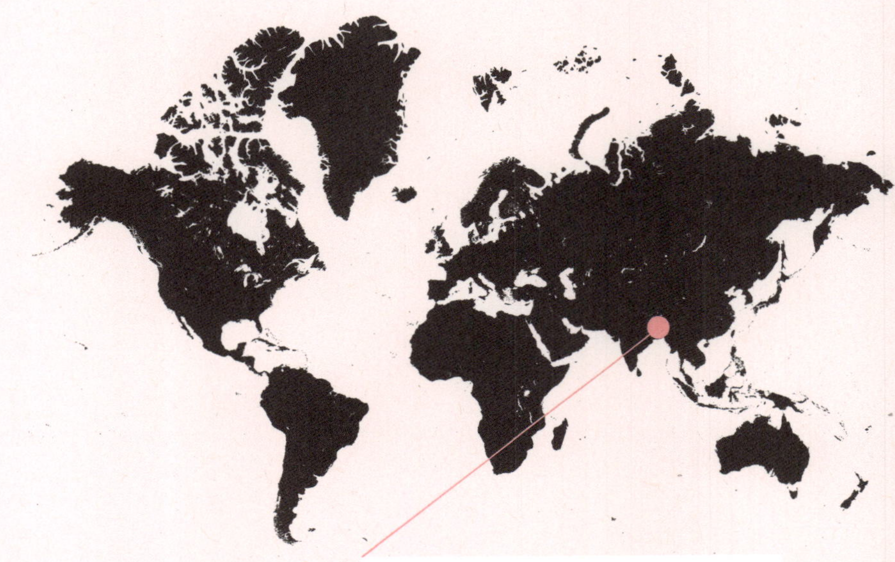

Bangladesh
Population: 165,000,000

It is estimated that up to 50 million people will need to be evacuated to higher ground in the near future. An estimated 6.5 million people have already been displaced due to the climate crisis.

Source: Climate Central (www.climatecentral.org)

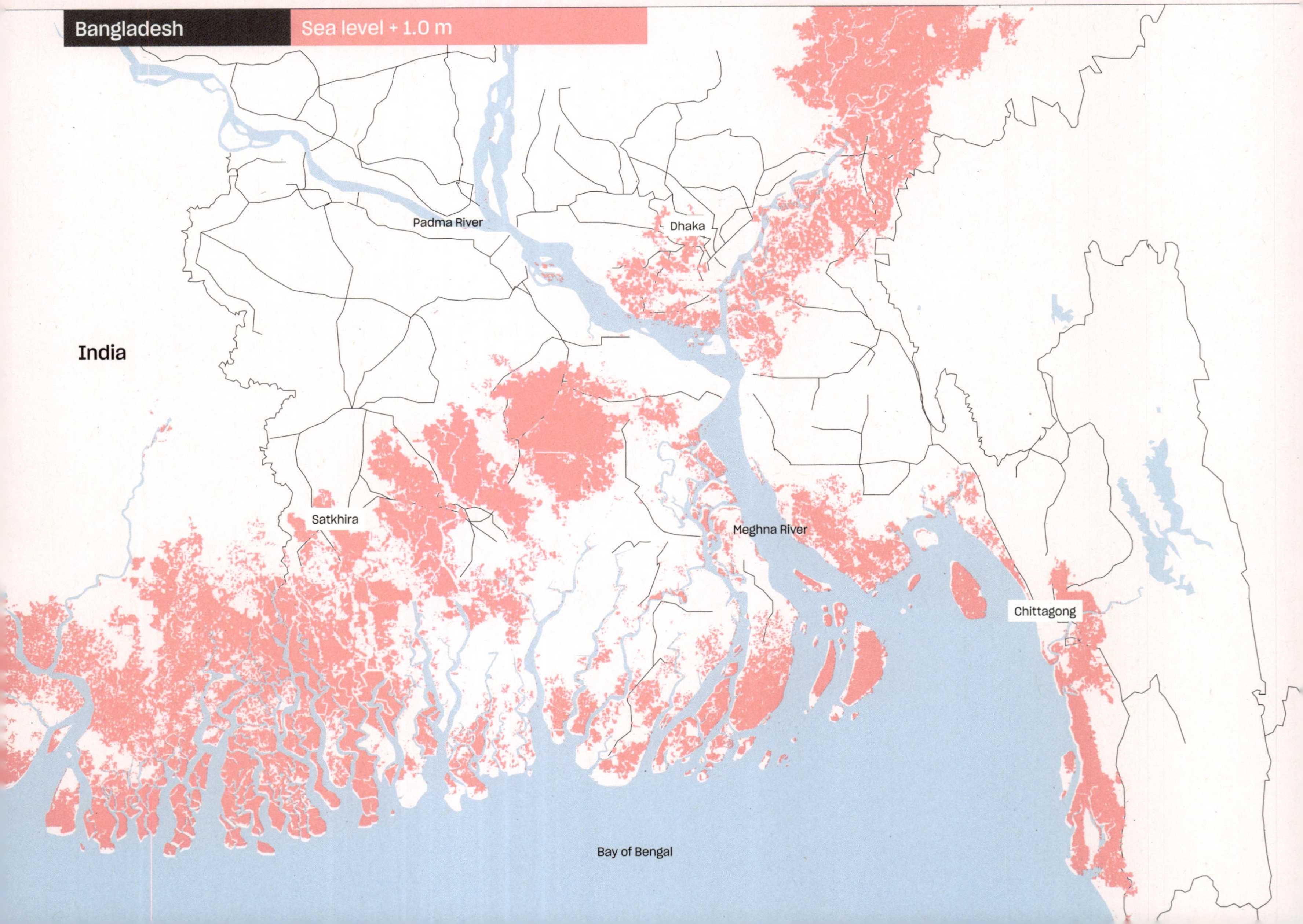

2020, the entire coastal belt became submerged when unusually high tides coincided with high rainfall. More than 100,000 people were marooned, even though there were no extreme weather events in the Bay of Bengal. Azad Kabir, the in-charge officer of the Karamjol Wildlife Rescue Center in Sundarban, said that he never had seen water at that level in the 10 years he had been working there.

Climate refugees in Bangladesh are on the rise as well, even though Bangladesh has taken several initiatives to protect its coastal population from the risks of natural hazards. According to World Bank Group (2013), Bangladesh has continued investment to reduce vulnerability since the 1960s. So far, the country has established 2,130 cyclone shelters, 139 polders (areas of low-lying land that have been reclaimed from bodies of water and are protected by dykes), 2,900 water control structures for drainage and improved early warning systems. Many polders were constructed along the coastal belt to provide protection to the country's infrastructure and food production system. But the polders and embankments were not built high enough to hold back ever-larger flood events and storm surges. When saline water overtopped the polders and submerged the land, roads and communities they were designed to protect, the water was difficult to drain. It caused prolonged waterlogging, which made it miserable for the entire community and hard to source drinking water. It severely impacted agricultural and livestock resources. People became homeless and lost their livelihoods because they could no longer farm the salt-laden soil or access potable drinking water. The prolonged waterlogging behind polders and embankments led to a large number of migrations, particularly after Cyclone Aila on 26 May 2009.

COASTAL ZONES AND DISTRICTS

During monsoon season, upstream operators open flood gates in hydroelectric projects and barrages across the mighty Ganga-Brahmaputra-Barak Basin. The rising waters converge on the low-lying delta of Bangladesh, making life much harder in general. Bangladesh is geomorphologically dominated by the Ganges-Brahmaputra-Meghna River System, which flows into the Bay of Bengal, forming

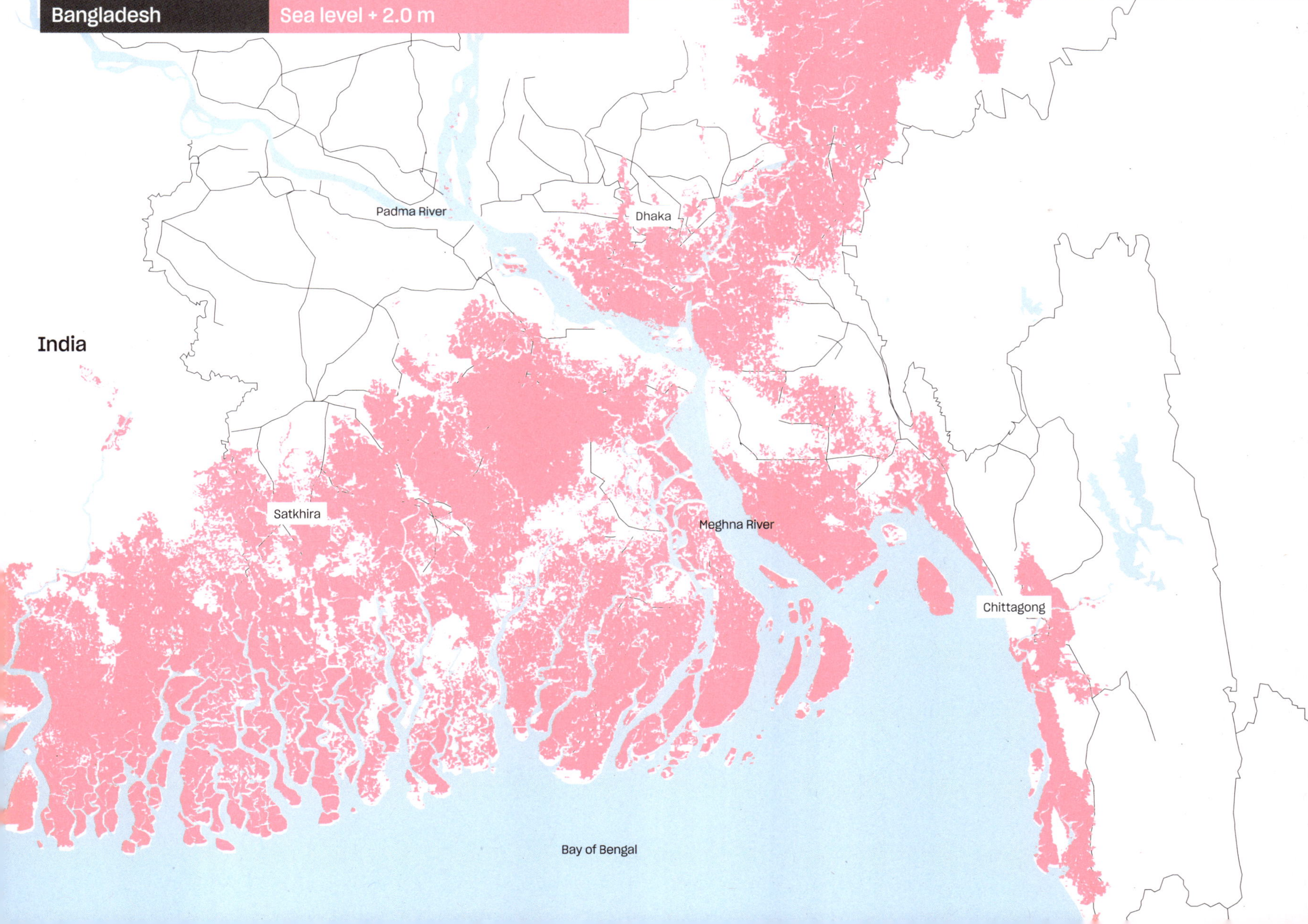

the main discharge passage of the largest active delta in the world.

The large coastal zone covers approximately 32% of the land area and is home to 29% of the total population (35 million people). The zone is administered by 19 coastal districts: Jessore, Narail, Gopalganj, Shariatpur, Chandpur, Satkhira, Khulna, Bagerhat, Pirojpur, Jhalokati, Barguna, Barisal, Patuakhali, Bhola, Lakshmipur, Noakhali, Feni, Chittagong and Cox's Bazar (Ahmad, 2019). The coastal areas of Bangladesh contain the world's largest unbroken coastline, which is 710 km long and composed of the interface of various ecological and economic systems, including mangroves (the largest single block of tidal halophytic mangrove forest in the world covers 6,017 km^2 of Bangladesh), tidal flat, estuaries, sea grass, about 70 islands, accreted land, beaches, a peninsula, rural settlements, urban and industrial areas and ports (Ahmad, 2019).

But the coast of Bangladesh is vulnerable to both surges from the ocean from the downstream and discharges of the Ganga, Brahmaputra and Barak River Systems from the upstream. The attempt to contain upstream rivers with dams and barrages while trying to hold back rising sea levels and storm surges with polders and embankments downstream is neither a feasible nor a sustainable solution for the nation.

FLOOD MANAGEMENT AND THE DELTA PLAN

To meet the increased water management challenges caused by the climate crisis, the government recently adopted *Bangladesh Delta Plan 2100*, a 100-year plan that addresses water-related issues in the Bengal Delta. The plan sounded exciting, purporting to offer a long-term vision for water governance, including comprehensive water management from the upstream to the downstream. But in reality, it offered nothing new, only more polders and embankments in the coastal zone. About 80 large-scale projects have already been adopted under *Bangladesh Delta Plan 2100*. About 60 of them are exclusively related to water resource management, but they have not considered previous experiences and lessons learned from the long-term adverse impacts of structural solutions in Bangladesh.

Flooding is a natural process that, on the one hand, causes damages and losses, but, on the other hand, provides fertility and renews croplands. However, the way that natural flooding has been perceived, managed and governed in Bangladesh so far needs to be rethought. For various reasons, we have not been successful in dealing with flood management in a sustainable way.

To understand this, it requires a comprehensive and in-depth review of our past initiatives and policy approaches to dealing with flood. Over time, upper riparian countries unilaterally built water regulatory projects in Bangladesh that aggravated the flood situation in the monsoon season and caused scarcity of water during the lean period. Mighty rivers like the Padma (Ganges) and vast natural wetlands like the Chalan Beel dry up during the non-rainy season. As a result, the volume of water towards the southern part of the country lessens at the downstream during winter. It also enhances the salinity intrusion towards the mainland.

During the British era, to channel water from the Ganges River to the Bhagirathi River in West Bengal, India, a large barrage on the Ganges was conceptualized for better functionality of Kolkata Port. But its feasibility study underscored the possible adverse impacts to its downstream. Therefore, the British government did not proceed with building it. However, post-British era, when India and East Pakistan (as Bangladesh was known initially, from 1947 to 1971) were formed, India went ahead with the same project, building the Farakka Barrage. Farakka completely destroyed the Padma River System in Bangladesh.

Chalan Beel was a 500 km^2 wetland located where the Ganges meets the Brahmaputra. First, the railway constructed by the British destroyed Chalan Beel. Later, the Dhaka-Rajshahi Highway and population pressure left Chalan Beel completely ineffective in blocking incoming seawater. Those complex changes led to further changes by way of setting up sluice gates and embankments in Bangladesh. Therefore, the natural flooding in the delta changed its characteristics significantly due to the engineered approach to water management.

Such in-depth review has been offered by Dr. Nazrul Islam in several of his publications, in particular his recent book, *Rivers and Sustainable Development: Alternative Approaches and Their Implications*, published by Oxford University Press in 2020. Earlier, he presented a shorter version of this review in his 2018 book, *Bangladesh Delta Plan 2100: A Review*, which he wrote in response to *Bangladesh Delta Plan 2100*. According to a press release published in *The Daily Star* on 14 June 2018, Dr. Islam expressed the following view:

It may be noted that Bangladesh already has the disappointing experience of foreign-led water development strategy. In the fifties, a commission led by Mr. Krug, the former US Interior Secretary was invited to Bangladesh to advice on its water development strategy. Following that commission's report, in the early sixties, the International Engineering Co. (IECO) prepared a master plan, containing a list of forty-seven projects. The water development efforts of the country for the next six decades have basically been directed toward implementation of the projects of that MasterPlan. The basic philosophy behind the Master Plan has been the 'Cordon Approach' to rivers, according to which the flood-plains and tidal plains are to be cordoned off from surrounding rivers by constructing uninterrupted embankments. This philosophy ignores the fundamental fact that in a delta the floodplains (also the tidal plains) are organically connected with the river channels, so that these two cannot be separated without causing harm to both. Foreign consultants who dominated the Krug commission and IECO were mostly from countries in which deltas are not that important (such as in the United States), so that they did not quite understand the reality of a delta.

The ill effects of the Cordon approach are now everywhere to see. It has led to promotion of below-flood-level settlement, aggradation of river bed, subsidence of flood and tidal plains, deterioration of the water bodies inside the floodplains, disruption of open capture fisheries, diminution of waterways, reduced recharge of groundwater table, and most importantly ubiquitous waterlogging. The Dhaka-Narayanganj-Demra (DND) project - the showcase of the Cordon approach - provides an example of these effects. Meanwhile, the promised surface water irrigation potential of the Cordon projects generally failed to materialize, and expansion of irrigation was almost completely taken over by tube-well irrigation, for which Cordons were mostly unnecessary, if not a hindrance.

INFRASTRUCTURE AND INDUSTRIALIZATION

Many large-scale infrastructure and industrialization projects in the coastal areas of Bagerhat, Barguna, Patuakhali, Chittagong and Cox's Bazar are heading aggressively forwards. Human settlements, agricultural lands, wetlands, floodplains, small rivers, riverbanks, foreshores and canals are being replaced by heavy industrial development. The acquisition of land and the conversion of these landscapes to factories, power plants, roads and rooftops, coupled with the climate crisis, is leading to mass migration and destruction of the natural water-based network and the fishing, salt and tourism industries.

Because of these industrializations, the UNESCO World Heritage Centre recommended that the Sundarbans World Heritage Site be added to the World Heritage in Danger list. The World Heritage Committee, the governing body of the World Heritage Centre, requested that the state party not proceed with the construction of large-scale industry before conducting a strategic environmental assessment.

Since the formation of the Bangladesh Economic Zones Authority in 2010, the vision of the economic development of the country has been based on industrial growth, facilitated by navigational transportation using the coast and rivers. It prioritizes specific navigational routes which are not feasible without massive and continuous dredging operations to enable ships to supply power plants with fuel and provide raw goods to factories and export finished goods. Before these massive industrial developments and dredging plans are implemented, a comprehensive, inclusive, science-based and transparent impact assessment is needed. Doubling down and using more of the same approaches that have failed in the past will likely have an even more harmful impact on the ecology, livelihoods and people of Bangladesh.

A PAUSE FOR REFLECTION

We have to reclaim the delta to keep development efforts sustainable. COVID-19 has caused a global pause in economic development. It gives us a much-needed opportunity to review and reorganise development plans to be pro-deltaic, green and sustainable. If the government mindset makes a positive shift towards an open approach and sustainability, it can be implemented through the newly formed Delta Governance Council. They can engage people and experts in the policymaking and implementation processes, re-orienting *Bangladesh Delta Plan 2100* away from unsustainable development that worsens the impacts of the climate crisis to green projects like renewable energy that power a more sustainable future.

Bangladesh is the only country in the world where all of the rivers have been declared living entities by the High Court. It designated management oversight to the National River Conservation Commission (NRCC), directing the government to make NRCC an independent body. By empowering the NRCC and implementing court orders to protect the rivers and wetlands, the government can better mitigate the ongoing climate catastrophe. It will take a much-needed philosophical and policy shift away from the failed cordon approach to a living delta approach. Protection of the riverbeds, foreshores, riverbanks, mangrove forests and floodplains in Bangladesh is vital to survive the climate crisis. We can do it by adopting comprehensive, inclusive, transparent, data- and science-guided short-term, mid-term and long-term plans and projects. •

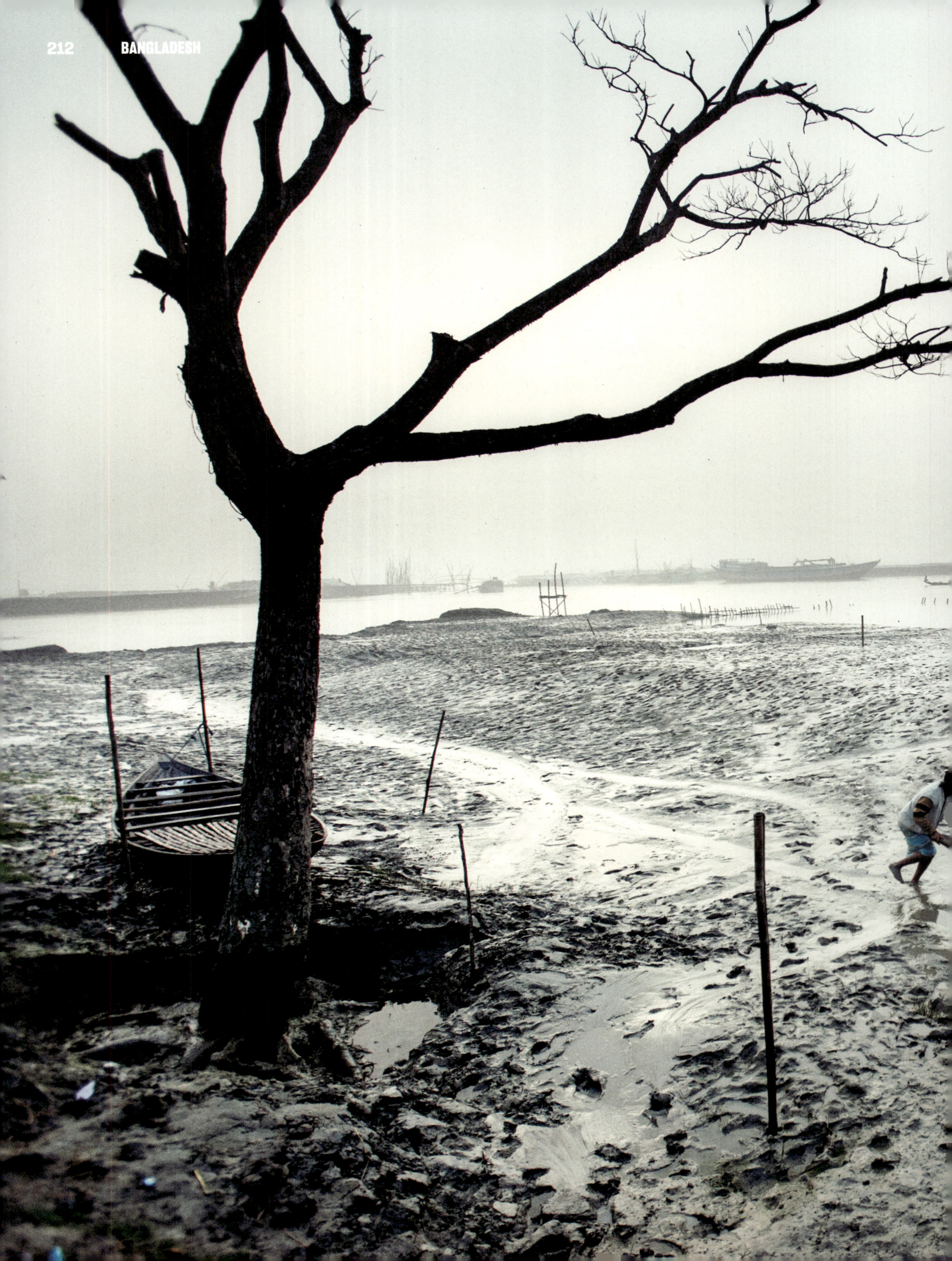

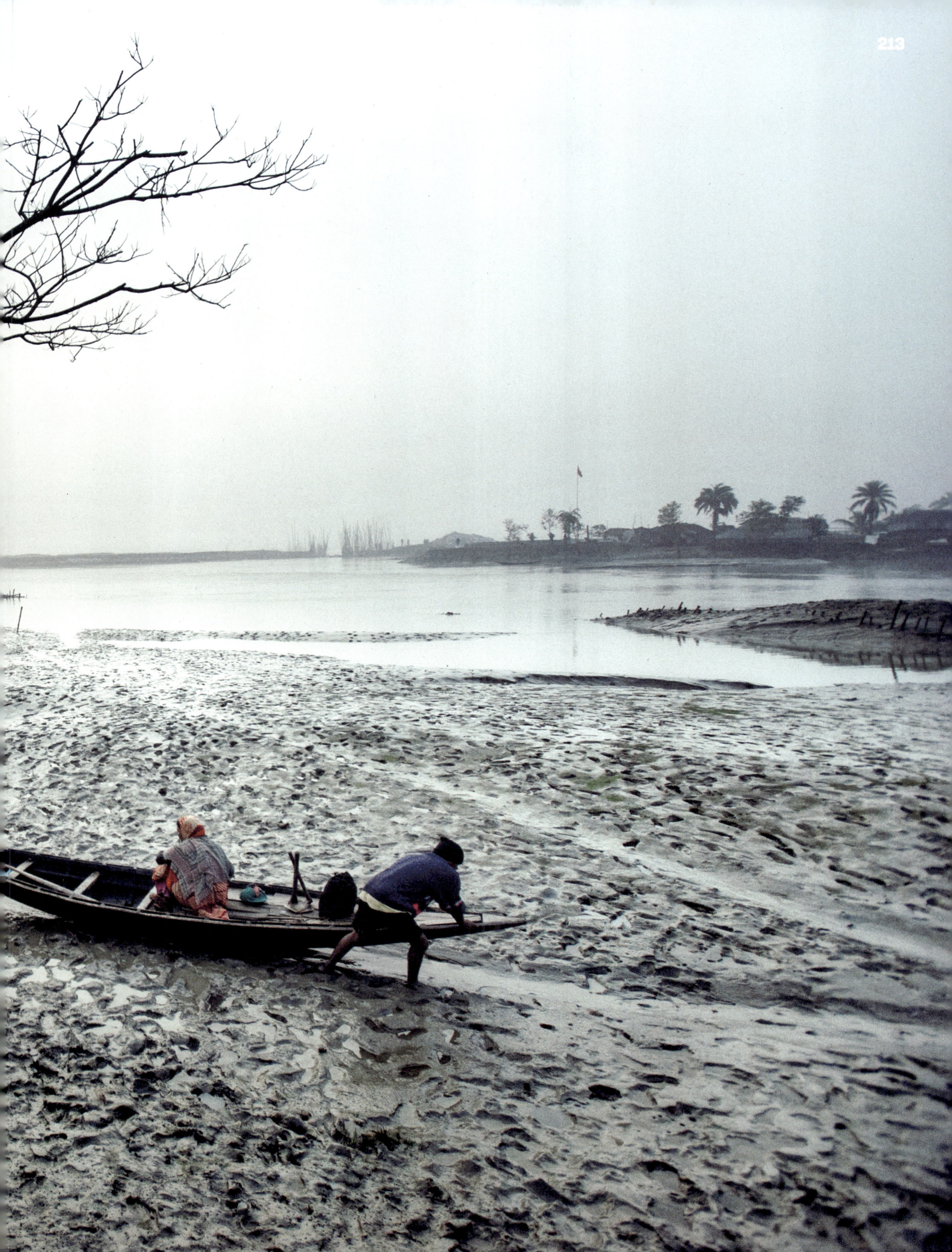

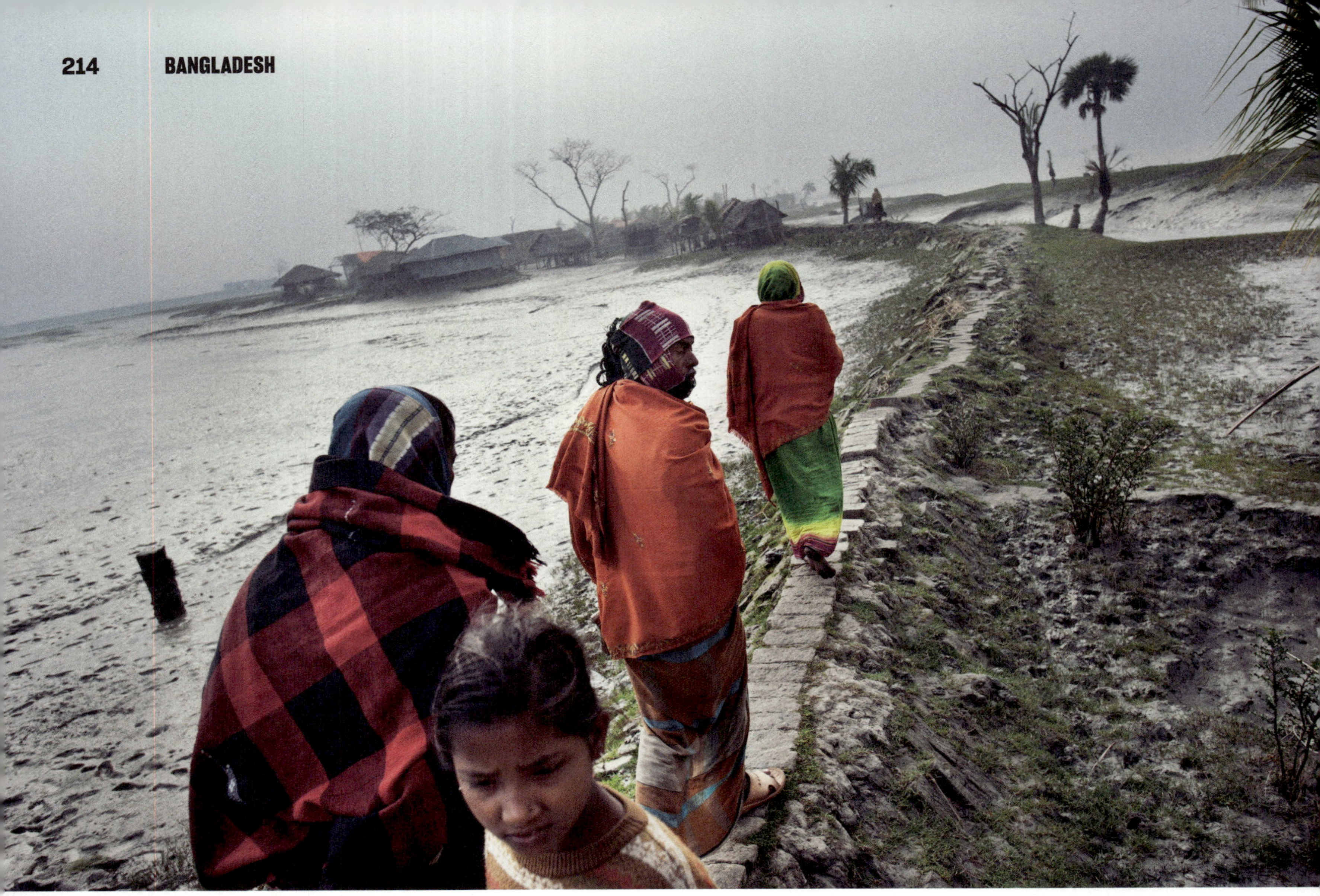

A family walks over a makeshift levee
at Bainpara, at low tide. Some houses
remain, but most were swallowed by
Cyclone Aila.

< Pages 212-213: Drowned land at low
tide. Some houses remain, most were
swallowed by cyclone Alia. After the
water receded the water never re-
turned to its old levels and all farmland
was lost.

The village of Choto Jaliakhali. The
village relocated to be safe from
the water, but there is no electricity.

A mother and her daughter at
Bainpara, their former village.

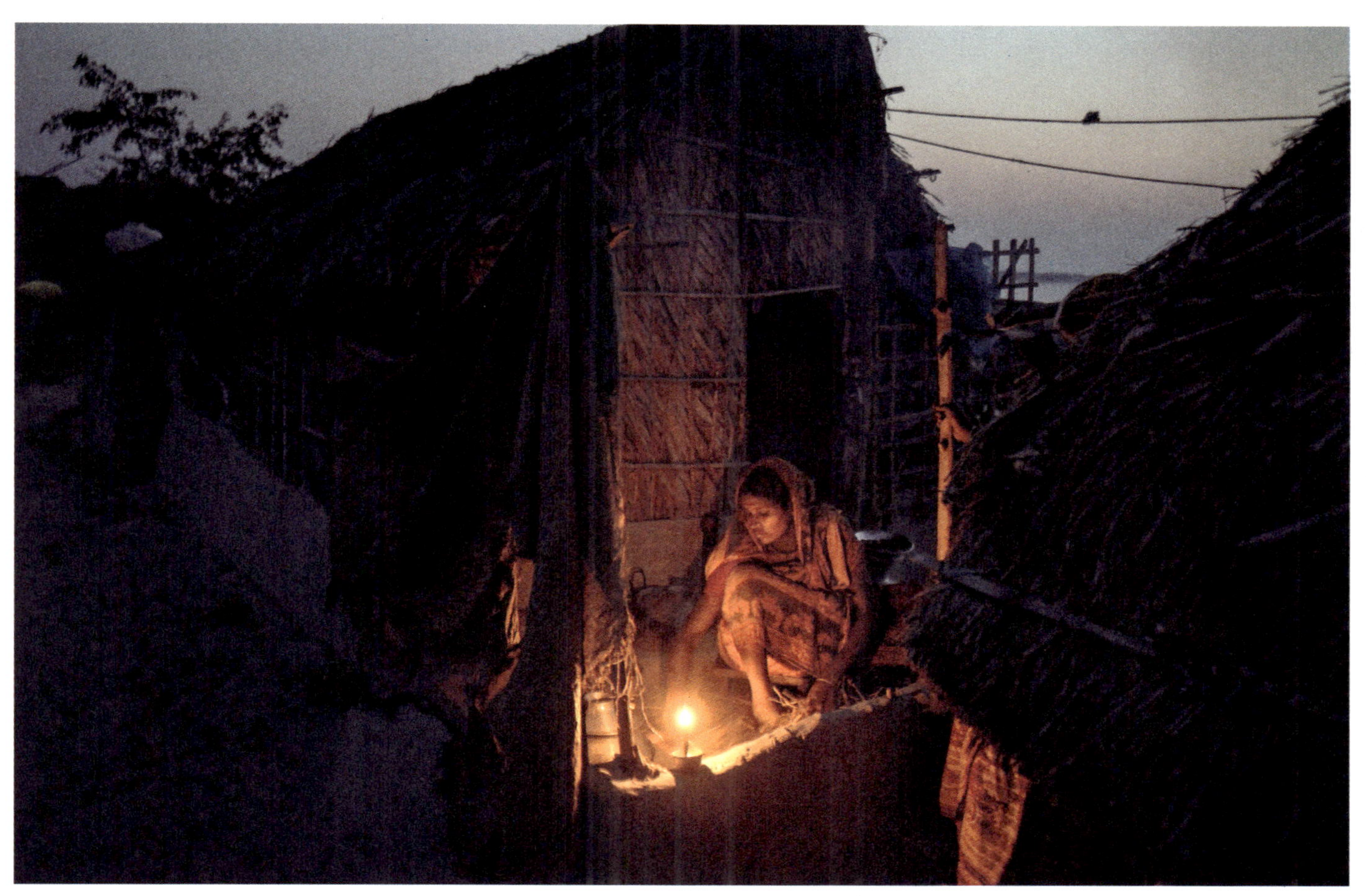

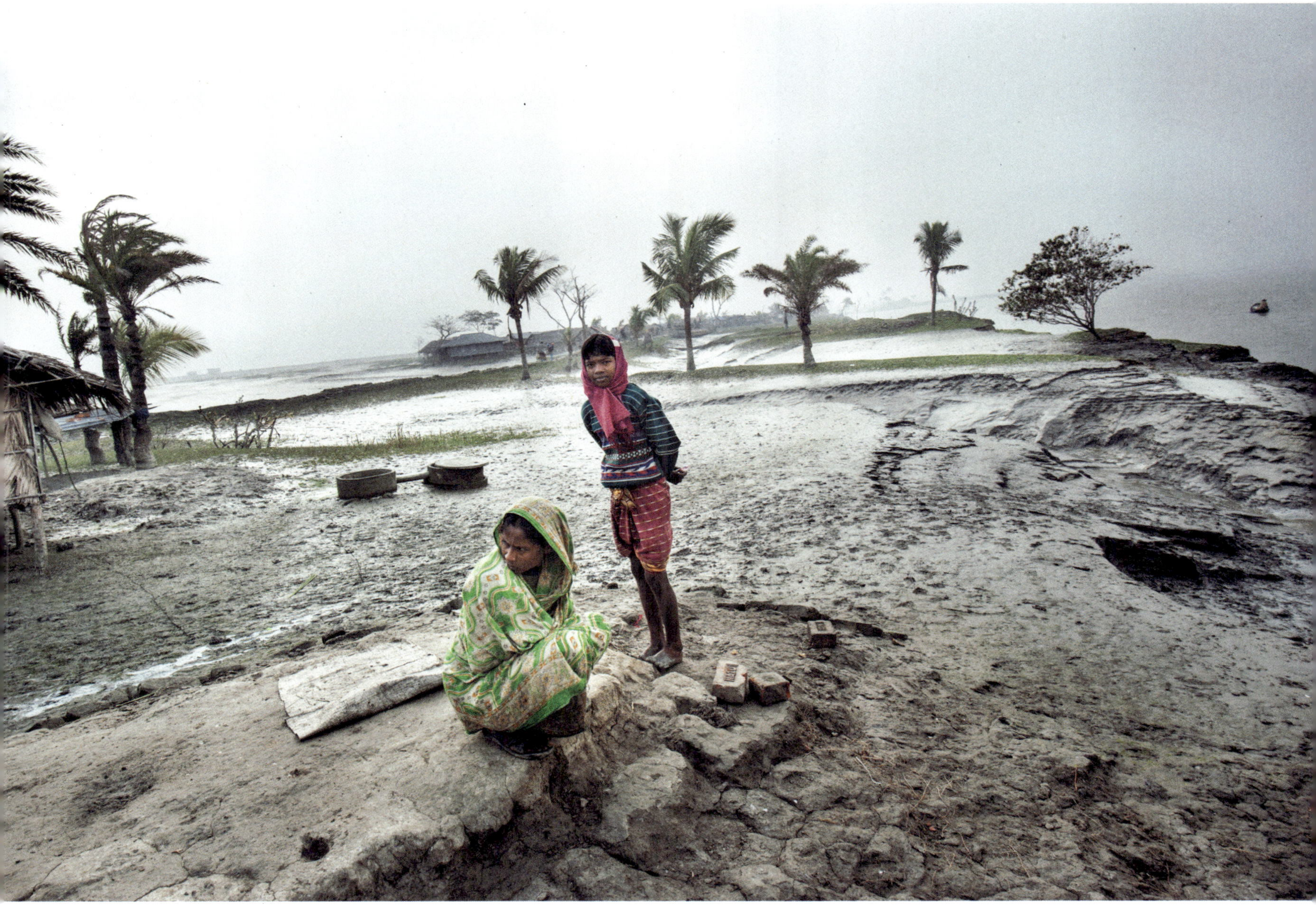

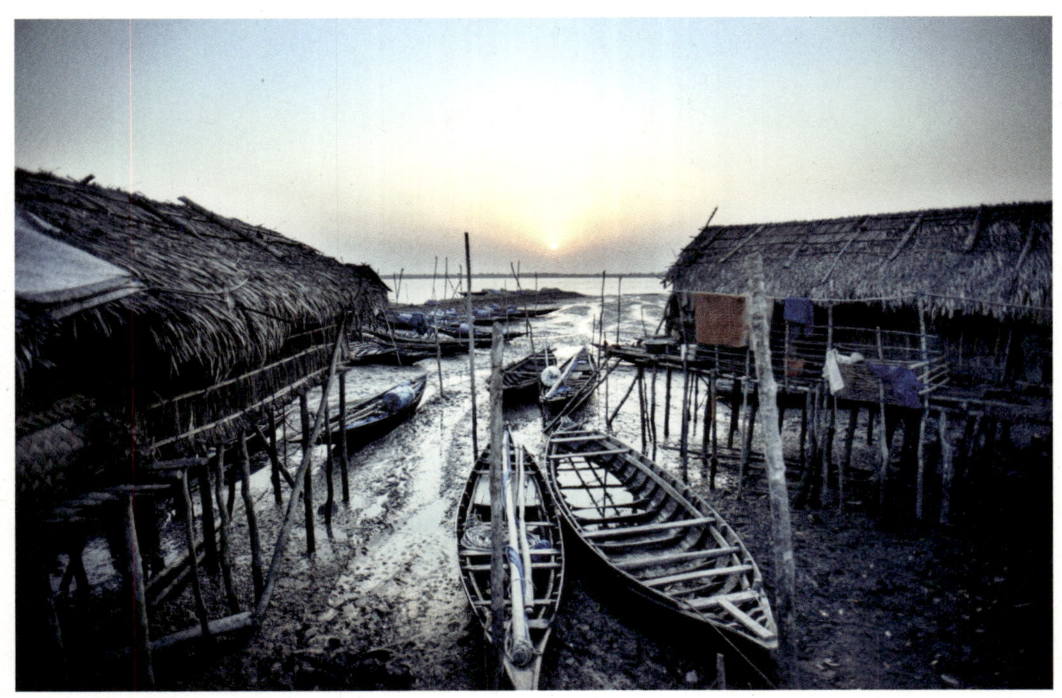

< The village of Kala Bogi, where Mohammed Fazar Gazi lives. He has relocated nine times due to the rising waters and knows the next time is coming soon.

A family walks towards the levee through drowned land at Bainpara, at low tide. At high tide, they are cut off by the water.

> People working in Katakhali to close the breaches in the levees caused by Cyclone Aila.

A woman sits next to her house in the village of Gunari. Tens of thousands of people live in makeshift huts on narrow makeshift levees following Cyclone Aila.

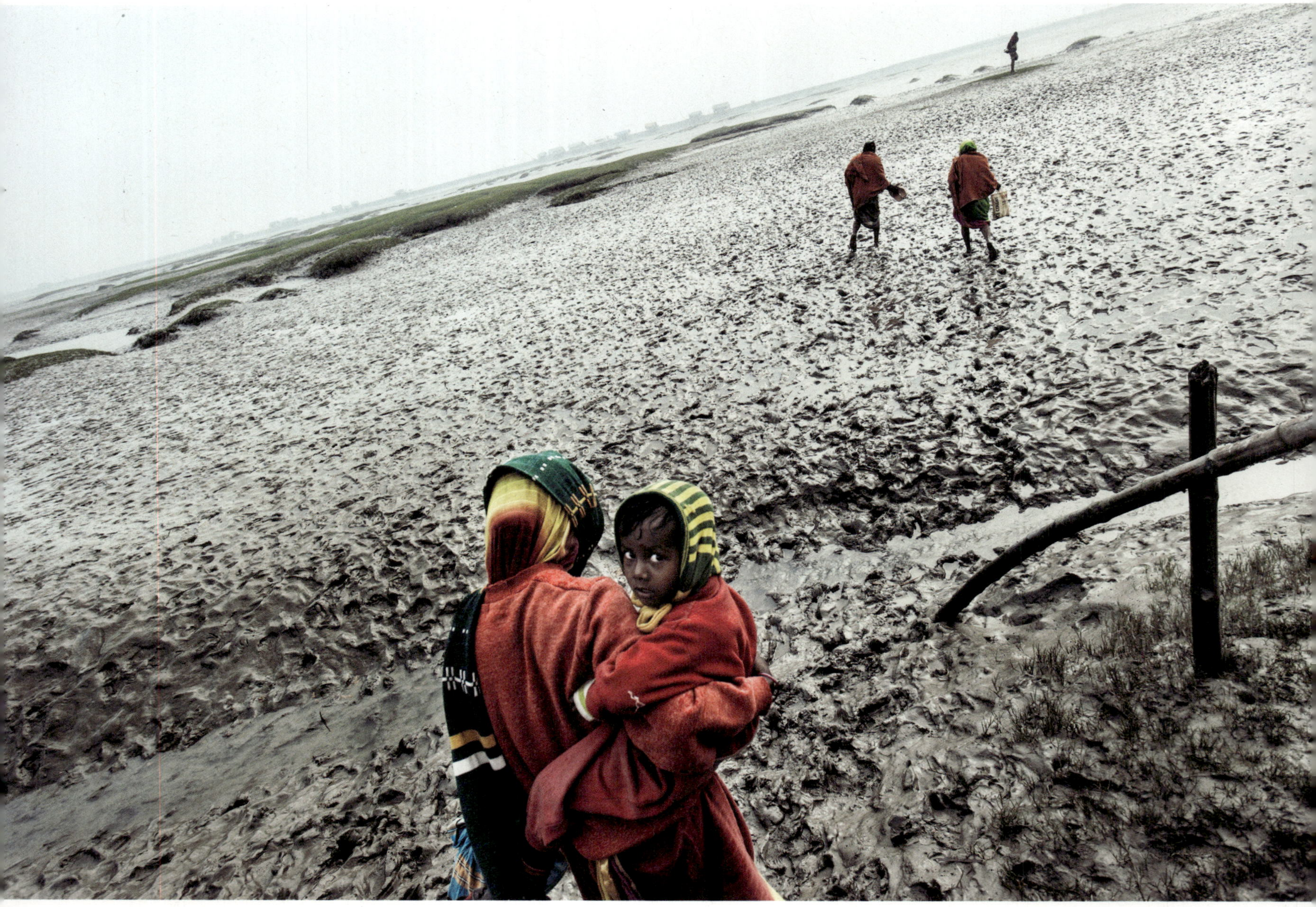

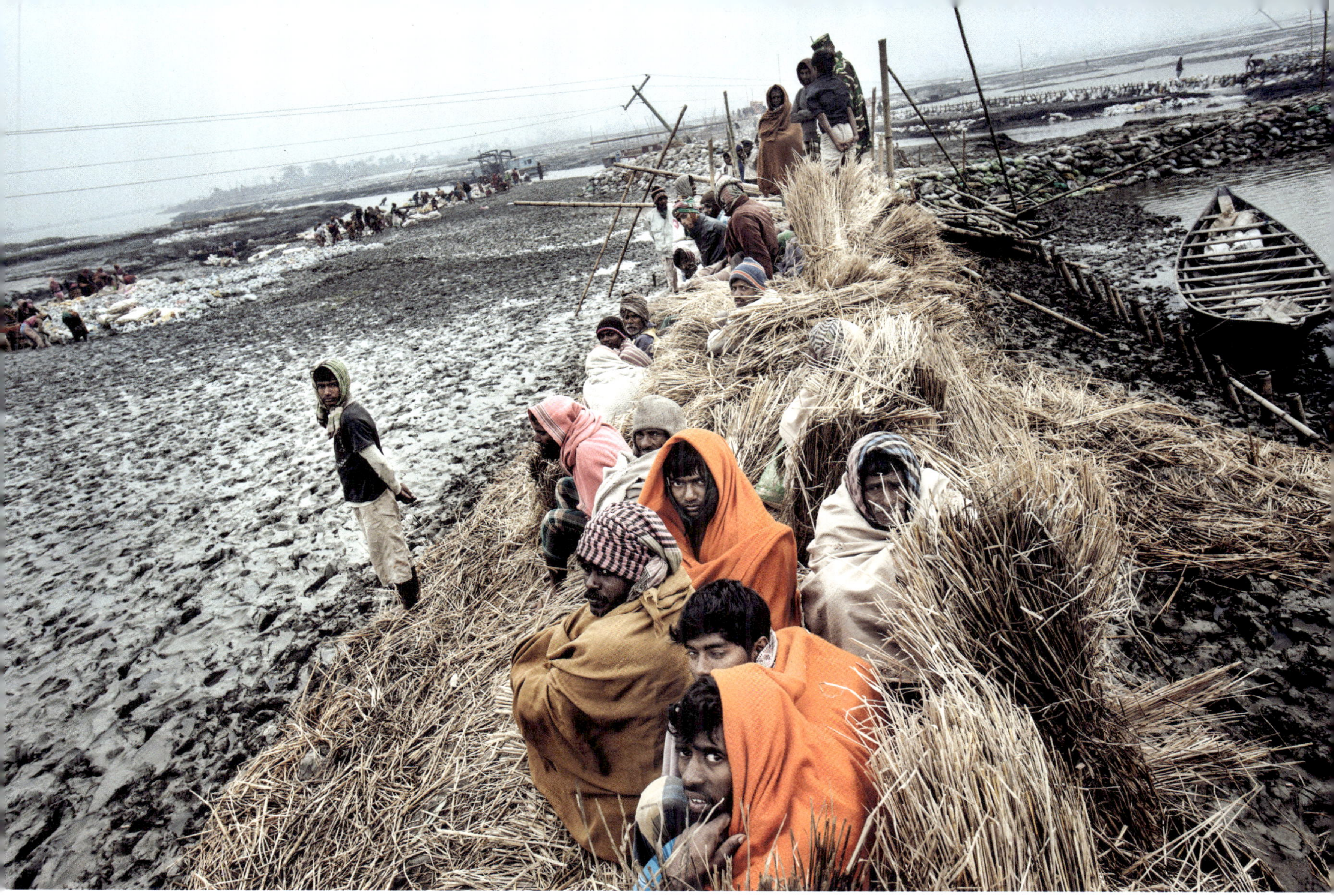

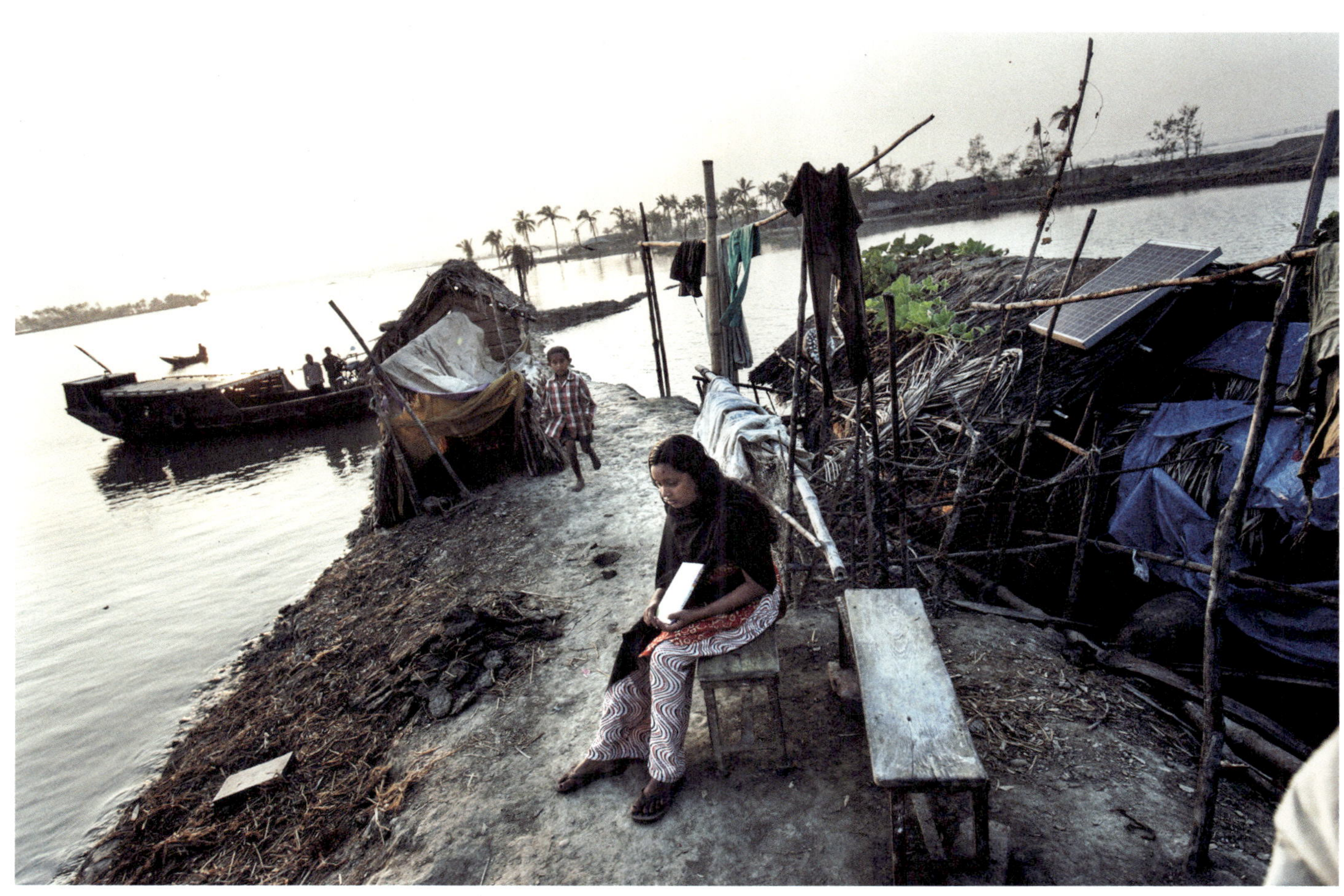

Thousands of people working in
Katakhali to close the breaches in the
levees caused by Cyclone Aila and to
heighten and reinforce the levees.

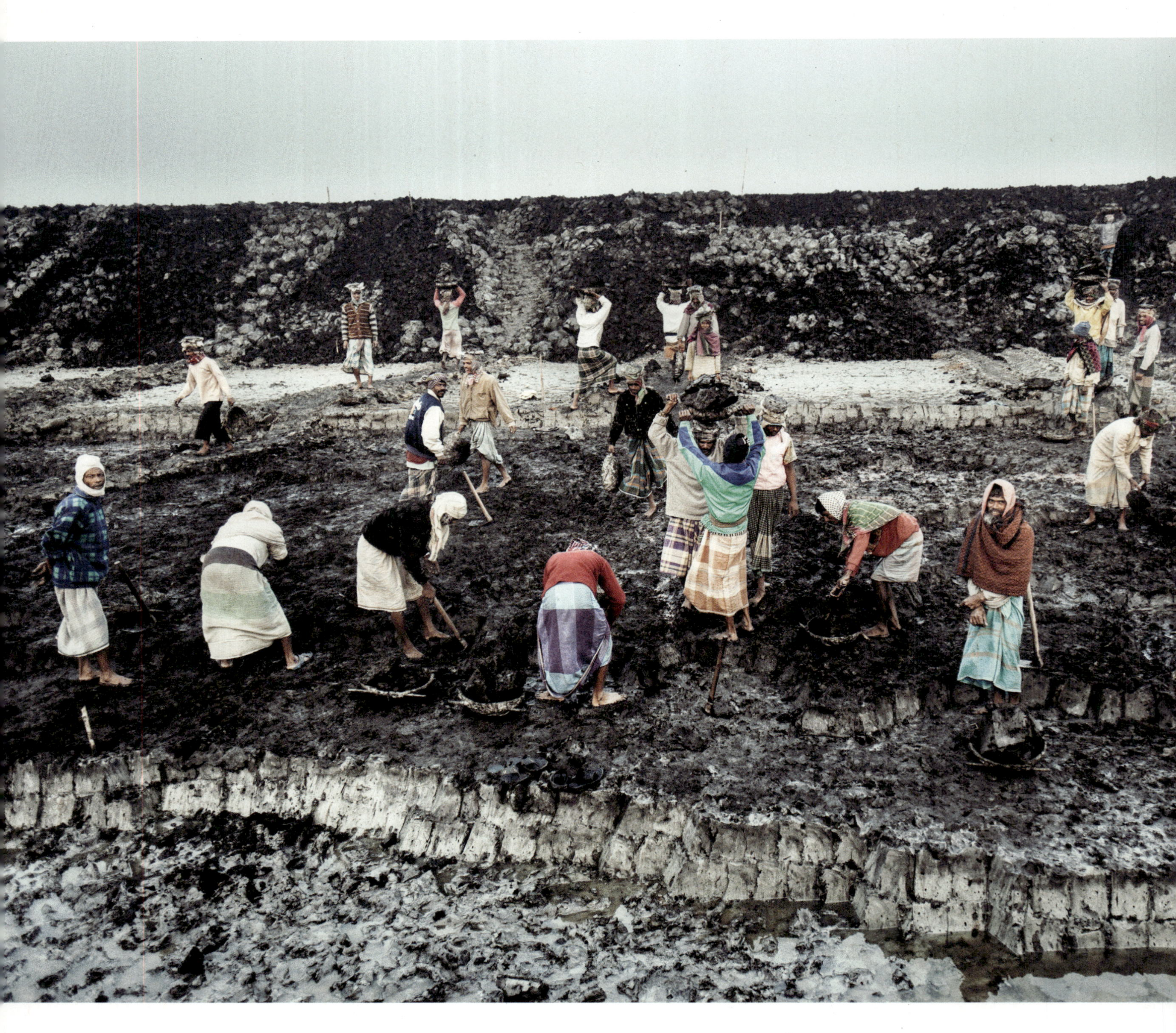

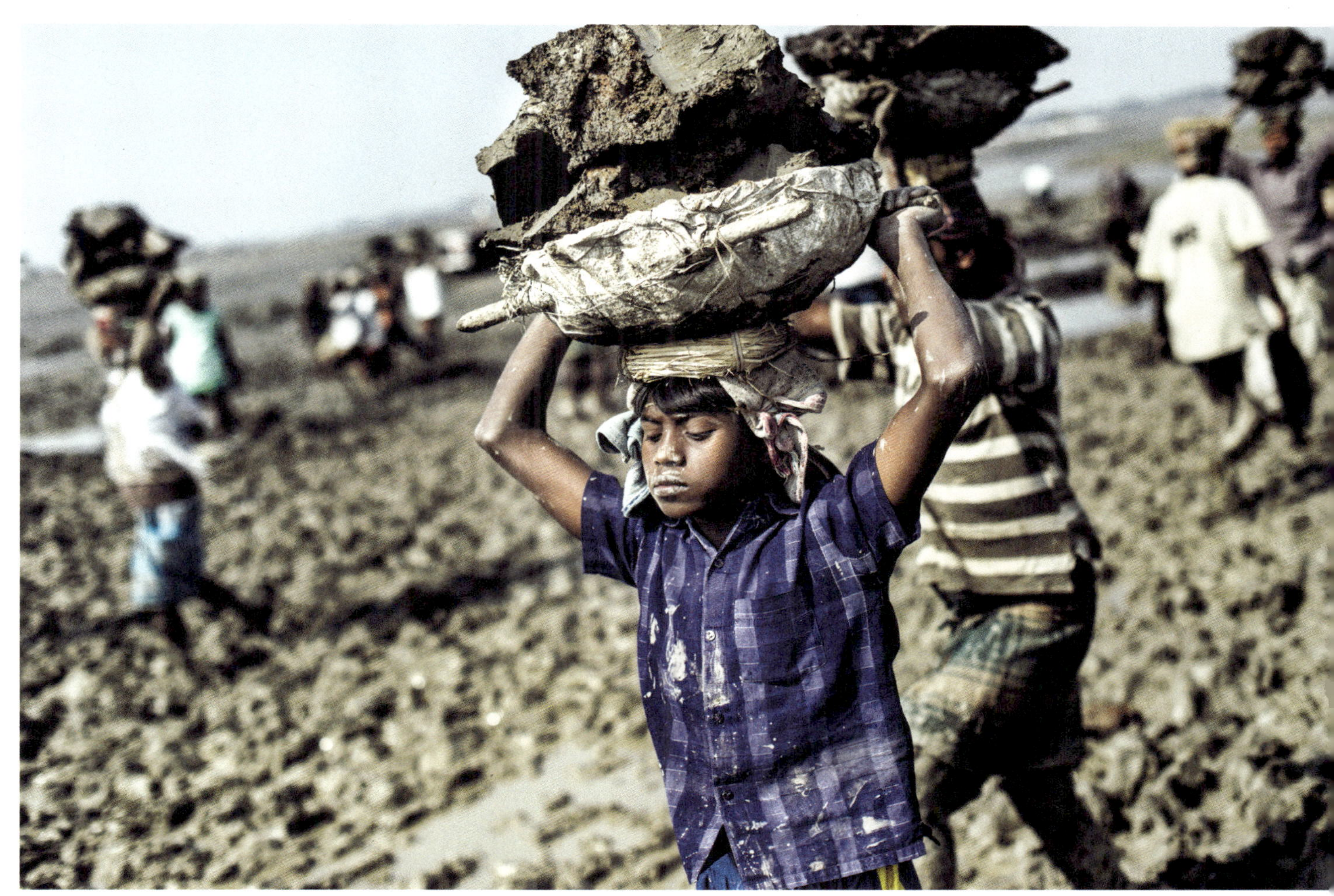

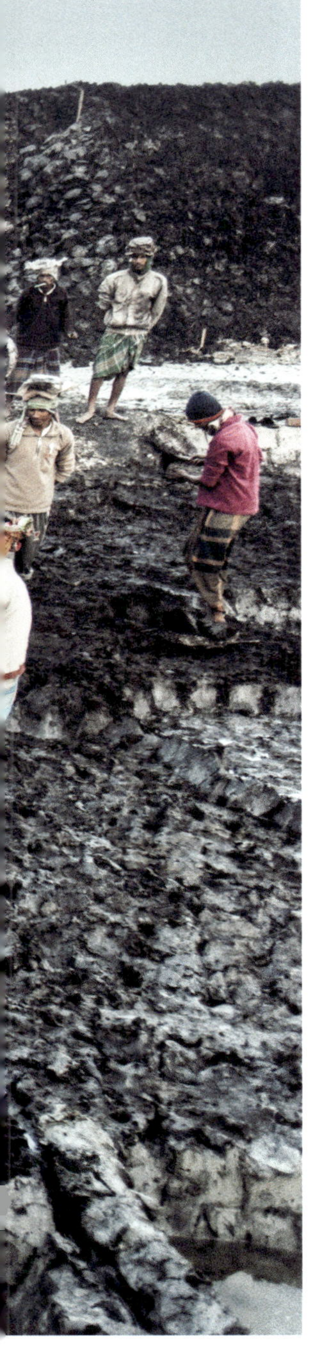

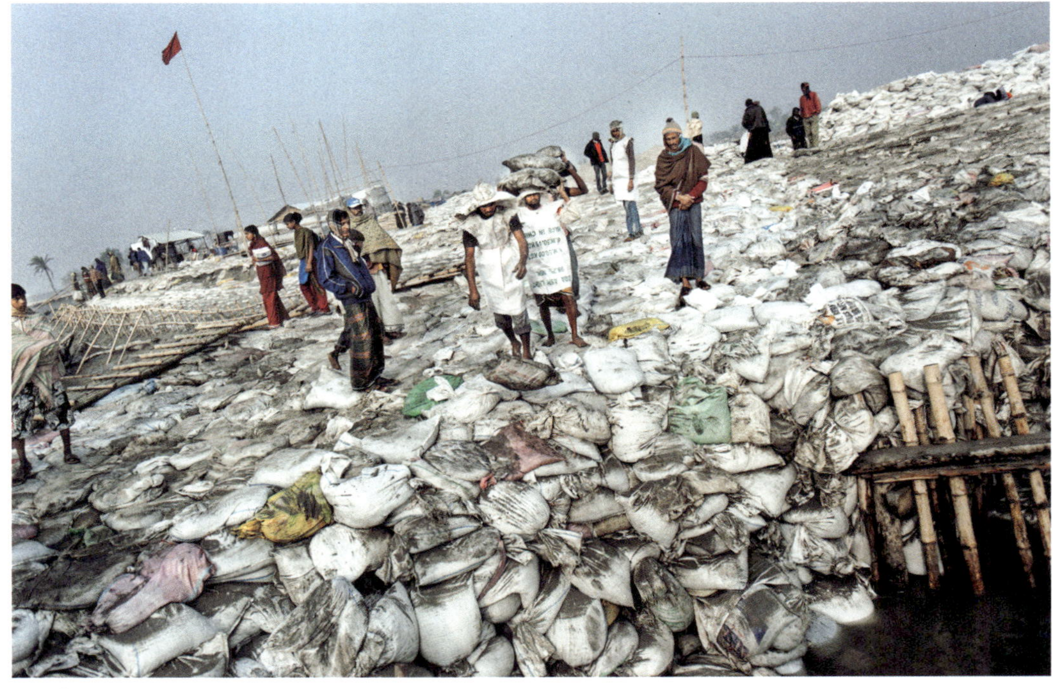

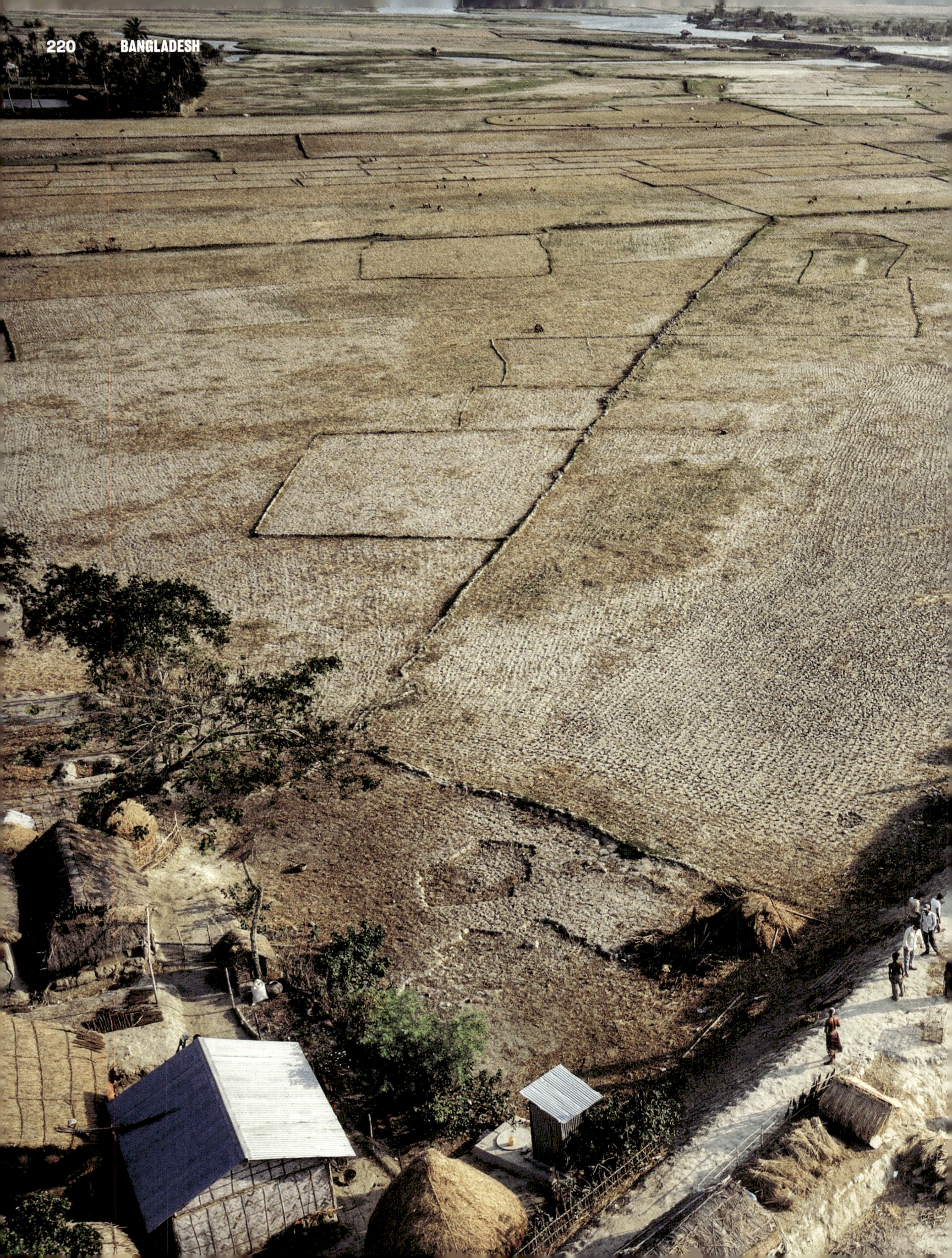

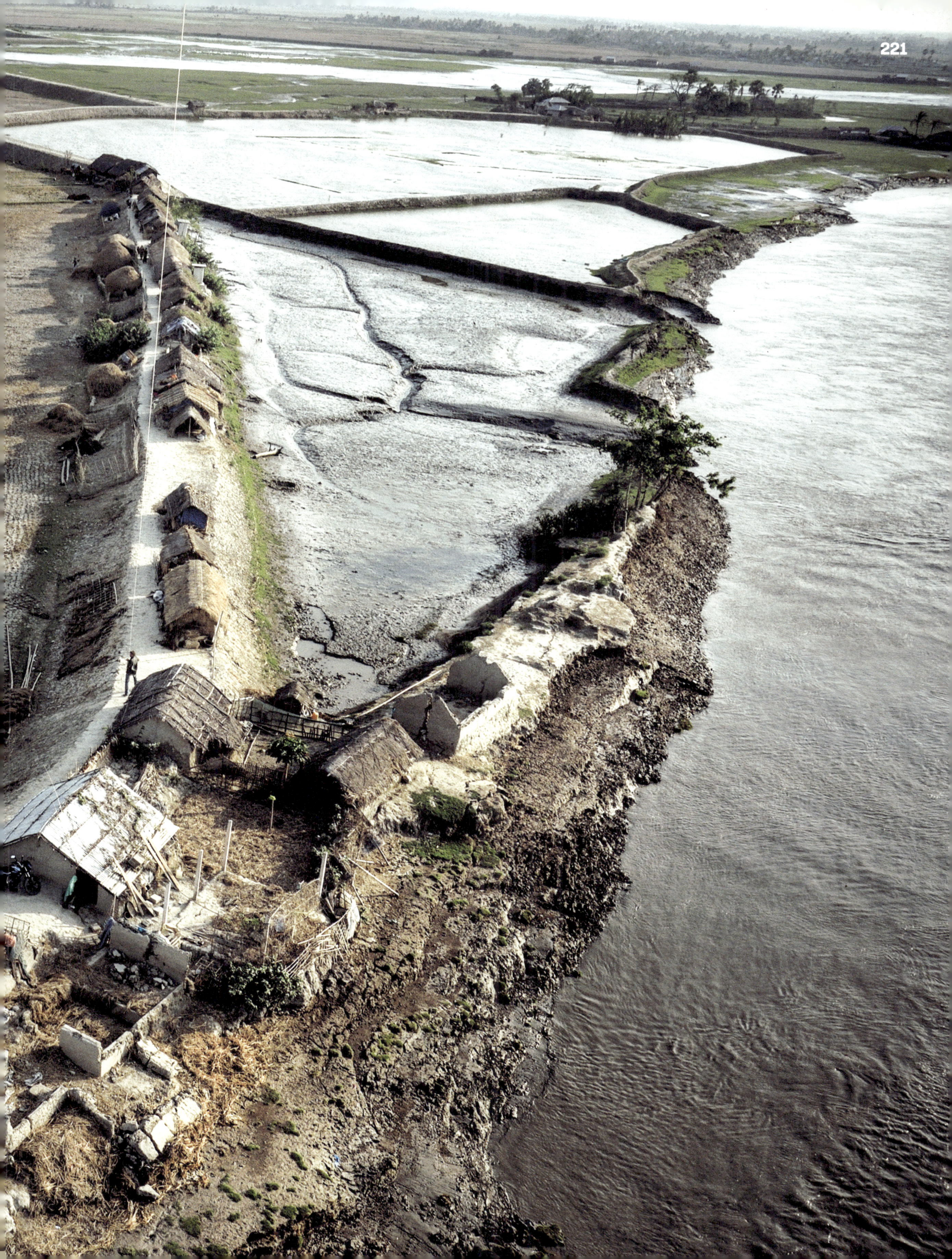

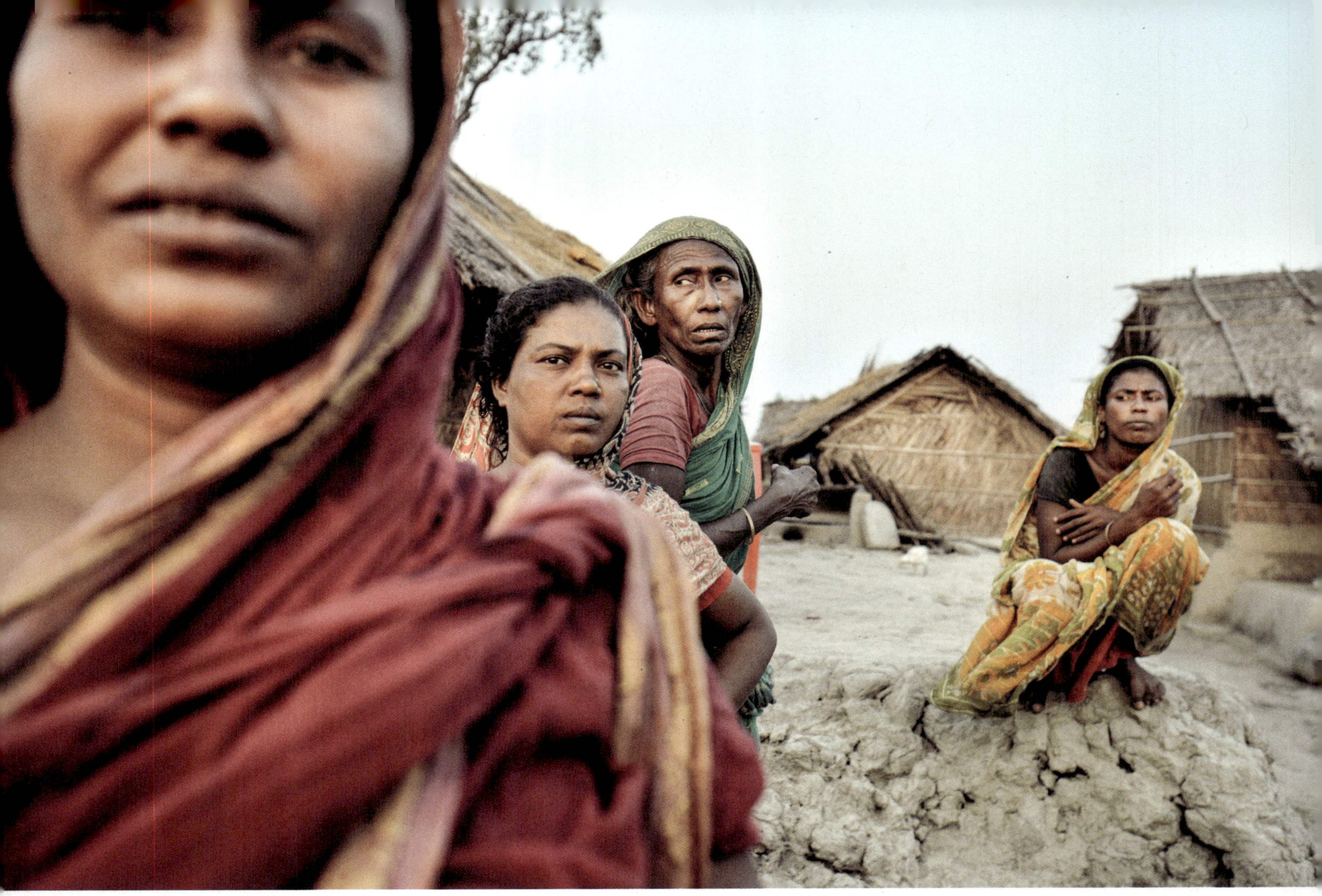

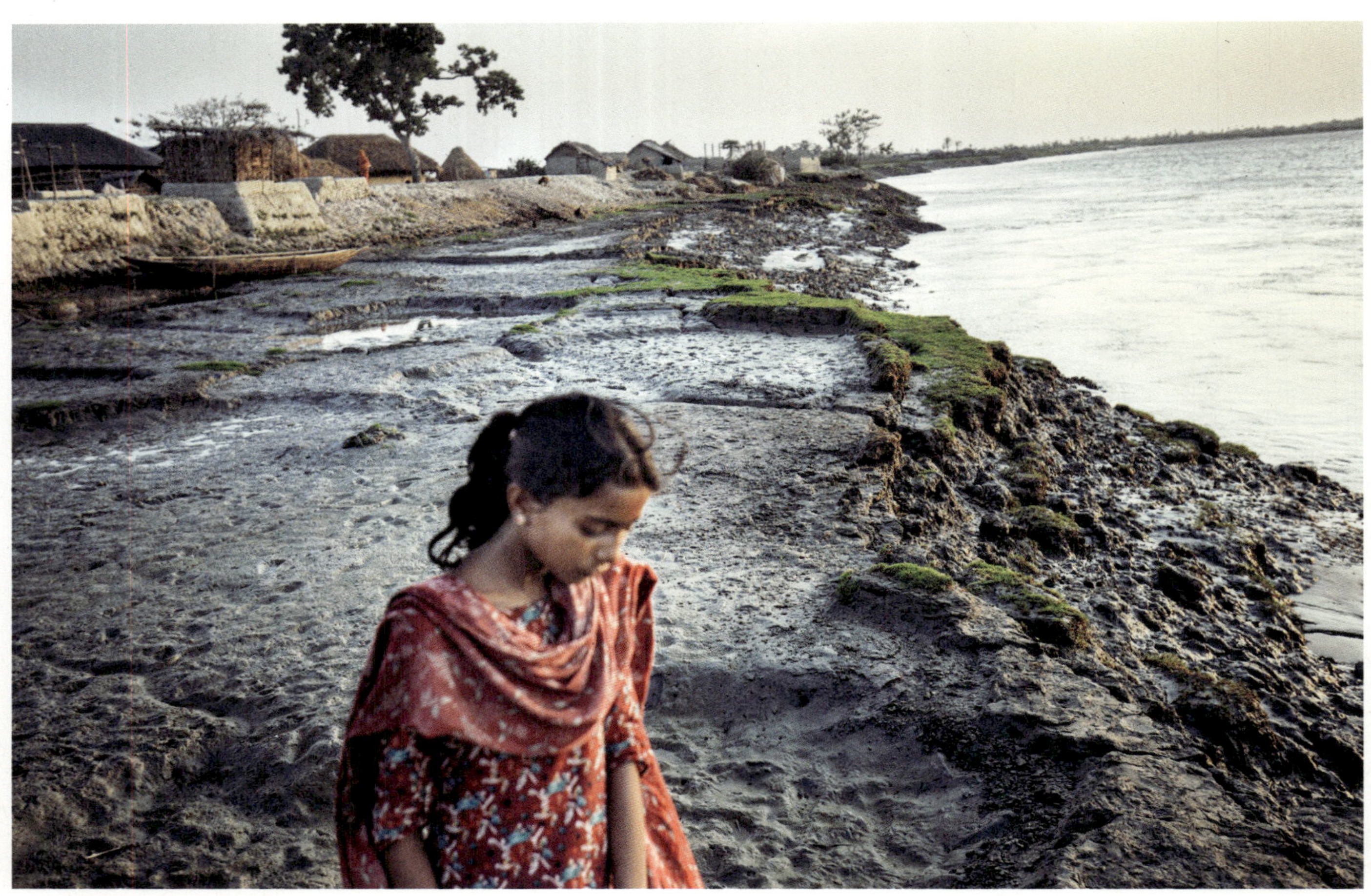

< Pages 220-221: The village of
Choto Jaliakhali.

The village of South Gunari, home of
Jahnara Begun (38; far left). Her two
children have moved to Dhaka; the
boy works in a brick factory, the girl
in a biscuit factory.

Choto Jaliakhali is the home of Nasrin
Khatun (13) and her grandmother
Guljan Biwi (53). Although the village
has already relocated, the embankment is eroding away very quickly
and will force the people to move
again. Nasrin's parents and brother
have moved to Chittagong.

A man stands next to what was once
was his land in the village of Gunari.
Tens of thousands of people live in
makeshift huts on narrow makeshift
levees following Cyclone Aila.

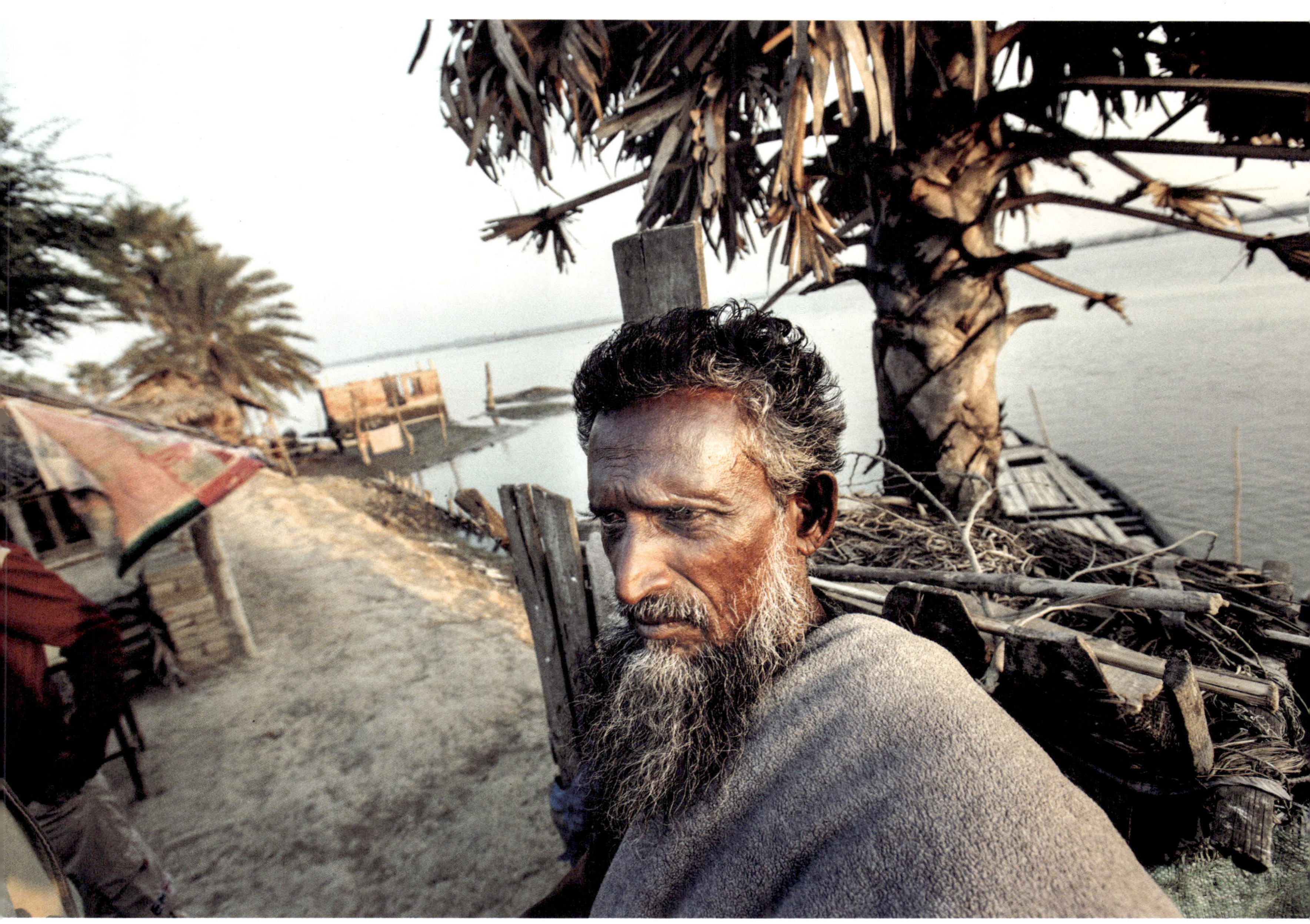

Choto Jaliakhali, the festival Charak Puja is celebrated by Hindu communities a few days before the Bengali new year celebration, which is on 15 April, according to the Indian calendar.

The village of Nalian. Tens of thousands of people live in makeshift huts on narrow makeshift levees.

Carpenters working on a boat in the village of Kala Bogi.

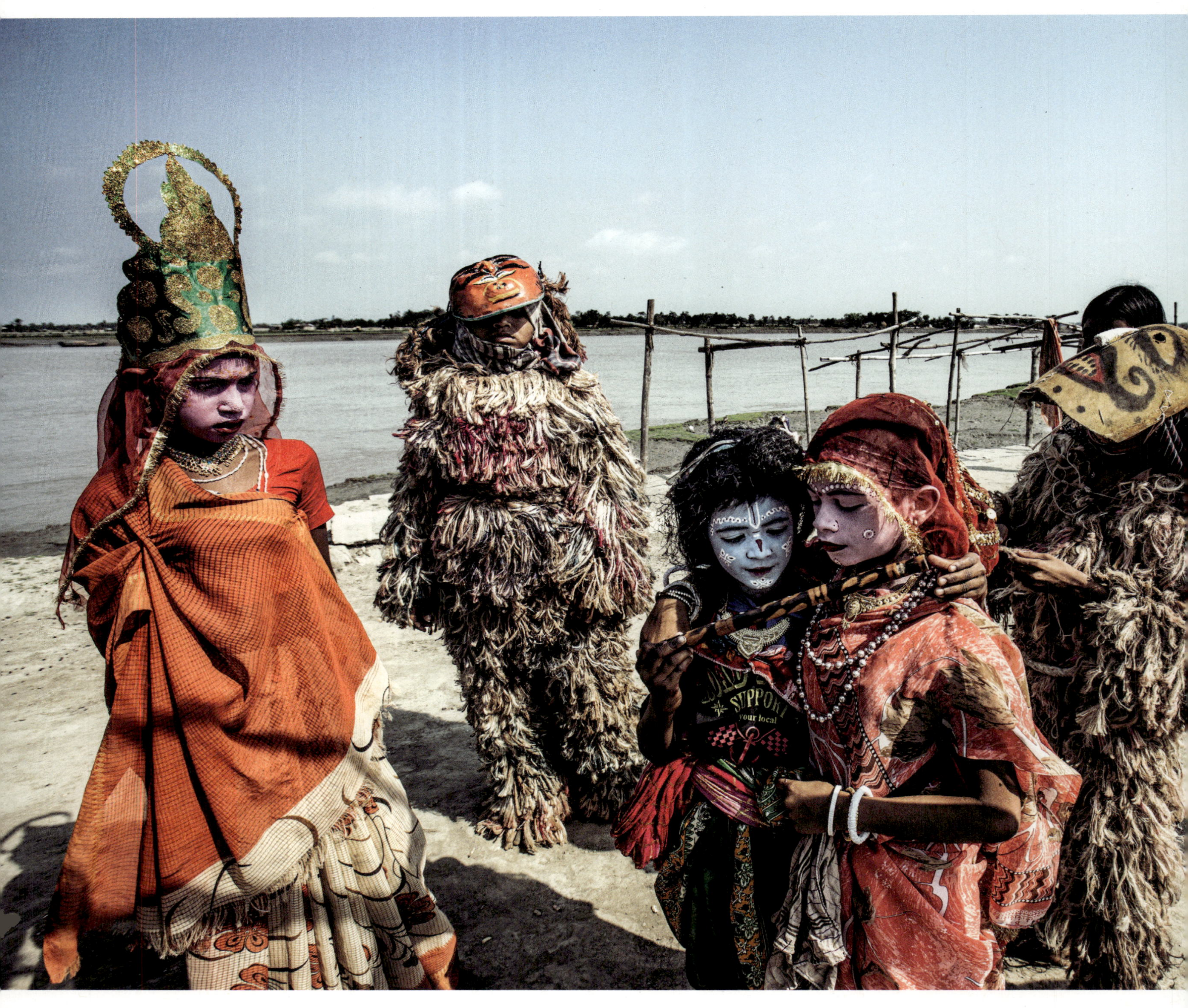

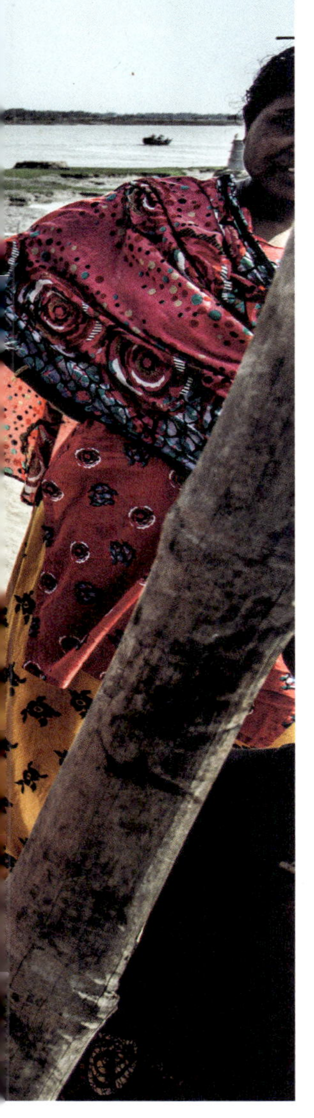

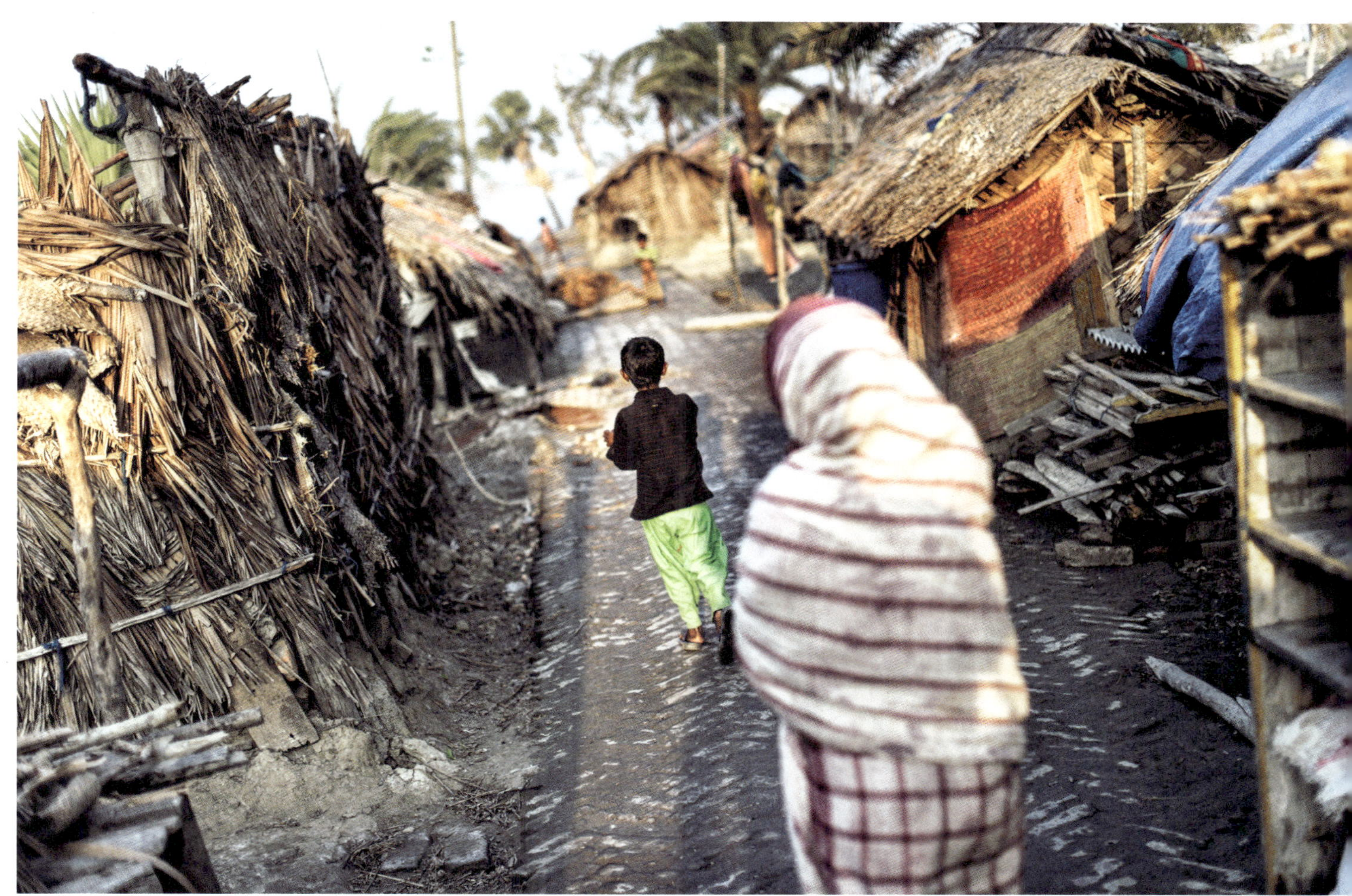

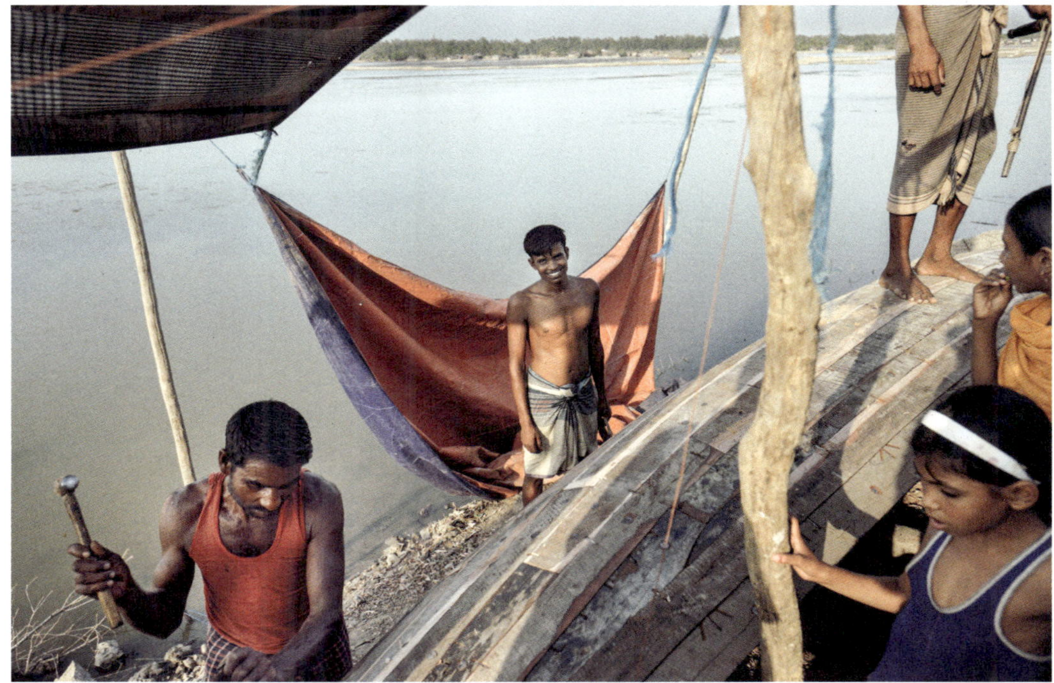

Sandwip Island is located in the Bay of Bengal, north of Chittagong. The island has a population of approximately 350,000 and is eroding away at a fast pace.

Abdul Baten (65) and his wife Hasne (55) have nine children and have moved five times because of the rising waters. Their most recent move was five months before, and they are already looking for where to relocate next.

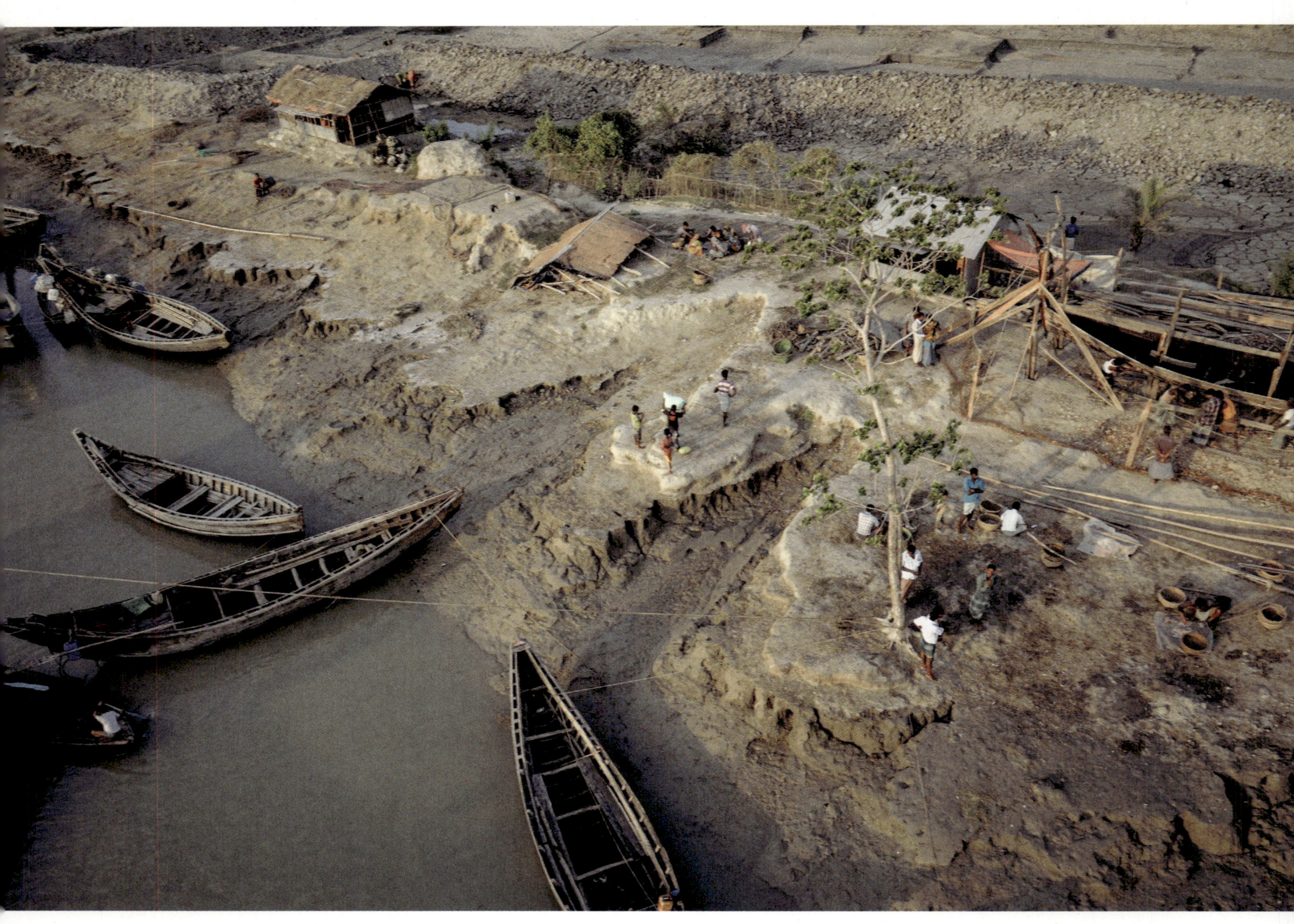

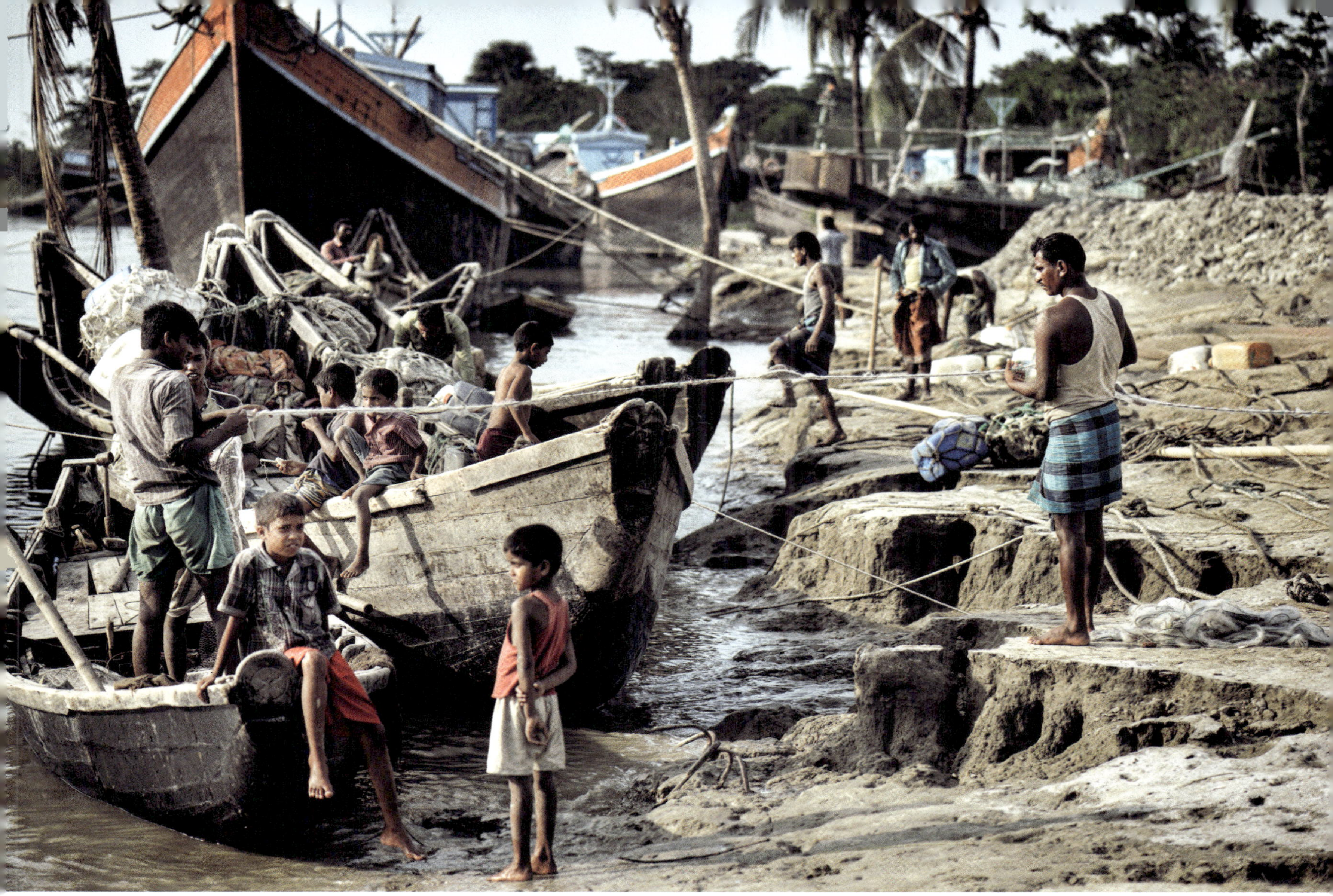

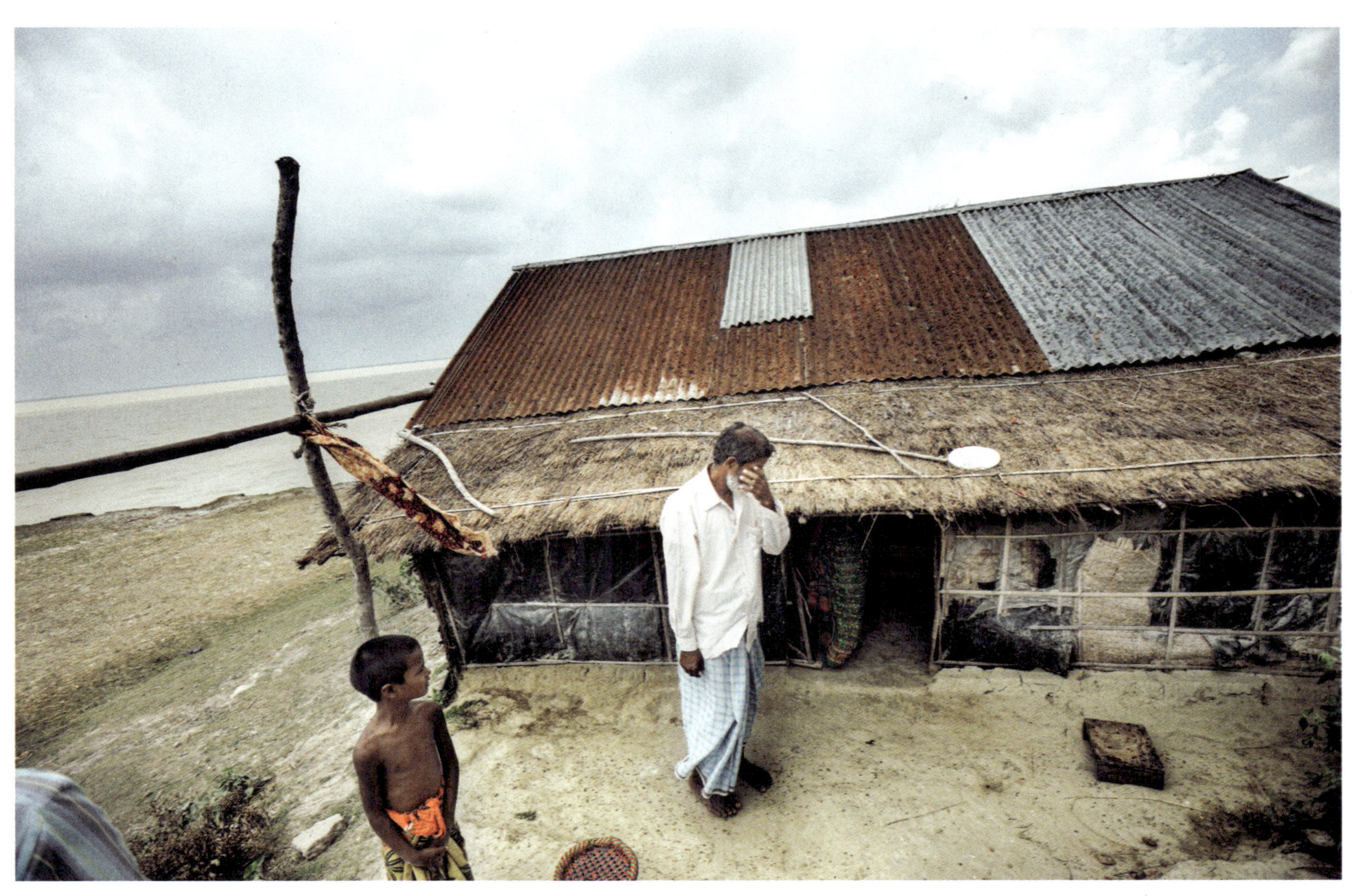

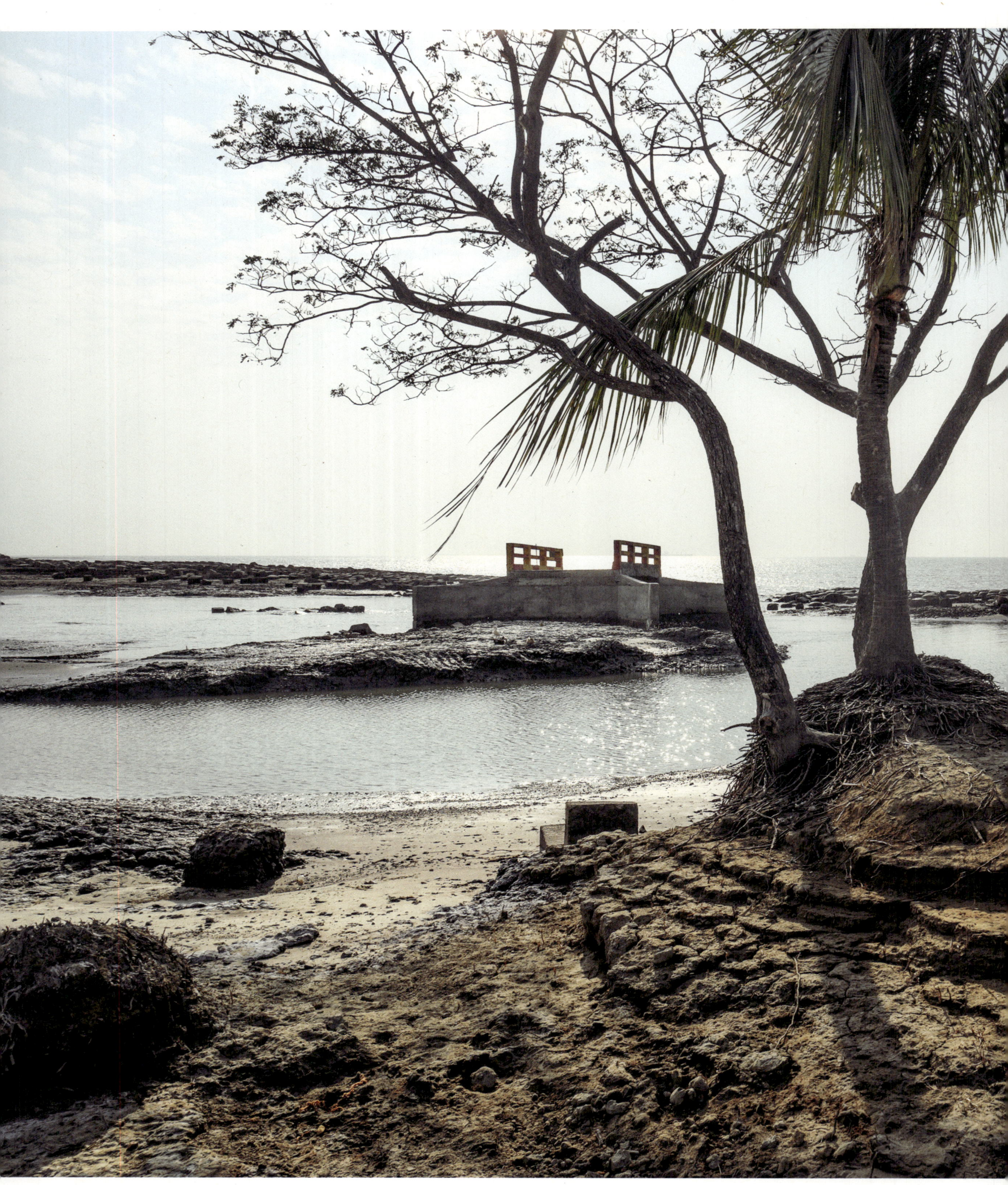

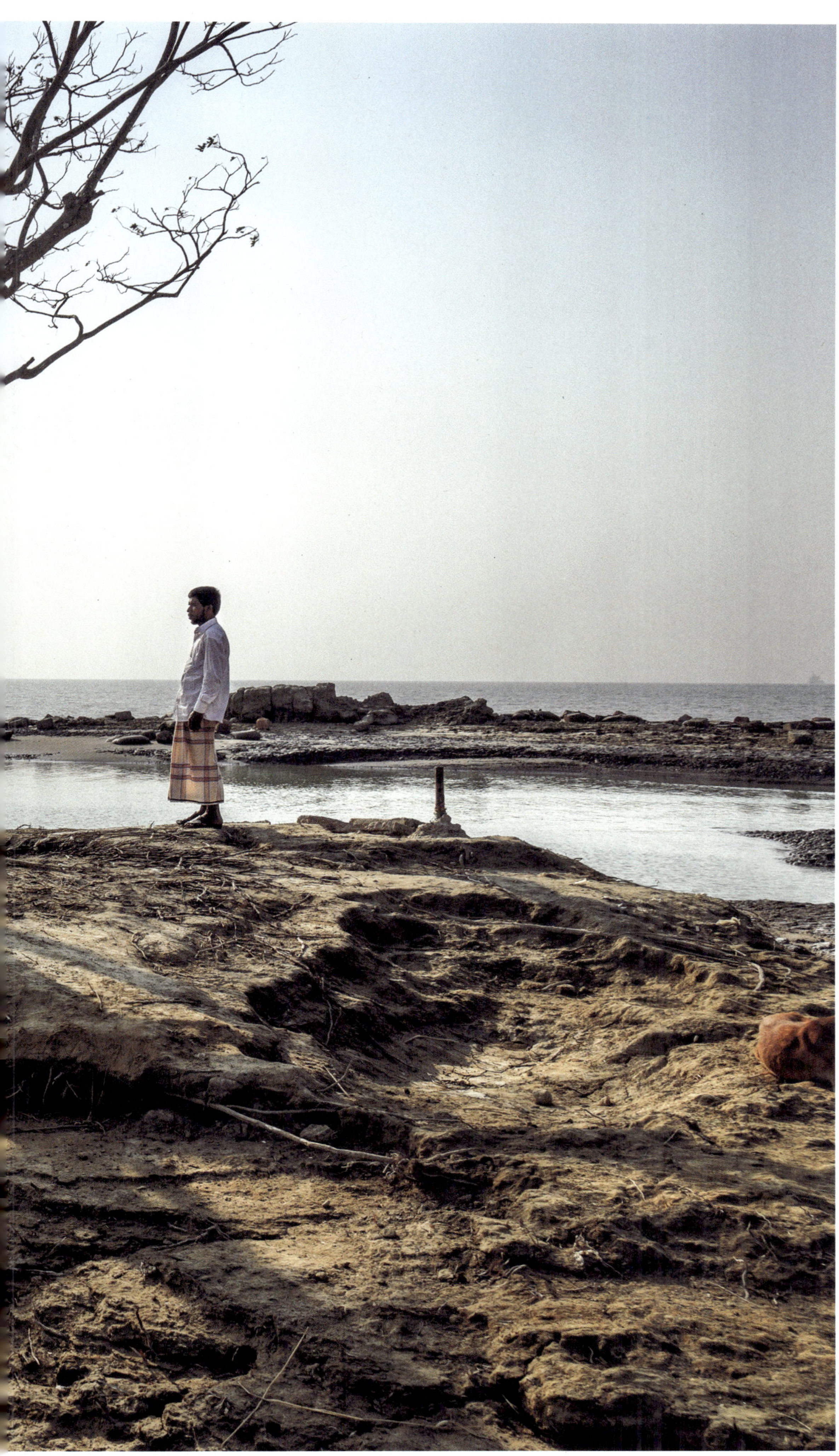

Coastal erosion and flooding on Kutubdia, an island in the Bay of Bengal with 145,000 inhabitants that is eroding at a fast pace. Due to sea level rise, the island will probably become submerged (also on pages 230 – 231).

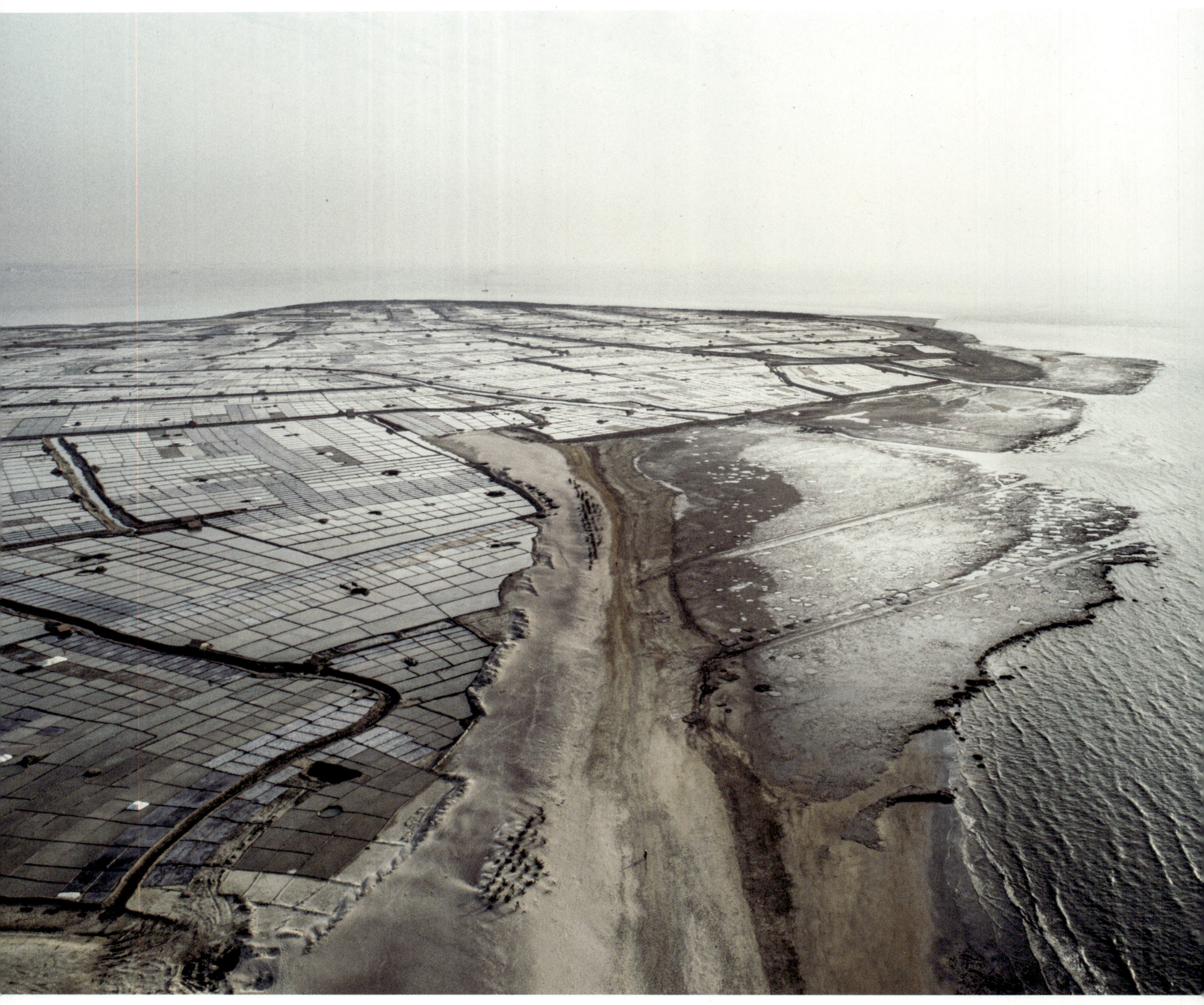

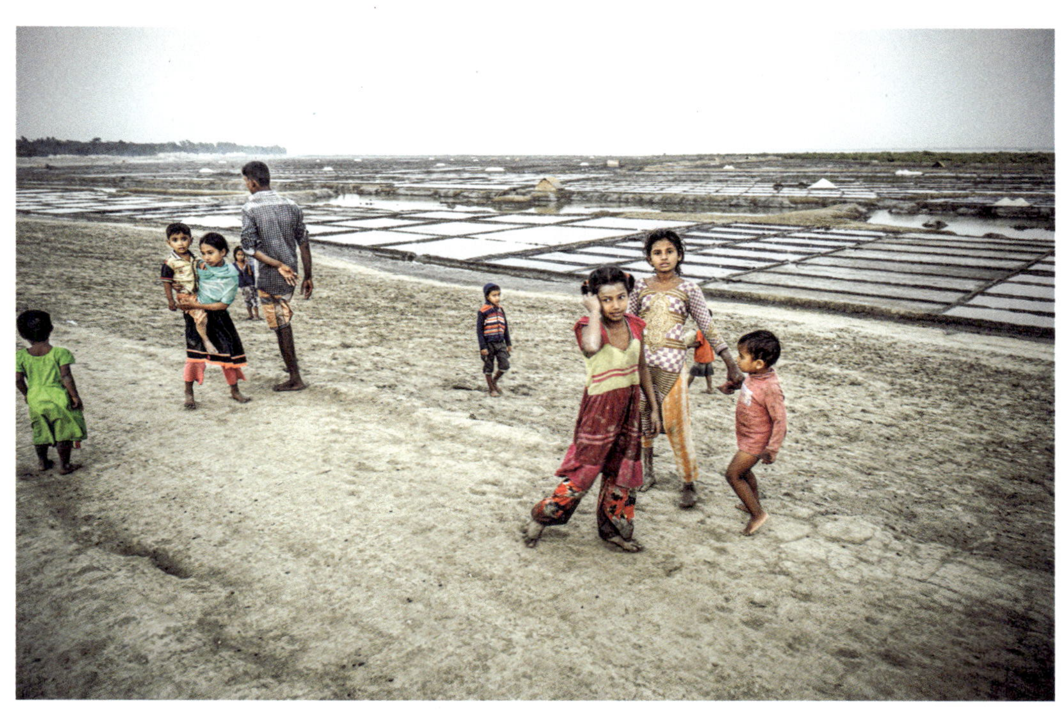

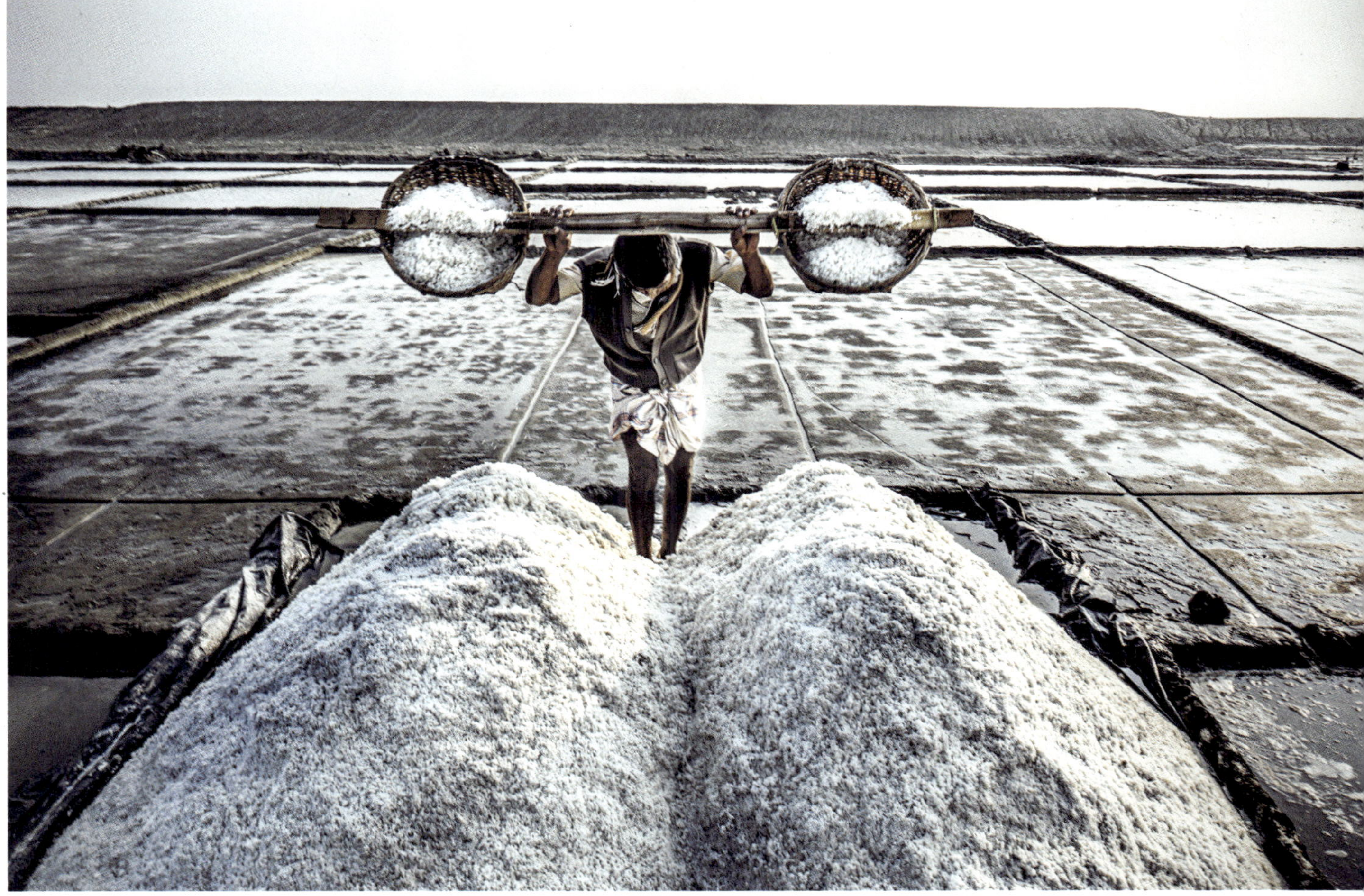

Low tide on the island of Kutubdia.
The island is being fortified by sea
walls, but lack of funds and time have
had a negative impact on progress.

Gulshan is a district in Dhaka where
Mamun Sardar (13) lives with his sister
Rabeya Khatun (15). He works in a
brick factory, she in a biscuit factory.
They are from the village of South
Gunari, in the delta, their family's
home.

Mobarak Gazi (30) is the son of
Mohammed Fazar Gazi, who still lives
in Dacope. They moved from Dacope
in 2009. They came to the Chittagong
Hill Tracts, the only hills in Bangladesh,
because they lost everything; now
they grow rice and other crops. They
rent the house from a landlord and
hope that in the future they will have
their own land and house.

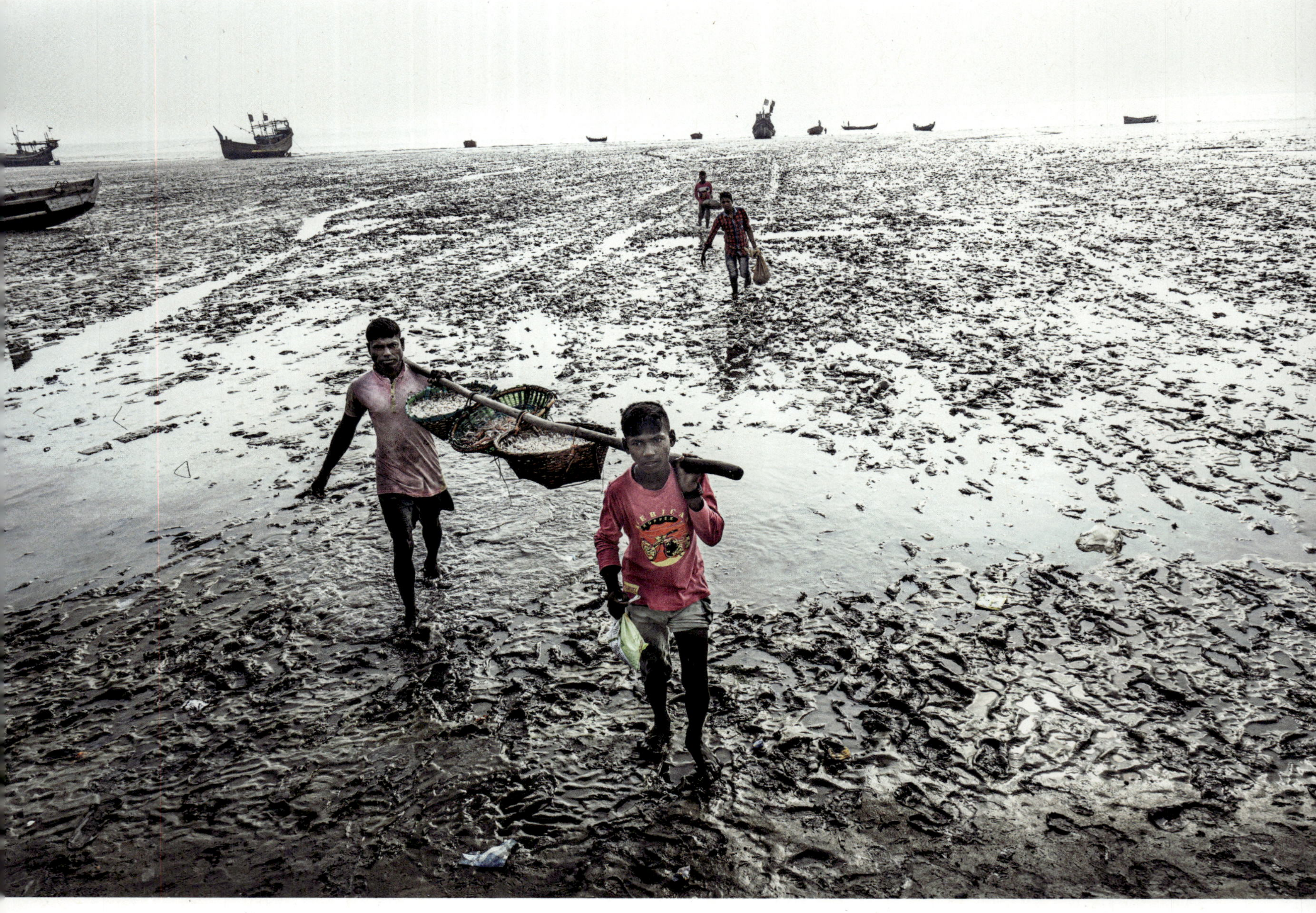

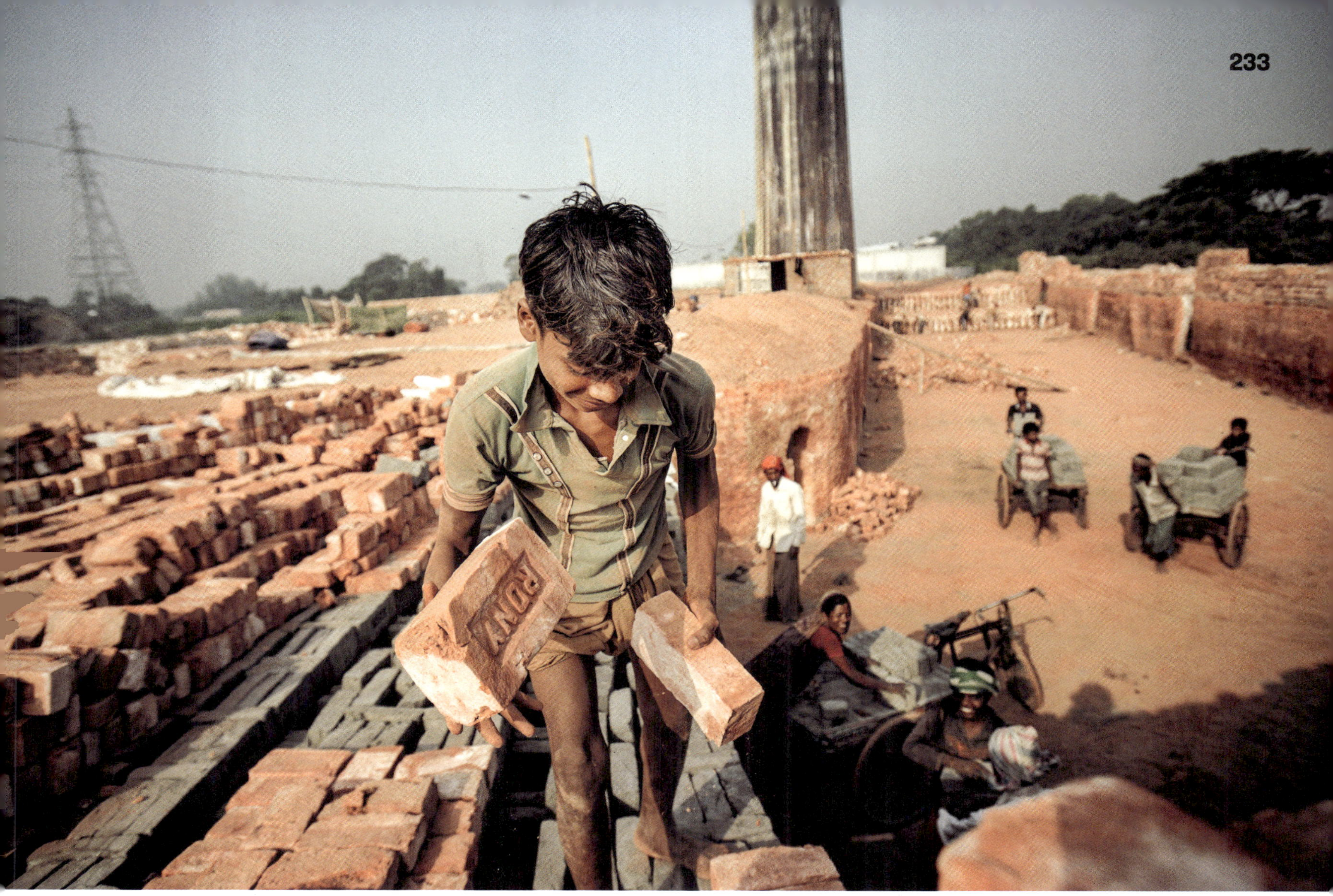

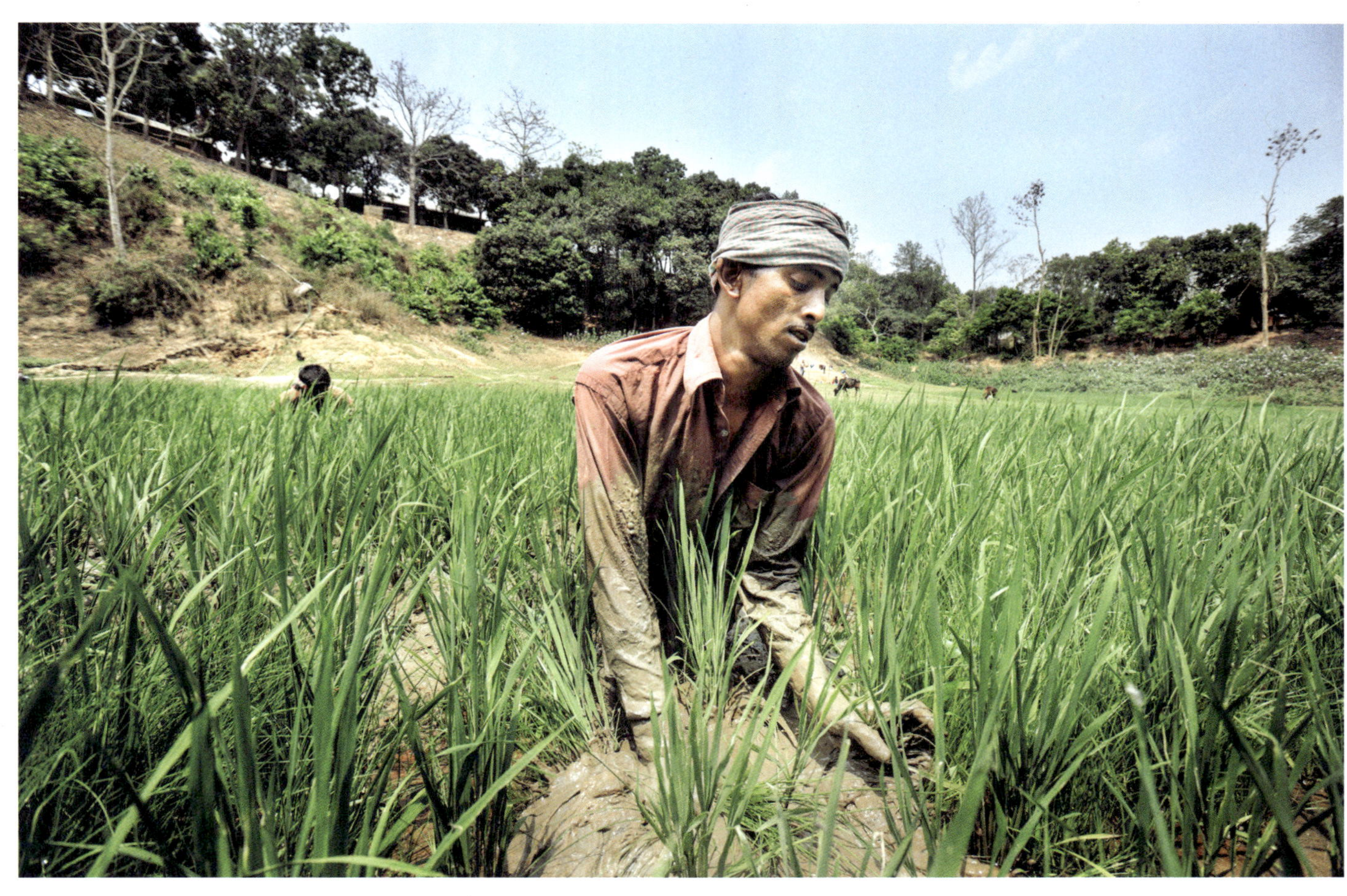

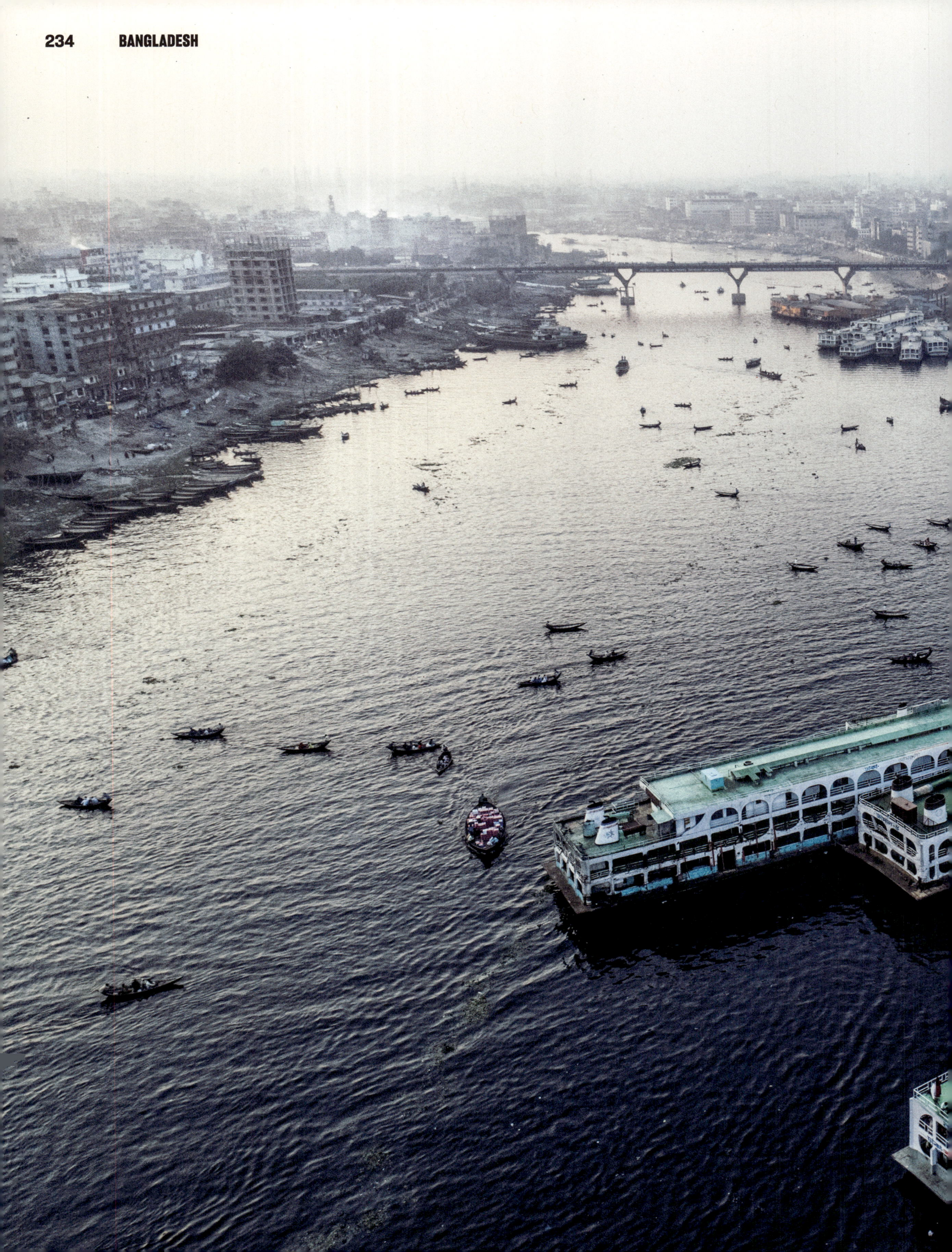

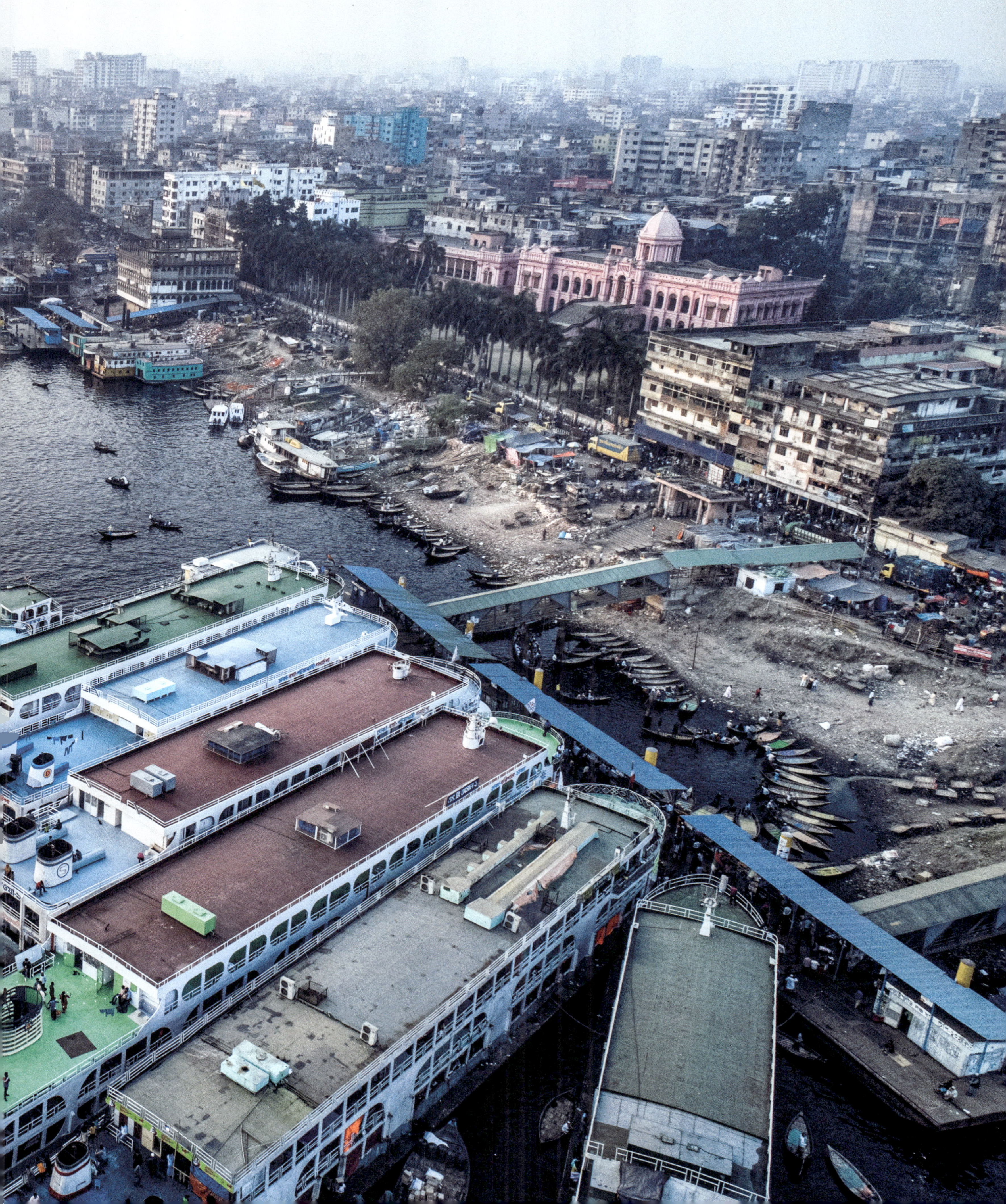

< Pages 234-235: Sadarghat is the main port in Dhaka. Many ships from here navigate the route to the delta.

Every day, many people arrive at the port of Dhaka by boat from the delta, migrating in the hope of a better life. Many lost their livelihoods due to frequent flooding.

A woman brings soil to the shore, which is being used to reclaim land to expand the area.

Karail is a slum area in Dhaka next to Lake Banali, a former river arm. Many of the people who live here have come from the delta, where they lost their livelihoods due to frequent flooding.

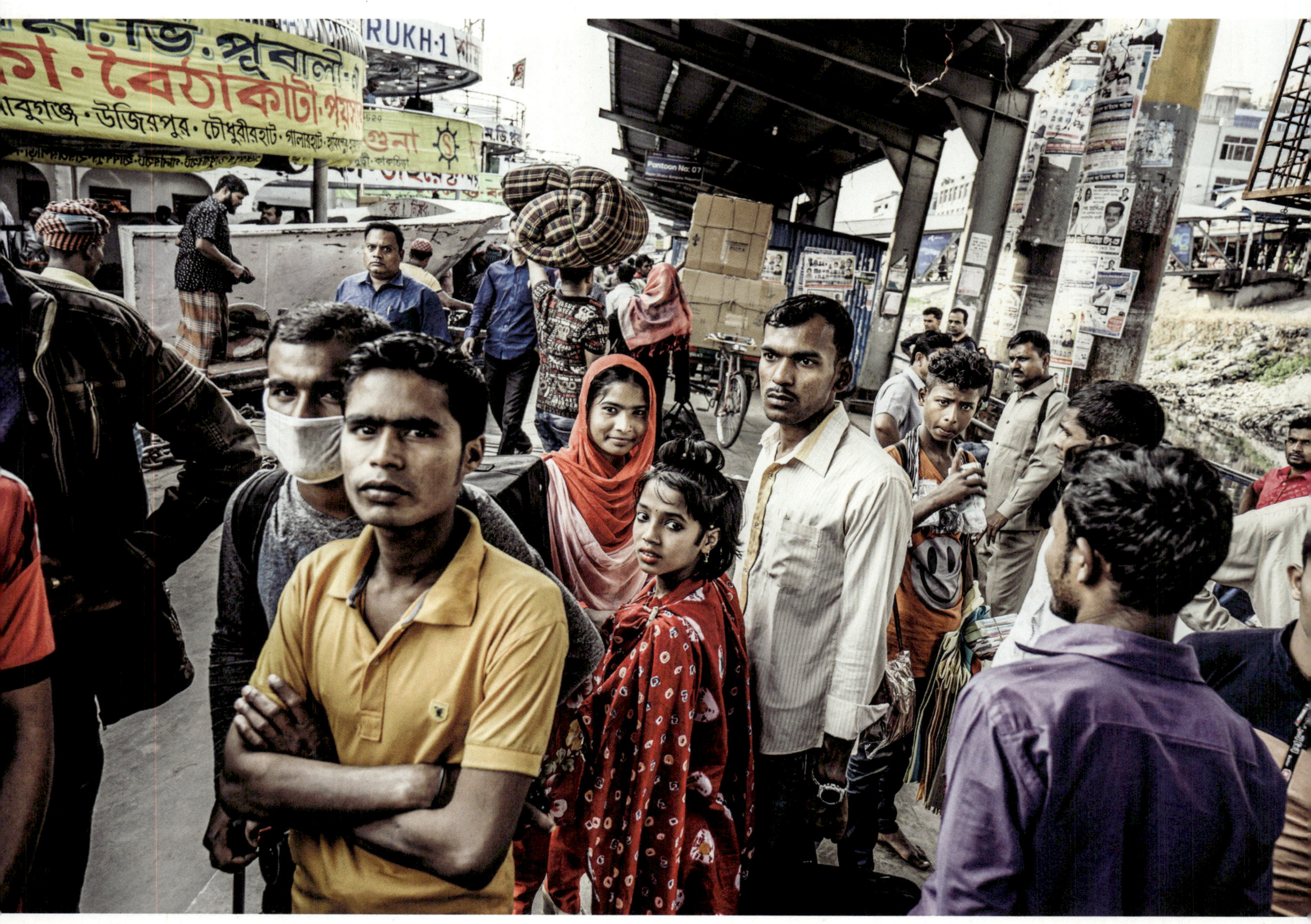

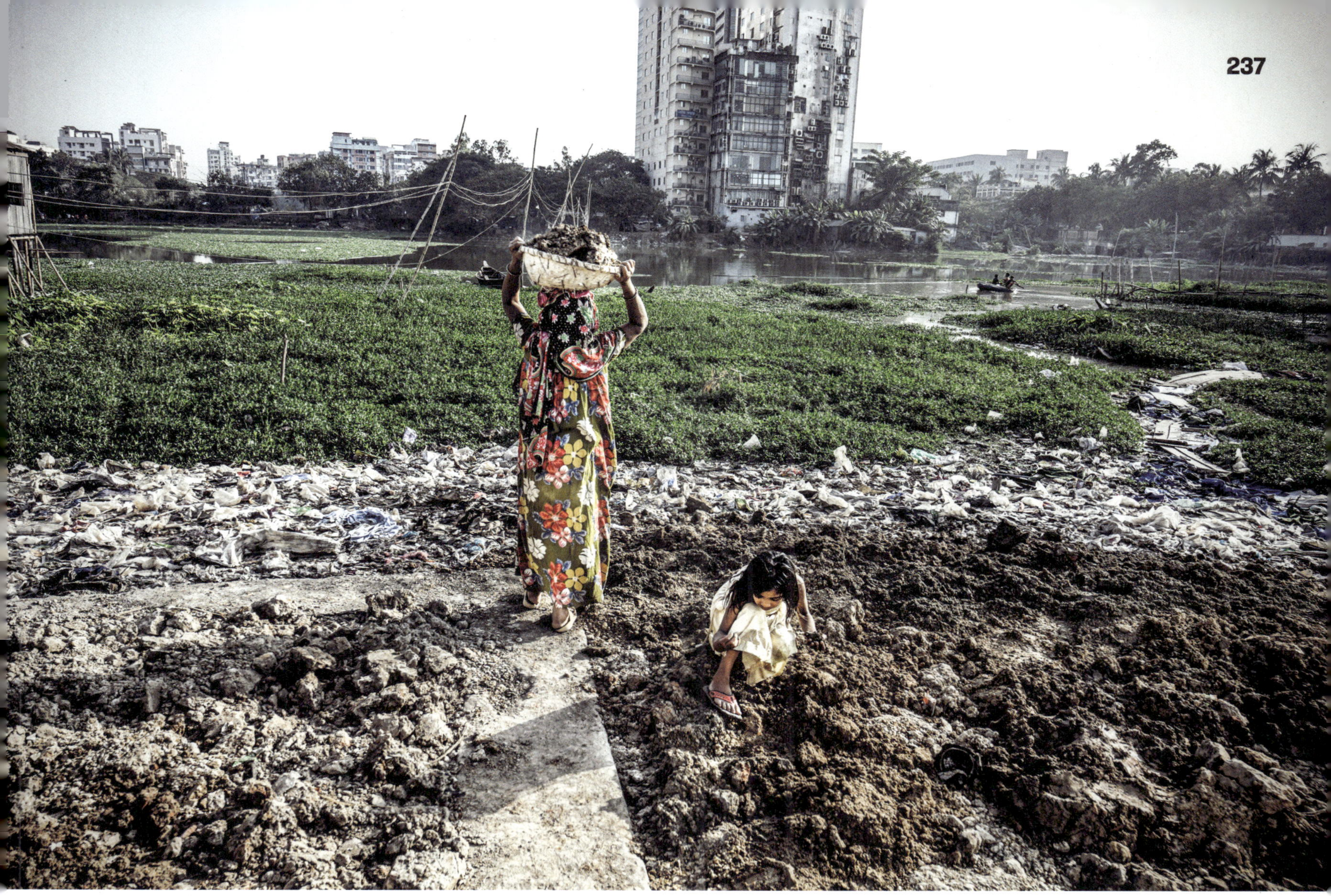

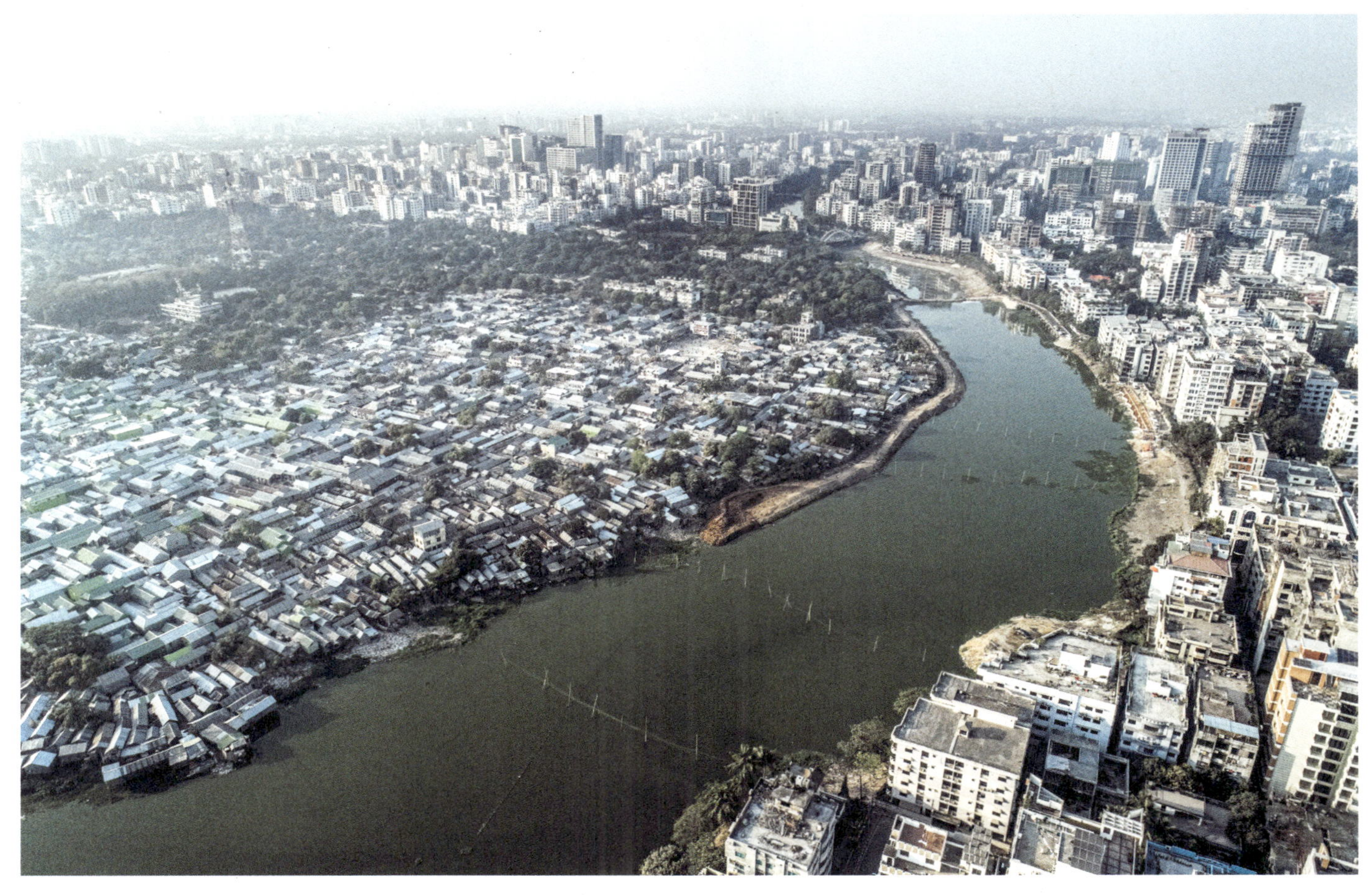

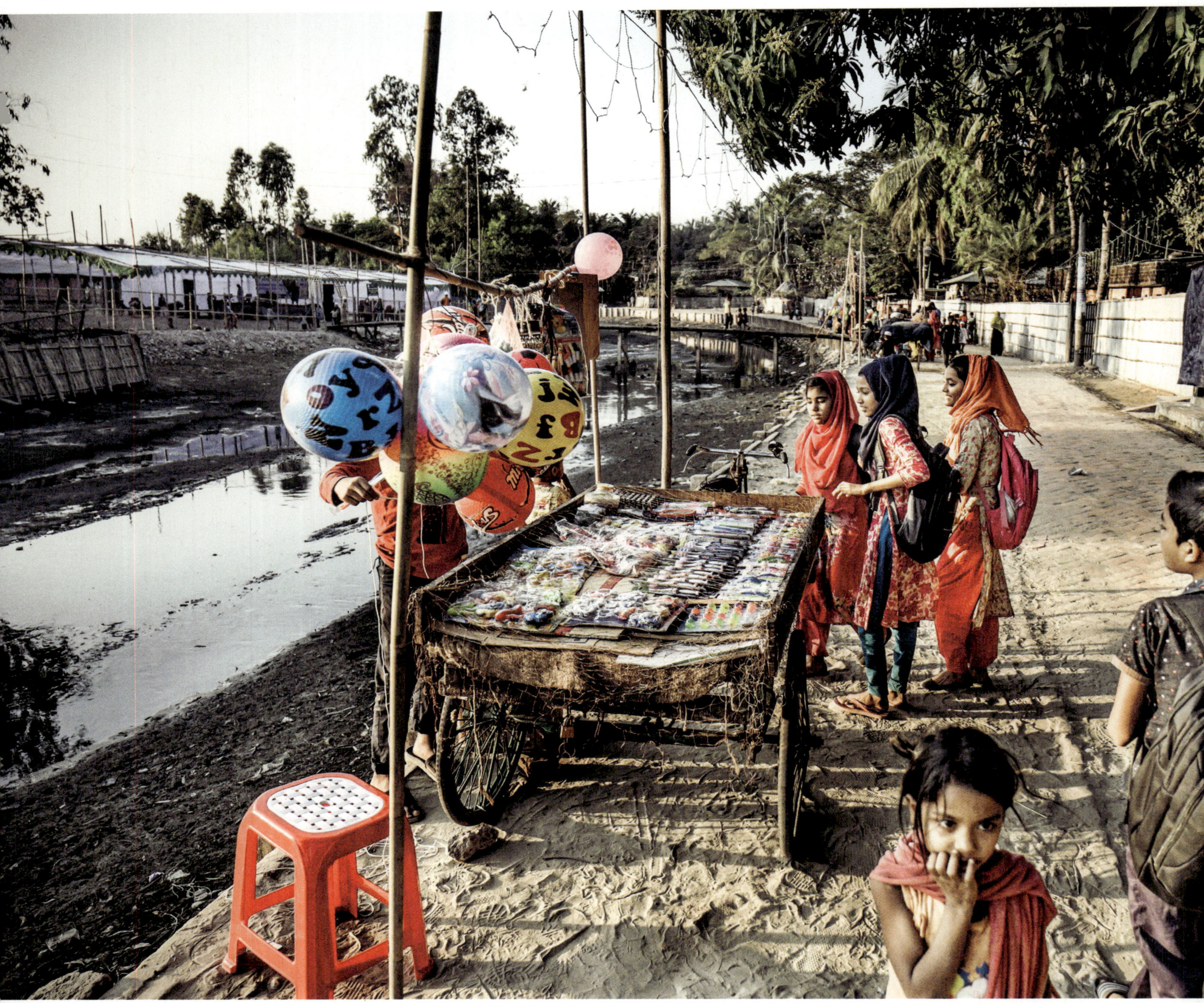

An area north of Cox's Bazar is the
home of many former inhabitants of
the island of Kutubdia. They have left
the island because they have lost their
livelihoods due to frequent flooding.

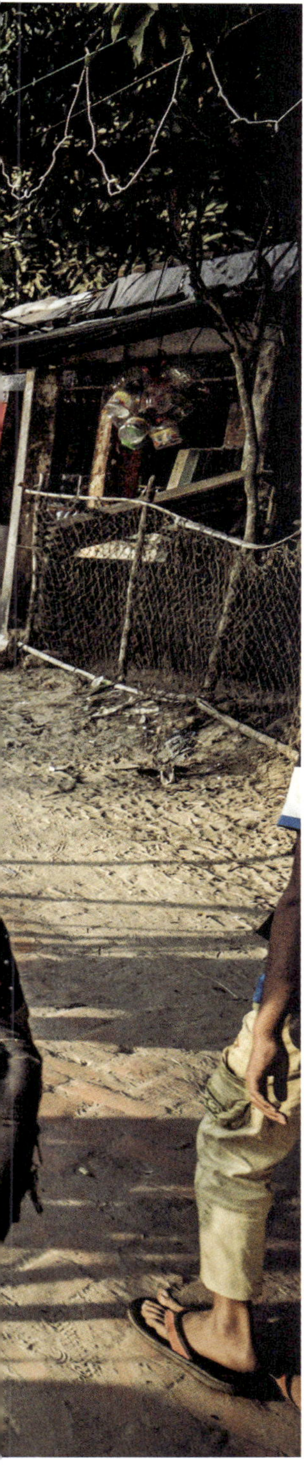

A community in Chittagong which houses many refugees from islands like Kutubdia who have fled the frequent flooding. The community is located next to the river mouth and is therefore also experiencing frequent flooding.

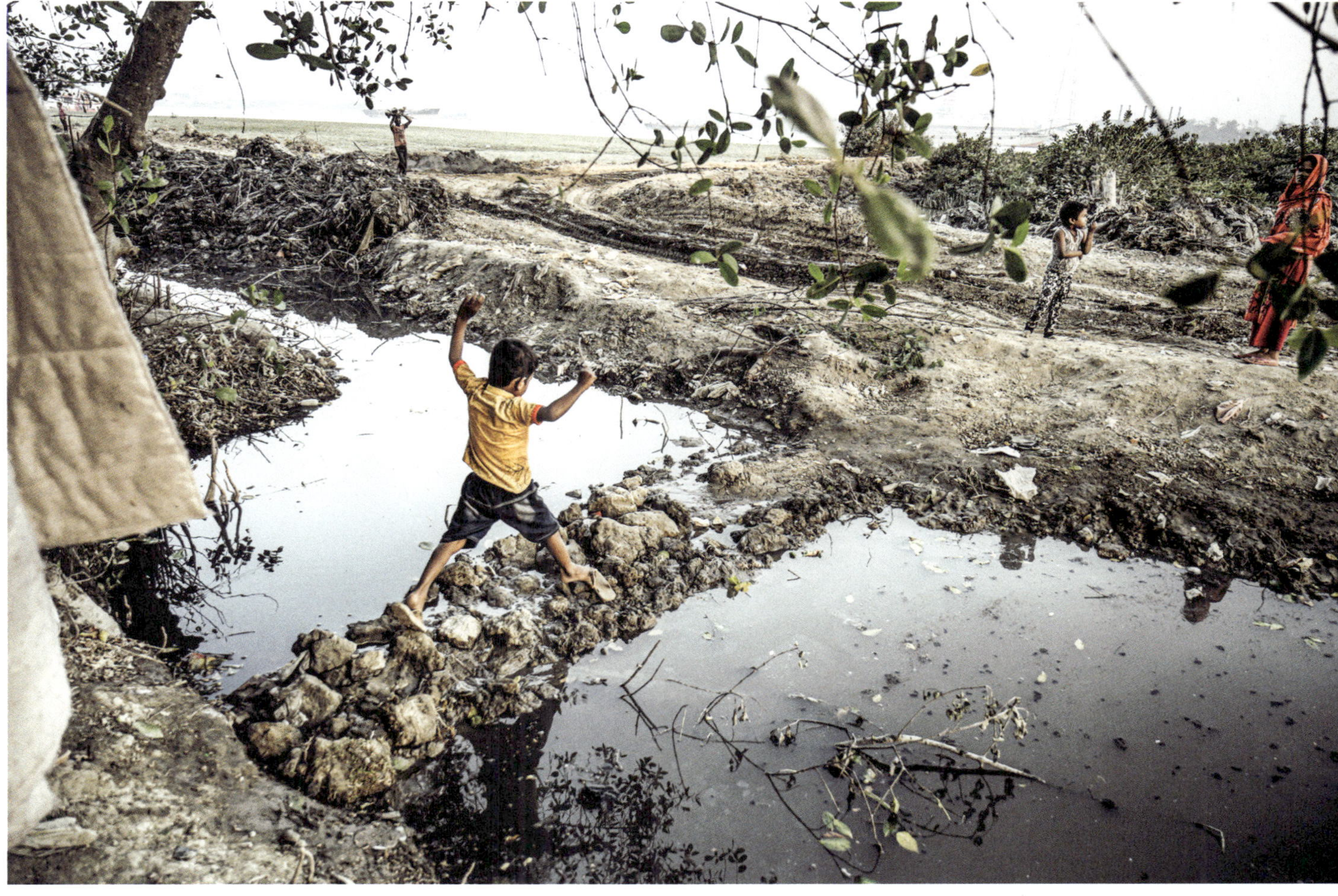

A group of women, all climate refugees, on their way to a protest. Their community in Chittagong is under threat of being demolished for city development.

A community in Chittagong which houses many climate refugees from islands like Kutubdia who have fled the frequent flooding.

A fire at a community in Chittagong, which houses many climate refugees from islands like Kutubdia who have fled the frequent flooding, destroyed many houses. The city is planning to demolish the area for development, and many residents believe the fire was not an accident, but a way to evict them.

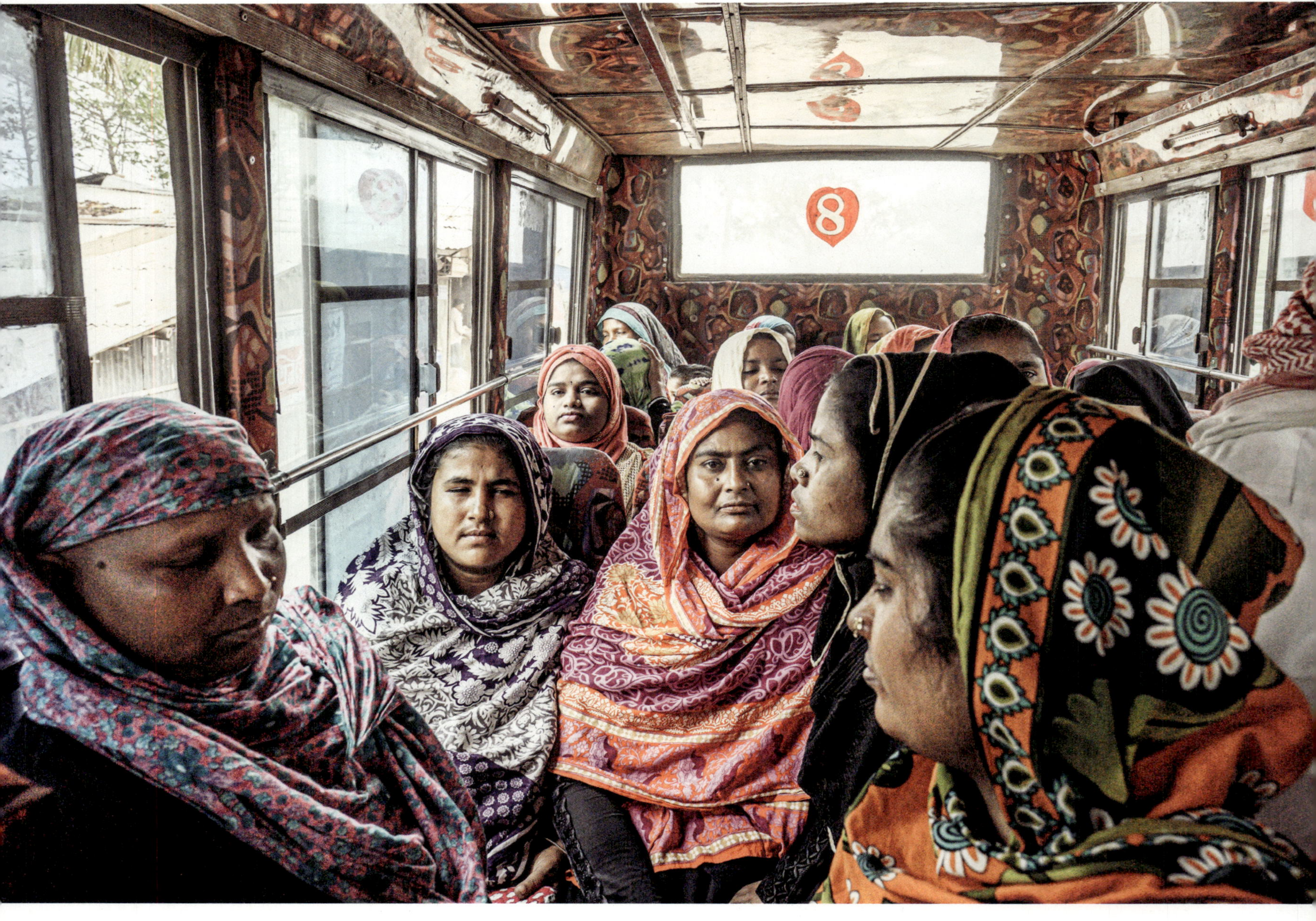

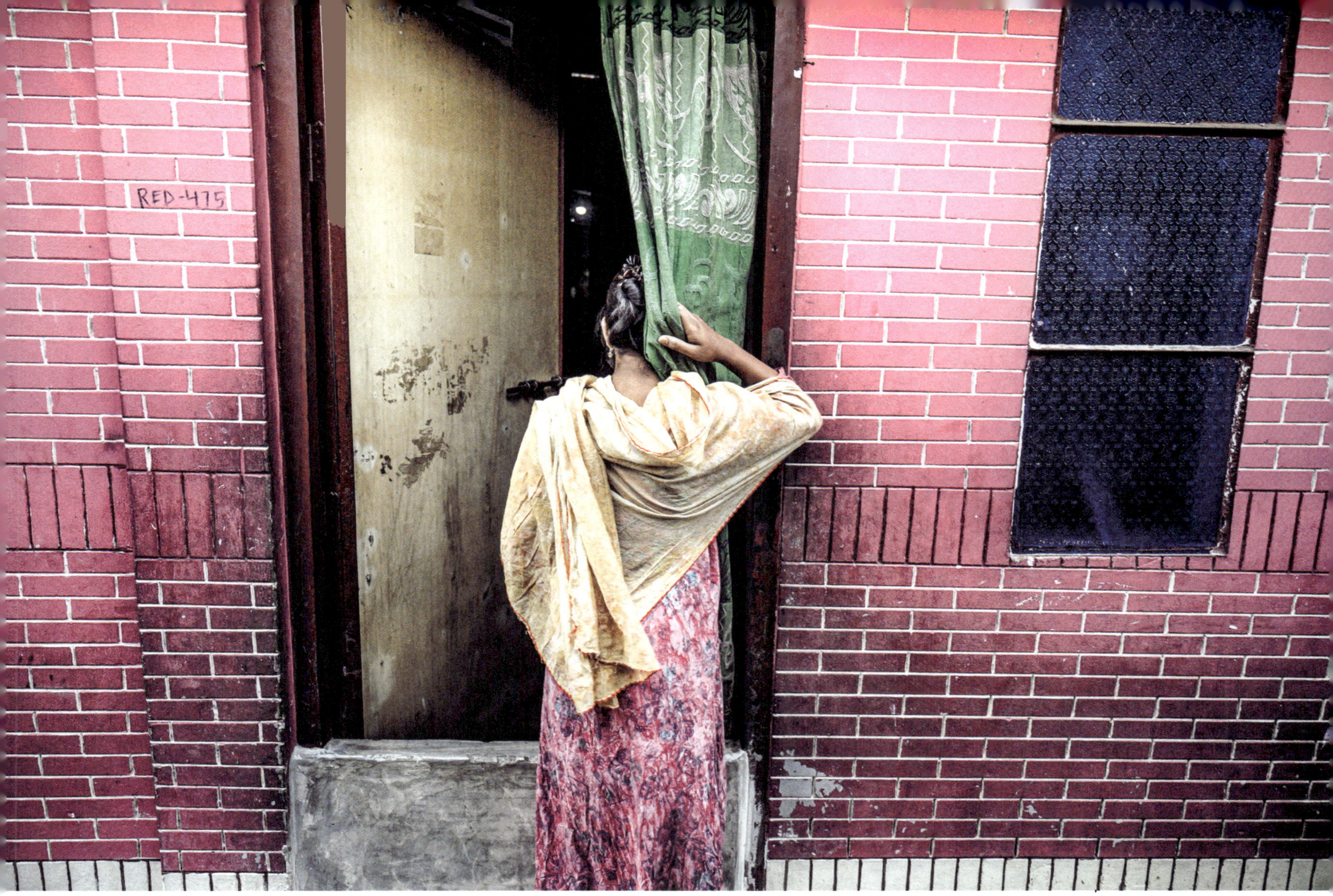

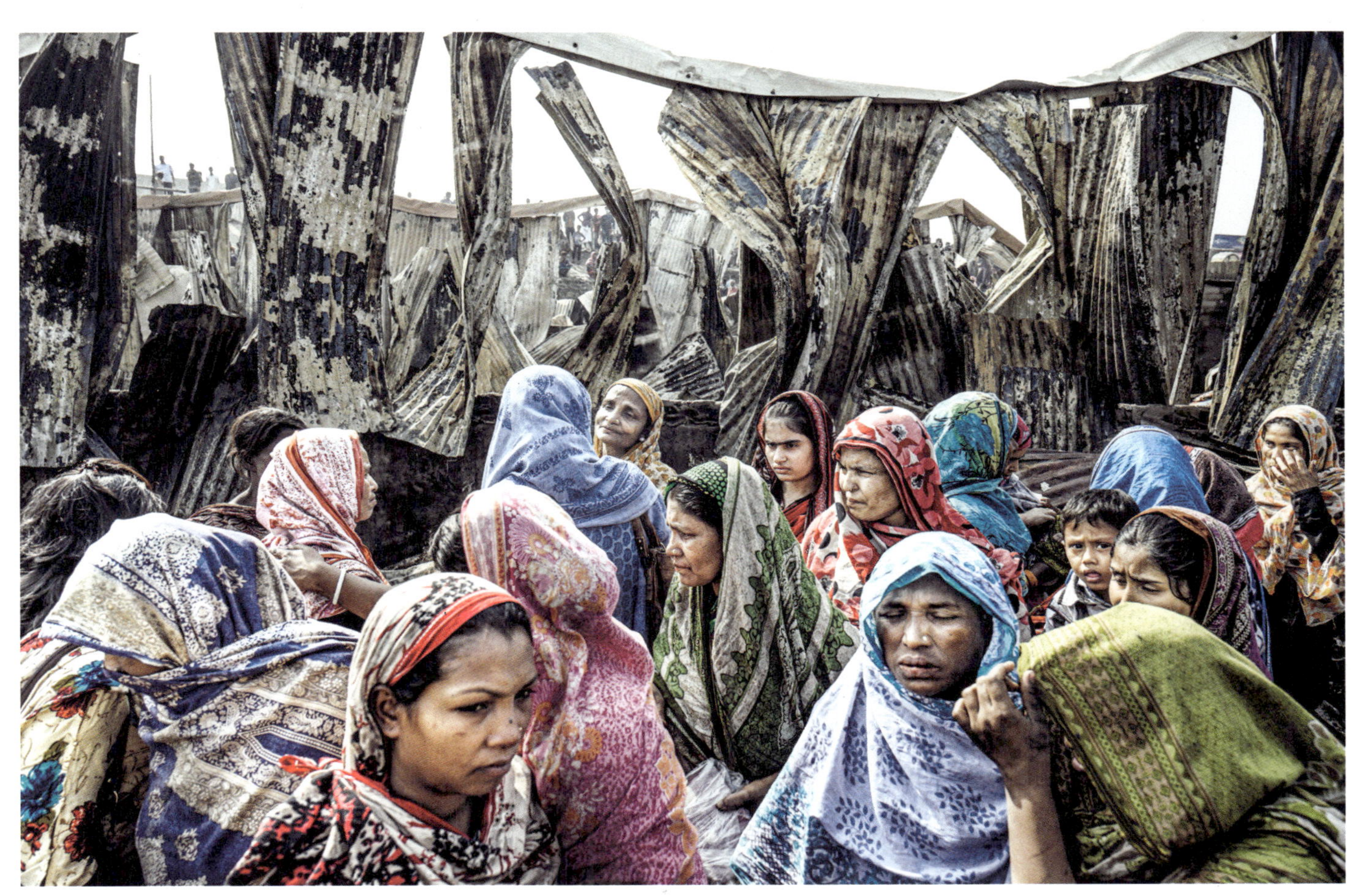

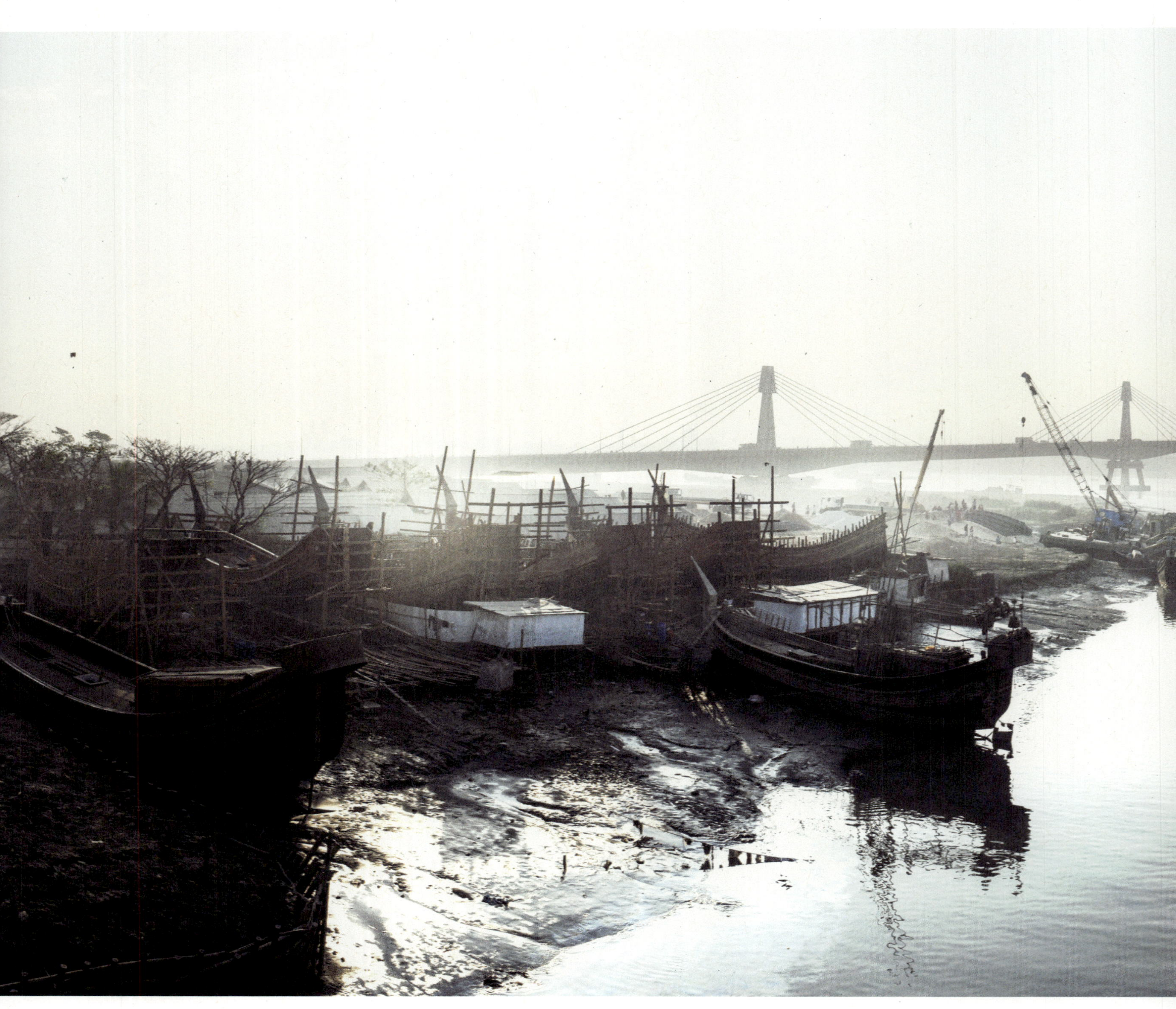

The Karnaphuli River at the city of
Chittagong, a major port in the Bay of
Bengal and a city that is threatened by
sea level rise and coastal erosion.

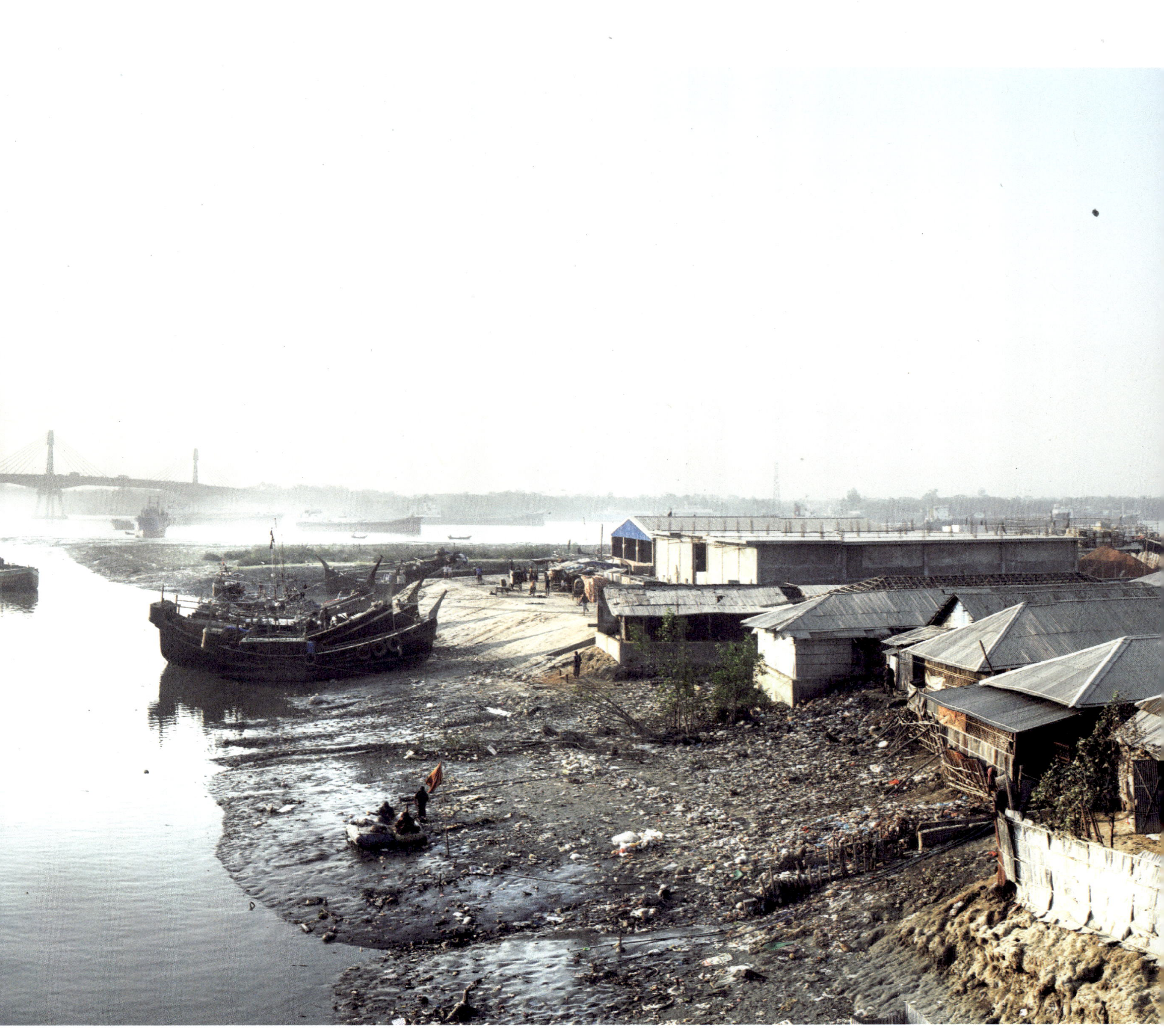

THE NETHERLANDS
Marjan Minnesma

The Netherlands could face a sea level rise of 1 m to 3 m by the end of the century if the world doesn't meet the agreement to reduce temperatures by 1.5 °C, preferably 2 °C. A 2-3 m rise could mean that part of the country will need to be given up, and major cities like Amsterdam and Rotterdam might have to relocate. The Dutch know how to protect their low-lying country, but time is running out - and still, the Dutch government doesn't want to be a 'front-runner' in tackling the climate crisis, it seems.

The name of our country, 'the Netherlands', says it all - itmeans 'the low-lying lands'. About one-quarter of the country is situated below sea level. Roughly 59% would be under water without our dykes and our water management system. Thus, a large part of the country would regularly have serious problems due to storms or excess water coming down the country's vast rivers on their way to the ocean.

The Dutch people started fighting the water hundreds of years ago, building dykes and implementing a very detailed and ingenious system of water management which has made the Netherlands famous. Our water management system is currently run by 21 regionally organised water authorities, which cooperate when needed in a very efficient way. They each have their own tax collection system and use this money to protect the people in their region. About 12,000 people work for the water authorities, and they protect 18,000 km of dykes.

Whereas there were many major floods in the Netherlands before the 20th century, over the last century, there have hardly been any major disasters. The last time was in 1995, when 250,000 residents were evacuated due to a fear of flooding, but in the end, the dykes were strong enough. Because of this very solid water management system, we have forgotten that water was not always our friend, and we seem to ignore that the climate crisis is changing the climate system much more quickly than expected, at a pace that should worry us.

The warmer oceans and the melting of Greenland and Antarctica are resulting in an increase in sea level rise. In the 1990s, global sea level rise was on average 2.5 mm per year. The most recent *Special Report on the Ocean and Cryosphere in a Changing Climate*, produced by the International Panel on Climate Change, reported a rise of 5 mm per year. The rise is speeding up, and the question is, can we build higher dykes quickly enough, if the sea level also keeps rising so quickly?

THE STEEPER THE CURVE, THE FASTER THE SEA LEVEL IS RISING.

In 2007, the ministries started to realise that they might need to scale up their activities to protect the Netherlands from rising sea water. They installed a new 'Delta Commission', tasked with giving advice to the Dutch government on protecting the country from flooding over the very long term, while ensuring it remains an attractive place to live, work, recreate and invest. In September 2008, the Commission delivered their report, which stressed that the climate crisis is a fact and that the sea level will probably rise even faster than previously expected. The report predicted a sea level rise of 0.65 m to 1.30 m by 2100, and of 2 m to 4 m by 2200, also noting that the country's soil is declining due to other factors. This was a shock to many people. Unfortunately, this also resulted in much critique. Other scientists, not part of the Commission, stated that the worst-case scenario of a 1.30 m sea level rise in this century was absurd and even 'activistic'. The Commission stressed that the worst-case scenario was of course not the most probable scenario, but

Global mean sea level rise

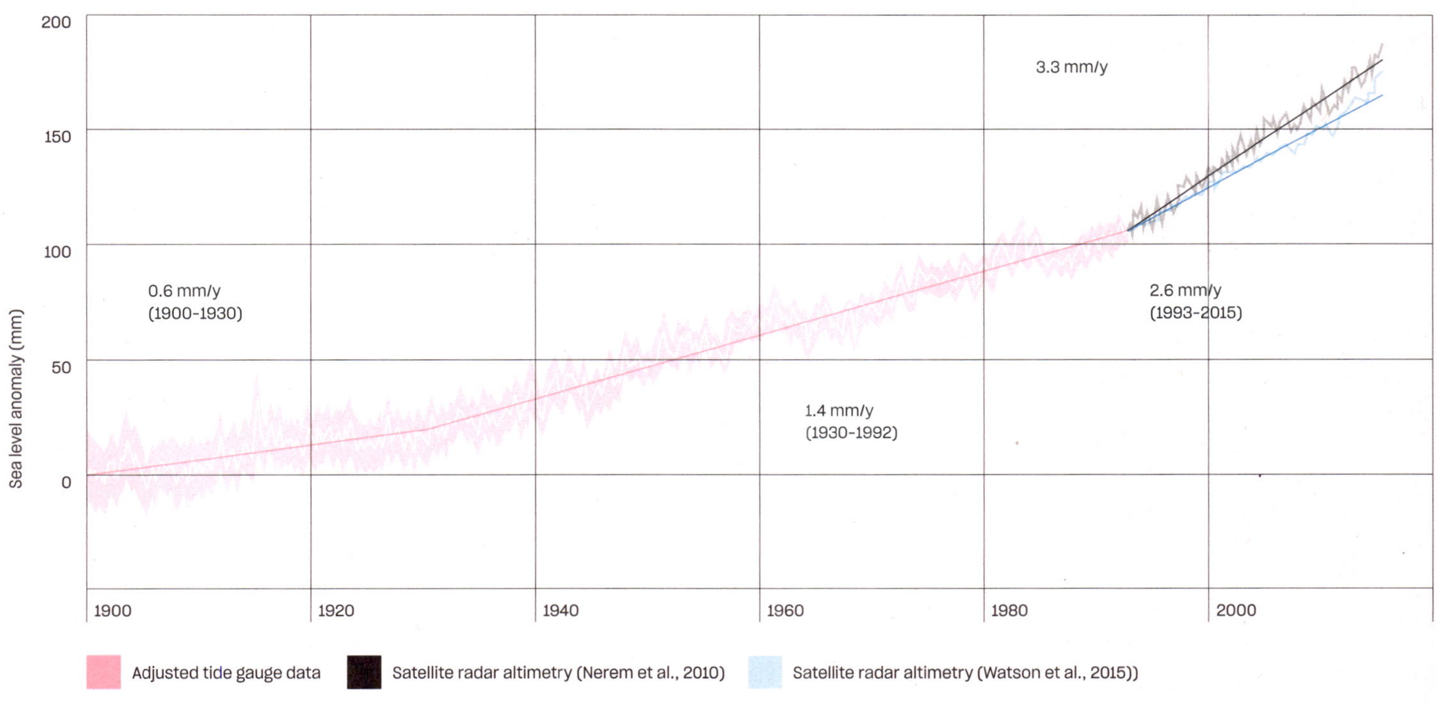

Source: Oppenheimer et al., in press

that we still needed to anticipate this much-quicker sea level rise and - better safe than sorry - to plan for the worst.

The Commission recommended initiating a new 'Delta Programme' and investing €1 billion to €1.5 billion per year (0.2% to 0.3% of gross national income) to improve the country's dykes and overall water system. Based on the Commission's recommendations, the Dutch government did initiate the Delta Programme, reserving €1 billion per year for almost 100 years to protect the country against water, a move probably not seen in many other countries.

THE WORST-CASE SCENARIO GETS WORSE

Ten years later, it appears that this 'worst-case scenario' of a 1.30 m sea level rise was actually optimistic. The national research centre for meteorology and climate, the Royal Netherlands Meteorological Institute (KNMI), reported in a study on the worst-case scenario that sea level rise this century might even be between 2.5 m and 3 m. Again, this is not the most likely scenario, but it shows that within 10 years, the worst-case scenario went from 1.30 m to 2.5 m or 3 m. A big leap in only 10 years!

This does not seem to worry the Dutch people. It is not a major topic on radio or TV, and it is not a political topic that we should act much faster. We are so convinced that we are the best water managers in the world (which we might be), that we have forgotten that there are physical realities that might prevent us from acting in time. Building new and bigger dykes takes 30 to 40 years. We need to plan ahead. It is not something we can speed up in just a few years. If the climate crisis is not taken more seriously, we might simply act too late. If the worst-case scenario becomes real, we will have a major problem. A few groups of scientists know this and have started to look at alternatives, ranging from constructing extremely high dykes and floating homes, to moving to Germany. Ten years ago, they thought that this was an absurd idea; nowadays it is one of the options being considered for the future.

A NEW PROBLEM: DROUGHT

In the meantime, it has become clear that it is not only too *much* water that creates a problem; it is also long periods of drought that should concern us. In a country in which almost 20% of the surface is covered by water, you do not expect problems from having too little water. Until 2018, most people in the Netherlands did not think that our country could ever not have enough water. However, the summer of 2018 was extremely dry and a difficult period for farmers on sandy ground, as well as for many nature areas. Our dykes suddenly appeared to not be as robust as we had thought. When a dyke dries out, it loses some necessary capacity and might not function as it should. The water authorities started to check on dykes more often and strengthened them for more dry periods. The economy was also impacted, as there was not enough water coming down the rivers, and many boats could not travel to Germany and back.

The next year, 2019, was slightly better (less dry, but not really wet either), whereas the first half of 2020 was extremely dry again. For months, there was hardly any rain. Again, farmers and nature faced the most challenges. People start to realise that these two dry years might be not just isolated incidents, but the 'new normal'. This requires more action, in the cities as well, but certainly for farmers, to keep water within our country longer and to store it more cleverly. Currently, all water entering the Netherlands is moved to the North Sea as quickly as possible - removed from the country. In the near future, people will need to store more water around their homes. Farmers should store water on their farms, and the water authorities should store more water in a variety of waterways, lakes and other places.

LAGGING BEHIND

Looking back at the last 30 years, since we started working on and signed the United Nations Framework Convention on Climate Change (UNFCCC), we must conclude that the Netherlands once were front-runners in preventing climate change but are now among the laggards of Europe. What happened? At the end of the 1980s, we were among the few countries that put the climate crisis on the agenda and argued in favour of a worldwide convention on the climate crisis. However, in the last 20 years, we have talked a lot, but we have not taken rigorous action, even though we should be among the first to take action and understand that we will have a serious problem when Greenland and Antarctica melt. Instead, we are among the 20% of countries with the highest CO_2 emissions in absolute terms, and within that group of high emitters, we are among the top 10 countries with the highest emissions per capita. We are not at all a small country with small emissions. We are a small country that is extremely fossil fuel oriented, with one of the largest natural gas fields in Europe and large harbours that import enormous amounts of oil, gas and coal. We have built a large chemical industry based on this fossil fuel infrastructure, and we have a large steel factory close to the coast, which is our biggest emitter. We are a big emitter.

And that should change quickly. How can we ever demand that other countries act to decrease their greenhouse gas emissions, when we ourselves are among the biggest emitters?

SPURRING ACTION

For me, this was the reason that, in 2012, I started urging our government to take more action on the climate crisis. The Netherlands has signed numerous documents within the context of the UNFCCC agreeing that industrial countries should reduce their emissions by 25% to 40% by 2020, to have a fair chance at avoiding a 2 °C warmer climate on average worldwide, as compared to 1850 (the beginning of the Industrial Revolution). Urgenda, the NGO I founded in 2007, wrote a letter to our government demanding a 40% reduction by 2020. The government agreed that the climate crisis was a major risk, with potentially enormous consequences in this century, but indicated 'that they did not want to be a front-runner' in climate action, as it might hurt the country's industry and economy. This is ironic when you realise that we were at that time one of the laggards in the European Union, with only Luxembourg using less renewable energy than the Netherlands (measured as a percentage of total energy used). Currently, even Luxembourg uses more renewable energy than the Netherlands.

When the government refused to take additional action, Urgenda decided to take the Dutch government to court for breaching their duty of care to protect the Dutch citizens from the dangerous consequences of the climate crisis. It was clear that more action was needed. And although the government admitted that the climate crisis was a clear threat (we did not debate that at all), they refused to take more action. And that is typical for many countries in the world. They refuse to act and do not behave as if climate change is a crisis, which it is.

Urgenda asked the Dutch citizens to join the case, so we ended up in court with almost a thousand co-plaintiffs, who supported us all the way to the Supreme Court. The *Urgenda Foundation* v. *State of the Netherlands* case became famous, because we won three times in a row: at the District Court in June 2015 (six months before the Paris Agreement!), at the Court of Appeal in October 2018 and finally at the Supreme Court in December 2019.

SOLUTIONS AND SUPPORT

As a result, the Netherlands were required to reduce their greenhouse gas emissions by at least 25% from 2020 onwards, compared with 1990. This is not enough to avoid a 1.5 °C warming compared to 1850, but it does force the government to speed up their actions and get in line with their own target for 2030 of a 49% reduction in emissions.

The court case was unusual for Urgenda, as our mission is more to take action ourselves, not demand that others take action. More in line with our typical projects, we have been working on a plan with 54 different solutions that the government could still take, even in 2020, to achieve the goal of a 25% reduction in emissions by the end of the year. We launched the plan together with 800

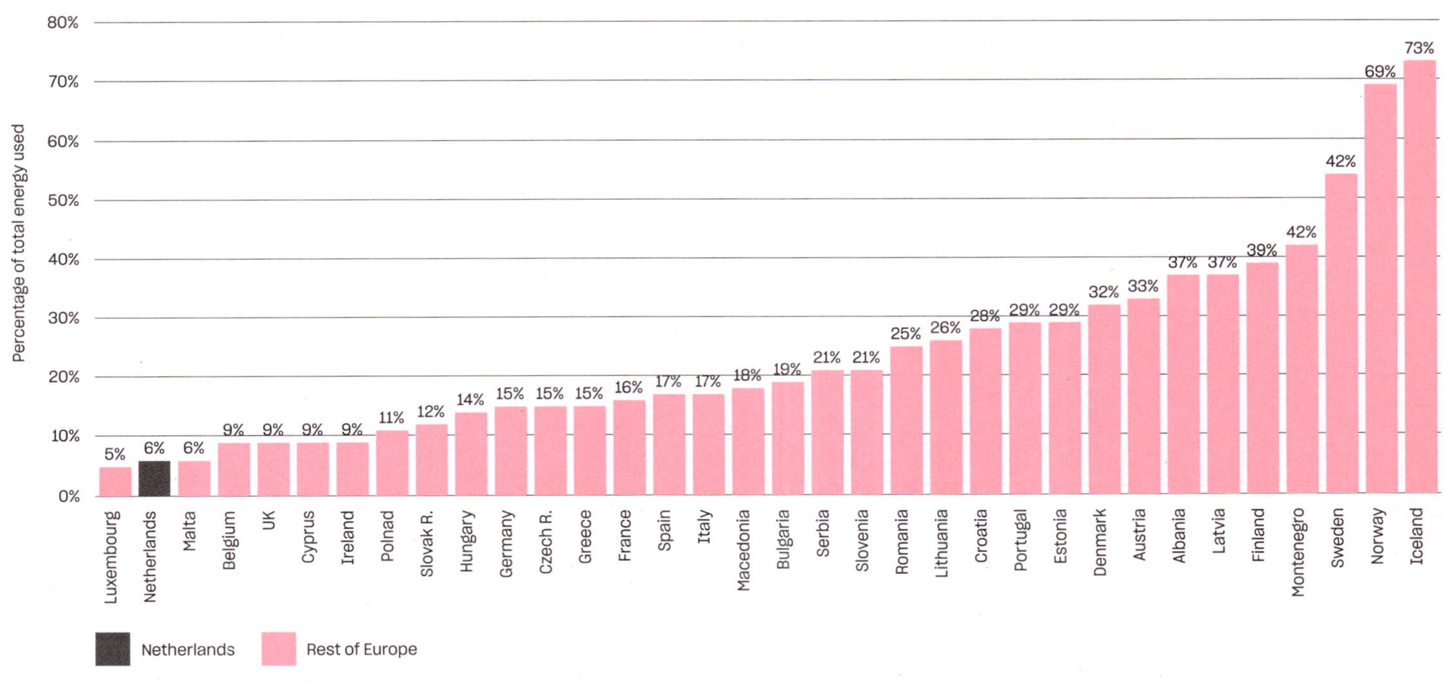

Renewable energy usage

Source: Urgenda, courtesy of Mathijs Bouman, @mathijsbouman

Map of the world showing countries resized according to their total CO_2 emissions, 2009

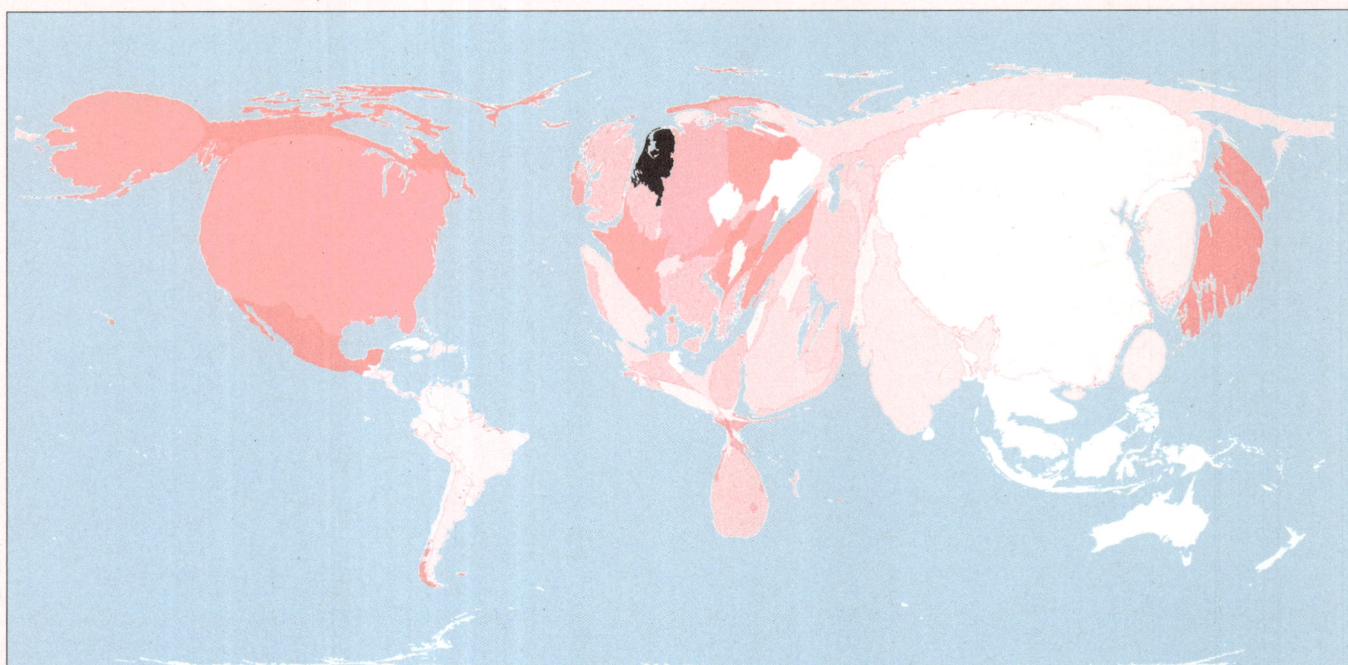

Source: IWR (2009) & UNFCC (2007). Map created by Benjamin Hennig, Sasi Research Group, University of Sheffield – www.viewsoftheworld.net

different organisations, which also offered to help the government in meeting their goal of reducing emissions. We explained that the government could either take these 54 small steps to reach their goal or combine a few larger steps, such as closing down coal-fired power plants. And we let them know we were prepared to help.

In the end, the government chose to close one old coal-fired power plant at the beginning of 2020, and they allocated a large budget to the coal sector to close down a second one. The remaining coal-fired power plants (three of them brand new, having opened in 2015 and 2016) face legislation that forces them to keep the plant at its technical minimum, meaning running on coal at a maximum of 25% to 33% of capacity. In addition to these actions for steeply reducing the use of coal, the government also agreed to carry out 30 of the 54 measures in our action plan for 2020. About €3 billion was set aside for this - a necessary step towards a much bigger reduction in 2030.

Urgenda has also worked out a plan, with input from many experts, for the Netherlands to move to a society that would run on 100% renewable energy in 2030. We used the open-source Energy Transition Model, which is also used by the government and many other actors in society. The first version of this plan was issued in 2012, when there were still 18 years to go until 2030. The last update was in 2019, when just over 10 years were left. As most

climate scientists and activists know, 10 years is extremely short, and under normal conditions it is not very likely that this goal will be reached. Still, it is the only way to have a reasonable chance at avoiding a 1.5 °C warming compared to 1850, and it is what all industrial countries should aim for. We have learnt from COVID-19 how much we can change in a very short time and what steps we are prepared to take when we realise that we are in a serious crisis. It is time we finally realise that the climate crisis is an even bigger crisis. We can easily afford to take the necessary steps, and we will end up with cleaner air and improved health, while also slowing the current loss of biodiversity. Why wait? Let's act! •

In 1953, the Netherlands was struck by a major flood. A combination of a very heavy 'northwesterly' storm and a spring tide caused major parts of the southwest to flood, and nearly 1,900 people died. The country realised it wasn't as safe from the sea as it had thought.

An ambitious plan was developed to protect the Netherlands: open sea arms in the southwest were closed off from the sea, and all dykes were brought to 'delta heights', which meant that a storm surge like the one in 1953 could statistically happen only once every 10,000 years.

But a new report by Deltares, published in 2018, indicates that the Netherlands could face a sea level rise of 1 m to 3 m by the end of the century.

It's thought that the country could handle a 1 m or possibly 2 m sea level rise if additional funding is made available, but a 3 m rise could mean that part of the country will need to be given up and that major cities like Amsterdam and Rotterdam might have to relocate.

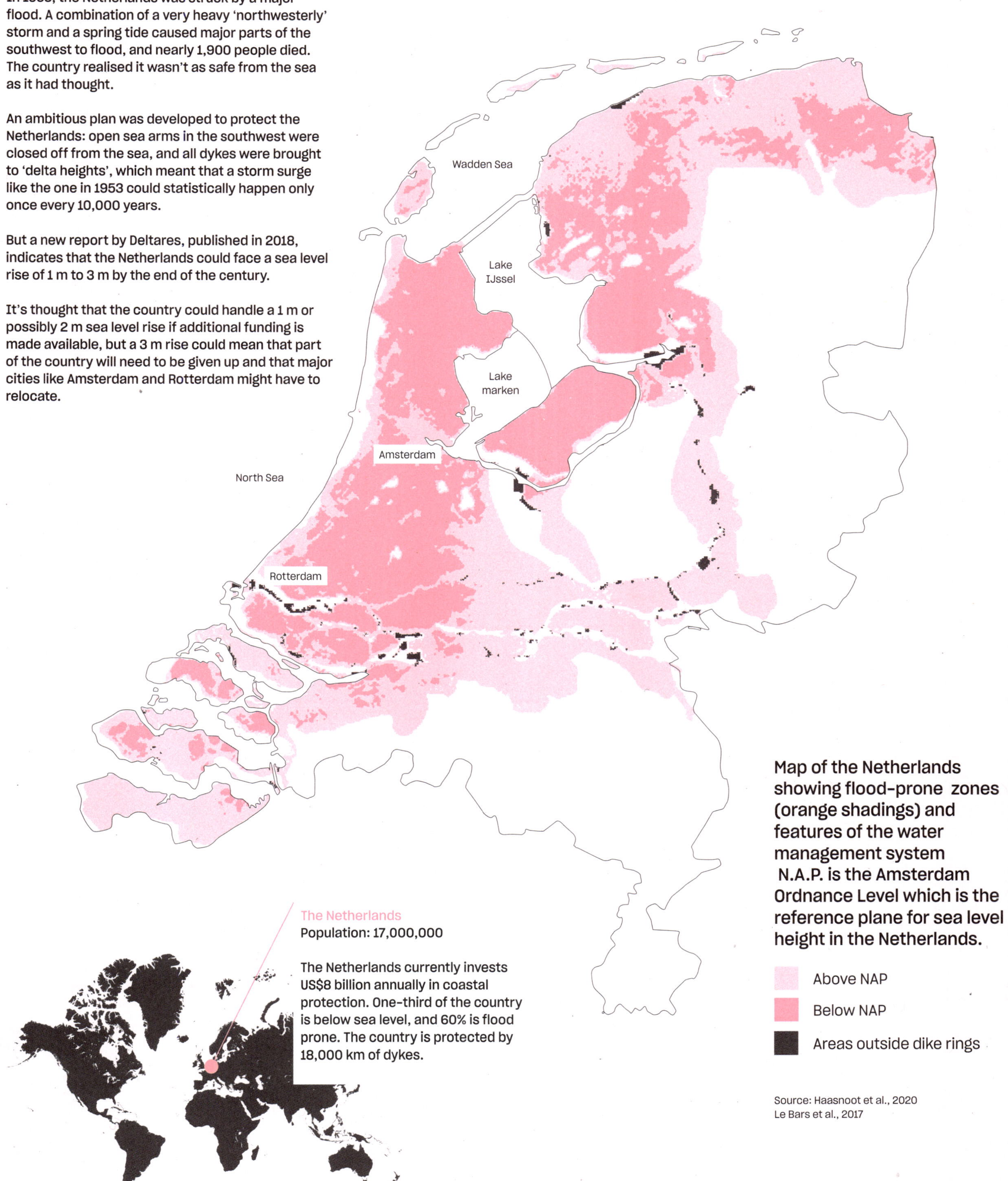

Map of the Netherlands showing flood-prone zones (orange shadings) and features of the water management system N.A.P. is the Amsterdam Ordnance Level which is the reference plane for sea level height in the Netherlands.

Above NAP

Below NAP

Areas outside dike rings

The Netherlands
Population: 17,000,000

The Netherlands currently invests US$8 billion annually in coastal protection. One-third of the country is below sea level, and 60% is flood prone. The country is protected by 18,000 km of dykes.

Source: Haasnoot et al., 2020
Le Bars et al., 2017

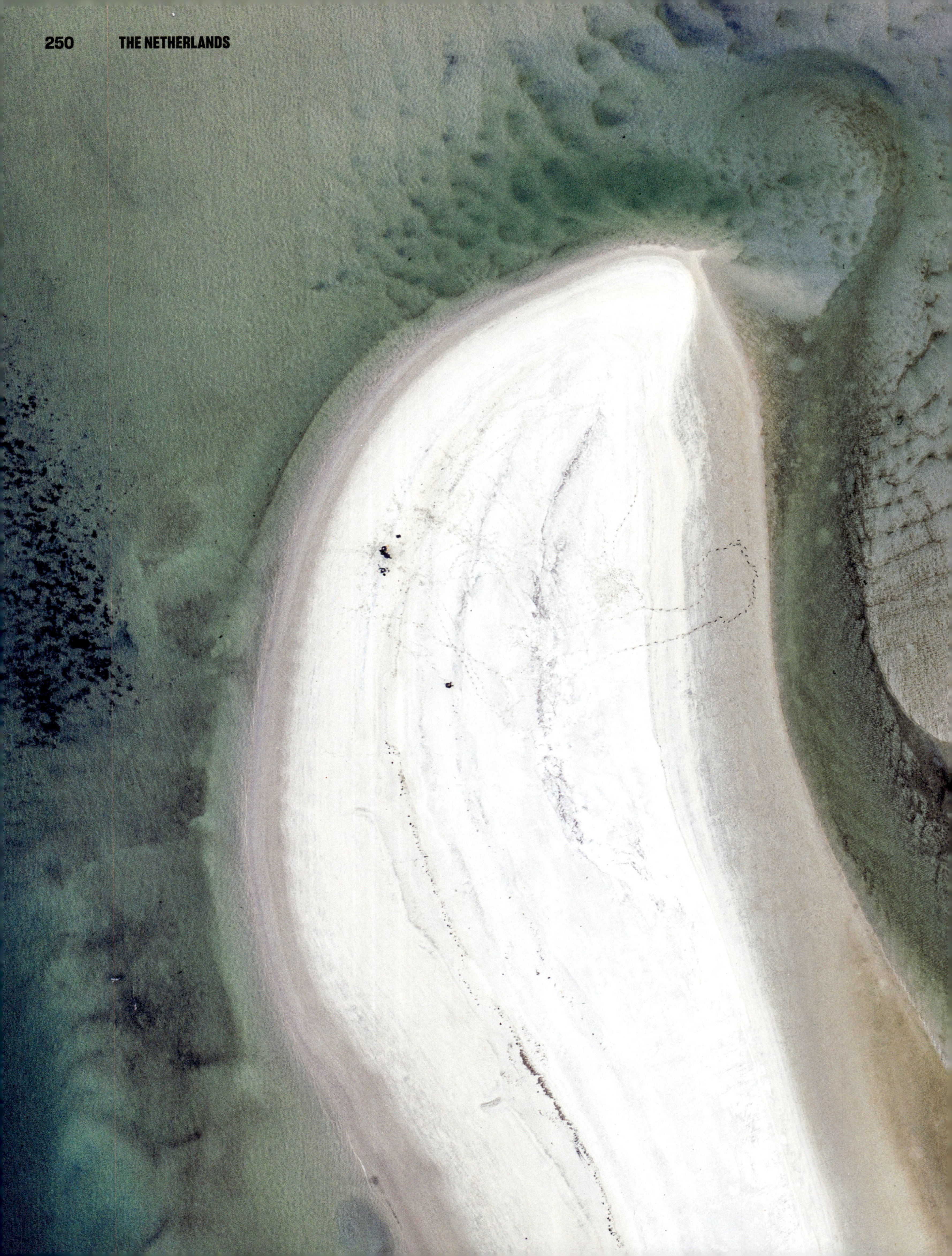

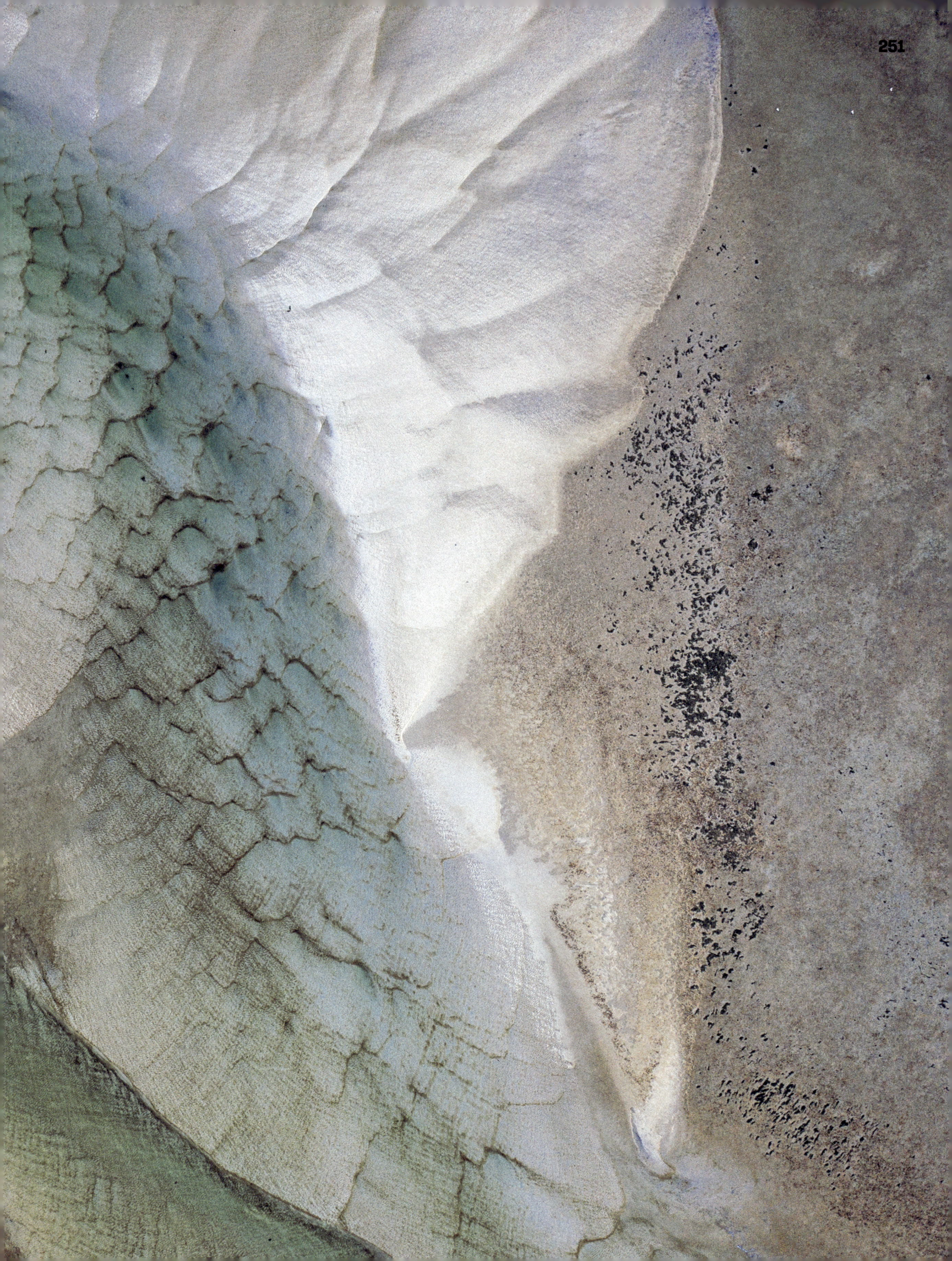

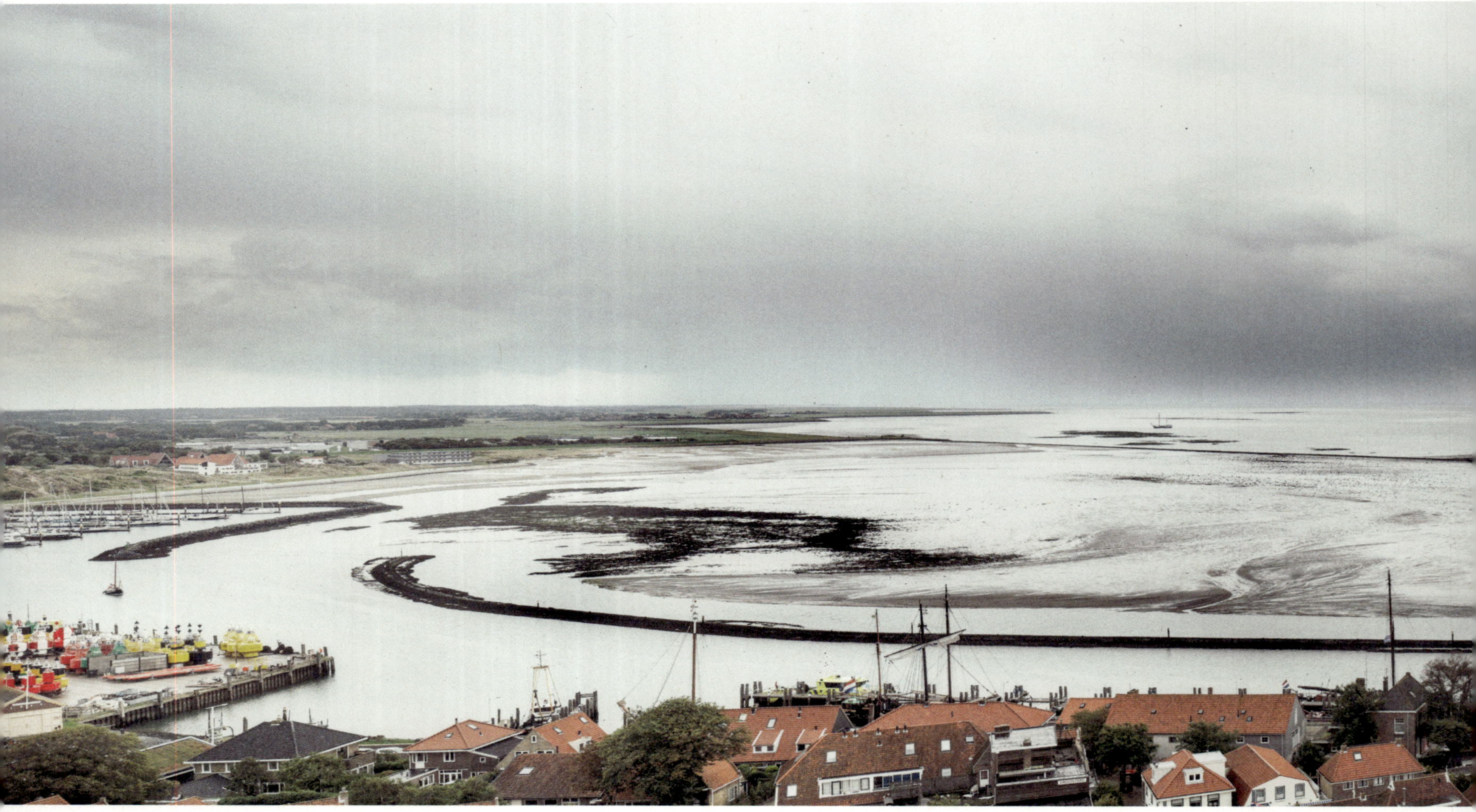

< Pages 250-251: Sandbanks becoming visible at low tide west of Terschelling. It is predicted that the Wadden Sea estuary will be permanently flooded in the future, which will have a big impact on the islands, which function as storm barriers.

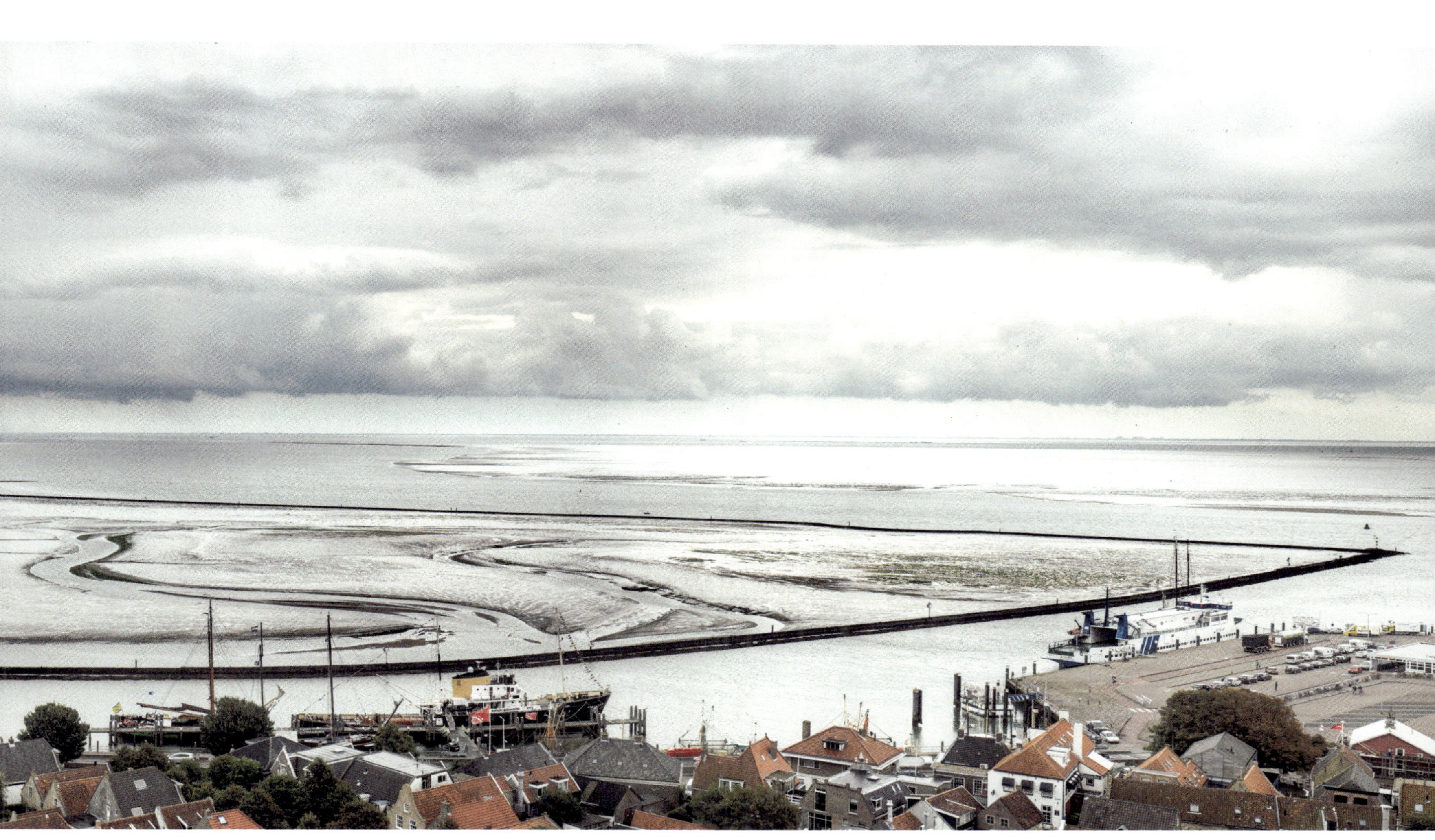

The village and port of West-Terschelling.

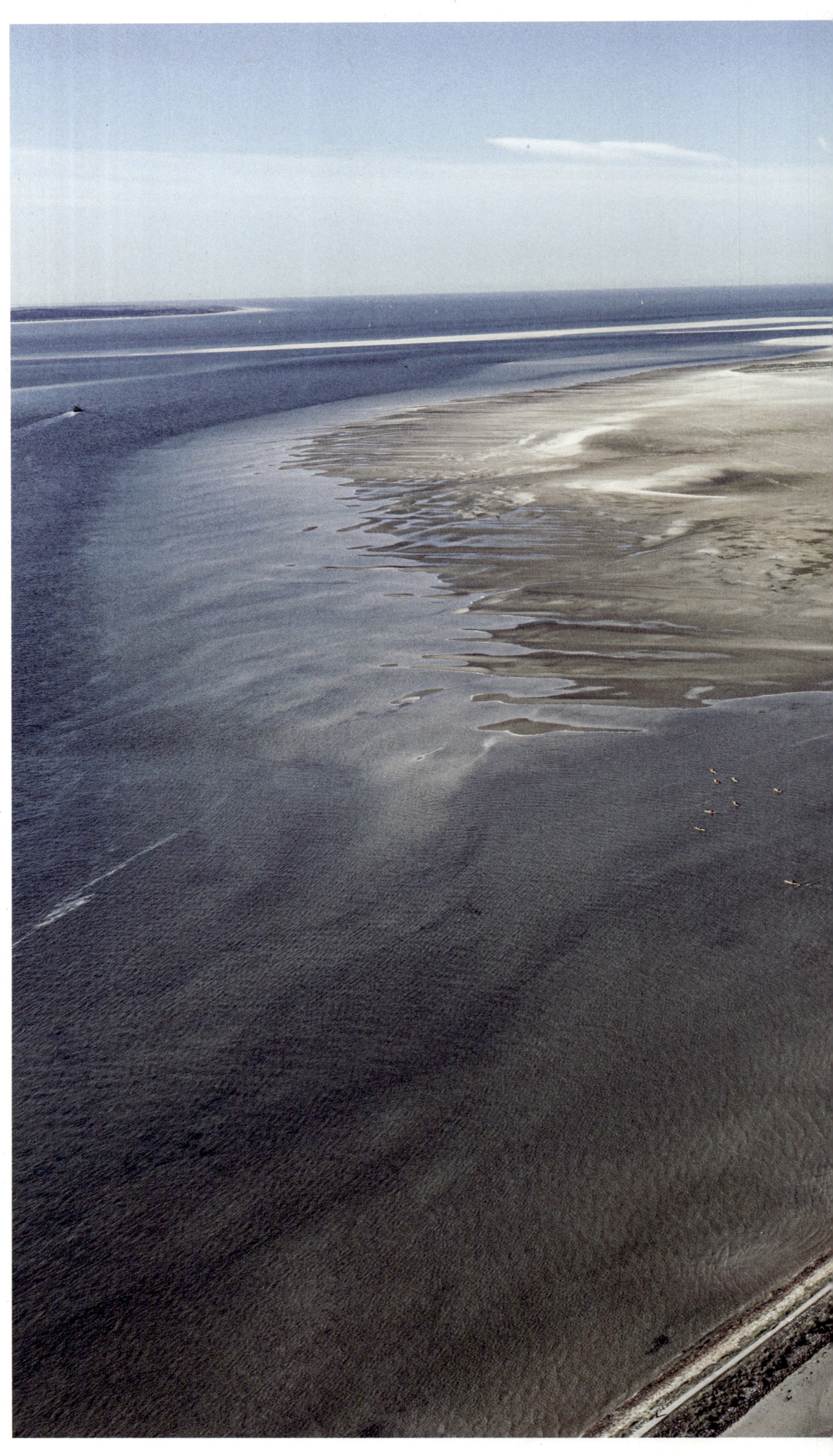

The village of West-Terschelling is considered the outer dyke and therefore
not protected from storm surges
and future sea level rise. The local,
provincial and national governments
are arguing over who will pay for it.

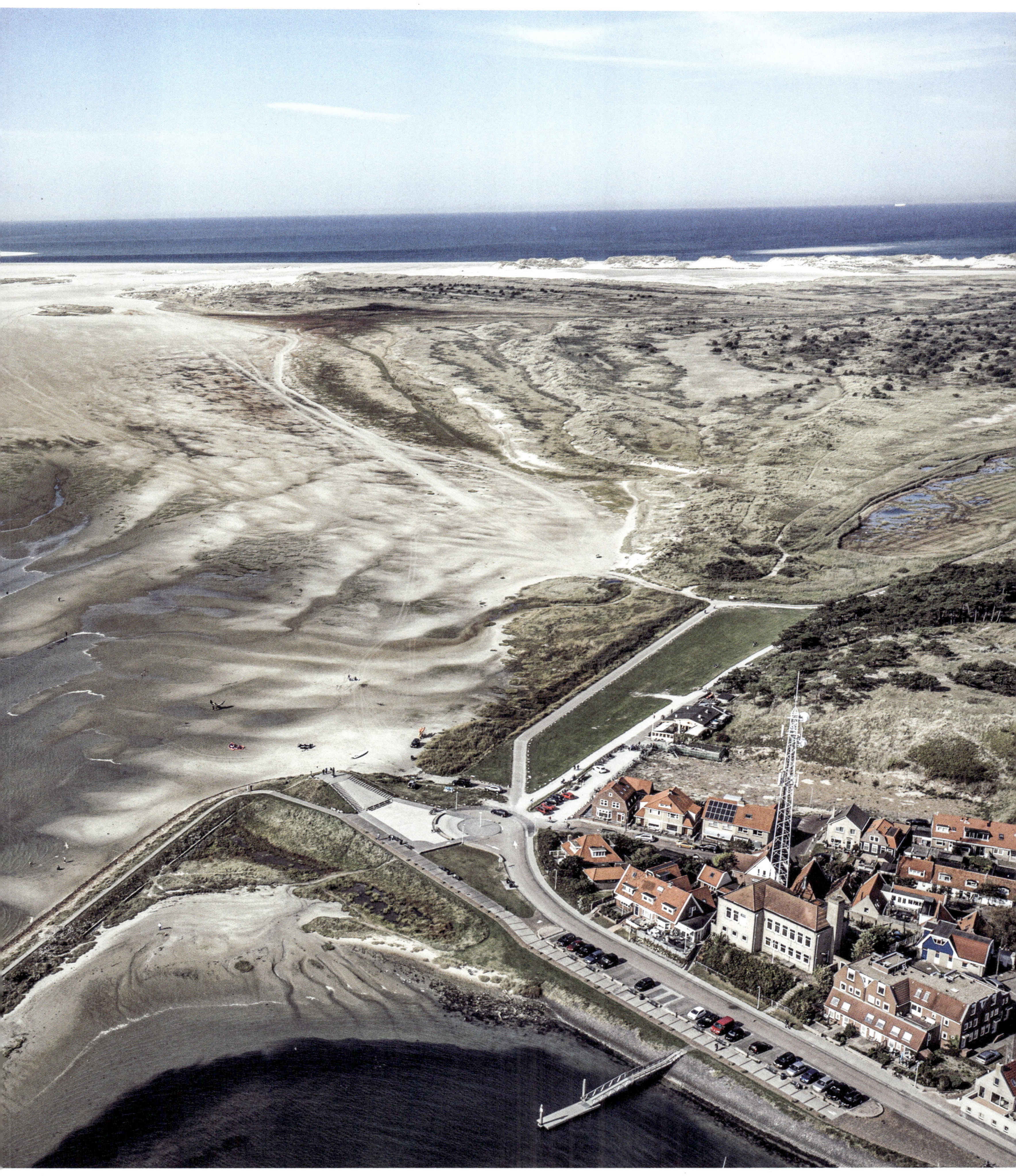

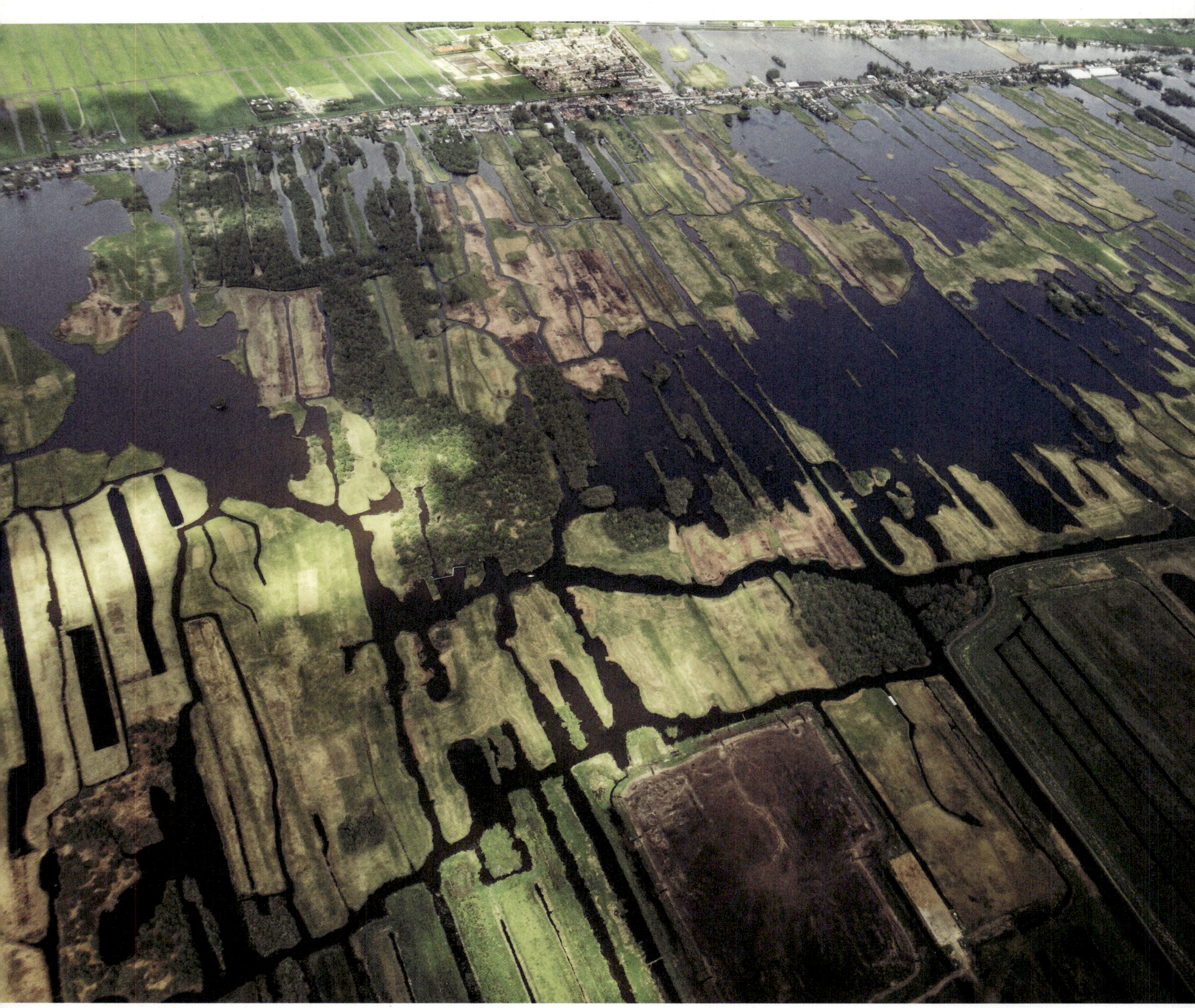

Waterland is a very low-lying region which has seen several floods over the centuries. The land is essentially peat, with many canals crossing the landscape.

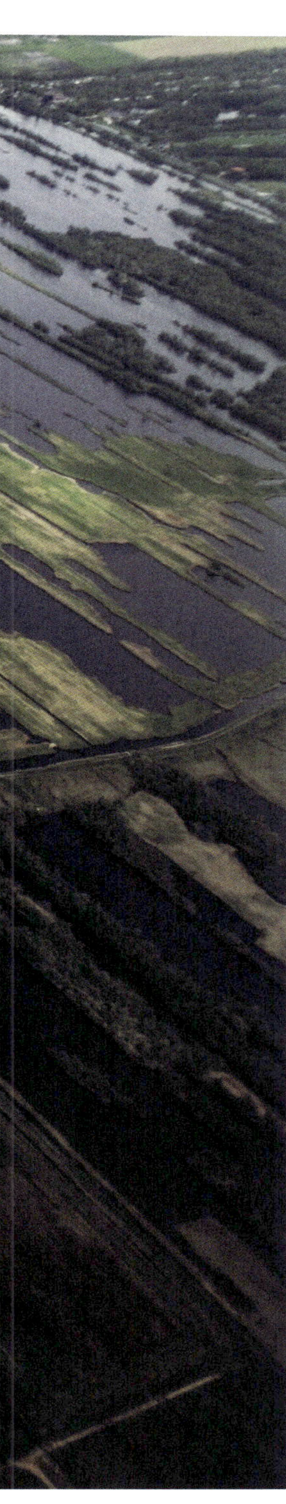

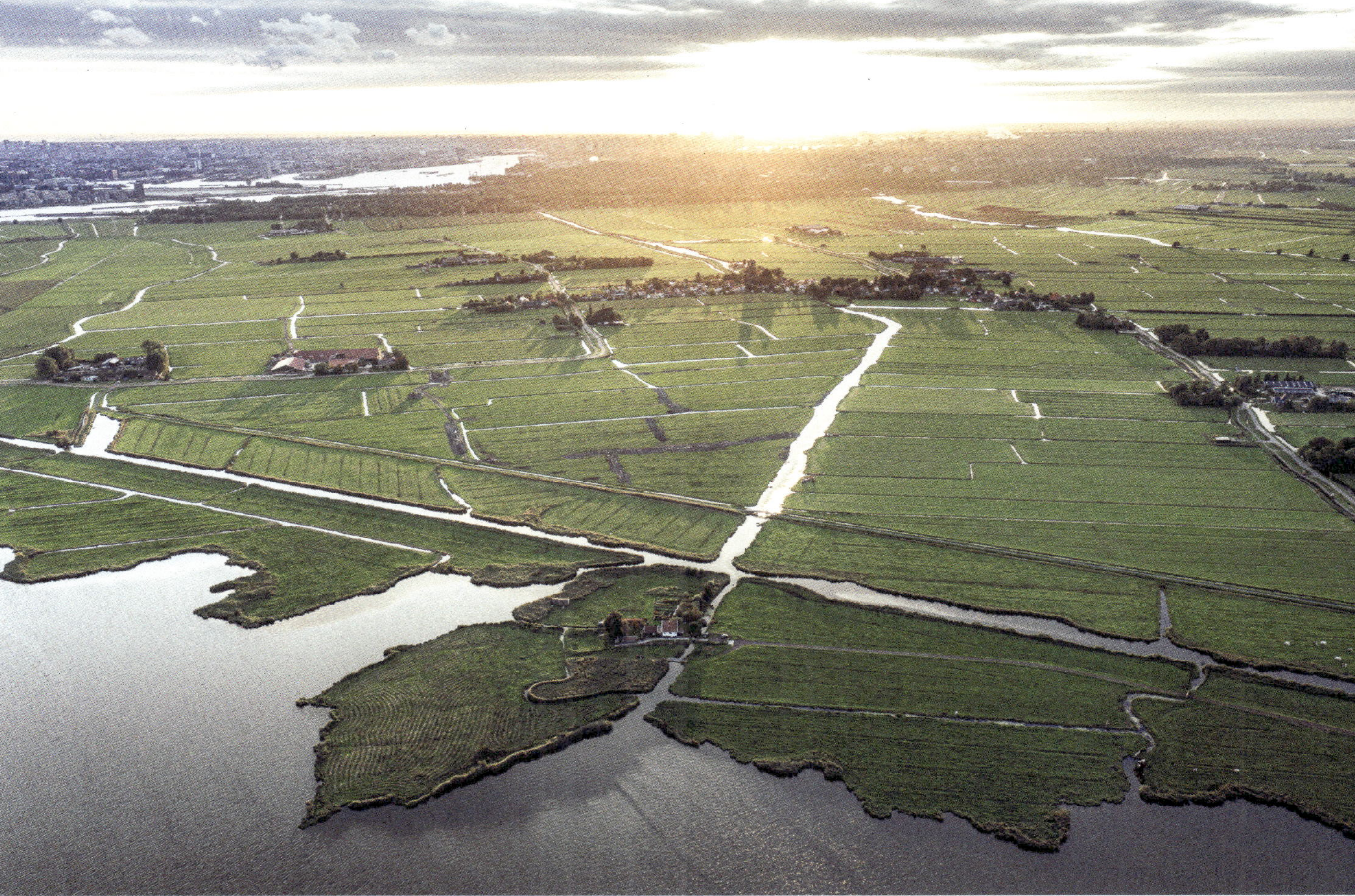

An area called Waterland, between Durgerdam and Uitdam. In the background is the city of Amsterdam.

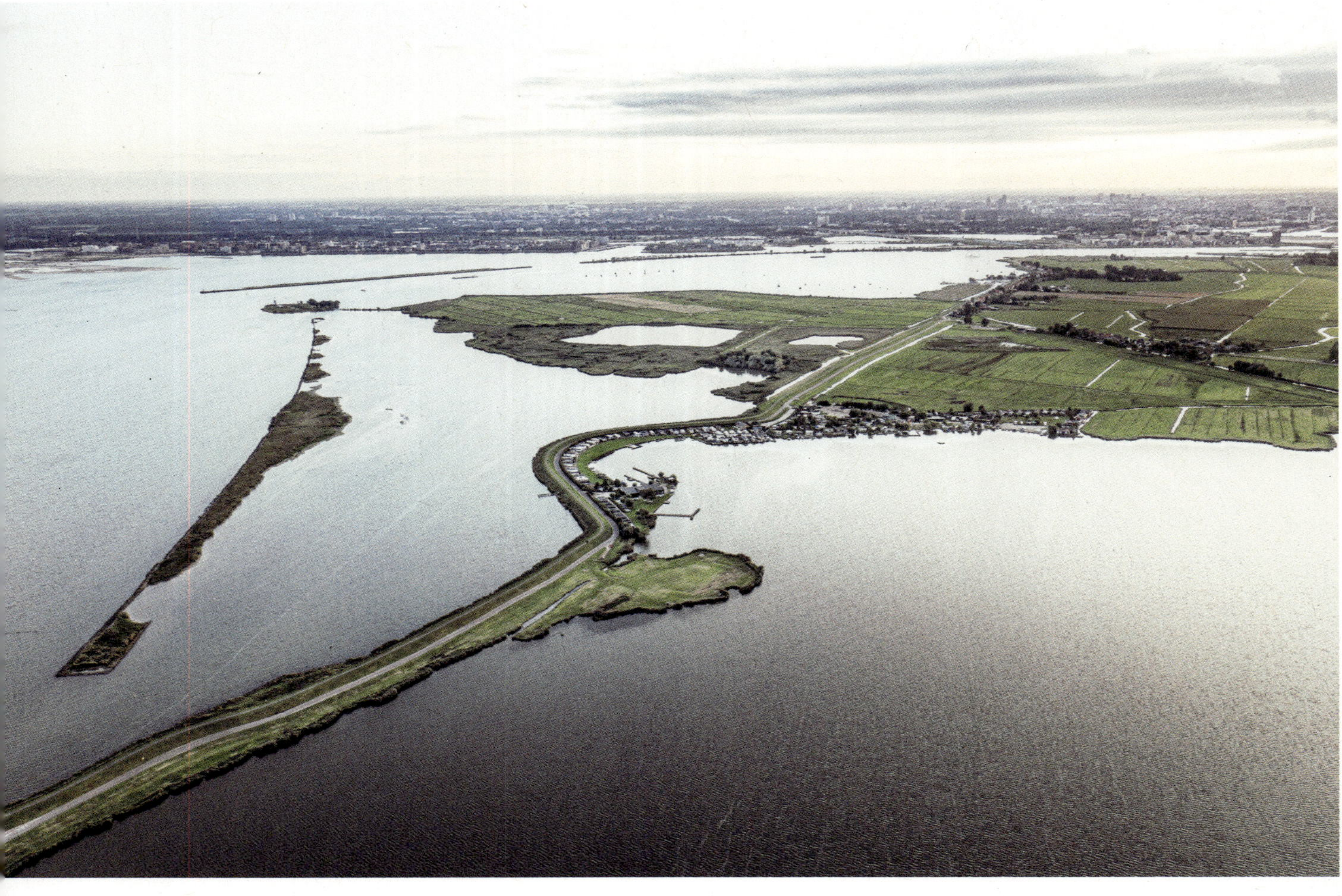

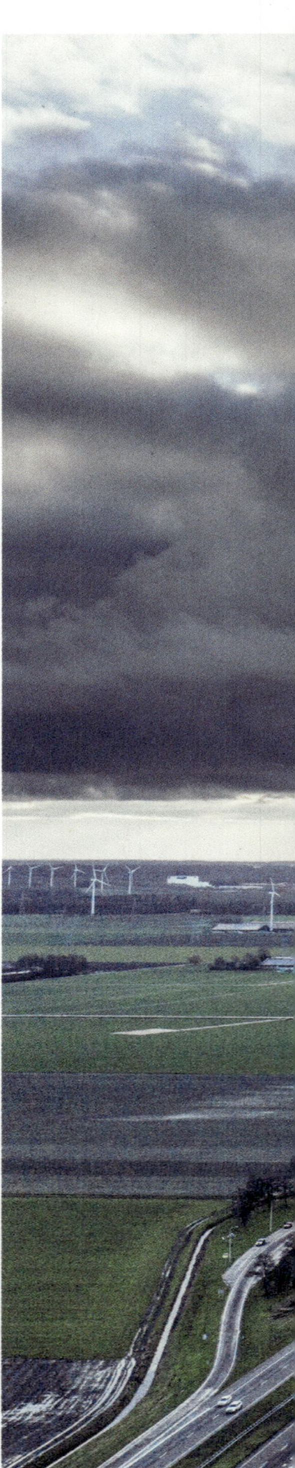

Waterland, between Durgerdam and Uitdam. In the background are the entrance to Amsterdam and the newly reclaimed land of IJburg, on which the city of Amsterdam is expanding. The sea dyke running along the Markermeer is being reinforced in the near future.

Flevoland is a polder, 3 to 5 m below sea level, reclaimed from the sea in the 1960s and '70s. It is one of the most productive agricultural lands in the world but is vulnerable to future sea level level rise.

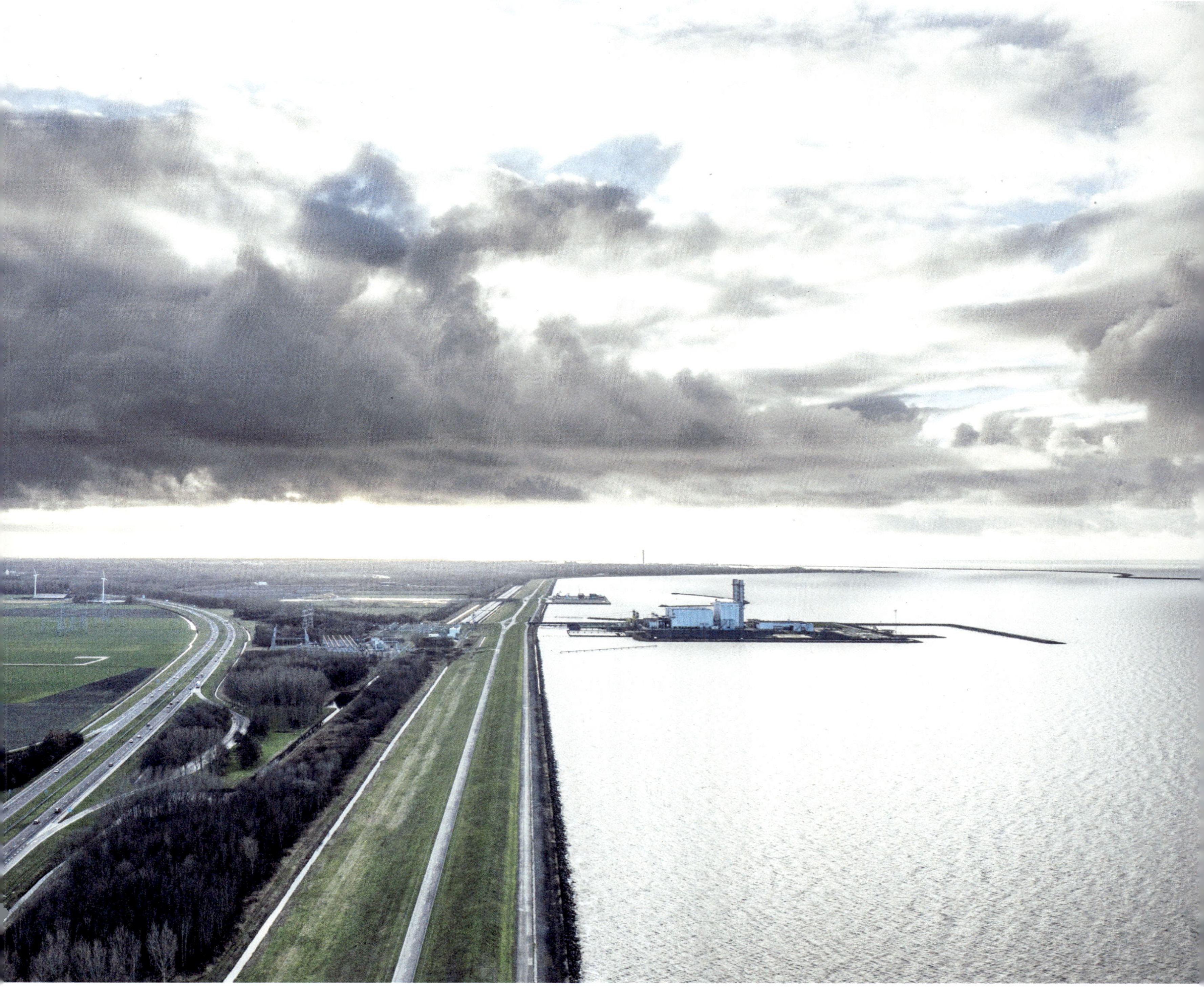

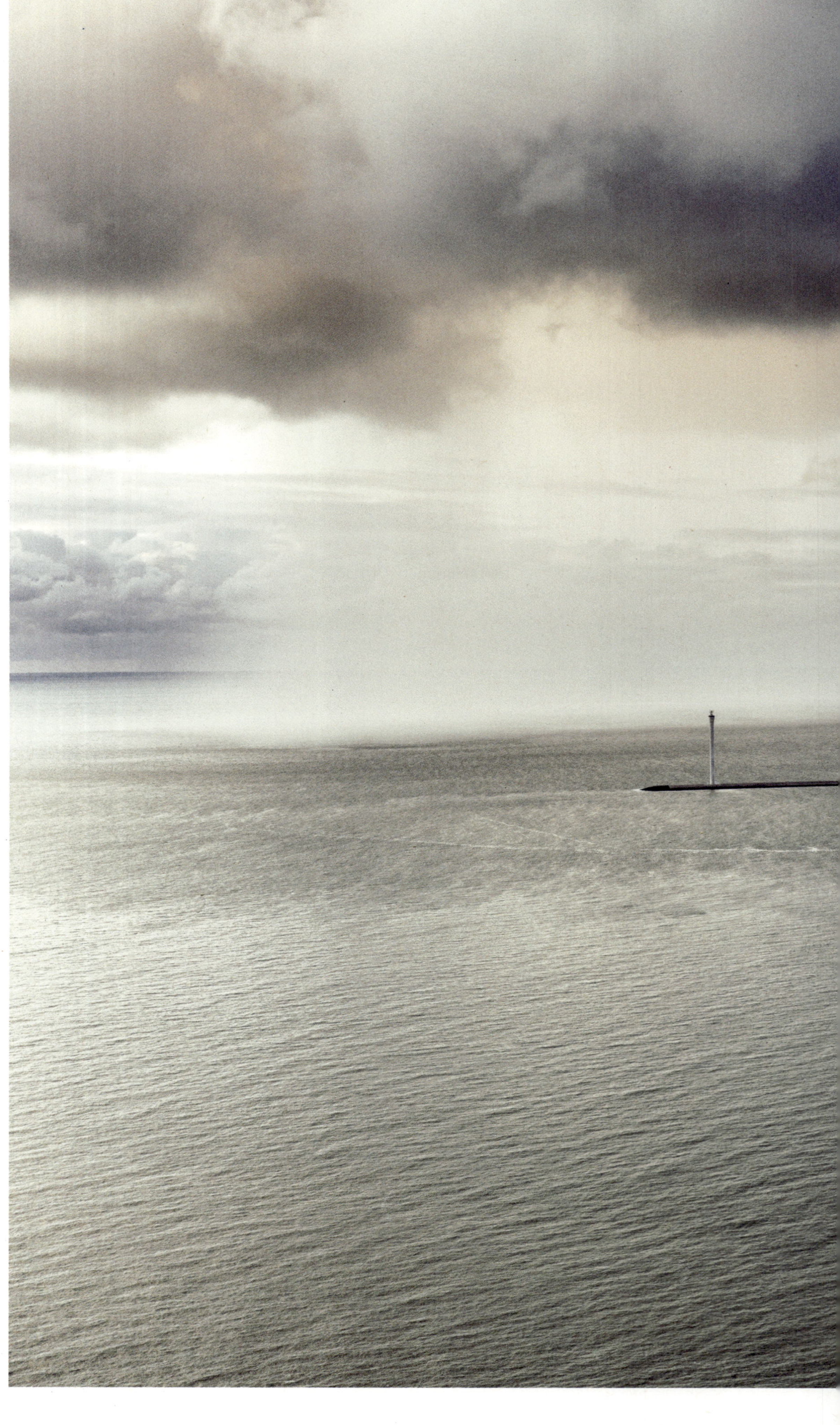

The Oosterscheldekering (Eastern Scheldt Storm Surge Barrier) was the most complex project of the Delta Works, which were built after the flood of 1953. It can close for a storm surge but will otherwise remain open to ensure the flora and fauna remain intact. It is a sophisticated piece of engineering that the Dutch remain proud of today, in a sea where currents are strong. The question is whether this dam will be high enough and remain open most of the time in the future.

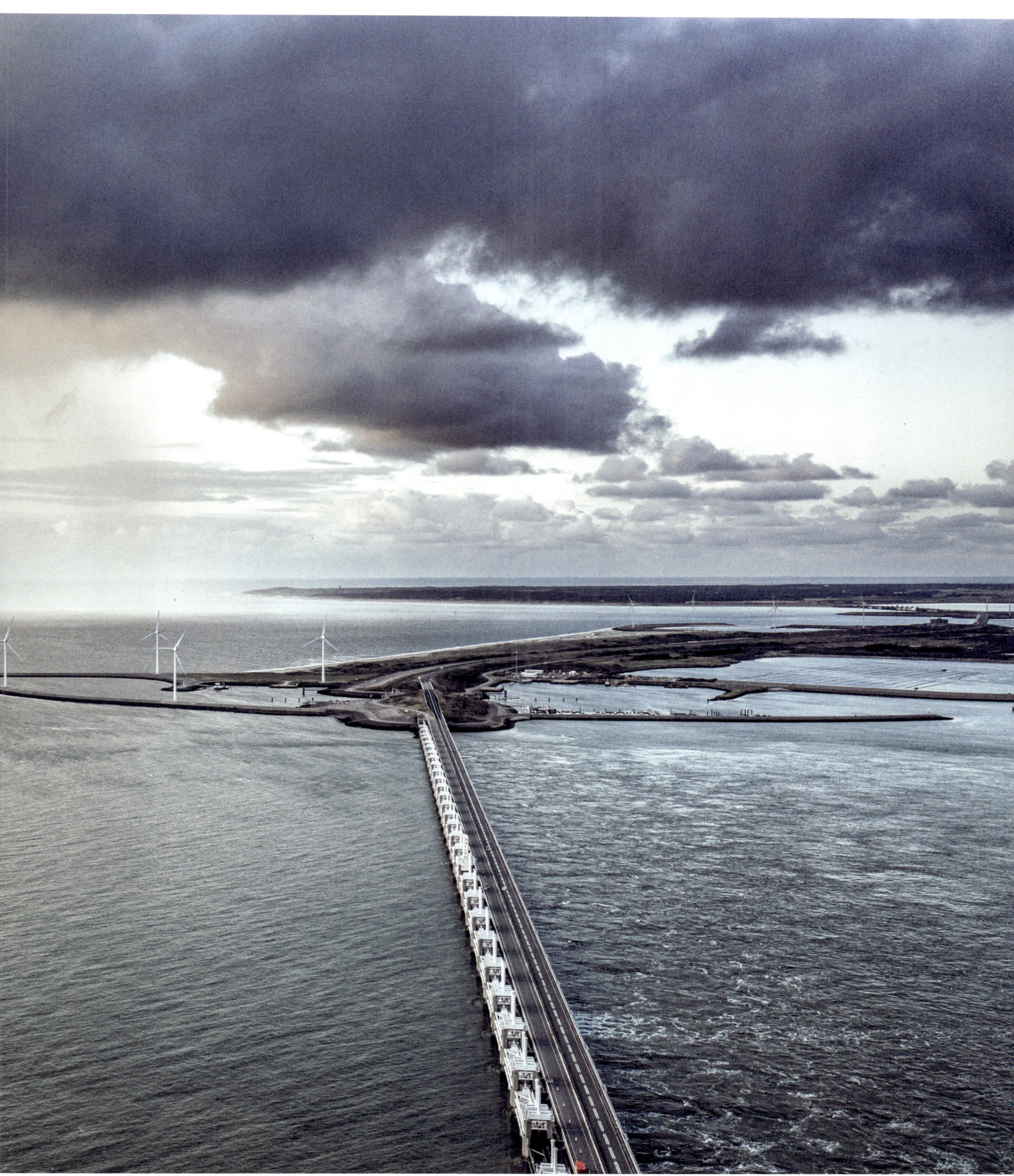

The Afsluitdijk dam and causeway completed in 1932 closed off the Zuiderzee (now called the IJsselmeer), a former inner sea, from the Wadden Sea (North Sea). It stopped the water from being tidal and threatening the coastal zones. It also enabled the Dutch to reclaim much-needed land (polders). In April 2019, an ambitious project was initiated to broaden the base of the dyke, heighten it and reinforce it to protect this part of Holland from future sea level rise.

The Zandmotor ('sand engine') is located between The Hague and the port entrance of Rotterdam. A sand bank was created by depositing 12 million m³ of sand, which naturally supplements sand to the beaches. The beaches are essential to protecting the coasts from eroding.

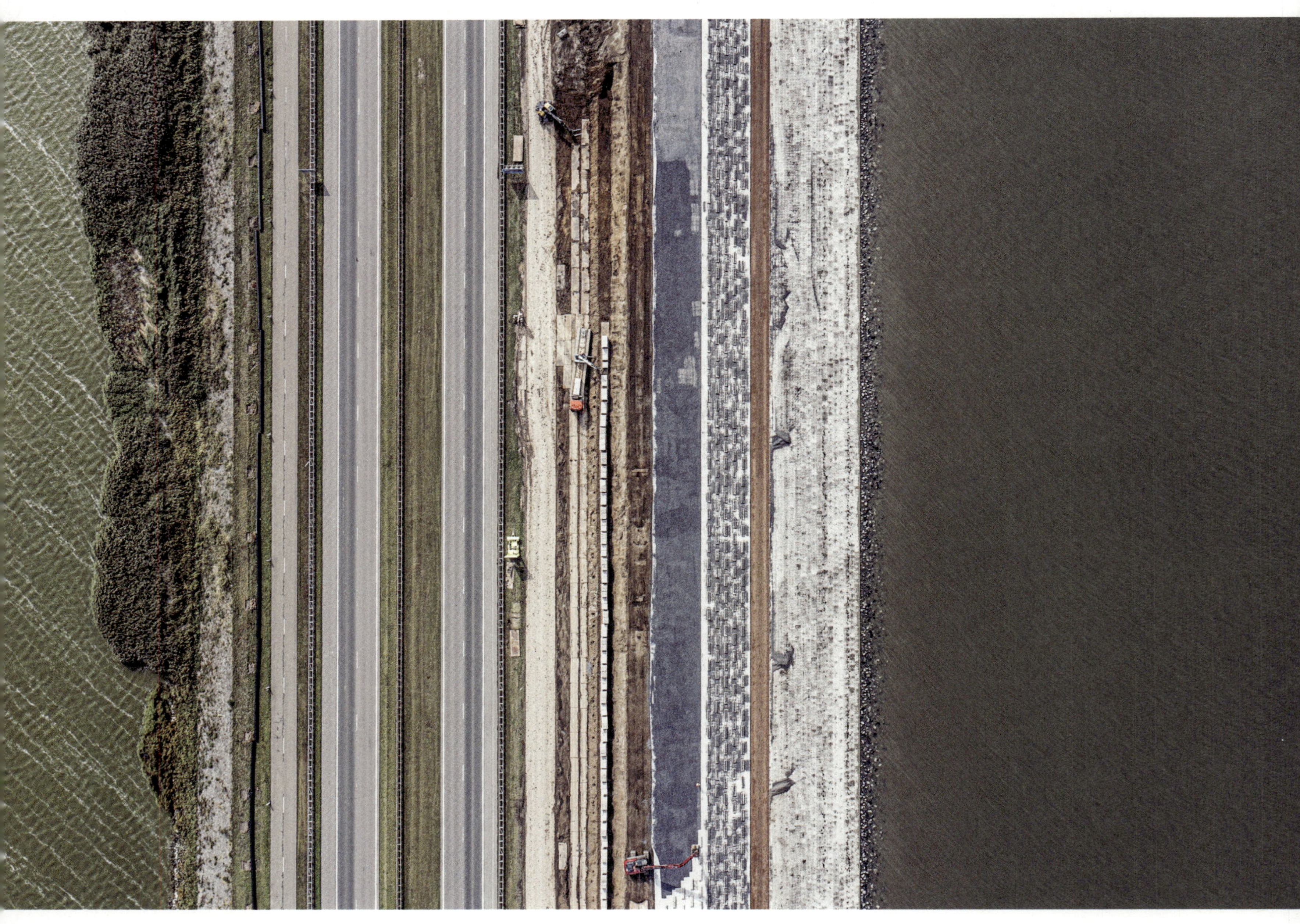

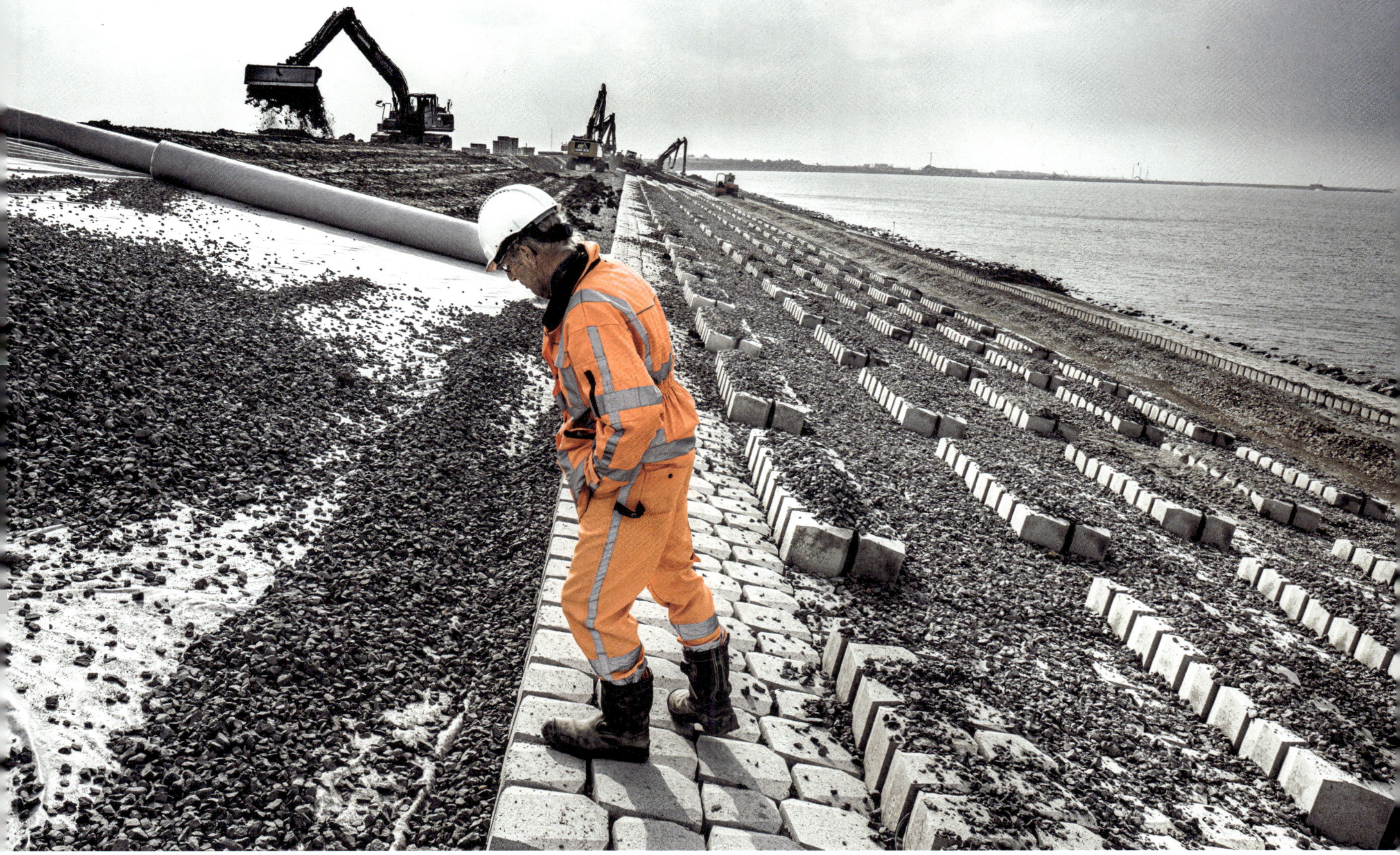

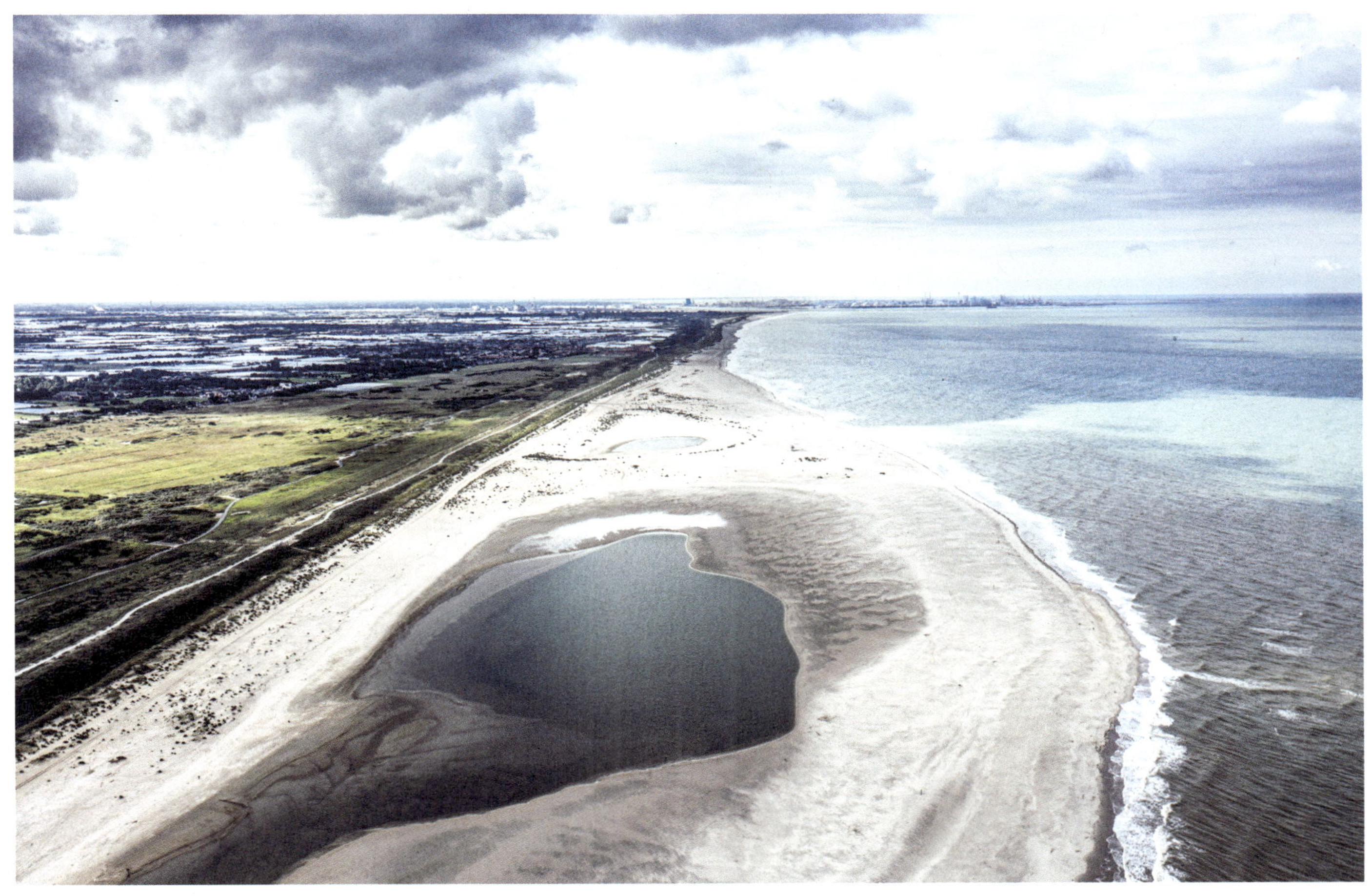

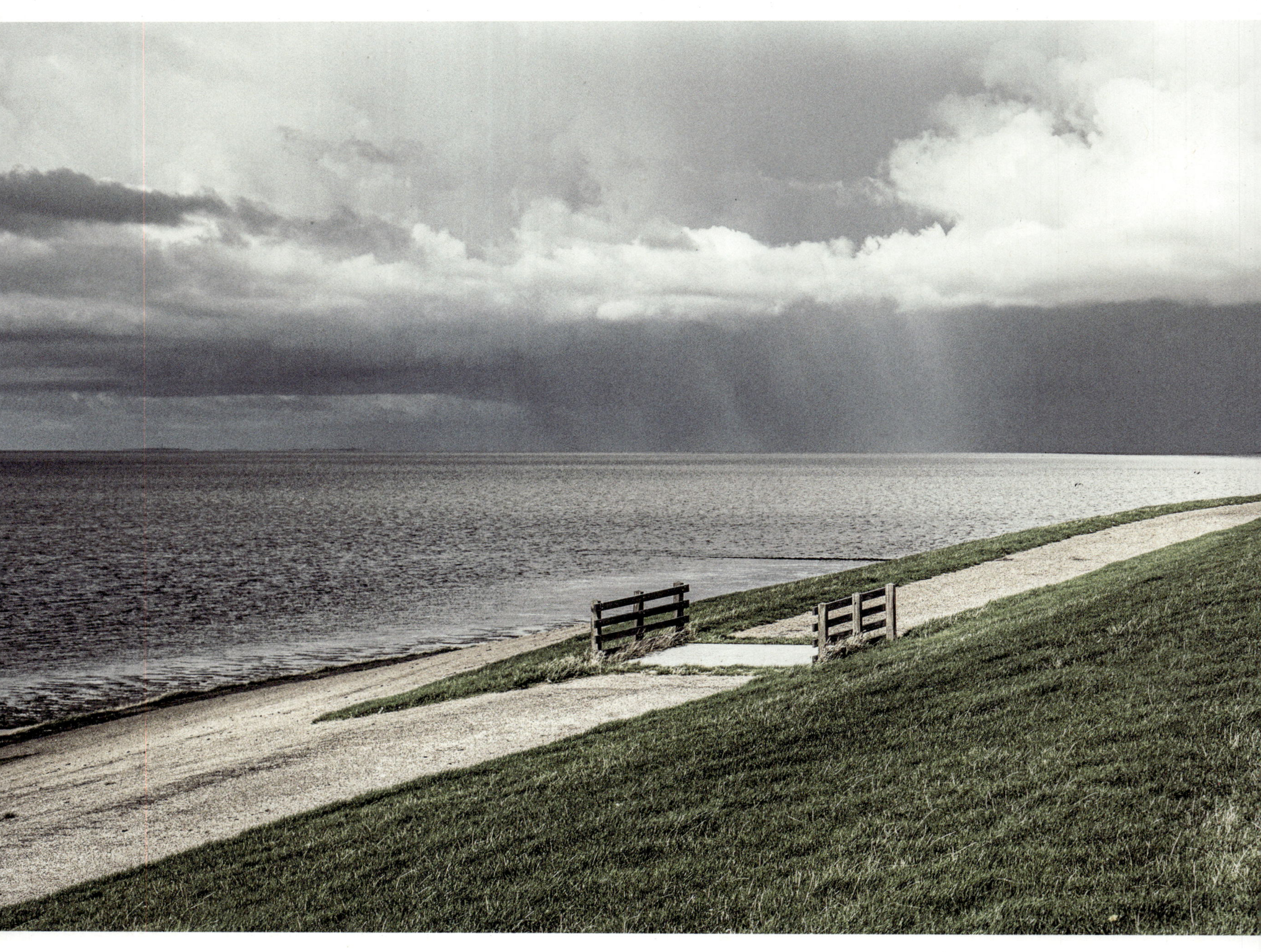

A sea dyke on the the Frisian coast and
the Wadden Sea north of Franeker.

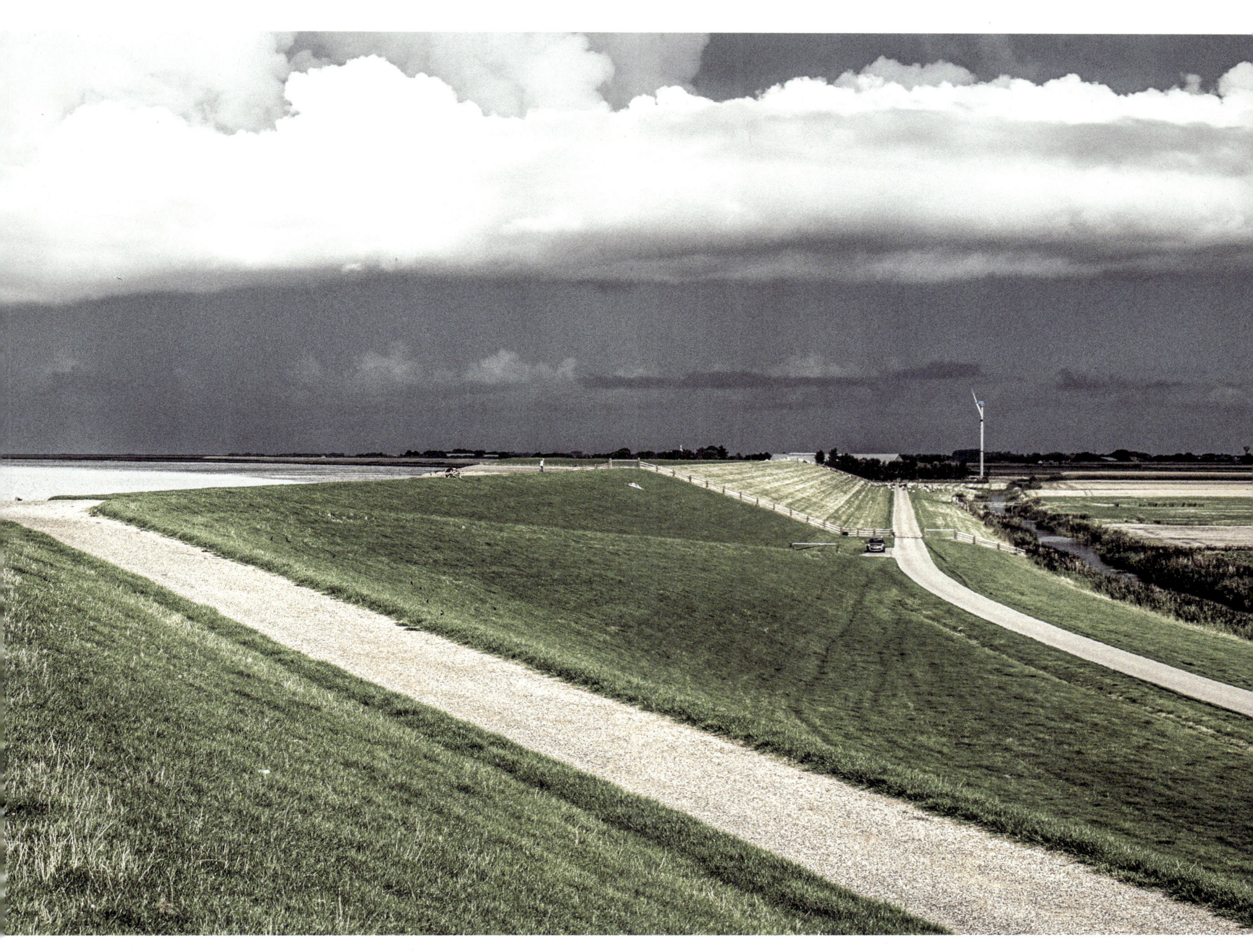

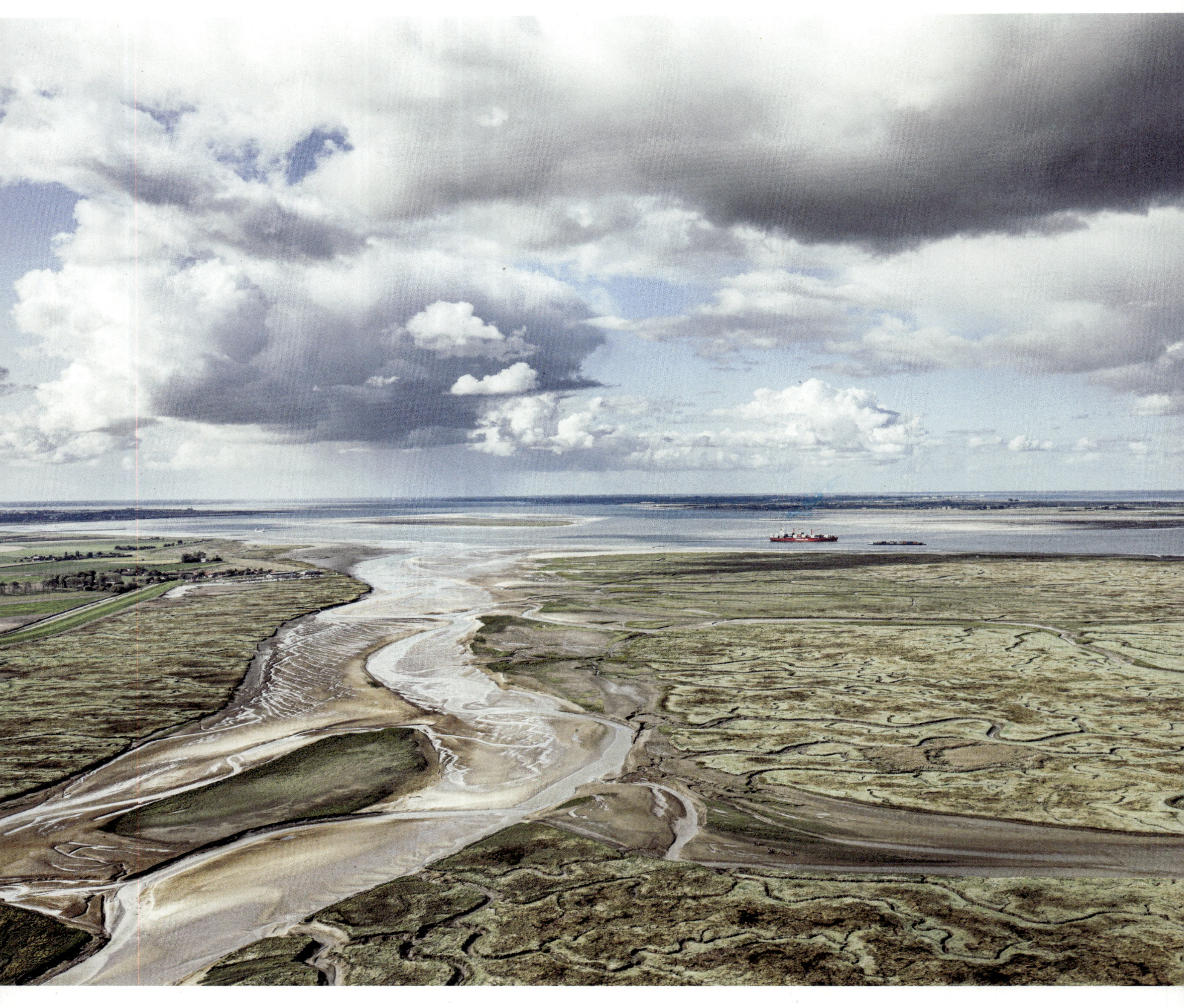

'The Drowned land of Saeftinghe' located in Zeeland on the island of Zeeuws Vlaanderen. In the back a containership on the the Westerschelde, the only open sea arm in the Southwest of the Netherlands and the entrance of the Schelde, the entrance of the port of Antwerp. This is an area that remains very vulnerable to storm surges and future sea level rise.

Hubrecht Janse is a farmer in Wolphaartsdijk, a village in the province of Zeeland, in the southwest of the Netherlands. He is growing regular corn, but also samphire, a great vegetable which grows in saline soil. It is believed that farmers in the Netherlands and other coastal regions will need to adapt when the sea level rises and the land becomes saline. Certain potatoes and even strawberries can be grown in saline soil.

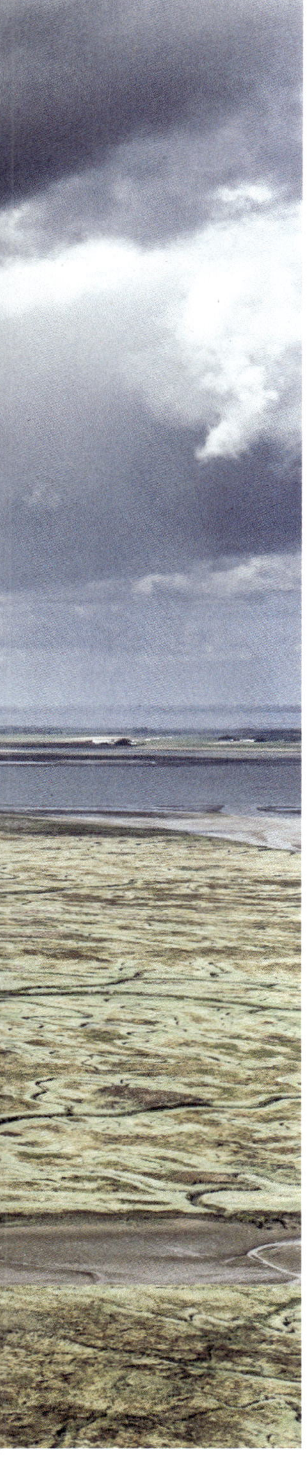

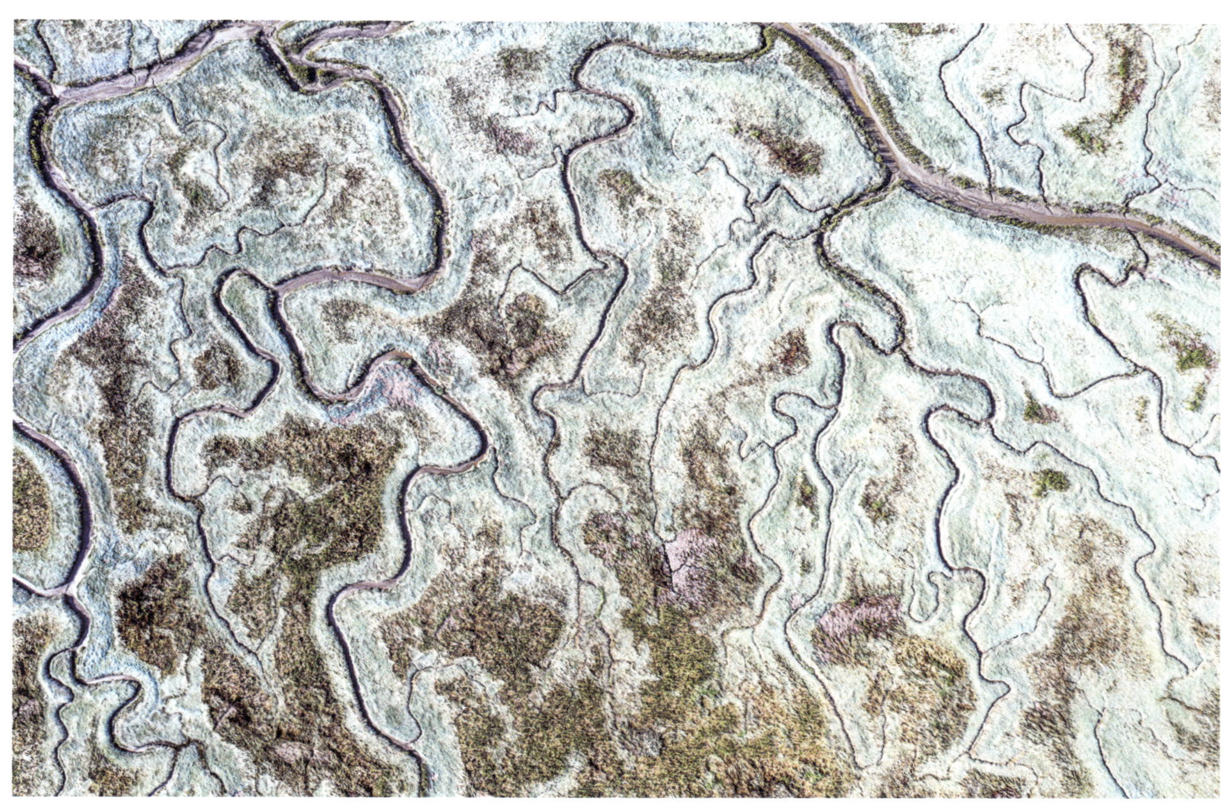

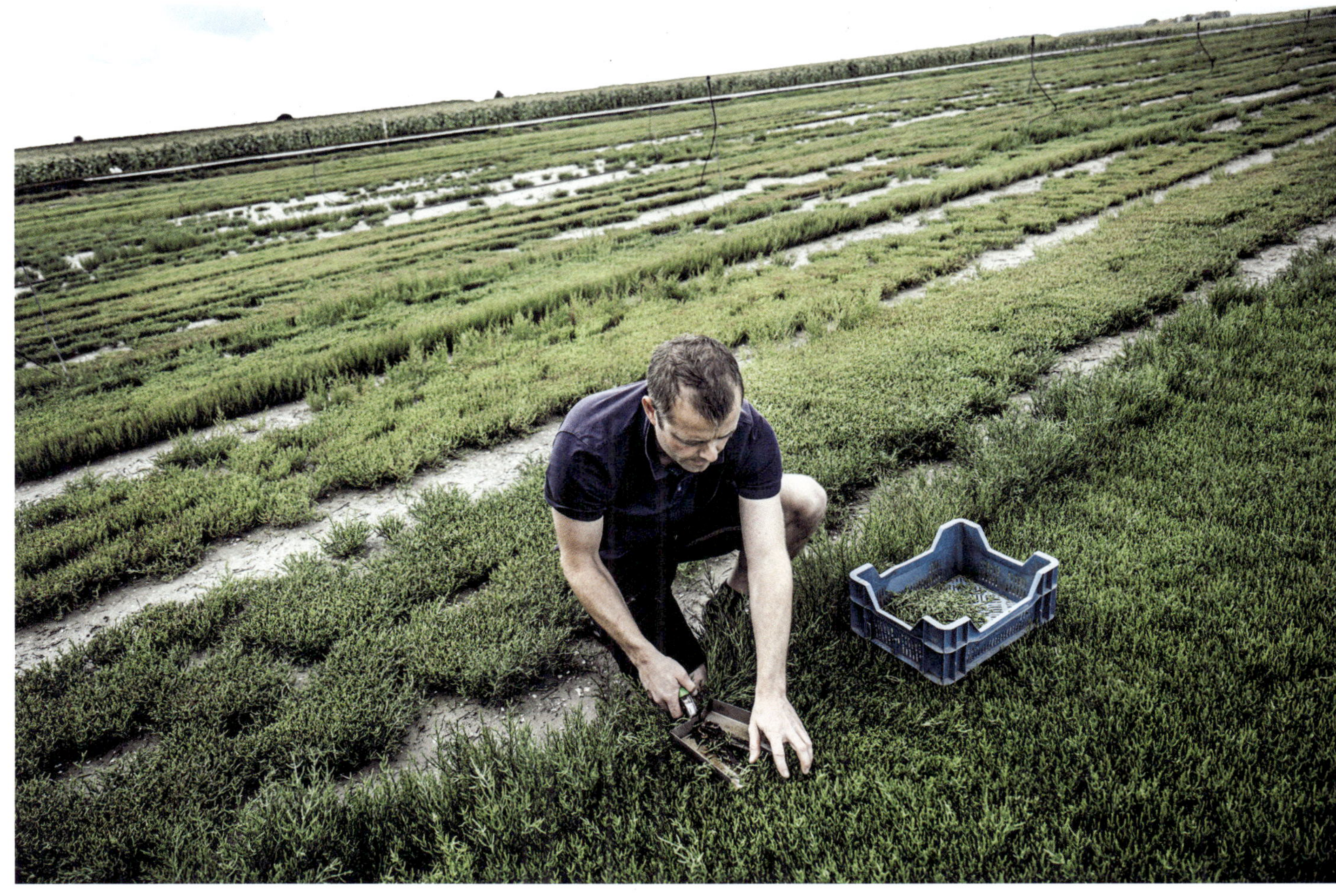

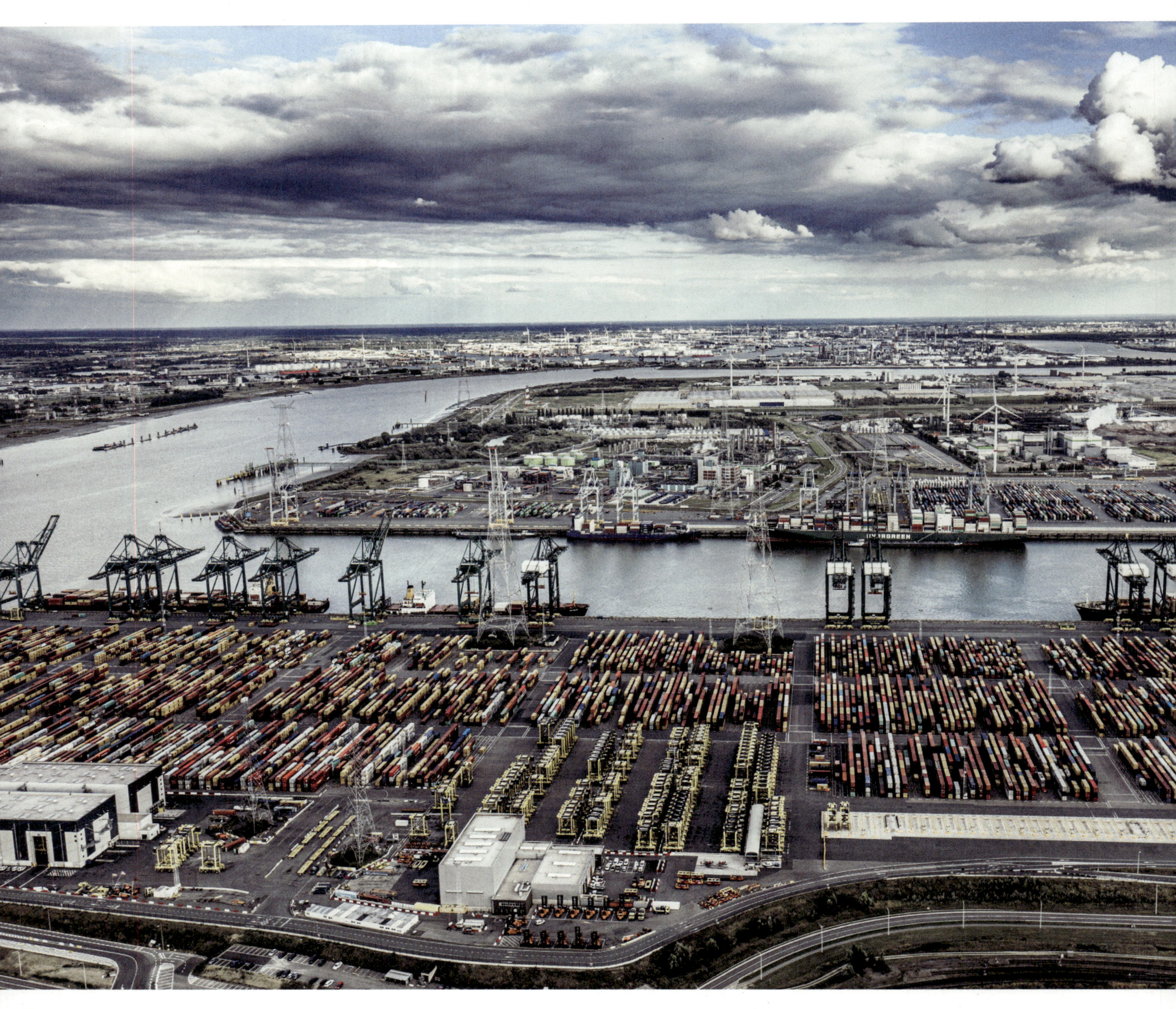

The new container terminals at the port of Antwerp, which has a direct connection with the Westerschelde (Western Scheldt), the only open sea arm in the southwest of the Netherlands.

The Maeslant Barrier was the last part of the engineering of the Delta Works. It protects the port and city of Rotterdam from 4 m storm surges when it is closed. If the sea level rises in the future, the barrier will have to be closed most of the time and will cut off Rotterdam from the sea, which is its economic lifeline. In the background are the Maasvlakte, Rotterdam's most important container terminals, which are not protected from storm surges and the rising sea level.

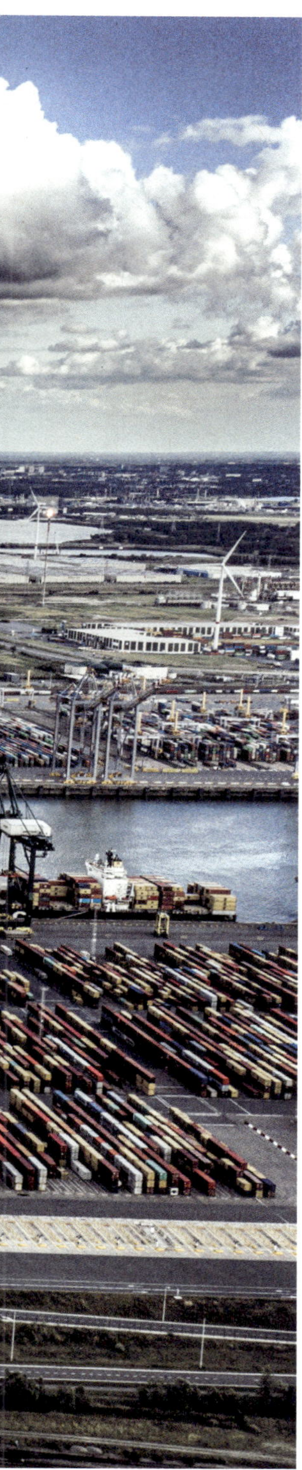

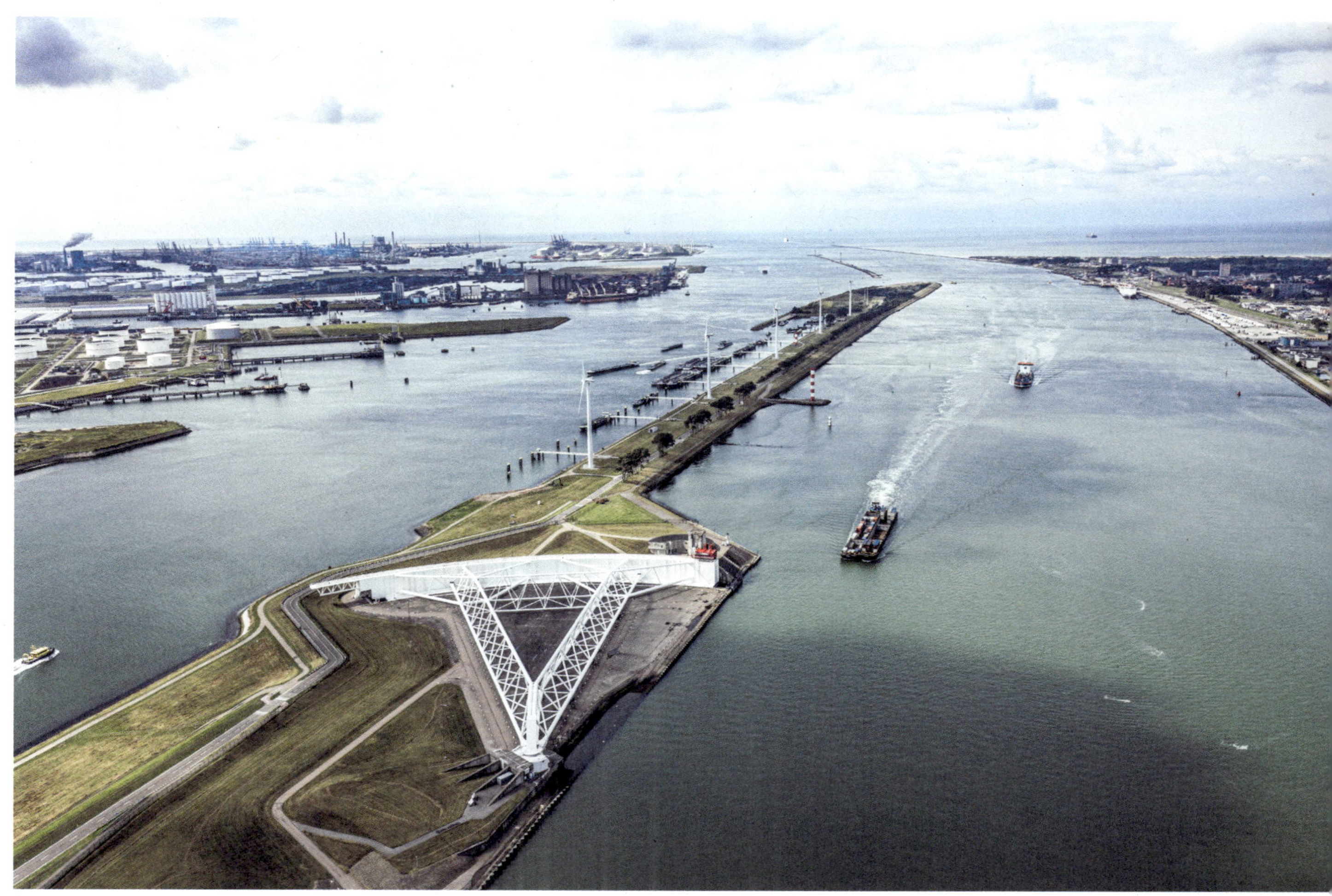

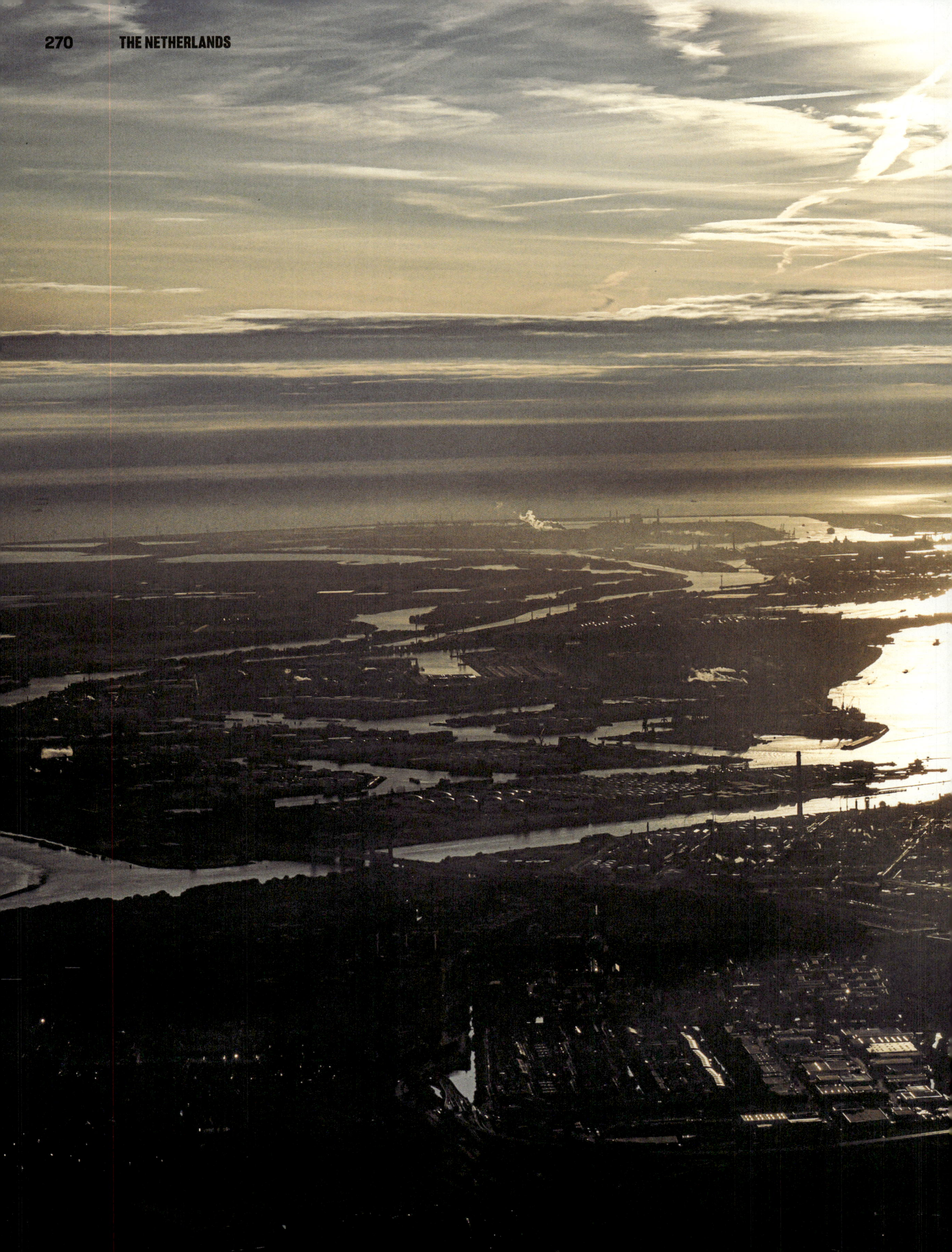

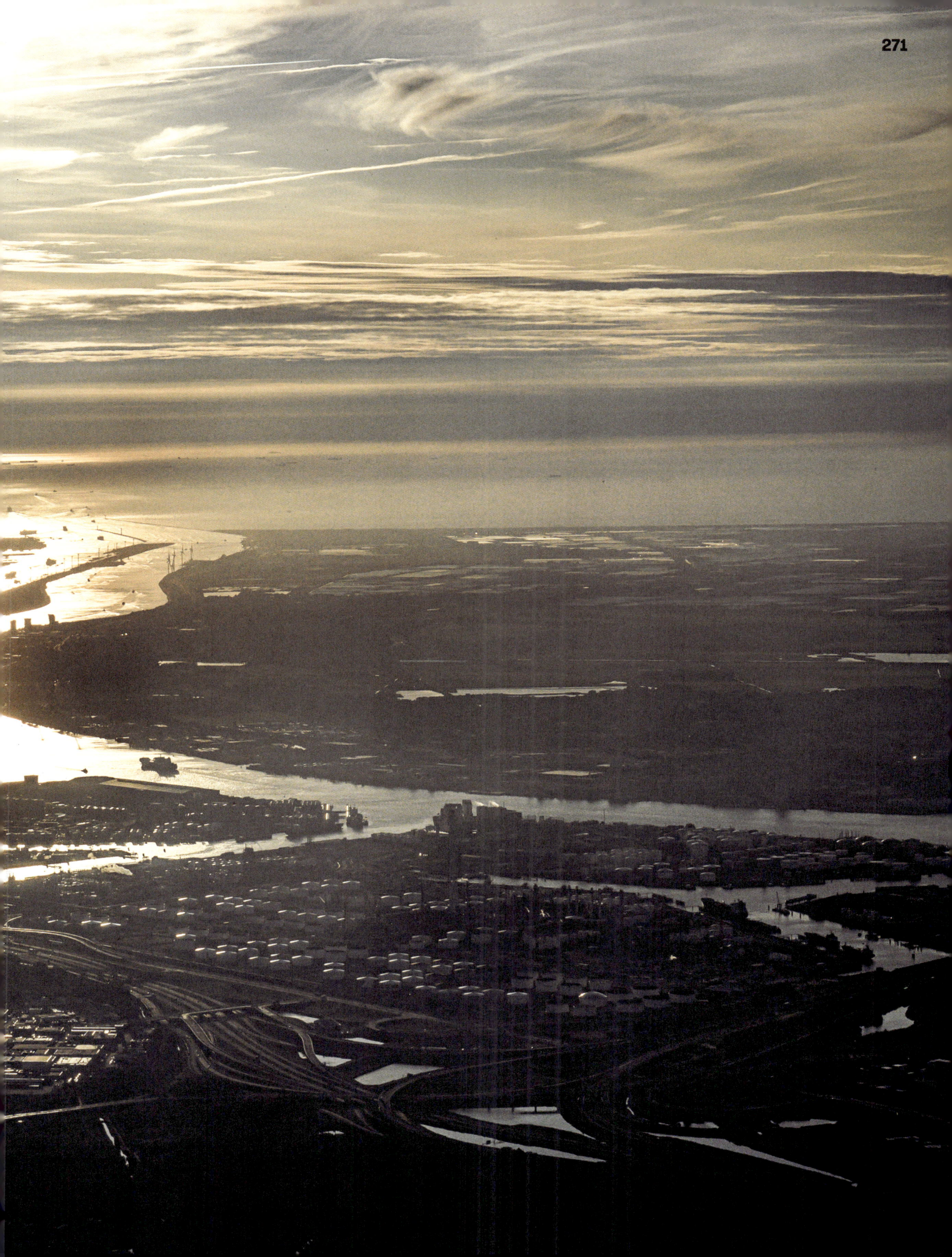

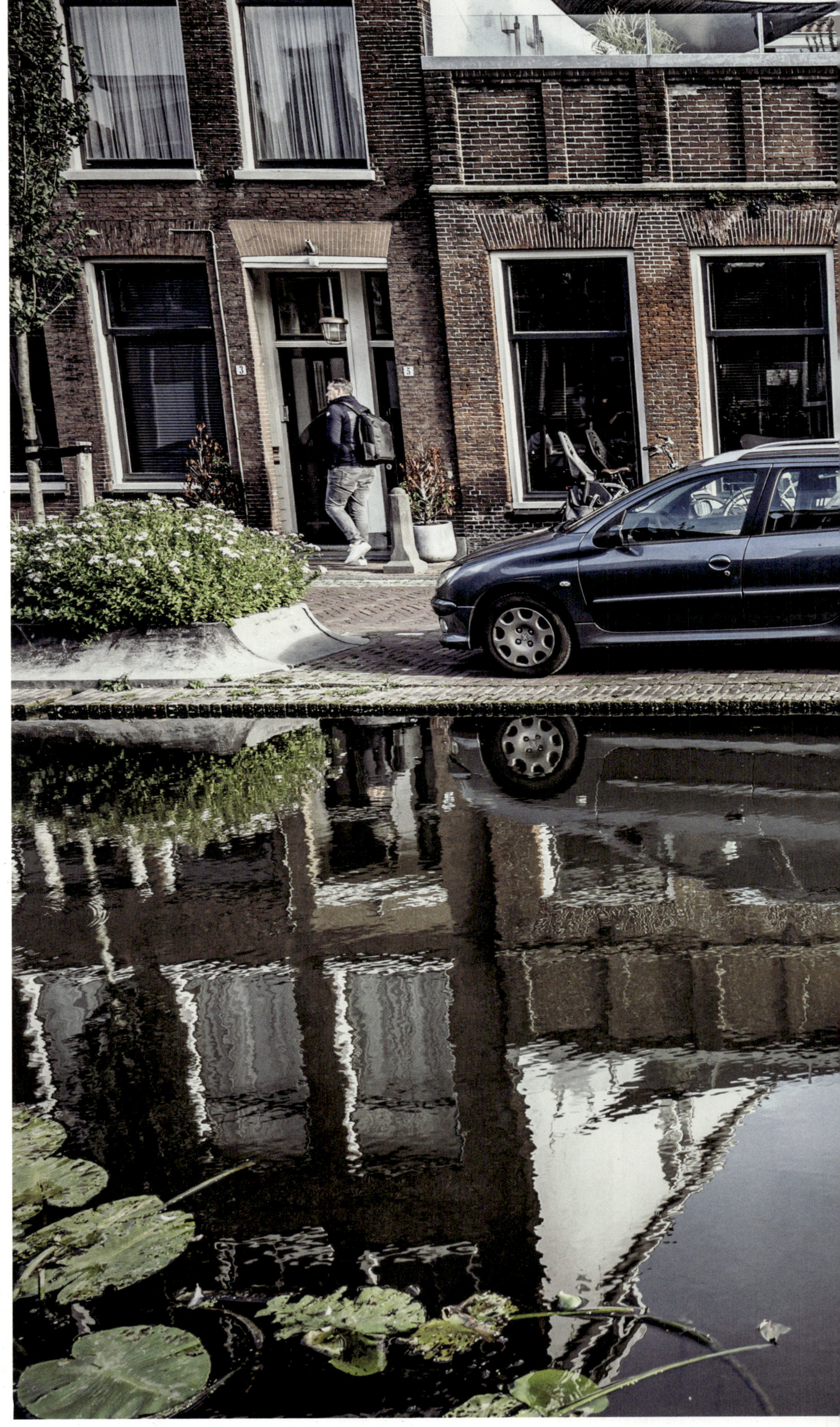

< Pages 270-271: The entrance of the port of Rotterdam, the largest port in Europe, is an open connection to the sea, not protected behind locks like Amsterdam.

The 1,000-year-old city of Gouda, like most part of Western Holland, was built on peatland and has flourished because of the still tidal river, the Hollandse IJssel. Ever since it was built, the city has been slowly sinking - peat is like a sponge. It was thought this would stop at some point, but it hasn't. The solution would be to pump out water, but since the houses are built on stilts, they can't - if the tops of the stilts are exposed to oxygen, they will start to rot, and the buildings will collapse. Combined with a rising sea level, the future of the city looks bleak.

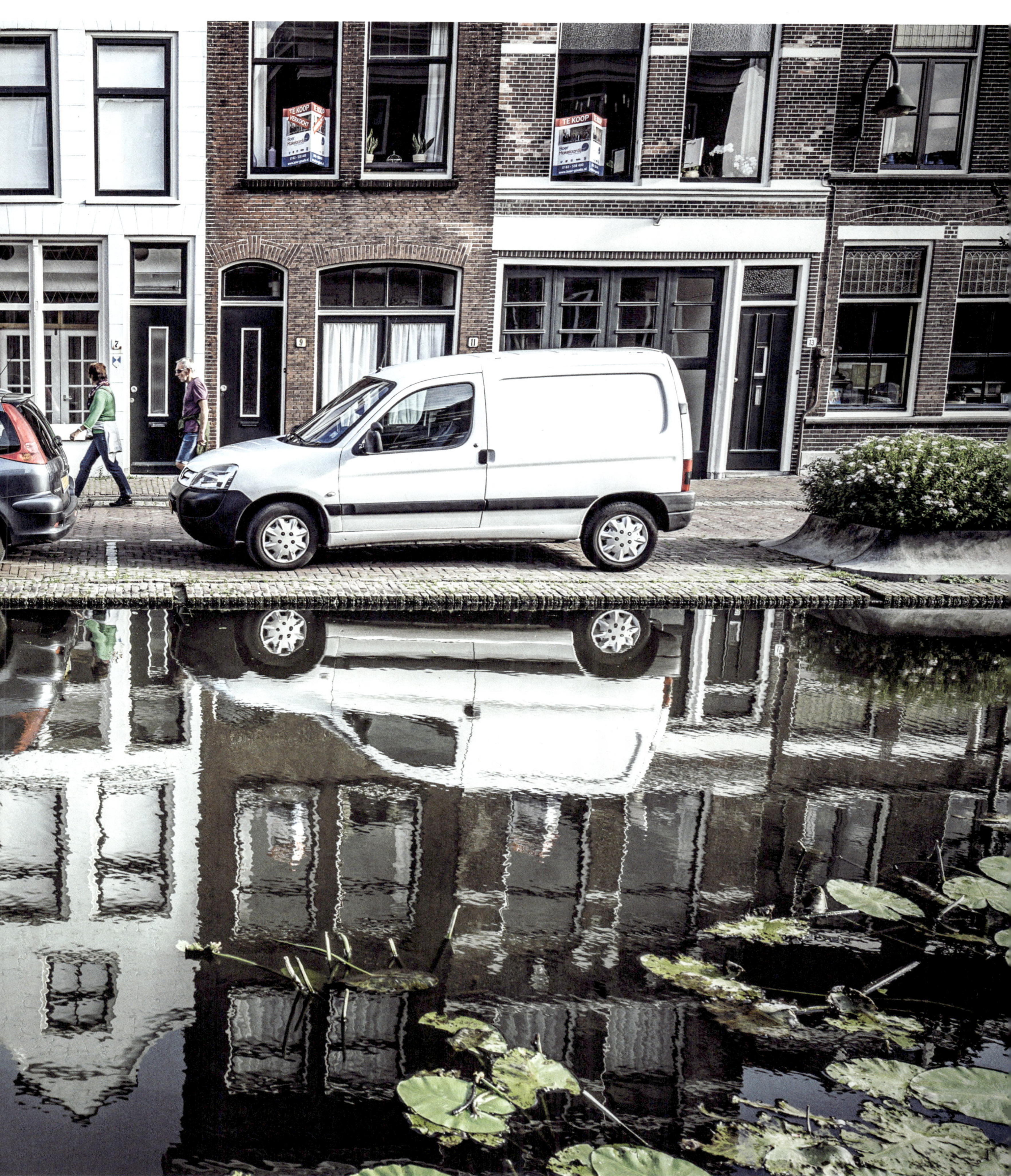

On 8 January 2019, the Netherlands was hit by a heavy northwesterly storm and a spring tide, a situation which equalled the circumstances of the flood of 1953. This time, the Netherlands was much better protected against storm surges, so the damage was minor. Nevertheless, the island of Terschelling was hit hard, and the quays and harbour of West-Terschelling flooded. This has happened before, but nowadays it happens much more frequently. The village is not protected, and future sea level rise will endanger the low-lying parts. It's predicted that the Wadden Sea estuary, which is mostly dry at low tide, will change to an open sea at low tide. This will have a serious impact on the islands and the environment, and predictions are that the islands will cease to exist in the long term.

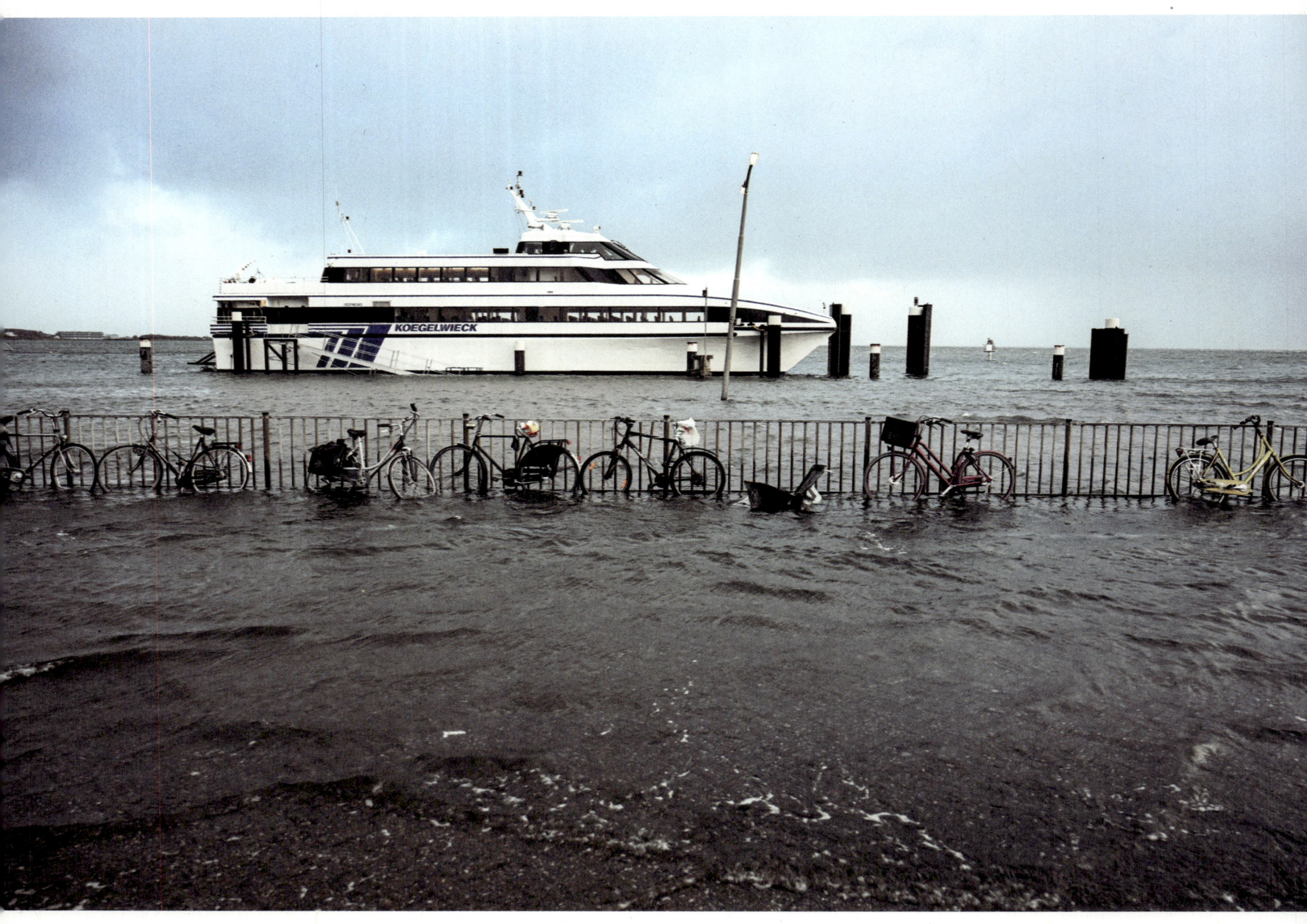

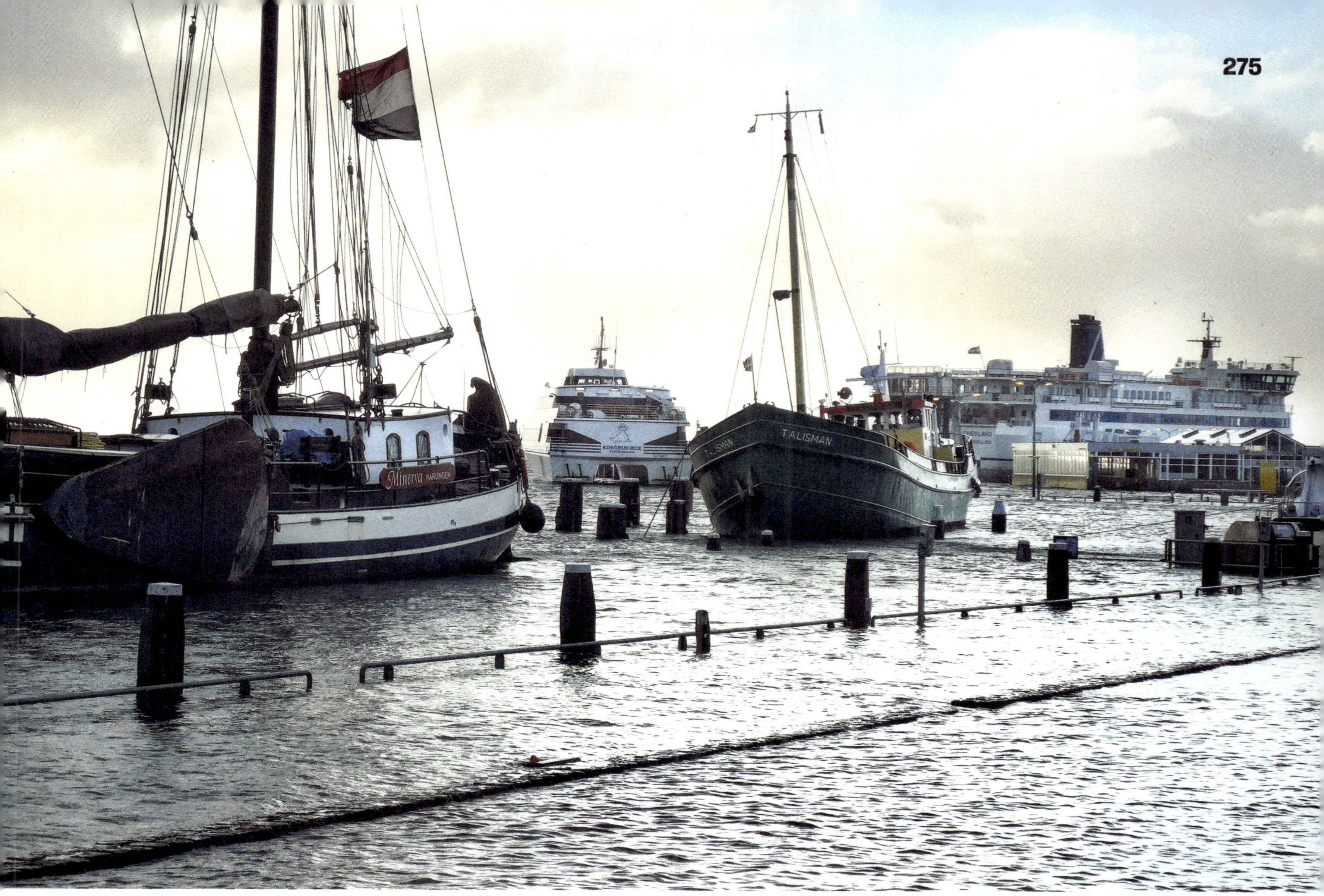

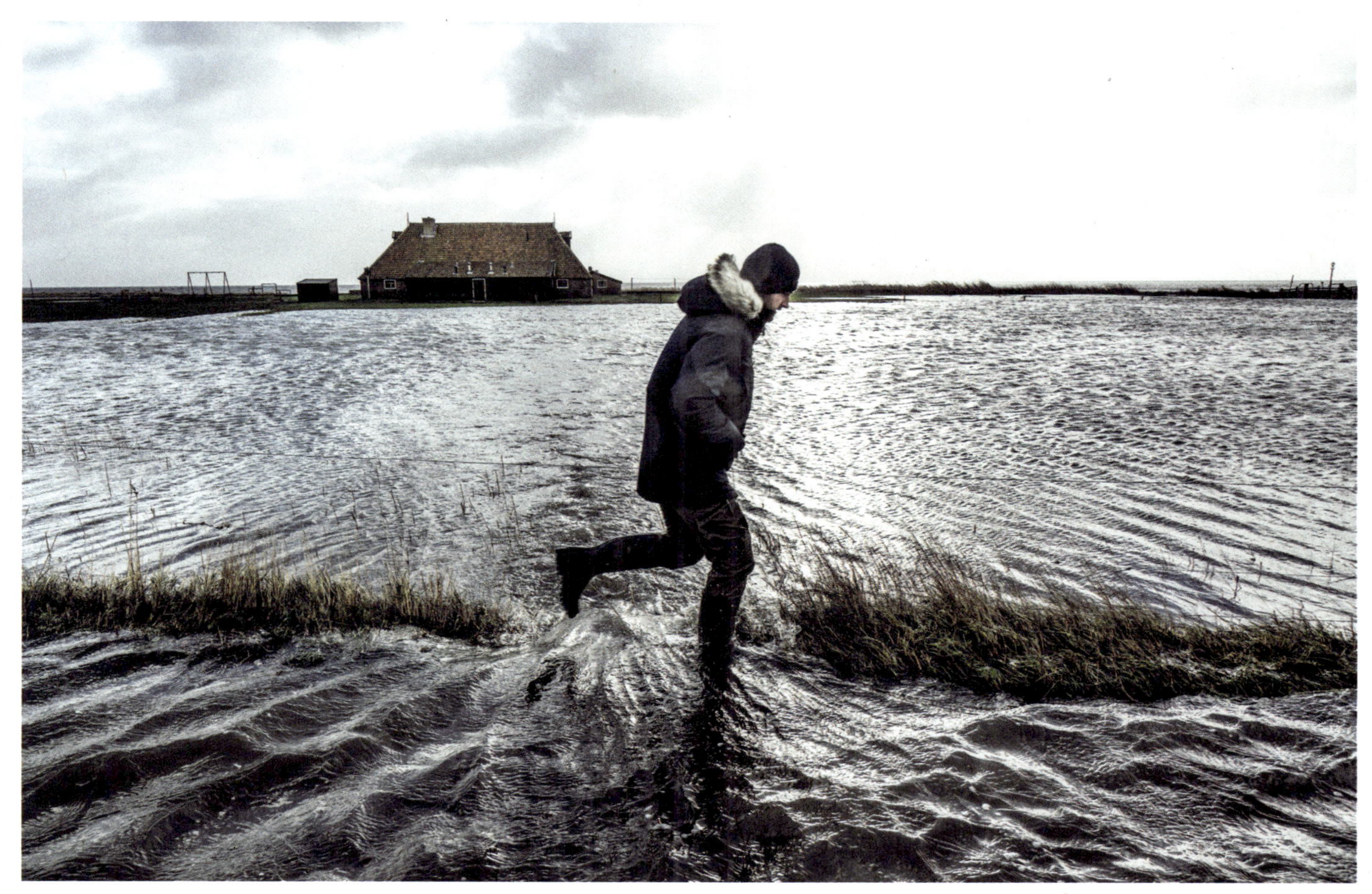

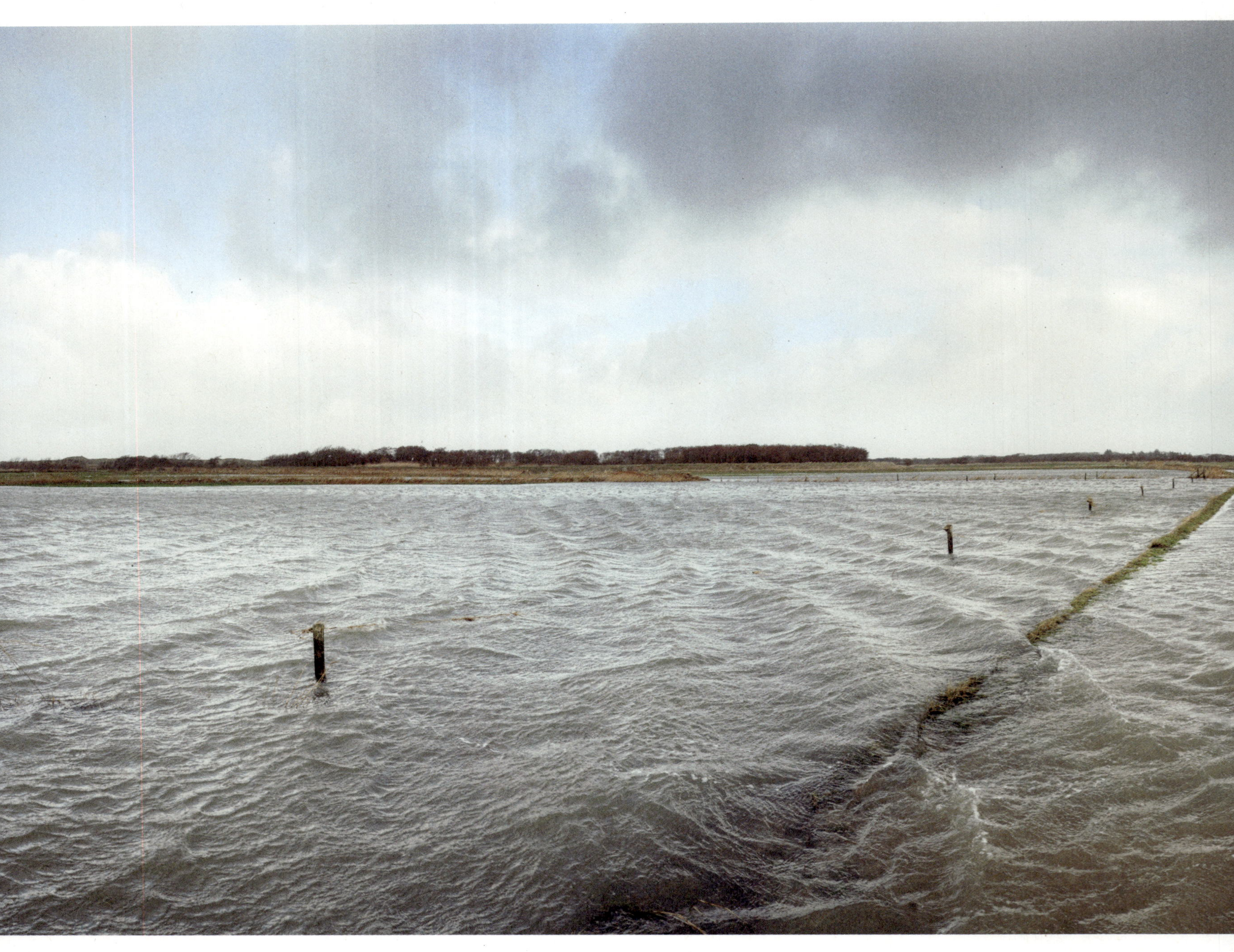

The Wierschuur at the east of
Terschelling is inaccessible due to
flooding.

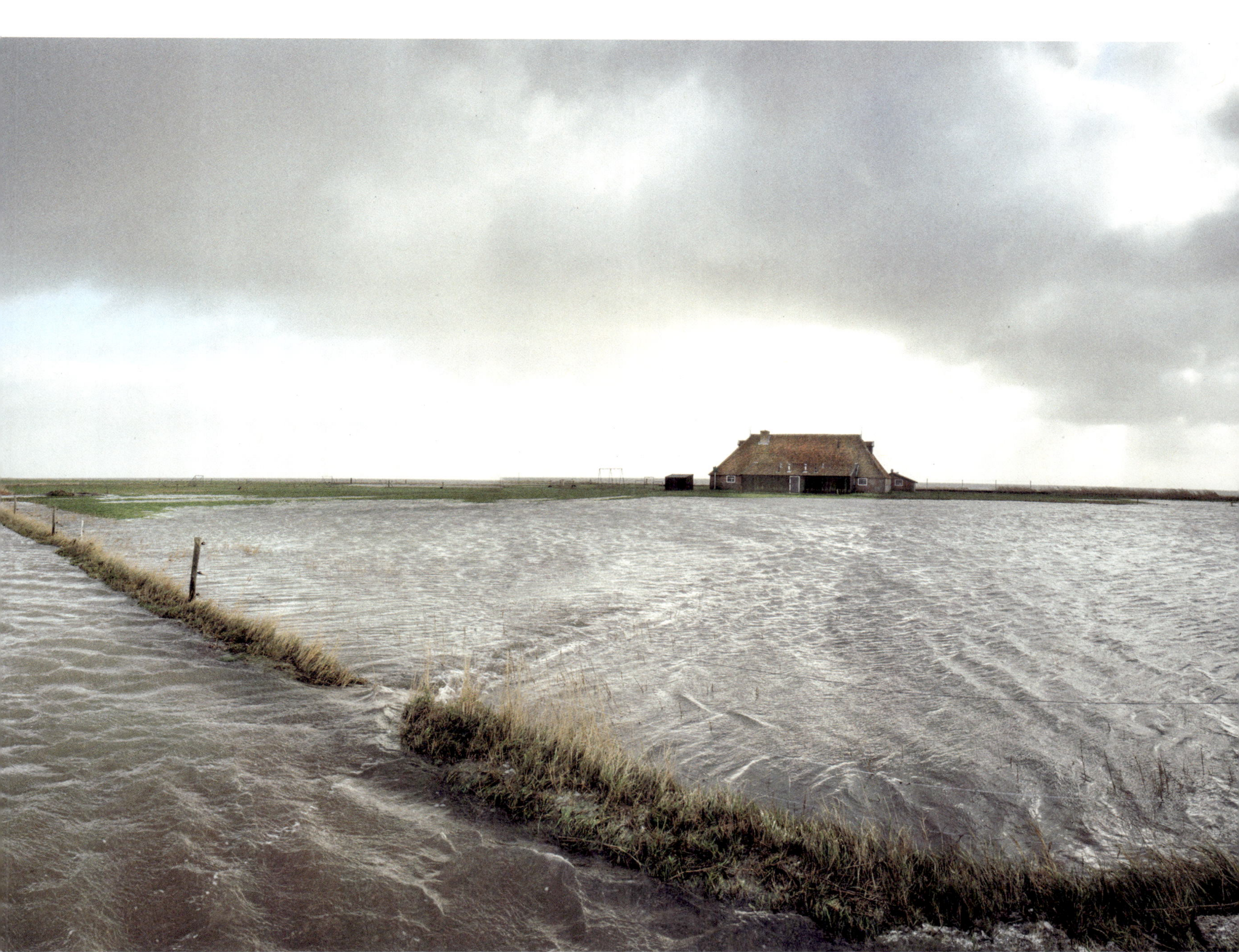

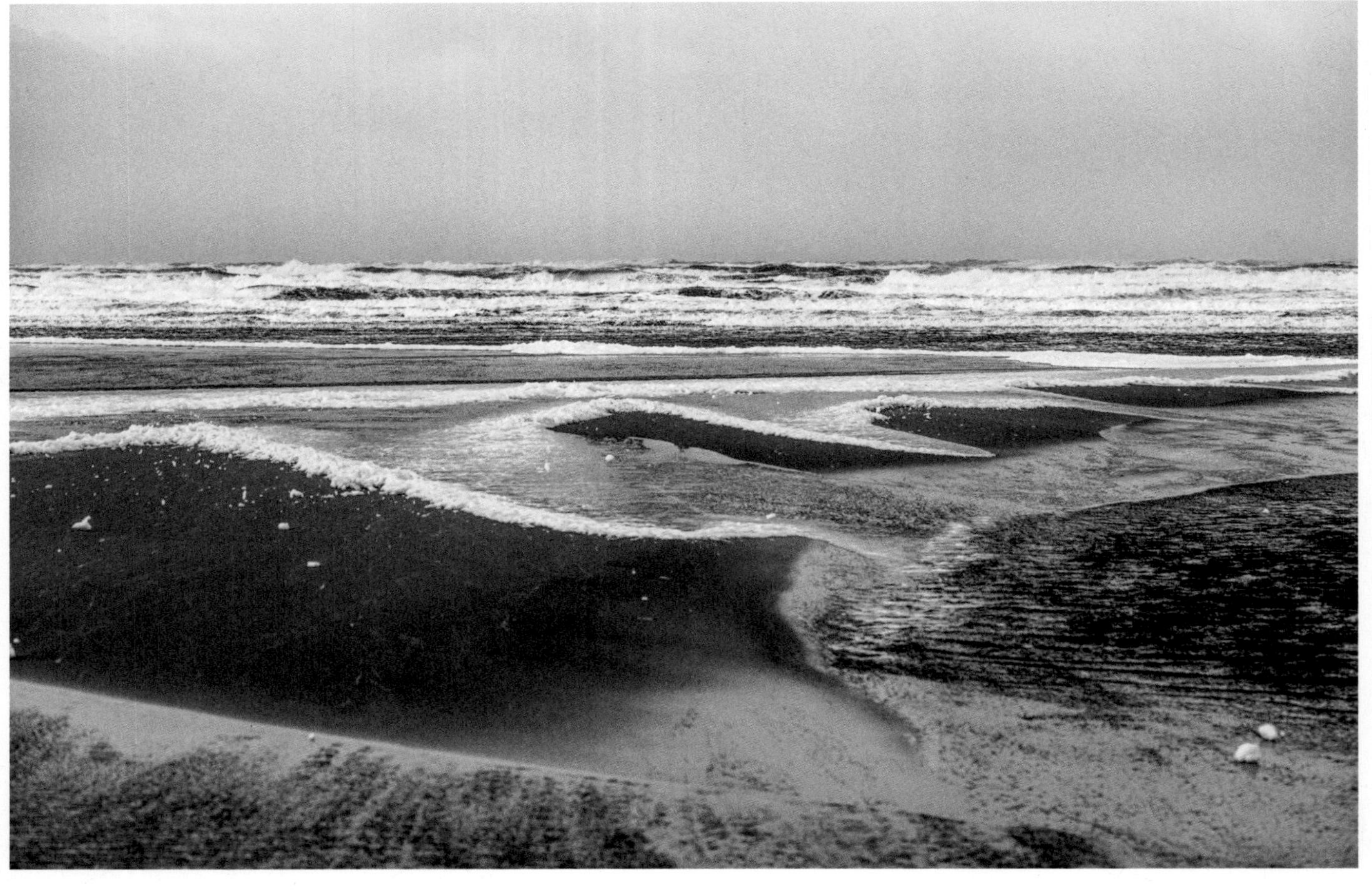

Kadir van Lohuizen
Photographer P5

Kadir van Lohuizen has covered conflicts in Africa and elsewhere but is probably best known for his long-term projects on the 'seven rivers of the world', rising sea levels, the diamond industry and migration in the Americas. In *Wasteland*, he investigated the mismanagement of waste in six megacities and the Arctic. He has covered the issue of rising sea levels for a number of years. Van Lohuizen has received numerous prizes and awards in photojournalism and is a frequent lecturer and photography teacher. He is based in Amsterdam, the Netherlands.

Henk Ovink
Special Envoy for International Water Affairs, the Netherlands, and Sherpa to the UN High Level Panel on Water P9

Henk Ovink was appointed in 2015 by the Dutch cabinet as the first special envoy for international water affairs. As an ambassador for water, he is responsible for advocating for water awareness around the world; initiating and leading research; building institutional capacity; forging coalitions among governments, multilateral organisations, the private sector, academia and NGOs; and initiating innovative approaches to address the world's pressing needs in relation to water. One of his flagship initiatives is the global partnership Water as Leverage, focused on climate and water action for resilient cities.

Ovink served on then-president Obama's Hurricane Sandy Rebuilding Task Force in the aftermath of Hurricane Sandy. He led the long-term innovation, resilience and rebuilding efforts and developed and led Rebuild by Design, the task force's groundbreaking challenge for innovative climate action.

Before joining the task force, Ovink was both acting director general of spatial planning and water affairs and director of national spatial planning for the Netherlands. He teaches at the Harvard Graduate School of Design, the London School of Economics and the University of Groningen. His latest book, *Too Big: Rebuild by Design: A Transformative Approach to Climate Change*, written together with Jelte Boeijenga, explores Ovink's climate and water work for the Obama Administration.

Dorthe Dahl-Jensen
Arctic scientist, Copenhagen and Greenland P18

Professor Dahl-Jensen's major scientific achievements have been in leading ice core drilling and the subsequent analysis of ice core data in conjunction with models to determine the past climate and how it has affected the Greenland ice sheet. She has led several international deep drilling projects, such as the North Greenland Ice Core Project (NGRIP), North Greenland Eemian Ice Drilling (NEEM) and the East Greenland Ice-core Project (EGRIP). In addition, she has led large research projects funded by grants from the Danish National Research Foundation, the European Research Council, the Seventh Framework Programme (FP7), the Villum Investigator Programme and the Canada Excellence Research Chairs Program. This research has led to numerous achievements and is also very well cited, making Dahl-Jensen a Thomson Reuters Highly Cited Researcher.

Jeff Goodell
Author and contributing editor, *Rolling Stone* P54

Jeff Goodell's most recent book, *The Water Will Come: Rising Seas, Sinking Cities, and the Remaking of the Civilized World,* was a *New York Times* Critics' Top Book of 2017 as well as one of the *Washington Post's* 50 Notable Works of Nonfiction in 2017. He is the author of five previous books, including *Big Coal: The Dirty Secret Behind America's Energy Future* and is a contributing editor at *Rolling Stone*, where he has covered the climate crisis for more than a decade. As a commentator on energy and climate issues, he has appeared on NPR, MSNBC, CNN, CNBC, ABC, NBC, Fox News and *The Oprah Winfrey Show*. He is a Senior Fellow of the Atlantic Council and a 2020 Guggenheim Fellow.

Elliot Brown
Resident Guna Yala, Panama P90

Elliot Brown is a native of the Gardí Sugdub community in the Guna Yala region of Panama. He obtained his bachelor's degree in biology with a specialty in environmental biology at the University of Panama's School of Biology, Faculty of Exact and Natural Sciences and Technology. Currently, he is a master's student in the management and evaluation of environmental impact studies at the Specialized University of the Americas. Brown is a researcher, associated with the Yaguará Panamá Foundation, a program for the conservation of wild cats, especially jaguars, and a volunteer biologist in the technical commission of the neighbourhood relocation project in Gardí Sugdub.community of Gardí Sugdub.

Amalinda Savirani
Associate professor, Gadjah Mada University,
Yogyakarta, Indonesia   P118

Amalinda Savirani works in Gadjah Mada University's
Department of Politics and Government, serving as
department head until the end of December 2020. She
obtained her PhD at the University of Amsterdam in
2015, majoring in political anthropology. Her research
interests in recent years have been focussed on urban
politics, particularly urban citizens' movement in relation
to housing, rights of livelihood and political linkages
between movement and democratic means such as
elections (Savirani & Aspinall, 2017; Savirani & Saidi
2017). Her previous publication was about 'publicness'
in the discourse between policymakers and civil society
organisations on the Jakarta Bay reclamation project
(Savirani, 2017).

Anote Tong
Former president, Kiribati   P146

Twice nominated for the Nobel Peace Prize, the winner
of the Sunhak Peace Prize and a 2012 Hillary Laureate,
His Excellency Anote Tong, former president of Kiribati, is
a world-renowned leader in the battles against climate
change and for ocean conservation. Under his leadership, the government of Kiribati created the Phoenix
Islands Protected Area, a 408,250 km² UNESCO World
Heritage Site off limits to fishing and extractive industries. Initiatives such as this, typical of Tong's vision and
his ability to balance political reality with humanitari-
an and ecological idealism, have earned him the Peter
Benchley Ocean Award from the Blue Frontier Campaign.

Since leaving office, he has continued to speak about
the realities of climate change: the urgency of the
issues, the complexity of the causes and the possible
solutions and the stark simplicity of the consequences
should we fail to act. Tong been awarded an honorary
doctorate in engineering from Pukyong National University in South Korea to add to his BSc from the University
of Canterbury in New Zealand and his master's in economics from the London School of Economics.

Sharif Jamil
Activist and leader, Bangladesh   P206

Sharif Jamil is a well-recognized environmental activist
at the national, regional and international levels. He
possesses over two decades of experience in activism,
during which he has led numerous campaigns to protect
and promote the rights of the environment in Bangladesh, even at the risk of his own life.

Currently, Jamil is the Buriganga Riverkeeper and the
coordinator of Waterkeepers Bangladesh, a programme
of the Waterkeeper Alliance, a global not-for-profit
network committed to ensuring clean, drinkable water.
He is also the general secretary of Bangladesh Poribesh
Andolon, the leading environmental rights organisation in Bangladesh. Jamil has been recognized for his
tireless commitment to activism. He was awarded Joke
Waller-Hunter Fellowship in 2007 as a prospective leader
in the environmental movement of developing countries.
He was also recognized as one of the 20 global Waterkeeper Warriors by the Waterkeeper Alliance during its
20th anniversary in 2019.

Marjan Minnesma
Co-founder and director, Urgenda, a Dutch organisation
for innovation and sustainability   P244

Marjan Minnesma studied business administration,
philosophy and law. She worked for the Dutch government in Central Europe on energy efficiency and
renewable energy projects for five years. She has been
campaign director for Greenpeace Netherlands and
worked for 10 years at different universities on large
international projects. Minnesma founded Urgenda, an
organisation dedicated to innovation and sustainability
and making required transitions happen, together with
Professor Jan Rotmans from the Dutch Research Institute For Transitions at Erasmus University. Urgenda has
worked as a front-runner in developing solutions to the
climate crisis, from organising the first collective buying
initiative for solar panels to the introduction of the first
electric car produced in series to the Netherlands in
2008. However, as quicker action was needed, Urgenda
brought the government to court, together with almost
900 citizens, in the famous Urgenda climate case, which
Urgenda won - all the way up to the Supreme Court -
forcing the government to take increased action.

Water
Henk Ovink · P9

Allmendinger, P., & Haughton, G. (2009). Soft spaces, fuzzy boundaries, and metagovernance: The new spatial planning in the Thames Gateway. *Environment and Planning* A, 41(3), 617-633.

Global Commission on Adaptation. (2019). *Adapt now: A global call for leadership on climate resilience.* https://cdn.gca.org/assets/2019-09/GlobalCommission_Report_FINAL.pdf

Guterres, A. (2020, April 14). *"This is a time for science and solidarity".* United Nations. https://www.un.org/en/un-coronavirus-communications-team/time-science-and-solidarity

Hallegatte, S., Rentschler, J., & Rozenberg, J. (2019). *Lifelines: The resilient infrastructure opportunity.* Washington, DC: World Bank Group. http://hdl.handle.net/10986/31805

Ligtvoet, W., Bouwman, A., Knoop, J., de Bruin, S., Nabielek, K., Huitzing, H., Janse, J., van Minnen, J., Gernaat, D., van Puijenbroek, P., de Ruiter, J., & Visser, H. (2018). *The geography of future water challenges.* The Hague, NL: PBL Netherlands Environmental Assessment Agency. https://www.clingendael.org/sites/default/files/2018-04/The-geography-of-future-water-challenges_pdf.pdf

Ovink, H., & Boeijenga, J. (2018). *Too big: Rebuild by design: A transformative approach to climate change.* Rotterdam, NL: nai010 publishers.

Rebuild by Design. (n.d.). *Hurricane Sandy design competition.* http://www.rebuildbydesign.com/our-work/sandy-projects

UNESCO. (n.d.). *Local and Indigenous Knowledge Systems (LINKS).* https://en.unesco.org/links

UNESCO World Water Assessment Program. (2020). *The United Nations world water development report 2020: Water and climate change.* https://en.unesco.org/themes/water-security/wwap/wwdr/2020

World Economic Forum. (2020). *The Global Risks Report 2020.* https://www.weforum.org/reports/the-global-risks-report-2020

World Health Organization. (n.d.). *Climate change and human health.* https://www.who.int/globalchange/global-campaign/cop21/

World Health Organization & UN-Water. (2014). *Investing in water and sanitation: Increasing access, reducing inequalities* [UN-Water global analysis and assessment of sanitation and drinking-water (GLAAS) - 2014 report]. https://www.unwater.org/publications/investing-water-sanitation/ t

World Water Atlas. (n.d.). *Methodologies: Water as Leverage programme.* https://www.worldwateratlas.org/curated/water-as-leverage

Panama
Elliot Brown · P90

Displacement Solutions. (2014, July). *The peninsula principles in action: Climate change and displacement in the autonomous region of Gunayala, Panama* [Mission report]. http://displacementsolutions.org/wp-content/uploads/Panama-The-Peninsula-Principles-in-Action.pdf

Lazrus, H., & Arenas, C. (2020). Islands on an angry Earth: Climate change, disasters, and implications for two island communities. In A. Oliver-Smith & S. M. Hoffman (Eds.), *The angry Earth: Disaster in anthropological perspective* (pp. 359-370). Routledge.

Indonesia
Amalinda Savirani · P118

Bakker, M., Kishimoto, S., & Nooy, C. (2017). *Social justice at bay: The Dutch role in Jakarta's coastal defence and land reclamation.* Both Ends, Centre for Research on Multinational Corporations (SOMO), and Transnational Institute (TNI). https://www.bothends.org/uploaded_files/document/1LR_Social_justice_at_bay_A4.pdf

Batubara, B., Warsilah, H., Wagner, I., & Salam, S. (2020). *Maleh dadi Segoro: Krisis sosial-ekologis kawasan pesisir Semarang-Demak* [Our land becomes sea: Social and ecological crisis in the Semarang-Demak area, Central Java]. Lintas Nalar.

Burhaini Faizal, E. (2016, April 21). Almost all reclamation in Indonesia illegal, claims environmental group. *The Jakarta Post.* https://www.thejakartapost.com/news/2016/04/21/almost-all-reclamation-in-indonesia-illegal-claims-environmental-group.html

Deltares. (2015). *Sinking cities: An integrated approach towards solutions.* https://www.deltares.nl/app/uploads/2015/09/Sinking-cities.pdf

Dutch Water Sector. (2016, March 24). *Royal Boskalis to construct five new islands for coastal city Makassar, Indonesia.* https://www.dutchwatersector.com/news/royal-boskalis-to-construct-five-new-islands-for-coastal-city-makassar-indonesia

Holliani Cahya, G. (2020, February 26). Climate change cause of Greater Jakarta floods, BMKG says. *The Jakarta Post.* https://www.thejakartapost.com/news/2020/02/26/climate-change-behind-2020-floods-that-displaced-thousands-in-jakarta-agency-says.html

Indonesia witnesses 1,928 disasters in 2020 so far. (2020, August 31). *The Jakarta Post.* https://www.thejakartapost.com/news/2020/08/31/indonesia-witnesses-1928-disasters-in-2020-so-far.html Kahfi, K. (2020,

January 4). New normal: Climate crisis to bring extreme rainfall more often, scientists say. *The Jakarta Post.* https://www.thejakartapost.com/news/2020/01/03/new-normal-climate-crisis-to-bring-extreme-rainfall-more-often-scientists-say.html

KuiperCompagnons. (n.d.). *The Great Garuda to save Jakarta.* https://www.kuipercompagnons.nl/ru/work/urban_planning_and_design/the_great_garuda_to_save_jakarta/

LBH Jakarta. (2017). *Seperti puing: Laporan penggusuran paksa di wilayah DKI Jakarta tahun 2016* [Like debris: Forced eviction report in Jakarta 2016]. https://www.bantuanhukum.or.id/web/seperti-puing-laporan-penggusuran-paksa-di-wilayah-dki-jakarta-tahun-2016

Lin, M.M., & Hidayat, R. (2018, August 12). Jakarta, the fastest-sinking city in the world. *BBC Indonesian.* https://www.bbc.com/news/world-asia-44636934

Mariani, E. (2016, May 20). Cheaper solution for subsidence offered. *The Jakarta Post.* https://www.thejakartapost.com/news/2016/05/20/cheaper-solution-subsidence-offered.html

Partners for Water. (2020, April). *Country update: Indonesia.* https://www.netherlandswaterpartnership.com/sites/nwp_corp/files/2020-04/Country-Update-Indonesia-April-2020.pdf

Pramita, D. (2015, September 17). *Korban Reklamasi Teluk Jakarta, ratusan nelayan jadi pemulung* [Victims of Jakarta Bay Reclamation, hundreds of fishermen become scavengers]. *Tempo.co.* https://metro.tempo.co/read/701337/korban-reklamasi-teluk-jakarta-ratusan-nelayan-jadi-pemulung/full&view=ok

Savirani, A., & Wilson, I. (2018, April 28). Distance matters: Social housing for the poor. *Inside Indonesia.* https://www.insideindonesia.org/distance-matters-social-housing-for-the-poor

Simone, A. M. (2004). People as infrastructure: Intersecting fragments in Johannesburg. *Public Culture,* 16(3), 407-429. https://muse.jhu.edu/article/173743/summary

Stepputat, F., & van Voorst, R. (2016). *Cities on the agenda: Urban governance and sustainable development* (DIIS Report 2016:04). Danish Institute for International Studies. http://hdl.handle.net/10419/144739

ter Braak, B. (Ed.). (2016). *Indonesia - The Netherlands integrated approach of future water challenges.* The Netherlands Water Partnership (NWP) for the Dutch Government.

World Bank Group. (2016, January 8). *Keeping Indonesia's capital safer from floods.* https://www.worldbank.org/en/news/feature/2016/01/08/keeping-indonesias-capital-safer-from-floods

Bangladesh
Sharif Jamil   P206

Ahmad, H. (2019). Bangladesh coastal zone management status and future trends. *Journal of Coastal Zone Management, 22(1)*, Article 1000466.

Bodrud-Doza, Md., Khan, R.M., & Asad, A. (2020, August 12). Trapped population: The combined impact of climate change and COVID-19. *The Business Standard*. https://tbsnews.net/thoughts/trapped-population-combined-impact-climate-change-and-covid-19-118429

Darby, M. (2017, August 14). What will become of Bangladesh's climate migrants? *Climate Home News*. https://www.climatechangenews.com/2017/08/14/will-become-bangladeshs-climate-migrants

McDonnell, T. (2019, January 24). Climate change creates a new migration crisis for Bangladesh. *National Geographic*. https://nationalgeographic.com/environment/2019/01/climate-change-drives-migration-crisis-in-bangladesh-from-dhaka-sundabans

McPherson, P. (2018, March 21). The dysfunctional megacity: Why Dhaka is bursting at the sewers. *The Guardian*. https://www.theguardian.com/cities/2018/mar/21/people-pouring-dhaka-bursting-sewers-over-population-bangladesh

Ministry of Disaster Management and Relief. (2015). *Comprehensive Disaster Management Programme Phase II*. https://www.preventionweb.net/files/globalplatform/entry_bg_paper~cdmpphaseii.pdf

Our cities not ready for climate migrants. (2019, October 24). *The Daily Star*. https://www.thedailystar.net/city/news/our-cities-not-ready-climate-migrants-1818049

UCAR Center for Science Education. (n.d.). Sea level change in Bangladesh. https://scied.ucar.edu/image/sea level-change-bangladesh

Wernick, A. (2019, March 25). A climate migration crisis is escalating in Bangladesh. *The World*. https://www.pri.org/stories/2019-03-25/climate-migration-crisis-escalating-bangladesh

World Bank Group. (n.d.). Building resilience of the coastal population of Bangladesh to climate change. https://www.worldbank.org/en/news/feature/2013/06/26/bangladesh-building-resilience-coastal-population-climate-change

P8

Oppenheimer, M., Glavovic, B.C., Hinkel, J., van de Wal, R., Magnan, A.K., Abd-El-gawad, A., Cai, R., Cifuentes-Jara, M., DeConto, R.M., Ghosh, T., Hay, J., Isla, F., Marzeion, B., Meyssignac, B., & Sebesvari, Z. (in press). Sea level rise and implications for low-lying islands, coasts and communities. In H.-O. Pörtner, D.C. Roberts, V. Masson-Delmotte, P. Zhai, M. Tignor, E. Poloczanska, K. Mintenbeck, A. Alegría, M. Nicolai, A. Okem, J. Petzold, B. Rama, & N.M. Weyer (Eds.), IPCC special report on the ocean and cryosphere in a changing climate (chapter 4). Geneva: Intergovernmental Panel on Climate Change. https://www.ipcc.ch/srocc/chapter/chapter-4-sea-level-rise-and-implications-for-low-lying-islands-coasts-and-communities/

P14

Haasnoot, M., Kwadijk, J., van Alphen, J., Le Bars, D., van den Hurk, B., Diermanse, F., van der Spek, A., Oude Essink, G., Delsman, J., & Mens, M. (2020). Adaptation to uncertain sea level rise; how uncertainty in Antarctic mass-loss impacts the coastal adaptation strategy of the Netherlands. Environmental Research Letters, 15(3), Article 034007. https://doi.org/10.1088/1748-9326/ab666c

Le Bars, D., Drijfhout, S., & de Vries, H. (2017). A high-end sea level rise probabilistic projection including rapid Antarctic ice sheet mass loss. Environmental Research Letters, 12(4), Article 044013. https://doi.org/10.1088/1748-9326/aa6512

P15

See p. 8

P20

Top left:
Joughin, I., Smith, B., Howat, I., & Scambos, T. (2016). MEaSUREs multi-year Greenland ice sheet velocity mosaic, version 1. [https://nsidc.org/data/NSIDC-0670/versions/1?qt-data_set_tabs=2#qt-data_set_tabs] National Snow and Ice Data Center. https://doi.org/10.5067/QUA5Q9SVMSJG

Top right:
World Meteorological Organization. (2018). WMO statement on the state of the global climate in 2017 (WMO-No. 1212). https://library.wmo.int/doc_num.php?-explnum_id=4453

Bottom right:
NASA. (n.d.). Operation IceBridge. https://icebridge.gsfc.nasa.gov/

P23

NASA Jet Propulsion Laboratory. (n.d.). Greenland ice loss 2002-2016. https://gracefo.jpl.nasa.gov/resources/33/greenland-ice-loss-2002-2016/

P57

Climate Central (www.climatecentral.org)

P58

Climate Central (www.climatecentral.org)

P93

Climate Central (www.climatecentral.org)

P120

Dr. Heri Andreas, Faculty of Earth Sciences and Technology, Bandung Institute of Technology

P122

Climate Central (www.climatecentral.org)

P151

Climate Central (www.climatecentral.org)

P207

Dhaka University; Intergovernemntal Pannel of Climate Change (IPCC)

PP208 – 209

Climate Central (www.climatecentral.org)

P245

See p. 8

P 247

Data source: IWR (2009) & UNFCC (2007). Map created by Benjamin Hennig, Sasi Research Group, University of Sheffield - www.viewsoftheworld.net

P247

Urgenda, courtesy of Mathijs Bouman, @mathijsbouman

P248

IWR (2009) & UNFCC (2007). Map created by Benjamin Hennig, Sasi Research Group, University of Sheffield - www.viewsoftheworld.net

P249

See p. 14

Thank you to all of the staff and my colleagues at NOOR, Sanne Klap, Henk Ovink, (Ministry of Infrastructure and the Environment), Sarah Theerlynck and Niels Famaey (Lannoo), Jeroen Kummer and Simon Burer (Kummer & Herman), Teun van der Heijden and Sandra van der Doelen (Heijdens Karwei), Jenny Smets, Jeroen de Vries, Frank Ortmanns, Meike Ziegler, Stanley Greene, Chantal Bijkerk, Peter Glas (Delta Programme Commissioner), Michael Huijser, Tessa van den Dolder, Sarah Keijzer (National Maritime Museum, Amsterdam), NTR, Witfilm, Martijn Blekendaal, Pim Hawinkels, Benny Jansen, Clément Saccomani, Claudia Hinterseer, Evelien Kunst, Frank Zuidweg, Yvonne Schaefers (Nikon Europe), Catherine Beltrandi and Sophie Loran (UNEP), Carlos Arenas, Scott Leckie (Displacement Solutions), David Furst, Magdalena Herrera, Rimon Rimon, Shahidul Alam, Fondation Carmignac, VPRO, Doke Romeijn, Peter Kuipers Munneke, Tinus Kramer, Onno van der Wal, Mike Kamber, Maggie Steber, MaryAnne Golon, Nina Alvarez, Adam Lyberth, Ng Swan Ti, Edy Purnomo, JanJaap Brinkman, Peter Letitre (Deltares), all my friends in Amsterdam and the Netherlands.

Also big thank you to all the people who allowed me to photograph them and who were willing to speak to me.

Photography:
Kadir van Lohuizen

Texts:
Kadir van Lohuizen
Henk Ovink
Dorthe Dahl-Jensen
Jeff Goodell
Elliot Brown
Amalinda Savirani
Anote Tong
Sharif Jamil
Marjan Minnesma

Book Design:
Kummer & Herrman

Translation (text Kadir van Lohuizen):
Lisa Holden

Image Editing:
Kadir van Lohuizen and Jeroen Kummer

Lithography:
Colour and Books, Sebastiaan Hanekroot

Infographics and figures:
Redesigned by Kummer & Herrman based
on multiple sources

Copy editing and proofreading:
Amy Haagsma

Paper:
Interior: 135 grs. Nautilus Classic, cover:
350 grs. Nautilus Classic

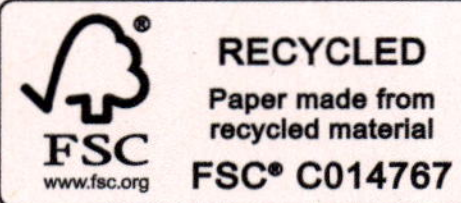

Sign up for our newsletter with news about
new and forthcoming publications on art,
interior design, food & travel, photography and
fashion as well as exclusive offers and events.

If you have any questions or comments
about the material in this book, please do
not hesitate to contact our editorial team:
art@lannoo.com

© Lannoo Publishers, Belgium, 2021
D/2021/45/16- NUR 652/653
ISBN: 9789401473590
www.lannoo.com

Every effort has been made to trace copyright
holders. If, however, you feel that you have
inadvertently been overlooked, please contact
the publishers.